MODELING METHODS FOR MARINE SCIENCE

This textbook on modeling, data analysis, and numerical techniques for marine science has been developed from a course taught by the authors for many years at the Woods Hole Oceanographic Institution.

The first part of the book covers statistics: singular value decomposition, error propagation, least squares regression, principal component analysis, time series analysis, and objective interpolation. The second part deals with modeling techniques: finite differences, stability analysis, and optimization. The third part describes case studies of actual ocean models of ever-increasing dimensionality and complexity, starting with zero-dimensional models and finishing with three-dimensional general circulation models. Throughout the book hands-on computational examples are introduced using the MATLAB programming language and the principles of scientific visualization are emphasized.

Modeling Methods for Marine Science is a textbook for advanced students of oceanography on courses in data analysis and numerical modeling. It is also an invaluable resource as a reference text for a broad range of scientists undertaking modeling in chemical, biological, geological, and physical oceanography.

DAVID M. GLOVER is a Senior Research Specialist in the Department of Marine Chemistry and Geochemistry at Woods Hole Oceanographic Institution. He is the author or co-author of 67 published articles, book chapters and abstracts. Dr. Glover's research uses satellite data, model results, and shipboard data to elucidate the mechanisms and processes by which the oceans play a major role in the maintenance of the global climate.

WILLIAM J. JENKINS is a Senior Scientist in the Department of Marine Chemistry and Geochemistry at Woods Hole Oceanographic Institution. He has published 84 peer-reviewed journal and book articles. Dr. Jenkins is the Director of the National Ocean Sciences Accelerator Mass Spectrometry Facility (NOSAMS). In 1983 he received the Rosenstiel Award in Oceanographic Science from the University of Miami and in 1997 he received the Henry Bryant Bigelow Award in Oceanography from the Woods Hole Oceanographic Institution. Dr. Jenkins' interests include studying tracers as applied to oceanic physical, chemical, biological, and geological processes; air–sea and ice–water exchange of gases; ocean biological productivity and its controls; radiogenic and primordial noble gas isotopes in the sea, atmosphere, lakes, ground waters, sediments and rocks; climatic changes in the ocean and its effects on biogeochemical systems; and radiocarbon and the global carbon cycle in the past 60,000 years.

SCOTT C. DONEY is a Senior Scientist in the Department of Marine Chemistry and Geochemistry at Woods Hole Oceanographic Institution. He has authored or co-authored more than 160 peer-reviewed journal and book articles. He was awarded the James B. Macelwane Medal from the American Geophysical Union in 2000 and an Aldo Leopold Leadership Program Fellowship in 2004. He has traveled extensively, lending his expertise to a number of national and international science programs, most recently as inaugural chair of the Ocean Carbon and Biogeochemistry (OCB) Program. He has also testified before both the US House of Representatives and the US Senate. His research interests include marine biogeochemistry and ecosystem dynamics, ocean acidification, the global carbon cycle, climate change, and the intersection of science and policy.

MODELING METHODS
FOR MARINE SCIENCE

DAVID M. GLOVER

WILLIAM J. JENKINS

and

SCOTT C. DONEY

Woods Hole Oceanographic Institution

CAMBRIDGE
UNIVERSITY PRESS

32 Avenue of the Americas, New York NY 10013-2473, USA

Cambridge University Press is part of the University of Cambridge.

It furthers the University's mission by disseminating knowledge in the pursuit of education, learning and research at the highest international levels of excellence.

www.cambridge.org
Information on this title: www.cambridge.org/9780521867832

First published 2011

A catalogue record for this publication is available from the British Library

ISBN 978-0-521-86783-2 Hardback

This book is dedicated to
Tina, Susan, and Andrea.
They have endured our preoccupations with
loving support, rare good humor,
and infinite patience.

Contents

Preface

If you are a student of science in the twenty-first century, but are not using computers, then you are probably not *doing* science. A little harsh, perhaps, and tendentious, undoubtedly. But this bugle-call over-simplification gets to the very heart of the reason that we wrote this book. Over the years we noticed, with increasing alarm, very gifted students entering our graduate program in marine chemistry and geochemistry with very little understanding of the applied mathematics and numerical modeling they would be required to know over the course of their careers. So this book, like many before it, started as a course – in this case, a course in modeling, data analysis, and numerical techniques for geochemistry that we teach every other year in Woods Hole. As the course popularity and web pages grew, we realized our efforts should be set down in a more formal fashion.

We wrote this book first and foremost with the graduate and advanced undergraduate student in mind. In particular, we have aimed the material at the student still in the stages of formulating their Ph.D. or B.Sc. thesis. We feel that the student armed with the knowledge of what will be required of them when they synthesize their data and write their thesis will do a much better job at collecting the data in the first place. Nevertheless, we have found that many students beyond these first years find this book useful as a reference. Additionally, many of our colleagues in the ocean sciences, broadly defined (chemical, biological, geological, and yes, even physical), find this book a useful resource for analyzing or modeling data.

Readers will find this book to be self-contained inasmuch as we introduce all of the concepts encountered in the book, including bringing the reader up to speed on ocean science and physics. Consequently, prerequisites for this book are few. However, exposure to linear algebra, statistics, and calculus sometime in the reader's past will be helpful, but not absolutely required. Additionally, this book uses MATLAB^{TM} as its computational engine and some programming in MATLAB^{TM} is required; for that reason exposure to programming concepts will be helpful as well. We have chosen MATLAB^{TM} (rather than some other mathematics and statistics package) because we find it subsumes arcane details (e.g. data formats) without concealing the process of analysis. There are a number of very useful MATLAB^{TM} m-files in this book (some written by us, some donated), which we have made available at http://www.cambridge.org/glover, the web page the publisher maintains for this book. These m-files are working, practical examples (i.e. code that runs), and each chapter contains detailed problems sets that include computer based assignments and solutions. A fair amount of MATLAB™ instruction occurs throughout the book and in Appendix A, which we call *Hints and Tricks*, so familiarity with MATLAB™ will be helpful but not required as well.[1]

We teach our course in a one-semester blitz divided into three parts. And, yes, taking the course is a little like drinking from a fire hose, but we feel that there is something beneficial about the Zen-like concentration required. The first part of the book deals with the mathematical machinery of data analysis that generally goes under the heading of *statistics*, although strictly speaking some of it is not really statistics (e.g. principal component analysis). The second part deals with the techniques of modeling that we choose to cover in this book: finite differences, stability analysis, and optimization. The third part of this book deals with case studies of actual, published models, of ever increasing dimensionality and complexity, starting with zero-dimensional models and finishing with three-dimensional general circulation models. Our goal is to instill a good conceptual grasp of the basic tools underlying the model examples. We like to say the book is correct, but not mathematically rigorous. Throughout the book the general principles and goals of scientific visualization are emphasized through technique and tools. A final chapter on scientific visualization reviews and cements these principles.

This book makes a very nice basis for a one- or two-semester course in data analysis and numerical modeling. It begins with data analysis techniques that are not only very useful in interpreting actual data, but also come up again and again in analyzing model output (computa). This first "third" of the book could also be used in a one-semester data analysis course. It begins with an introduction to both MATLAB and singular value decomposition via a review of some basic linear algebra. Next the book covers measurement theory, probability distributions, and error propagation. From here the book covers least squares regression (both linear and nonlinear) and goodness of fit (χ^2). The next analysis technique is principal component analysis which begins with covariance, correlation, and ANOVA and ends with factor analysis. No data analysis course would be complete without a treatment of sequence data starting with auto- and cross-correlation, proceeding through Fourier series and transforms, and optimal filtering, and finishing up with, of course, the FFT. We finish the data analysis third of the book with a chapter on gridding and contouring techniques from simple nearest neighbor methods to objective interpolation (kriging).

The middle third of this book is the transitional segment of any course that attempts to bring together data analysis techniques and numerical modeling. However, this portion of the book can also be used as part of a more traditional course on numerical modeling. We begin with integration of ordinary differential equations and introduce some simple but useful zero-dimensional models. At this point we pause for a chapter and present a tutorial on model building, practical things one needs to consider no matter how simple or complex the problem. We then demonstrate how the parameters in such models can be optimized

---

[1] MATLAB is a registered trademark of The MathWorks, Inc., of Natick, MA 01760, USA. In order to avoid the appearance of crowded redundancy we are dropping the TM from the name, but when we write MATLAB we are referring to the trademarked product.

with respect to actual data. When the problems become too complex to be expressed as ordinary differential equations we use partial differential equations, and a discussion and practical introduction to the advection–diffusion equation and turbulence is presented. Next the concept of finite differences is developed to solve these complex problems. In the final chapter of this middle section we cover the important topics of von Neumann stability analysis (Fourier resurfaces), conservation, and numerical diffusion.

We find that, at this point, the students are primed and ready to tackle some "real" models. However, this final third of the book could be used to augment a modeling survey course, although, to get the most out of such a survey the students would need to be well versed in numerical modeling techniques. The book takes the reader through a series of models beginning with simple one-dimensional models of the ocean that rely heavily on lessons learned in the earlier, nonlinear regression section. There are also one-dimensional models of the upper sediment and a very thorough exposition of a one-dimensional, seasonal model of the upper ocean water column. This last section of the book transitions to two-dimensional gyre models and culminates with a chapter on three-dimensional, general circulation models. Up to this point all of these models have been "forward" models. The final third of the book wraps up with "inverse" models. Here we introduce the concepts of inversion and data assimilation and return to the lessons learned from the singular value decomposition chapter at the beginning of the book. It is followed by three-dimensional inversions involving two-dimensional slices of the ocean.

There are certainly many "mathematical methods" books on the market. But this book is the only one we know of that attempts to synthesize the techniques used for analyzing data with those used in designing, executing, and evaluating models. So, where on one's bookshelf does this volume fit in? It goes in that gap that exists between your copies of Stumm and Morgan's *Aquatic Chemistry* and Broecker and Peng's *Tracers in the Sea* on one side, and Pedlosky's *Geophysical Fluid Dynamics* and Wunsch's *The Ocean Circulation Inverse Problem* on the other. Although we are, and our examples reflect this, oceanographers, we feel that scientists in other fields will find our explanations and discussions of these techniques useful. For while the density, pressure, and nature of the problem being analyzed and/or modeled may be vastly different from the ones commonly encountered in the pages of this book, the mathematics remain the same.

Over the years we have had a great deal of help (particularly from our students, who take great pride and pleasure in finding mistakes in our notes) in pulling together the information found between the covers of the book you hold in your hand. We thank each and every one of our students, friends, and colleagues who have contributed to the betterment of this work. However, at the end of the day, we take full responsibility for the accuracy of our work, and deficiencies therein are our responsibility.

1

Resources, MATLAB primer, and introduction to linear algebra

> "Begin at the beginning," the King said, very gravely, "and go on till you come to the end: then stop."
>
> *Lewis Carroll*

Welcome to *Modeling Methods for Marine Science*. The main purpose of this book is to give you, as ocean scientists, a basic set of tools to use for interpreting and analyzing data, for modeling, and for scientific visualization. Skills in these areas are becoming increasingly necessary and useful for a variety of reasons, not the least of which are the burgeoning supply of ocean data, the ready availability and increasing power of computers, and sophisticated software tools. In a world such as this, a spreadsheet program is not enough. We don't expect the reader to have any experience in programming, although you should be comfortable with working with computers and web browsers. Also, we do not require any background in sophisticated mathematics; undergraduate calculus will be enough, with some nodding acquaintance with differential equations. However, much of what we will do will not require expertise in either of these areas. Your most valuable tool will be common sense.

1.1 Resources

The activities of modeling, data analysis, and scientific visualization are closely related, both technically and philosophically, so we thought it important to present them as a unified whole. Many of the mathematical techniques and concepts are identical, although often masked by different terminology. You'll be surprised at how frequently the same ideas and tools keep coming up. The purpose of this chapter is threefold: to outline the goals, requirements, and resources of the book, to introduce MATLAB (and give a brief tour on how to use it), and to review some elements of basic linear algebra that we will be needing in subsequent chapters.

As we stated in the preface (you *did* read the preface, didn't you?), our strategy will be to try to be as "correct" as possible without being overly rigorous. That is, we won't be dragging you through any theorem proofs, or painful details not central to the things you need to do. We will rely quite heavily on MATLAB, so you won't need to know

1

how to invert a matrix or calculate a determinant. But you should know qualitatively what is involved and why some approaches may be better than others; and, just as important, why some procedures fail. Thus we will try to strike a happy medium between the "sorcerer's apprentice" on the one extreme and having to write your own FORTRAN programs on the other. That is, there will be no programming in this book, although you will be using MATLAB as a kind of programmer's tool. The hope is that the basic mathematical concepts will shine through without getting bogged down in the horrid details.

1.1.1 Book structure

The course of study will have four sections. The first section (Chapters 1–7) is aimed at data analysis and some statistics. We'll talk about least squares regression, principal components, objective analysis, and time-series approaches to analyzing and processing data. The second section (Chapters 8–12) will concentrate on the basic techniques of modeling, which include numerical techniques such as finite differencing. The third (Chapters 13–17) will consist of case studies of ocean models. The largest amount of time will be spent on 1D models (and to a lesser extent 2D and 3D models) since they contain most of the important elements of modeling techniques. The final section (Chapters 18–19) will discuss inverse methods and scientific visualization, which in some respects is an emerging tool for examining model and observational data. Throughout the text will be examples of real ocean and earth science data used to support the data analysis or modeling techniques being discussed.

1.1.2 World Wide Web site

We teach a course at the Woods Hole Oceanographic Institution using this text, and this text is supported by a web page (`http://www.cambridge.org/glover`), where the most up-to-date versions of our codes are available. The MathWorks also maintains a web presence (`http://www.mathworks.com/`) and versions of our work may be available there.

We draw upon a number of other sources, some from textbooks of a more applied mathematics background and others from the primary ocean literature. Davis's (2002) book is a primary example of the former, although it really only covers the first section of this book. Bevington and Robinson (2003) is such a useful book that we strongly recommend you obtain a copy (sadly it is not as inexpensive as it once was). The book by Press *et al.* (2007) is useful for the second section, although the Roache (1998) book is best for finite difference techniques (and with the latest edition it is happily no longer out-of-print). The third and fourth sections will rely on material to be taken from the literature. There is a list of references at the end of the book, and the other texts are listed only as supplemental references in case you (individually) have a need to delve more deeply into some aspects of the field.

1.2 Nomenclature

We could put a lot of quote marks around the various commands, program names, variables, etc. to indicate these are the things you should type as input or expect as output. But this would be confusing (is this an input? or an output?) plus it would be a real drag typing all those quote marks. So a word about nomenclature in this text. This book is being typeset with LaTeX, and we will try to use various fonts to designate different things in a consistent manner. From now on, when we mean this is something that is "computer related" (the name of a file, a variable in a MATLAB program, a MATLAB command, etc.) it will be in `simulated typewriter font`. If it is a downloadable link, it will be underlined as in `http://URL` or `filename`, at least on first introduction. If, however, we are referring to the mathematics or science we are discussing (variables from equations, mathematical entities, etc.), we will use *mathtext* fonts. In particular, scalar variables will be in simple italic font (x), vectors will be lowercase bold face (**b**), and matrices will be upper case bold face (**A**).

If you have read enough math books, you have learned that there is no universally consistent set of symbols that have unique meanings. We will try to be consistent in our usage within the pages of our book, but there are only 26 letters in our alphabet and an even smaller number of Greek characters, so some recycling is inevitable. As always, the context of the symbol will be your best guide in deciphering whether this is the λ from Chapter 4 (eigenvalues) or Chapter 13 (decay constants). On the first introduction of new jargon or names of things the text will be *italicized*.

1.3 A MATLAB primer

You can read this book, and benefit greatly, without ever touching a computer. But this is a lot like theoretical bicycle riding; there is no better way to learn than with hands on experience, in this case hands on a keyboard. MATLAB is a very powerful tool (there are others), and for showing practical examples it is more useful than pseudocode.

The best way to learn how to use MATLAB is to do a little tutorial. There are at least two places where you can get hold of a tutorial. The best is the *MATLAB Primer* (Davis and Sigmon, 2004), the seventh edition as of this writing. A second place is to use the online help available from your MATLAB command window. If you are using the MATLAB graphical user interface (GUI), pull down the help menu from the toolbar at the top of the command window and select "MATLAB help". This will start the online documentation MATLAB help browser. In the "Contents" tab on the left select the "MATLAB" level and go to "Printable Documentation (PDF)" to find a listing of manuals available to you. The first one, "Getting Started", is a good place to begin. If you are not using the command window GUI, type `helpdesk` at the MATLAB prompt, and this online documentation help browser will be launched. If you get an error with the `helpdesk` command we suggest you speak with your system administrator to learn how to configure your web browser or type `help docopt` if you are an experienced computer user. Whichever one you pick, just start at the beginning and work through the tutorial. It basically shows you how to do most things. Don't worry if some of the linear algebra and matrix math things don't make much sense yet.

Read the primer. We won't waste time repeating the words here, except to point out a few obvious features.

- MATLAB is case sensitive; a variable named A is not the same as a. Nor is `Alpha` the same as `ALPHA` or `alpha` or `AlPhA`. They are all different.
- Use `help` and `lookfor`. If you don't know the name of a command, but, for example, want to know how to make an identity matrix, type `lookfor identity`. You will then be told about `eye`. Use `help eye` to find out about the `eye` command. Note that MATLAB capitalizes things in its help messages for EMPHASIS, which confuses things a little. Commands and functions are *always* in lower case, although they are capitalized in the help messages.
- Remember that matrix multiplication is not the same as scalar or array multiplication; the latter is designated with a "dot" before it. For example C=A*B is a matrix multiplication, whereas C=A.*B is array multiplication. In the latter, it means that the elements of **C** are the scalar product of the corresponding elements of **A** and **B** (i.e. the operation is done element by element).
- The colon operator (:) is a useful thing to learn about; in addition to being a very compact notation, it frequently executes much, much faster than the equivalent `for ... end` loop. For example, j:k is equivalent to [j, j+1, j+2, ..., k] or j:d:k is equivalent to [j, j+d, j+2*d, ..., j+m*d] where m = fix((K-J)/D). There's even more: the colon operator can be used to pick out specific rows, columns, elements of arrays. Check it out with `help colon`.
- If you don't want MATLAB to regurgitate all the numbers that are an answer to the statement you just entered, be sure to finish your command with a semicolon (;).

MATLAB has a "scripting" capability. If you have a sequence of operations that you routinely do, you can enter them into a text file (using your favorite text editor, or better yet, MATLAB's editor) and save it to disk. By default, all MATLAB script files end in a .m so that your script (or "m-file") might be called `fred.m`. You can edit this file with the MATLAB command `edit fred`; if the file does not exist yet, MATLAB will prompt you, asking if you wish to create it. Then, you run the script by entering `fred` in your MATLAB window, and it executes as if you had typed in each line individually at the command prompt. You can also record your keystrokes in MATLAB using the `diary` command, but we don't recommend that you use it; better to see A.1 in the appendix of this book. You will learn more about these kind of files as you learn to write functions in MATLAB.

You can load data from the hard drive directly into MATLAB. For example, if you had a data file called `xyzzy.dat`, within which you had an array laid out in the following way:

$$
\begin{array}{cccc}
1 & 2 & 3 & 4 \\
5 & 6 & 7 & 8 \\
9 & 0 & 1 & 2
\end{array}
$$

then you could load it into MATLAB by saying `load xyzzy.dat`. You would then have a new matrix in your workspace with the name `xyzzy`. Note that MATLAB would object (and rightly so, we might add) if you had varying numbers of numbers in each row, since that doesn't make sense in a matrix. Also, if you had a file named `jack.of.all.trades` you would have a variable named `jack` (MATLAB is very informal that way). Note that if you had a file without a "." in its name, MATLAB would assume that it was a "mat-file", which is a special MATLAB binary format file (which you cannot read/modify with an editor). For example, `load fred` would cause MATLAB to look for a file called `fred.mat`. If it doesn't find it, it will complain. But first, make sure MATLAB is looking in the correct file directory, which is an equivalent way of saying make sure the file is in MATLAB's PATH.

You can save data to disk as well. If you simply type `save`, MATLAB saves everything in your workspace to a file called `matlab.mat`. Don't try to read it with an editor (remember it's in binary)! You can recover everything in a later MATLAB session by simply typing `load`. You can save a matrix to disk in a "readable" file by typing `save foo.dat xyzzy -ascii`. In this case you have saved the variable `xyzzy` to the file `foo.dat` in ASCII (editable) form. You can specify more than one variable (type `help save` to find out more). Remember the `-ascii`, because nobody but MATLAB can read the file if you forget it.

You can even read and write files that are compatible with (shudder!) Excel. There are a number of ways to do this. For example, to read an Excel file you can use `A=xlsread('filename.xls')`, and the numeric data in `filename.xls` will be found in the MATLAB variable `A`. The `xlsread` function has a number of other capabilities; to learn more simply type `help xlsread`. MATLAB even has a function that will tell you things about what is inside the Excel file, for example `SheetNames`; to learn more type `help xlsfinfo`. Also, the MATLAB functions `csvread` and `csvwrite` facilitate transferring data to and from Excel; do a `help csvread` to have MATLAB explain how to use these functions.

A final word about the MATLAB code presented in this book. As we write (and rewrite) these chapters we are using MATLAB release R2007b (depending whether we upgraded recently). To the best of our knowledge, all of the examples and programs we provide in this book are compatible with R2007b (version 7.5). As time goes on, some of our code will undoubtedly become incompatible with future MATLAB release X. To deal with this eventuality we have decided to make our material available on web pages (http://www.cambridge.org/glover) instead of the more static CD-ROM media.

1.4 Basic linear algebra

A scalar is a single number. A vector is a row or column of numbers. You can plot a vector, for example $[3, 7, 2]$, which would be an arrow going from the origin to a point in three-dimensional space indicated by $x = 3$, $y = 7$, and $z = 2$. A matrix may be thought of as a bundle of vectors, either column or row vectors (it really depends on what "physical reality" the matrix represents). If each of the vectors in a matrix is at right angles to all of its mates, then they are said to be *orthogonal*. They are also called *linearly independent*. If the lengths of the vectors, as defined by the square root of the sum of the squares of its components (i.e. $\sqrt{x^2 + y^2 + z^2}$), are also 1, then they are said to be *orthonormal* (ortho – at right angles, normal – of unit length). For example, a vector $[1/\sqrt{2}, 0, 1/\sqrt{2}]$ has a length of 1, as does $[1/\sqrt{3}, 1/\sqrt{3}, 1/\sqrt{3}]$ and $[0, 0, 1]$.

Before we start, there are some simple rules for matrix manipulation. First, you can only add or subtract matrices of the same size (same number of rows and columns). The one exception in MATLAB to this is when you add or subtract scalars to/from a matrix. In that case the scalar is added/subtracted from each element of the matrix individually. Second, when you multiply matrices, they must be *conformable*, which means that the left matrix must have the same number of columns as the right matrix has rows:

$$\begin{bmatrix} 1 & 2 & 4 \\ 5 & 8 & 7 \end{bmatrix} \times \begin{bmatrix} 3 & 9 \\ 2 & 4 \\ 1 & 8 \end{bmatrix} = \begin{bmatrix} 11 & 49 \\ 38 & 133 \end{bmatrix} \tag{1.1}$$

Matrix multiplication is non-commutative. That is, in general, $\mathbf{A} \times \mathbf{B}$ is not the same as $\mathbf{B} \times \mathbf{A}$ (in fact, the actual multiplication may not be defined in general). The algorithm for matrix multiplication is straightforward, but tedious. Have a look at a standard matrix or linear algebra text to see how matrix multiplication works. Even though you won't actually be doing matrix multiplication by hand, a lot of this stuff is going to make more sense if you understand what is going on behind the scene in MATLAB; Strang (2005) is a good place to start.

1.4.1 Simultaneous linear equations

The whole idea of linear algebra is to solve sets of simultaneous linear equations. For example, consider the following system of equations:

$$\begin{aligned} x + 2y + z &= 8 \\ 2x + 3y - 2z &= 2 \\ x - 2y + 3z &= 6 \end{aligned} \tag{1.2}$$

which can be represented as:

$$\mathbf{Ax} = \mathbf{b} \tag{1.3}$$

where **A**, the matrix of the coefficients, looks like:

$$\mathbf{A} = \begin{pmatrix} 1 & 2 & 1 \\ 2 & 3 & -2 \\ 1 & -2 & 3 \end{pmatrix} \tag{1.4}$$

The column value vector (**b**) representing the *right-hand side* (RHS) of the equation system is:

$$\mathbf{b} = \begin{pmatrix} 8 \\ 2 \\ 6 \end{pmatrix} \tag{1.5}$$

Note that the matrix **A** contains the coefficients of the equations, **b** is a column vector that contains the knowns on the RHS, and **x** is a column vector containing the unknowns or "target variables".

You enter **A** into MATLAB the following way:

```
A = [ 1 2 1; 2 3 -2; 1 -2 3];
```

(the array starts and ends with the *square bracket*, and the rows are separated by semicolons; also we've terminated the statement with a semicolon so as not to have MATLAB regurgitate the numbers you've just typed in.) You can enter **b** with:

```
b=[8; 2; 6]
```

Finally, the column "unknown" vector is:

$$\mathbf{x} = \begin{pmatrix} x \\ y \\ z \end{pmatrix} \tag{1.6}$$

(Don't try to enter this into MATLAB, it's the answer to the question!)

Now how do you go about solving this? If we gave you the following scalar equation:

$$3x = 6 \tag{1.7}$$

then you'd solve it by dividing both sides by 3, to get $x = 2$. Similarly, for a more general scalar equation:

$$ax = c \tag{1.8}$$

you'd do the same thing, getting $x = c/a$. Or more appropriately $x = (1/a) \times c$, or put slightly differently, $x = inv(a) \times c$. Here, we've said that $inv(a)$ is just the *inverse* of the number.

Well, you can do a similar thing with matrices, except the terminology is a little different. If you enter these data into MATLAB, then you can solve for the values of the three variables (x, y, and z) with the simple statement x=A\b. This is really equivalent to the statement x=inv(A)*b. Now check your answer; just multiply it back out to see if you get b by typing A*x. The answers just pop out! This is simple. You could do this just as easily for a set of 25 simultaneous equations with as many unknowns (or 100 or 10^4 if you have a big computer).

But what did you just do? Well, it's really simple. Just as simple scalar numbers have inverses (e.g., the number 3 has an inverse; it's 1/3), so do matrices. With a scalar, we demand the following:

```
scalar * inv(scalar) = 1
```

```
i.e.:   3 * 1/3 = 1
```

So with a matrix, we demand:

```
matrix * inv(matrix) = I
```

Here we have the matrix equivalent of "1", namely I, the *identity matrix*. It is simply a square matrix of the same size as the original two matrices, with zeros everywhere, except on the diagonal, which is all ones. Examples of identity matrices are:

$$
\begin{matrix} 1 & 0 \\ 0 & 1 \end{matrix} \qquad \begin{matrix} 1 & 0 & 0 \\ 0 & 1 & 0 \\ 0 & 0 & 1 \end{matrix} \qquad \begin{matrix} 1 & 0 & 0 & 0 \\ 0 & 1 & 0 & 0 \\ 0 & 0 & 1 & 0 \\ 0 & 0 & 0 & 1 \end{matrix}
$$

Well, you get the idea. Oh, by the way, a matrix must be square (at the very least) to have an inverse, and the inverse must be the same size as the original. Note that like its scalar little brother, "1", the identity matrix times any matrix of the same size gives the same matrix back. For example, A*I=A or I*A=A.

OK, now try this with the matrix you keyed into MATLAB. Type A*inv(A) (you are multiplying the matrix by its inverse). What do you get? You get the identity matrix. Why are some "ones" represented as "1" and some by "1.000"? Also, you sometimes get 0.0000 and −0.0000 (yeah, we know, there's no such thing as "negative zero"). The reason is that the computation of the matrix inverse is not an exact process, and there is some very small roundoff error (see Chapter 2, Section 2.1.5). Which means that "1" is not exactly the same as "1.00000000000000000000000", but is pretty darn close. This is a result of both the approximate techniques used to compute the inverse, and the finite precision of computer number representation. It mostly doesn't matter, but can in some special cases. Also try multiplying the matrix A by the identity matrix, A*eye(3) (the identity matrix of rank 3). What is *rank*? Keep reading!

Finally, let's do one more thing. We can calculate the *determinant* of a matrix with:

```
d=det(A)
```

The determinant of a matrix is a scalar number (valid only for square matrices) and gives insight into the fundamental nature of the matrix. If it is zero, there may be trouble ahead. We will run into the determinant in the future. We won't tell you how to calculate it, since

MATLAB does such a fine job of doing it anyway. If you're interested, go to a matrix math text. Anyway, now calculate the determinant of the inverse of A with:

```
dd=det(inv(A))
```

(See how we have done two steps in one; MATLAB first evaluates the `inv(A)` then feeds the result into `det()`.) Guess what? `dd = 1/d`. Before we do anything else, however, let's save this matrix to a new variable AA with:

```
AA=A;
```

Now that you see how solving linear equations works, try another set of equations:

$$x + 2y + z = 8$$
$$2x + 3y - 2z = 2 \tag{1.9}$$
$$3x + 5y - z = 10$$

i.e., you enter,

```
A = [1 2 1; 2 3 -2; 3 5 -1]
b = [8; 2; 10]
x = A\b
```

Whoops! You get an error message:

```
Warning: Matrix is close to singular or badly scaled.
         Results may be inaccurate. RCOND = 1.171932e-017
```

(Note that your result for RCOND might be a little different, but generally very small.)

What happened? Well, look closely at the original set of equations. The third equation is really not very useful, since it is the sum of the first two equations. It is not linearly independent of the other two. You have in effect two equations in three unknowns, which is therefore not solvable. This is seen in the structure of the matrix. The error message arises when you try to invert the matrix because it is *rank deficient*. If you look at its rank, with `rank(A)`, you get a value of 2, which is less than the full dimensionality (3) of the matrix. If you did that for the first matrix we looked at, by typing `rank(AA)`, it would return a value of 3.

Remember our friend the determinant? Try `det(A)` again. What value do you get? Zero. If the determinant of the inverse of A is the inverse of the determinant of A (get it?), then guess what happens? A matrix with a determinant of zero is said to be *singular*.

1.4.2 Singular value decomposition (SVD)

Common sense tells you to quit right there. Trying to solve two equations with three unknowns is not useful ... or is it? There are an infinite number of combinations of x, y, and z that satisfy the equations, but not every combination will work. Sometimes it is of value to know what the range of values is, or to obtain some solution subject to some other (as yet to be defined) criteria or conditions. We will tell you of a sure-fire technique to do this, *singular value decomposition*. Here's how it works.

You can split a scalar into an infinite number of factors. For example, you can represent the number 12 in the following ways:

- 2×6
- 3×4
- 1×12
- $2 \times 3 \times 2$
- $1 \times 6 \times 2$
- 0.5×24
- 1.5×8
- $1\frac{2}{3} \times 7.20$
- $24 \times 1 \times 2 \times 0.25$

and so on. You can even require (constrain) one or more of these factors to be even, integer, etc. Well, the same is true for matrices. This leads to a host of different techniques called *decomposition*. You'll hear of various ones, including "LU", "QR" and so on. Their main purpose is to get the matrices in a form that is useful for solving equations, or eliminating parts of the matrix. They are all used by MATLAB but the one you should know about is SVD (which is short for singular value decomposition). It is nothing magic, but allows you to break a matrix down into three very useful components. The following can be "proved" or "demonstrated" with several pages of matrix algebra, but we will just make an assertion. We assert that for any matrix (**A**) of dimension $N \times M$ (N rows, M columns), there exists a triple product of matrices:

$$\mathbf{A} = \mathbf{USV}' \tag{1.10}$$

where:

U is column orthonormal (i.e. each column as a vector is orthogonal to the others and of "unit length" – sum of squares of elements is 1) and of size $N \times M$,

S is a diagonal matrix of dimension $M \times M$, whose diagonal elements are called *singular values*. These values may be zero if the matrix is *rank deficient* (i.e. its rank is less than the shortest dimension of the matrix),

V is an orthonormal square matrix of size $M \times M$. In linear algebra and MATLAB **V**′ means **V** *transpose* where you swap rows and columns, i.e. if

$$\mathbf{G} = \begin{pmatrix} 1 & 2 & 3 \\ 4 & 5 & 6 \\ 7 & 8 & 9 \end{pmatrix} \quad \text{then} \quad \mathbf{G}' = \begin{pmatrix} 1 & 4 & 7 \\ 2 & 5 & 8 \\ 3 & 6 & 9 \end{pmatrix} \tag{1.11}$$

Note that there are two ways of defining the size of the matrices, you may come across the other in Strang's (2005) book, but the results are the same when you multiply them out. The actual procedure for calculating the SVD is pretty long and tedious, but it always works regardless of the form of the matrix. You accomplish this for the first matrix in MATLAB in the following way:

```
[U,S,V]=svd(AA,0);
```

That's how we get more than one thing back from a MATLAB function call; you line them up inside a set of brackets separated by commas on the *left-hand side* (LHS) of the equation. You can get this information by typing help svd. Note also that we have included a, 0 after the AA. This selects a special (and more useful to us) form of the SVD output. To look at any of the matrices, simply type its name. For example, let's look at S by typing S. Note that for the matrix AA, which we had no trouble with, all three *singular values* are non-zero.

```
S =

    5.1623         0              0
         0    3.0000              0
         0         0         1.1623
```

Now try it with the other, troublesome matrix:

```
[U,S,V]=svd(A,0);
```

and after typing S, you can see that the lowest right-hand element is zero. This is the trouble spot!

```
S =

    7.3728         0              0
         0    1.9085              0
         0         0         0.0000
```

Now, we don't need to go into the details, but it can be proven that you can construct the matrix inverse from the relation inv(A) = V*W*U', as we would write it in MATLAB, where W is just S with the diagonal elements inverted (each element is replaced by its inverse). For a rank deficient matrix (like the one we had trouble with), at least one of the diagonal elements in S is zero. In fact, the number of non-zero singular values is the *rank* of the matrix. Thus if you went ahead and blindly inverted a zero element, you'd have an infinity. The trick is to replace the inverse of the zero element with zero, not infinity. Doing that allows you to compute an inverse anyway.

We can do this *inversion* in MATLAB in the following way. First replace any zero elements by 1. You convert the diagonal matrix to a single column vector containing the diagonal elements with:

```
s=diag(S)
```

(note the lower and upper case usage). Then set the zero element to 1 with:

`s(3)=1`

Then invert the elements with:

`w=1./s`

(note the decimal point, which means do the operation on each element, as an "array oper-ation" rather than a "matrix operation"). Next, make that pesky third element *really* zero with:

`w(3)=0;`

Then, convert it back to a diagonal matrix with:

`W=diag(w)`

Note that MATLAB is smart enough to know that you are handing it a column vector and to convert it to a diagonal matrix (it did the opposite earlier on). Now you can go ahead and calculate the best-guess inverse of A with:

`BGI=V*W*U'`

where `BGI` is just a new variable name for this inverted matrix. Now, try it out with:

`A*BGI`

Bet you were expecting something like the identity matrix. Instead you get:

```
 0.6667   -0.3333    0.3333
-0.3333    0.6667    0.3333
 0.3333    0.3333    0.6667
```

Why isn't it identity? Well, the answer to that question gets to the very heart of *inverse theory*, and we'll get to that later in this book (Chapter 18). For now we just want you to note the symmetry of the answer to `A*BGI` (i.e., the 0.6667 down the diagonal with a positive or negative 0.3333 everywhere else).

Now, let's get down to business, and get a solution for the equation set. We compute the solution with:

`x=BGI*b`

which is:

$$x = \begin{pmatrix} 0.7879 \\ 2.1212 \\ 2.9697 \end{pmatrix} \tag{1.12}$$

Do you believe the results? Well, try them with:

`A*x`

which gives you the original b! But why this solution? For example, x = [1 2 3]'
works fine too (entering the numbers without the semicolons gives you a row vector, and
the prime turns a row vector into a column vector, its transpose). Well, the short answer
is because, of all of the possible solutions, this one has the shortest *length*. Check it out:
the square root of the sum of the squares of the components of [1 2 3]' is longer than
the vector you get from BGI*b. The reason is actually an important attribute of SVD, but
more explanations will have to wait for Chapter 18.

Also, note that the singular values are arranged in order of decreasing value on page 11.
This doesn't have to be the case, but the MATLAB algorithm does this to be nice to you.
Also, the singular values to some extent tell you about the *structure* of the matrix.

Not all cases are as clear-cut as the two we just looked at. A matrix may be nearly singular, so that although you get an "answer," it may be subject to considerable uncertainties
and not particularly robust. The ratio of the largest to the smallest singular values is the
condition number of the matrix. The larger it is, the worse (more singular) it is. You can
get the condition number of a matrix by entering:

cond(A)

In fact, MATLAB uses SVD to calculate this number. And RCOND is just its reciprocal.

So what have we learned? We've learned about matrices, and how they can be used to
represent and solve systems of equations. We have a technique (SVD) that allows us to
calculate, under any circumstances, the inverse of a matrix. With the inverse of the matrix,
we can then solve a set of simultaneous equations with a very simple step, even if there is
no unique answer. But wait, there's more! Stay tuned ...

1.5 Problems

1.1. Download the data matrix A.dat and the column vector b.dat (remember to put
     them in the same directory in which you use MATLAB). Now load them into MAT-
     LAB using the command load A.dat and load b.dat. (Make sure you are in
     the same directory as the files!)

   (a) Now solve the equation set designated by Ax=b. This is a set of seven equations
       with seven unknowns. List the values of x.
   (b) What is the rank and determinant of A?
   (c) List the singular values for A.

1.2. Download A1.dat and b1.dat. Load them into MATLAB. Then do the following:

   (a) What is the rank and determinant of A1?
   (b) What happens when you solve A1*x=b1 directly?
   (c) Do a singular value decomposition, compute the inverse of A1 by zeroing the
       singular value inverses and solve for x.

2

Measurement theory, probability distributions, error propagation and analysis

> ... [When] you can measure what you are speaking about and express it
> in numbers, you know something about it; but when you cannot express
> it in numbers, your knowledge is of a meagre and unsatisfactory kind; it
> may be the beginning of knowledge, but you have scarcely in your
> thoughts advanced to the state of science, whatever the matter may be.
>
> *Lord Kelvin*

In science, it's not enough to perform a measurement or make an estimate and report a number. It is equally important that you quantify *how well* you know the number. This is crucial for comparing measurements, testing theories, debunking bad models, and making real-world decisions. In order to do this, you need to know something about basic measurement theory, and how different sources of uncertainty combine to influence your final results. In this chapter we will be going over the fundamental statistical concepts that lie underneath all that we do when analyzing our data and models. We discuss some important issues that lie at the heart of measurement theory (scales, errors, and roundoff) and look at the concept of probability distributions. Related to this is the very delicate question of whether, how, and why some experimental data should be rejected. In short, this chapter summarizes the science of knowing what you know and how well you know it.

2.1 Measurement theory

2.1.1 Systems of measurements (scales)

How we measure or characterize things in science depends on our objectives. Are we trying to itemize or list things, or are we trying to compare them? Are there some fundamental, underlying quantities that we are trying to compute or model? There are essentially four scales of measurements: nominal, ordinal, interval, and ratio.

- **Nominal scales** are classification of data into mutually exclusive categories of equal rank. For example, "These are forams, those are diatoms, and that's left over from my lunch."
- **Ordinal scales** are observations organized in some qualitatively consistent order, but the steps between ranks are not necessarily equal in size, or even quantifiable. For example,

on the Mohs scale of hardness, talc has a hardness of 2 and orthoclase a hardness of 6, but orthoclase isn't really three times harder than talc. Another example is the Beaufort wind scale.

- **Interval scales** are used when the data can be put on a scale with successive intervals that are constant, but without any absolute, non-arbitrary reference point. For example, consider the centigrade temperature scale: even though it is defined by the freezing and boiling points of pure water, seawater with a temperature of 20 °C doesn't really have twice the heat content of seawater with a temperature of 10 °C.
- **Ratio scales** are used for measurements like the interval scale, but for data that do have a natural zero point. Consider the Kelvin temperature scale, or the length of things. Another example would be dissolved oxygen concentrations; a sample with 300 μmol/kg oxygen concentration does indeed have twice as much oxygen as a sample of the same mass with 150 μmol/kg concentration.

With some thought, you can see that what you can do with a measurement depends on what type of scale you are using. Two measurements made in a nominal scale cannot sensibly be subtracted; e.g. subtracting forams from diatoms doesn't make sense. Similarly for ordinal scales, adding the Mohs hardness of talc to that of diamond doesn't mean anything. Differencing interval scale measurements, however, does make sense. Knowing the temperature difference in centigrade (an interval scale) between the surface of the ocean and the atmosphere at some location will tell you how much sensible heat is transferred (assuming you know the physics well enough).

2.1.2 Precision versus accuracy

Related to these scales in a general sense are the concepts of *precision* and *accuracy*. You would be surprised at how often the two terms are used interchangeably in the published literature. You'd almost think the authors were confused about it, but we will assume that they're just being casual about things. Think of precision as simply being a measure of how well you can repeat your measurements. The more often you can get numbers that are close to each other for the same measurement procedure measuring the same thing, the better the precision (i.e. the smaller the "error"). You could imagine that repeating the measurement many times with the same device would allow you to average the results and somehow obtain a more precise determination. Implicit in this argument is the asumption that small errors associated with each measurement should somehow average out to zero if you did it enough times. If the device were somehow flawed (e.g. the end of your ruler had been bitten off by a starving graduate student), your multiple results would still average to some more precise value but would be nonetheless wrong. Accuracy, on the other hand, is a measure of how close you have come to the *true* value of whatever it is you are trying to measure. It is clearly the more desirable attribute, but one that is inherently more difficult to determine and demonstrate experimentally.

2.1.3 Systematic versus random errors

The concepts of precision and accuracy are closely related to systematic and random errors. A systematic error is a reproducible discrepancy that makes your measurement different from the "true" value; it directly influences *accuracy*. No amount of repeated measurement using the same technique or apparatus will improve things. Random errors, however, are due to fluctuations in the environment that yield results that differ from experiment to experiment even though you believe that the conditions imposed by you were identical; these types of errors influence *precision*. This belief implies these variations are beyond our knowledge or control, but because they are random in nature, sufficient repeated measurement leads to improvement in some gradual fashion by averaging or cancelling out.

2.1.4 Significant figures and roundoff

Significant figures and roundoff are computational issues: both what you do, and what you have your computer do. The number of significant digits (figures) that you report implies the amount of precision you were able to accomplish, so pay careful attention to how many digits you report. Further, how many significant figures you report in your results is influenced by another type of error that creeps into experimental work, *roundoff error*. You may be surprised to know that the following numbers all have the same number of significant figures:

$$1234 \quad 12\,340 \quad 123.4 \quad 12.34$$
$$1001 \quad 10.10 \quad 0.000\,101\,0 \quad 100.0$$

The rule goes like this: if there is no decimal point then the left-most non-zero number is the most significant digit, and the right-most number is the least significant digit provided it is not zero. If there is a decimal point, then the most significant digit is found as above, but now the right-most number, even if it is zero, is the least significant number. The least significant number implies the *precision*, so you need to be honest with your reader when you report your data.

Rounding off is usually done to the results to set the number of significant digits you report. It implies the precision of the result, and it is something we do deliberately. For example, if you measure a cylinder with a ruler that has millimeters as its smallest markings, you wouldn't report its length as 34.347 82 mm. The same is true with computed values; no matter how many digits your calculator or computer displays, you should only include the appropriate number of significant figures. It gets even more complicated when you think about the fact that errors "propagate" or combine in complicated ways when you use data to compute something else. We'll deal with that later in this chapter.

The rules for rounding off go like this: the change in the digit that you are rounding off is governed by the digit(s) to its immediate right. If the digit immediately to its right is greater than 5 you round the least significant digit up one count. If it is less than 5 you leave it the same (you simply truncate). The special case of when the digit to the right is exactly 5 is governed by the following rule: round up if the least significant digit is odd, truncate if it is even. For example:

$$1.979 \pm 0.082 \quad \text{becomes} \quad 1.98 \pm 0.08$$
$$25.8350 \quad \text{becomes} \quad 25.84$$
$$15.9450 \quad \text{becomes} \quad 15.94$$

The reason that odd and even numbers are treated differently in the "exactly 5" case is that you want the net effect to average out in case of many repetitions. If you treated them both the same, you'd introduce a bias.

2.1.5 Computational roundoff and truncation

Computational roundoff is done by the computer and is something we all need to be aware of. Generally, numerical roundoff error comes in two varieties: machine roundoff error and algorithm truncation error. Machine roundoff results from the way numbers are represented inside the computer. All numbers are stored as binary numbers, consisting of clusters of eight bits (each bit is a binary entity that may take on only one of two values, 0 or 1) called bytes. Numbers are either stored as *integers* (i.e. whole numbers), or as *floating point*, which can range from very small (e.g. 1.672×10^{-24}) to very large (e.g. 6.025×10^{27}) and much more.

For integer operations, numerical roundoff is only a problem when the operation produces a number too large in magnitude for the machine to represent. This is often referred to as *overflow* (when too positive) or *underflow* (when too negative) error. Typically, computers (or more properly programming languages and compilers) represent integer-like entities in a variety of forms, including

Type	Number of bits	Minimum value	Maximum value
Unsigned byte	8	0	255
Signed byte	8	−128	127
Short integer	16	−32 768	32 767
Long integer	32	−2 147 483 648	2 147 483 647

Thus you have to pay attention to what kind of "counter" you use when you do repetitive calculations. If you have a "loop" that executes 60 000 times (which isn't very much nowadays) you don't want to count your steps with a short integer! The integer limitation is more a function of language than computer hardware, so you need to be aware of how it's represented. You might be inclined to say here, "If there's such a fuss with using integers, why bother at all?" Well, don't forget that integers are special: they are discrete (e.g. exactly 15, not 14.999 99 or 15.000 01), and they behave differently in division because you always round *down*: for example you have 8/3 = 2 in integer division, but 8/3=2.6667 (or 3 to one significant digit) in floating point division. You will find as you progress in computational programming that you'll need integers and their special arithmetic more than you expect.

This integer business is different than the ultimate machine accuracy (ϵ_m), which is the smallest number that can be added to 1 and still make a noticeable difference. It is a part of *floating point* limitation. Such roundoff error accumulates in computer computations as $\epsilon_m N$ or $\epsilon_m \sqrt{N}$ where N is the number of operations. Computers store floating point numbers in a specific way, and in this case we will look at a specific example of a 32-bit word. This merely means that the computer uses 32 bits to represent each floating point number, and it might look something like:

The sign bit	The exponent	The mantissa
1	10010110	11101001100101100100101

Because of the way that numbers are converted to binary notation, the first bit of the mantissa would always be a one. So computers actually treat this bit as implicit and "bit-shift left" the remaining bits to increase the precision of the number representation slightly. A 32-bit floating point representation is often referred to as *single precision*, and has an inherent precision (resolution) of approximately 1 part in 10^7. Single precision floating point is actually the default storage for most versions of FORTRAN, for example. This kind of precision sounds pretty good for most calculations, but the problem arises when repeated computations occur. Or more seriously, when you try to take the difference between two very large but similar numbers. This problem arises more frequently than you might think. For most computations, a 64-bit or *double precision* floating point representation is used, which has roughly double (in a logarithmic sense) the resolution, approximately 1 part in 10^{15}. You'll be relieved (at least *we* were) to know that this is the default storage format used by MATLAB.

Truncation error is another pitfall you may encounter. It's what happens when you sum only N terms of a series instead of an infinite number of terms. For example, every time you compute a trigonometric function (e.g. a sine, cosine, or tangent), this is obtained by summing an infinite series; not forever, but until the series converges within a predetermined resolution. Truncation error is also an important and fundamental issue in dealing with *finite difference* calculations, which is at the heart of many numerical models (see Chapter 12).

Typically, truncation error and numerical roundoff error do not interact in well designed algorithms. Nevertheless, some algorithms are unstable and these two errors can interact in an unhealthy fashion. An example of an unstable algorithm is given in the problems.

2.2 The normal distribution

2.2.1 Parent versus sample distributions

To start our discussion of distributions in general, we first consider the concept of a *parent population* and the *sample* drawn from it. The parent distribution is the distribution of measurements you would get if you took an infinite number of measurements from the

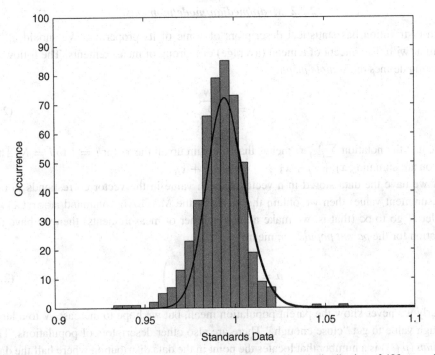

Figure 2.1 Two population distributions. The shaded histogram is the distribution of 480 secondary standard radiocarbon ratios measured over a period of time using an accelerator mass spectrometer. The smooth curve is the estimated parent population, computed using normpdf in MATLAB. Both have the same mean and standard deviation.

parent population (i.e. if you measured every possible sample). The sample distribution is the distribution of measurements you got when you decided to only make some limited number (N) of measurements of the parent population. Guess which one you're most likely to encounter in the real world?

If all (or practically most) of the systematic errors are corrected for, then the random errors will likely create a *normal* (also known as *Gaussian*) distribution of measurements. Figure 2.1 represents an example of a parent and sample distribution. The data are a compilation of measurements of secondary standards made on an accelerator mass spectrometer, each divided by their accepted values. The y-axis represents occurrence or number of counts, but if divided by the total counts it would represent probabilities. Plots of distributions with probability as the y-axis are known as *probability density functions* or PDF for short (not to be confused with a type of file that you can exchange between computers...). Putting histograms into this form is very useful (as we will see in a moment) because the area underneath these curves represents the probability of something; the "something" depends on the question you are asking. We should also point out that although most well behaved experiments produce approximately normal error distrbutions, not all error distributions are normal. There are other error distributions.

2.2.2 Mean/median/mode/moments

Each distribution has statistical descriptors of some of its properties. We should all be familiar with the concept of a mean (average) of a group of measurements. The following equation defines the *sample mean*:

$$\bar{x} = \frac{1}{N} \sum_{i=1}^{N} x_i \tag{2.1}$$

Note that the notation $\sum_{i=1}^{N} x_i$ means that you sum up all the x_i for $i = 1$ to $i = N$. That is, you are summing $x_1 + x_2 + x_3 + \cdots + x_{N-1} + x_N$.

If we have the data stored in a vector $\mathbf{x}$ (each value in the vector corresponds to one measurement value) then we obtain the mean by the MATLAB command mean(x). If we let N go to ∞ (that is, we make a huge number of measurements) then we have the equation for the *parent population* mean:

$$\mu = \lim_{N \to \infty} \left(\frac{1}{N} \sum_{i=1}^{N} x_i \right) \tag{2.2}$$

Clearly we never know the parent population mean, but we hope to increase N to a large enough value to get "close enough." There are also other descriptors of populations. The *median* ($\mu_{1/2}$) is a number that locates the point in the data distribution where half the data points are above that number and the other half are below. If we describe the probability of the ith element in a distribution of x having a value 27 as $P(x_i = 27)$, then for the median, we have

$$P\left(x_i < \mu_{1/2}\right) = P\left(x_i > \mu_{1/2}\right) = 0.5 \tag{2.3}$$

Don't forget that the real meaning of probability is that the sum of all possible exclusive states must be numerically 1. The definition above is not strictly correct, since there is a finite chance of a point that straddles the two possibilities. We're not going to split hairs here, though. The MATLAB command for this is median(x). And there is the concept of a *mode*, which represents the most probable value of $\mathbf{x}$:

$$P(\mu_{\max}) \geq P(x_i \neq \mu_{\max}) \tag{2.4}$$

MATLAB provides the mode(x) command with one important peculiarity: if all the values of $\mathbf{x}$ are different, then mode(x) returns the smallest value of $\mathbf{x}$.

There are two more descriptors of data that are very important: variance and standard deviation. The *variance* of data is given by:

$$s^2 = \frac{1}{(N-1)} \sum_{i=1}^{N} (x_i - \bar{x})^2 \tag{2.5}$$

and the MATLAB command is var(x). Note the vague similarity to Equation (2.1), with two notable exceptions: we're summing the squares of the deviations from the mean, and there is a $(N - 1)$ instead of a N in the divisor. The reason for the $(N - 1)$ is that we've used up *one degree of freedom* by defining the data in terms of the mean value. In a sense, by calculating things in terms of the mean, we've positioned the calculation in a favorable way. Another way to look at it is to think of the case where you have just one data point. The mean will fall exactly on that data point, and the sum will be zero. By subtracting 1 from $N = 1$, the statistic reduces to an undefined quantity (the ratio of two zeros). This makes sense since the whole idea of looking at mean-squared differences between points breaks down when you have fewer than two points. Thus the $N - 1$ is a way of keeping tabs on things.

The *standard deviation*, s, is given by the square root of Equation (2.5) and the MAT-LAB command is std(x). A word about nomenclature. When we are talking about the variance or standard deviation of the sample population we represent them as s^2 and s respectively, but when we are referring to the parent population we use σ^2 and σ respectively. Unfortunately this distinction is blurred in many math texts, and we commit the same *faux pas* ourselves in this very chapter! Since the parent population statistics are rarely known, perhaps you can forgive us for wanting to use such a cool symbol as σ.

We'll make mention of two more statistical descriptors of data, although we won't dwell long on their detailed use, since we argue that they are ultimately a little obscure, and not very robust. They are the skewness and kurtosis of the data. Whereas the mean and the variance are the first and second order moments of the data (i.e. they are the average of values raised to the first power and the average deviations from the mean raised to the second power), the *skewness* is related to the third moment (the average of deviations from the mean raised to the third power) and reflects how far its distribution is from being perfectly symmetrical.

$$Sk = \frac{\frac{1}{(N-1)} \sum_{i=1}^{N} (x_i - \bar{x})^3}{s^3} \tag{2.6}$$

where s is the standard deviation we've just defined. A perfectly normal (Gaussian distribution) would have a skewness of zero, one that is skewed to smaller values would have a negative skew, while one that has a positive skew would have more positive values (see Fig. 2.2).

Although some of us might think that *kurtosis* is caused by some sort of vitamin deficiency, it is actually the fourth moment of the distribution and reflects how "pointy" the distribution is.

$$K = \frac{\frac{1}{(N-1)} \sum_{i=1}^{N} (x_i - \bar{x})^4}{s^4} \tag{2.7}$$

Here, the normal distribution would have a kurtosis of 3. A "flatter" distribution would have a smaller kurtosis, and is referred to as *platykurtic*, and a pointier one would have a kurtosis greater than 3, and be called *leptokurtic*. Guess what: something intermediate, like

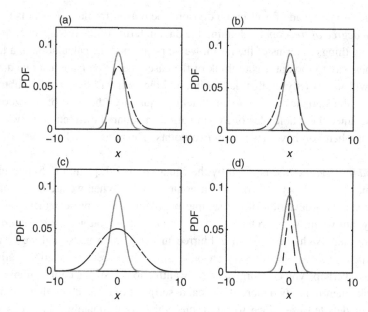

Figure 2.2 Here are some "theoretical" distributions (dashed curves) with (a) positive and (b) negative skewness, and with (c) smaller and (d) larger kurtosis. The light gray line corresponds to a comparable Gaussian curve with zero skewness and a kurtosis of 3. The areas of the dashed curves are not meant to equal the gray curves.

a Gaussian, is referred to as *mesokurtic*. We're sure that was on the tip of your tongue. You should be very careful, however, because some people define the "excess kurtosis" as the above value minus 3, so that a true Gaussian would have zero excess kurtosis, but these individuals can sometimes regard the term "Excess" as *excess* and leave it off, leading to some confusion if you're not paying close attention.

You may be tempted to explore the idea of looking at higher and higher moments of distributions as statistical diagnostics. This isn't a good idea, because the question of what they really mean becomes muddled. Such statistics are rather "non-robust" in that they depend on such high moments (powers of differences) that a small population of outliers tends to drive the result from the expected value. That is to say, the statistics are skittish. Also, the problem with these higher moments of distributions is that they are sums and differences of numbers raised to higher powers, so that computational and roundoff errors can accumulate rapidly. *Caveat emptor.*

2.2.3 *The normal (Gaussian) distribution*

There are quite a number of "distributions" of data in the literature, and we will survey some of the more important ones later in this chapter, but the one that you will often come across in this business is the *normal* (sometimes called *Gaussian*) distribution. We will

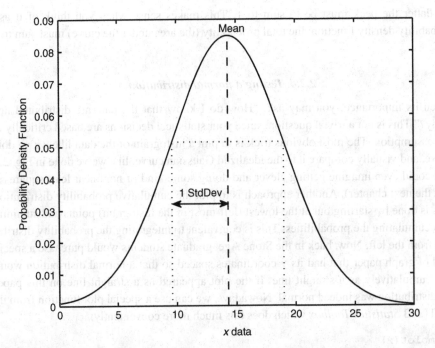

Figure 2.3 An example of what the distribution of a sample population with a mean of 14.2 and a standard deviation of 4.7 might look like if we collected lots and lots of data and divided the counts by the total number of counts to yield probabilities.

spend a fair amount of time discussing this particular distribution because it is so important to the fundamental, underlying assumptions of statistical analysis. Most importantly, the normal distribution describes the distribution of truly random errors that may accumulate in our observations. Just two parameters are needed to uniquely describe any normal distribution: the mean and the standard deviation. Imagine we had a sample population with a mean of 14.2 and a standard deviation of 4.7. Its probability density distribution might look like Fig. 2.3.

We can generate a plot like Fig. 2.3 in MATLAB with:

```
x=[0.0:0.1:30.0];
xpdf=normpdf(x,14.2,4.7);
plot(x,xpdf);
```

Now, the function normpdf in the code above is part of the MATLAB *Statistics Toolbox*, and generates the theoretical Gaussian curve characterized by the mean and standard deviation provided as arguments. The generic shape of a Gaussian is given by

$$P = P_0 e^{-(x-\bar{x})^2/2\sigma^2} \tag{2.8}$$

where the constant $P_0 = \frac{1}{\sigma\sqrt{2\pi}}$ is computed such that the area under the curve is exactly 1. It turns out that it is related to the width of the distribution, since the wider the distribution the flatter the peak must be to sum to 1. This makes sense when you think of it as a probability density function: the total probability (the area under the curve) must sum to 1.

2.2.4 Testing a normal distribution

Given its importance, you may ask, "How do I know that the data are distributed normally?" This is not a trivial question, since your statistical decisions are based critically on this assumption. The most obvious trick is to plot a histogram of the data like we've done above, and visually compare it to the idealized Gaussian curve, like we've done in Fig. 2.1. You could even imagine getting clever and doing some kind of nonlinear least squares fit (see the next chapter). Another approach is to plot the cumulative probability distribution. This is done by starting out at the lowest (left-most in the histogram) points and summing or accumulating the probabilities. This is equivalent to integrating the probability distribution from the left. Now, back in the Stone Age, graduate students would purchase a special kind of graph paper that had its y-coordinates spaced so that a normal distribution would plot cumulatively as a straight line. If the plot appeared as a straight line on this paper, the distribution was indeed normal. Nowadays, we can use a special plot function from the MATLAB *Statistics Toolbox* which does this much more conveniently:

```
normplot(x)
```

This produces a plot like Fig. 2.4 for our radiocarbon standards ratio data set shown in Fig. 2.1.

So what does this mean? The plot shows that over more than 90% of the range, the data overlap a straight line, and hence the data are distributed normally near the center of the data set. At the extremes, or "on the tails", the data diverge from normal. This means that there are "long tails" in the distribution. Or in other words extreme events are more probable than you would predict based on normal statistics. This, in fact, is the usual thing that you see with experimental data; extreme events occur infrequently that bias your data much more than typical fluctuations. These are also refered to as "fliers" or "outliers". Another, notable influence on data is human intervention: fliers may be rejected by analysts because they are unexpected, too improbable, or in small data sets, unduly influence means. We'll talk a little later about how and when to deal with outliers.

2.2.5 Standardization and normalization (Z-scores)

Since the normal distribution seems to describe the distribution of random errors that accumulate in our observations, we can best use the associated statistical machinery by putting our data into a *standard form*. Also, we may wish to make comparisons between two normally distributed populations. However, if the two populations are measured in different units, say temperature (°C) and nitrate concentration (μmol kg^{-1}), then such a comparison is difficult. We get around this problem by using the standard normal form of the data sets

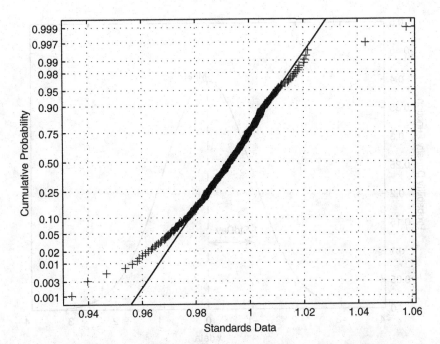

Figure 2.4 The cumulative probability distribution for the radiocarbon standards data in Fig. 2.1 compared with that of a normal distribution (the straight line).

(also known as the *Z-scores*). From each data point, the *Z*-scores are formed by subtracting the mean and dividing by the standard deviation for that population (or sample):

$$Z_i = \frac{x_i - \bar{x}}{s} \tag{2.9}$$

After this transformation the data are said to have been *standardized* (short for standard normal form) and Fig. 2.3 might look something like Fig. 2.5. On our website, we provide you with a MATLAB m-file called `standardiz`, which does this operation for you (be aware, `standardiz` operates column-wise so if x is a vector, make sure it is a column vector).

    Normalization differs from standardization. It deals with putting your data into a vector of unit length. If you consider that your data, any data, have a range of values, you can think of a mathematical transformation that will make the length of your data equal to one. For example:

$$\hat{x}_i = \frac{x_i}{\sqrt{\sum_i x_i^2}} \tag{2.10}$$

But the expression "to normalize one's data" gets used in all sorts of ways without knowing precisely what one is normalizing to, so its usage may be meaningless. We think that it's

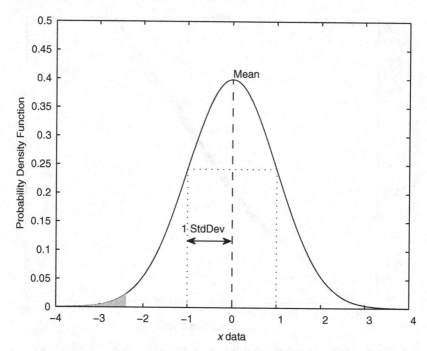

Figure 2.5 The data in Fig. 2.3 have been transformed to Z-scores. Note that the curve is now centered on zero and the width (standard deviation) is equal to one. Since some of the data points must lie below the mean, some of the Z-score must be negative. The gray area on the left is the area under the curve for Z-values less than -2.383, which in turn is associated with the probability of finding a sample with a Z-score less than -2.383 (see the next section).

better to stick with standardization or at the very least, be specific about to what (and for that matter, why) you are normalizing your data.

2.2.6 Calculating normal probabilities

What does knowing the PDF about the sample population tell us? First let's look at how we can make use of these distributions (in this particular case the normal distribution) by using the areas underneath the curves that represent probabilities. Suppose we had a sample population with $\bar{x} = 14.2$ and $s = 4.7$. We can ask ourselves the following question: "What is the probability of finding $x \le 3$?" By calculating the Z-score, we can look up the probability in standard statistics tables, or use MATLAB's `normcdf` function to calculate it for us. The Z-score is computed as:

$$ Z = \frac{3.0 - 14.2}{4.7} = -2.3830 \tag{2.11} $$

The probability of finding an $x = 3.0$ or smaller is just the area underneath the curve in Fig. 2.5 that we've shaded gray, integrated from $-\infty$ (negative infinity) to -2.3830. This yields P = 0.0086, a small probability indeed. Put another way, you have a chance of less than 1 in 100 of "seeing" a value quite so low. You can do this with normcdf in two ways:

normcdf(3.0,14.2,4.7)

or

normcdf(-2.3830,0,1)

Try it! In the first case we tell MATLAB the mean and standard deviation, but in the second case we use the standard normal form, and consequently the mean is always 0 and the standard deviation 1.

Likewise you can ask, "What is the probability of finding an $x = 20.0$ or larger?" In this case the Z-score would be 1.2340 and the area underneath the curve in Fig. 2.5 from $-\infty$ to 1.2340 is 0.8914. But remember that you want to know if x could be 20.0 or greater, which is represented by the area underneath the right-hand tail of the curve in Fig. 2.5. Since the total area underneath these curves ($-\infty$ to $+\infty$) is equal to one by definition, $P = 1 - 0.8914 = 0.1086$, or about 1 chance in ten. Similarly, by taking the appropriate differences you could ask, "What is the probability of finding x between 3.0 and 20.0?" The answer is 0.8828. In other words, if you know the mean and standard deviation of the sample population, you can easily calculate the probability of seeing any particular value assuming the actual distribution is truly normal.

2.3 Doing the unspeakable: throwing out data points?

Thinking about the probability discussion in the last section, consider the following: let's assume you've made 100 measurements of a value whose parent population mean is zero and whose parent standard deviation is 1 (in other words, it's Z-scored). Now in a sample of approximately 100 measurements with random errors, you will likely find one measurement that is more than 2.6 standard deviations away from the mean of the parent population. What does this mean? If you accept this "once in a hundred" data point in calculating your mean value, and since there's only one of them, your estimate will be off by 0.026 (2.6 standard deviations divided by 100). Since the uncertainty in your estimate is likely to be of order 0.1, this is almost significant, but not huge. You can probably live with it.

Suppose, however, that the measurement was much more difficult, and you could only manage to make 10 measurements in an experiment. Now the odds are that once in every 10th experiment, you're going to get one of those pesky "once in a hundred fliers", and it just happens to be *your bad luck* that you happen to hit that 10th experiment. Now your estimated mean is off by 0.26, which is roughly the same size as the calculated uncertainty of your estimated mean. What's worse is the effect that your flier has on your estimate of your standard deviation, since it is weighted by its square. What do you do? Furtively, you look around to see if anyone is paying attention, and then with a simple keyboard click ...

The problem becomes even more complicated for two important reasons. First, in order to know whether a data point is more than 2.6 standard deviations from the parent population mean you need to know what the parent population mean actually *is* and perhaps more importantly what the parent population standard deviation is. Since this is most likely what you were trying to find out in the first place, that's problematic. Second, data are not always exactly normally distributed. There's often some sporadic, external effect that contributes to a genuinely anomalous result. Examples include:

- there's some undetected power spike in the laboratory electrical supply;
- someone opens the door in the lab, causing an air pressure fluctuation, or temperature spike;
- there's a faulty solder-joint in your spectrometer that responds to occasional vibration;
- the ship takes a sudden roll;
- Jupiter aligns with Mars...

So let's face it, fliers *do* happen, and data fall abnormally off the beaten path. Moreover, should the threshold for data rejection be 2.6, 3.3, or 4 standard deviations, somewhere in between, or not at all? Well, not surprisingly, someone actually has already thought about this.

2.3.1 Chauvenet's criterion

Here's where you get a little common sense coming into play. What would you consider to be an offensive data point? If you made 100 measurements, you'd expect to find on average one point that was 2.57 standard deviations off. Tossing out such a point is actually a little severe, so you might argue that you'd opt for a slightly less stringent approach. If you had done the experiment twice, then the chances are that one of the two times, you'd catch a "once in *two* hundred flier", which means that it would be 2.81 standard deviations away. Thus your one 100-measurement experiment would be biased relative to the other 100-measurement experiment, and you ought to throw the point out that is 2.81 standard deviations away. This is the reasoning behind *Chauvenet's criterion for data rejection*. If you had made $2N$ measurements, then you would expect on average to have one measurement that would fall a distance z_c from the mean; that is, there will be one datum with a Z-score greater than $+z_c$ or less than $-z_c$. So if one or more points fall further from the mean than z_c away from the computed mean of your data, you can (and perhaps should) safely reject them.

So how do we go about this? Suppose you have made N measurements. You would then ask what Z-score would be associated with having *one* outlying point occuring in $2N$ measurements. (This is sometimes rephrased as having a half of a point occurring within N measurements, but that doesn't make a lot of sense.) You can do this either by looking it up in a table of normal probabilities (e.g. Table C.2 in Bevington and Robinson, 2003), or you can calculate it using MATLAB's `norminv` function. Bear in mind that there is a subtle trick in that the probability p you provide corresponds to the area under the curve between $+z_c$ and $-z_c$, so that $1 - p$ corresponds to *twice* the probability of finding a point greater than z_c above the mean or below the mean.

So here's how you might do it. Let's say you made $N = 100$ measurements. Then the probability that you're interested in is $1/(2N) = 0.005$, but since it's a *two-sided* or *two-tailed* test (the point could fall either above or below the curve), you're actually interested in $1 - p = 0.0025$ since the two tails must add up to 0.005. Thus you need to use a probability of $p = 0.9975$

```
zc=norminv(0.9975,0,1)
```

which returns a value of 2.807. Thus any points with a Z-score whose magnitude is greater than this value ought to be rejected. Try checking this and other values against the Bevington and Robinson table mentioned above. The numbers should agree reasonably well. This is often a good sanity check to ensure you have things right.

Now the truly clever researcher might be tempted to perform this rejection iteratively. That is, why not compute a mean and standard deviation, Z-score the data and reject the fliers, then compute an even better mean and standard deviation and do the same thing all over again, rejecting more data. The advice of all the statistical sages and texts is *do Chauvenet rejection only once in a given distribution*. If the data were normally distributed, and there weren't many fliers, you'll probably find that the second iteration will not yield any more rejectable points. If it does, then it suggests that your data may *not* be normally distributed. The philosophy is that filtering once is a valid thing to do, but iterative filtering may dramatically alter the data distribution in a fundamental way, invalidating the assumptions behind your statistical calculations, and leading to erroneous results. Moreover, you may accused of being a Chauvenet Chauvinist.

2.4 Error propagation

We have already seen how to calculate the standard deviation and some of its uses. But here is another and very important use, *the propagation of error*. Most often, we are interested in values derived from raw data. Suppose you measured two quantities and then used them to calculate a third. One question that will naturally arise is: "What's the error of the third number?" You can assume that it must be related to the errors in the first two numbers that you measured (with replicates or some other means of assessing their errors), but what is the mathematical relationship? Actually, you use the variance, the standard deviation squared.

2.4.1 The general equation

In the general case, if $y = f(a, b, \ldots)$, then:

$$\sigma_y^2 = \sigma_a^2 \left(\frac{\partial y}{\partial a}\right)^2 + \sigma_b^2 \left(\frac{\partial y}{\partial b}\right)^2 + \cdots + 2\sigma_{ab}^2 \left(\frac{\partial y}{\partial a}\right)\left(\frac{\partial y}{\partial b}\right) + \cdots \quad (2.12)$$

This is an important equation! Note that, as promised, we broke the notational rule and are using σ instead of s, but we can't resist. One of us (who shall remain anonymous) has had this equation written on the whiteboard in his office since 1996 to remind our students.

Let's take a simpler version and look at some specific examples. If $y = f(a, b)$ (just two variables), then:

$$\sigma_y^2 = \sigma_a^2 \left(\frac{\partial y}{\partial a}\right)^2 + \sigma_b^2 \left(\frac{\partial y}{\partial b}\right)^2 + 2\sigma_{ab}^2 \left(\frac{\partial y}{\partial a}\right)\left(\frac{\partial y}{\partial b}\right) \tag{2.13}$$

All of these terms should be familiar, with the possible exception of the σ_{ab}^2 term. This is referred to as the covariance (as opposed to the variance) between a and b.

$$\sigma_{ab}^2 = \langle (a - \bar{a})(b - \bar{b})\rangle \tag{2.14}$$

where the quantities between the $\langle\ \rangle$ represent what is known, in statistical circles, as "the expected value", which you can think of as a type of mean value. The main difference here is that the expected value is the probability weighted mean of all possible values:

$$\langle y \rangle = \sum_{i=1}^{N} P_i y_i \tag{2.15}$$

2.4.2 Assumptions regarding independence or orthogonality

The nice thing about this σ_{ab}^2 term is that if you can show that a and b are uncorrelated, σ_{ab}^2 becomes zero and drops out of the calculation. That is, if you imagine a and b randomly jittering around their mean values, and if on average they don't tend to be on "the same side" of their respective means at the same time, then the terms in the definition of σ_{ab}^2 nicely cancel out. As a simple example, suppose we had an equation for y that was calculated from a and b as follows:

$$y = pa + qb \tag{2.16}$$

Then we would calculate the uncertainty in y from the following:

$$\sigma_y^2 = \sigma_a^2 p^2 + \sigma_b^2 q^2 + 2\sigma_{ab}^2 pq \tag{2.17}$$

And if a and b were to be found uncorrelated, this would simplify to:

$$\sigma_y^2 = \sigma_a^2 p^2 + \sigma_b^2 q^2 \tag{2.18}$$

Just be sure to mind your ps and qs. Of course, things can be a lot more complicated, and Bevington and Robinson (2003) have a number of simpler examples worked in their Chapter 3. We will return to the very important and central concept of covariance in Section 2.8, where we will see how it is calculated, and how it acts as a descriptor of the structure "hidden" in data sets.

2.5 Statistical tests and the hypothesis

At its most fundamental level, the scientific method involves asking basic questions and attempting to answer them. Posing a question can often be reduced to construction of a hypothesis. You then design an experiment which can either refute the hypothesis, or accept it.

2.5.1 Hypothesis building and testing

Now here's a really big deal: we can test hypotheses by merely calculating a number! Of course, the hypothesis must be posed and the number calculated properly. So, there is one more subject to cover before we move on. That is the concept of hypothesis building and testing. The initial step in any statistical test is posing the appropriate hypothesis. Remember, in statistics (and in science in general) you can never prove anything, you can only disprove things. This being the case, the simplest case of demonstrating a difference is accomplished by posing the opposite as your hypothesis, i.e. that there is no difference. This is called the *null hypothesis* (H_o). The null hypothesis is accompanied by the alternate hypothesis (H_1). They are generally represented as:

$$H_o : \mu_o = \mu_1 \tag{2.19}$$

$$H_1 : \mu_o \neq \mu_1 \tag{2.20}$$

The null hypothesis (H_o) and the alternate hypothesis (H_1) should be mutually exclusive and all inclusive. That is to say, H_o should be specific and H_1 should be general. One can draw up a contingency table that describes how hypothesis testing might turn out. It is given as a two by two table (see Table 2.1).

In statistical tests α is known as the level of significance, and it must be chosen a priori before the test. It essentially amounts to the probability of committing a *type I error*. One must ask one's self, "How much of a chance am I willing to take of rejecting the null hypothesis when, in fact, it is correct?" If you are willing to take a 1 in 20 chance (a common level of risk) then your α level is 0.05. The type II error occurs with a probability of β and frequently is not known. To minimize the probability of committing a type II error, the null hypothesis is written with the intention of rejecting it. Often we talk about a level of *confidence*; it is given by:

$$P = 1 - \alpha \tag{2.21}$$

2.5.2 Example 1: testing a null hypothesis

Now we can test a null hypothesis. Suppose we had a sample population with a mean of $\bar{x} = 30.0$ and we wanted to test whether or not this mean was, within some statistical level of certainty, the same as or different from a parent population with a mean of $\mu = 14.2$. We

Table 2.1 *Hypothesis testing contingency table*

	Hypothesis is correct	Hypothesis is incorrect
Hypothesis is accepted	Correct decision	Type II error (β)
Hypothesis is rejected	Type I error (α)	Correct decision

can define a test statistic similar to the Z-scores we calculated earlier. It is a little confusing that the *test statistic* used for normal populations is the same letter as the Z-score (without the subscripted i), but you will see that they are calculated in such a similar manner that it is understandable that statisticians would make this choice. In fact, if you reflect on what you are doing (calculating how many standard deviations from the mean, i.e. zero, you are) this makes sense. We calculate:

$$Z = \frac{\bar{x} - \mu}{s_e} = \frac{\bar{x} - \mu}{\sigma\sqrt{\frac{1}{N}}} = \frac{30.0 - 14.2}{4.7\sqrt{\frac{1}{6}}} = 8.2345 \tag{2.22}$$

Because we "know" (we told you earlier) that $\mu = 14.2$ and $\sigma = 4.7$ (and we're telling you now that six independent sample populations were analyzed to make an estimate of the parent population mean), the null hypothesis (H_o) and alternate hypothesis (H_1) are:

$$H_o : \bar{x} = \mu \text{ or } H_1 : \bar{x} \neq \mu \tag{2.23}$$

Remember, a null hypothesis is an assumption of *no* difference.

Since we are only testing whether it is equal (or not equal), not greater than, or less than, this test is called a *two-tailed* test. The level of significance we have chosen a priori to be 0.05. So we only need find out what the standard score is for 0.025 probability in each tail, and that will be the *critical value*. If our test statistic exceeds that critical value then we can, with 95% confidence, reject the null hypothesis. Looking in the standard tables for normal distributions, we find the tails are marked by standard scores of ± 1.96, and since 8.2345 is larger, we can reject the null hypothesis. That's all there is to it. The basic concepts apply for all other statistical tests although some of the details may change.

2.5.3 Example 2: testing for a normal distribution

Our original foray into questioning whether a data distribution was truly normal by plotting the cumulative probability distribution was instructive, but dissatisfyingly non-quantitative. We could see that the data distribution deviated from normal on the tails, but it would have been helpful to be able to reject (or accept) the null hypothesis that the data were normal. The benchmark test for Normal distributions is the Kolmogorov–Smirnov test, which determines the maximum deviation between the cumulative distribution function for the data and that of a normal distribution with the same mean and standard deviation. This is then used to accept (if smaller than a critical value) or reject (if larger) the hypothesis of normalcy. The problem with the K–S test is that it requires that you know in advance the parent population mean and standard deviation, which you are not often (if ever) likely to know. The solution to this dilemma, you might have guessed, is to compute the mean and standard deviation from the sample population and hope for the best. This is basically what the Lilliefors test does. This test is implemented in the MATLAB *Statistics Toolbox* as

```
[H, P, L, C] = lillietest(x);
```

where **x** is the vector containing your data, *H* is the null hypothesis result (it's 1 if you *reject* the hypothesis) that the data are normally distributed, *P* is the p-statistic (probability), *L* is the Lilliefors statistic (maximum distance between the actual and idealized normal CDF), and *C* is the critical value that was used in the test (this number varies with the number of data points). The p-statistics cannot be directly calculated, and have been determined by Monte Carlo methods and tabulated. The values reported by the program are interpolated from the table and NaN is reported for data outside the range of the table. The default probability level for rejecting the null hypothesis is 0.05 (i.e. 95% confidence limit), but you can specify different levels. We would show you how this might work in the case of our example data set, but perhaps you'd like to practise this in the problem set?

2.6 Other distributions

We mention here a few of the distributions that you may encounter. There are many, many more and you're encouraged to consult the literature for others that you come across. Each has its place, and we offer you some solace at the end of section with the *central limit theorem* which, although it sounds abstract, is a powerful tool for converting data that are not normally distributed into the more familiar and useful Gaussian distribution.

2.6.1 Student's t-distribution

As mentioned above, there are lots of other distributions that you may encounter in your travels. Not all sample populations will approximate a normal distribution so closely that the Z-statistic will be applicable. This is especially true for small samples. The *Student's t-distribution* was developed especially for this case. This distribution is particularly applicable to sample populations that are small in size and is useful for performing statistical tests on the equality of means. You may be tempted to conclude that this test was developed specifically for *students* (hence the name) but this is not the case. It was named after a fellow who published under the name "Student". No, that wasn't his name, which was William Sealy Gosset. He was forced by his employer (an Irish brewery of considerable fame) to use this *nom de plume* to protect trade secrets.

This distribution is dependent upon one additional input variable: the *degrees of freedom* (v). In calculating the t-statistic you will need to calculate sample population means; for

every mean you need to calculate you can subtract one degree of freedom. You start with
as many degrees of freedom as you have data points. Then:

$$t = \frac{\bar{x} - \mu}{s_e} = \frac{\bar{x} - \mu}{s\sqrt{\frac{1}{N}}} \tag{2.24}$$

and

$$\nu = N - 1 \tag{2.25}$$

This is the t-statistic for what is referred to as a one-tailed test (the parent population is
known). Use MATLAB's `ttest` to perform this test.

When comparing the means of two sample populations you use the two-tailed version of
this statistic. Problem 2.4 at the end of this chapter shows how this works (see Davis, 2002,
for details and use MATLAB's `ttest2`). For small data populations (say around 10–12),
the Student's t-distribution tends to be a little broader than a normal distribution. As the
degrees of freedom (i.e. the number of data points) approach infinity, that is the number
of points becomes large, the t-distribution approaches normal. In practice, the differences
become very small when you have more than a few dozen points.

The Student's t-test is usually applied to two distributions that have equal variances. If
the variances differ, then a more general test called the *Welch's t-test* is applied.

2.6.2 The F-distribution

Whereas the Student's t is useful for testing the equality of the means of two distributions,
the *F-distribution* is useful for testing the equality of variances for two distributions. The
F-distribution is dependent upon two additional input variables: ν_1 and ν_2, the degrees of
freedom of the two sample populations. The F-statistic is simply:

$$F = \frac{s_1^2}{s_2^2} \tag{2.26}$$

where subscripts 1 and 2 again refer to the first and second population statistics. This test
statistic is used extensively in the *Analysis of Variance* (ANOVA), as we shall see later
(Section 2.8.1).

2.6.3 Poisson distribution

There is another distribution that you, as a geochemist or biologist, may find particularly
useful: the *Poisson distribution*. Here we interject a note of statistical coolness. Don't pro-
nounce Poisson like it looks. Poisson is pronounced "Pwah-Sohn" if you really want to
sound like you know what you're doing (without fishing for compliments, that is). The
Poisson distribution arises from random events whose probability of occurrence is independent of time. It is applicable to the statistics of counting discrete events, like the discrete
radioactive decays some of you will be using to measure radioisotopes. It arises from the

fundamental nature of radioactive decay. If you have a radioactive source that you are "counting" with a detector for some period of time, then the uncertainty associated with the N events that you observed is simply $\sqrt{N}$. That is, if you were to repeat the exact same experiment a number of times, the variance of the results would be N. This doesn't mean that you made an inaccurate measurement, but simply that there was an intrinsic variability in the number of events during that time window that would be reflected in varying results. The power of this knowledge is that you know precisely how well you have determined the count rate. What's more, you know how long you need to count to achieve some target precision.

For example, if you had counted 100 events, the uncertainty (one standard deviation) would be 10 counts, or 10%. If you had counted 10 000 events, then the uncertainty would be 100 counts, or 1%. Thus the more counts you have, the smaller (proportionately) your error becomes. This distribution is discrete, not continuous as the others are, and depends on one parameter, λ, which is both the mean and the variance.

2.6.4 Weibull distributions

You often see the *Weibull distribution* discussed in meteorology when describing the distribution of wind speed or precipitation, but it has uses in a broad range of other areas, e.g. in biology (survival and mortality statistics) and in industry (product reliability and Mean Time To Failure). It has the general characteristic of being zero for $x \leq 0$ and has a cumulative distribution function that asymptotically approaches unity as $x \to \infty$. These characteristics fit well to the idea of some scalar value such as wind speed or precipitation that can approach zero but never be negative, and which tends to have positive excursions (e.g. wind gusts). There is a whole class of functions that satisfy this requirement, with a general form of

$$f(x) = \frac{\gamma}{\sigma} \left(\frac{x - \mu}{\sigma} \right)^{(\gamma - 1)} e^{-\left(\frac{x-\mu}{\sigma} \right)^{\gamma}} \tag{2.27}$$

where the restriction is that you have $x \geq \mu$, and $\gamma, \sigma > 0$, but you will more likely encounter the *two-parameter form* which has $\mu = 0$

$$f(x) = \frac{\gamma}{\sigma} \left(\frac{x}{\sigma} \right)^{(\gamma - 1)} e^{-\left(\frac{x}{\sigma} \right)^{\gamma}} \tag{2.28}$$

This generalized form also includes the Exponential (*Gamma*) distribution as a special case for $\gamma = 1$, and approximates a Gaussian for $\gamma = 3.6$. Climatically, wind distributions appear to fit a Weibull distribution with $\gamma \approx 1.5$ (see, for example, Pavia and O'Brien, 1986).

Considering its importance in a wide variety of fields, it is not surprising that the MATLAB *Statistics Toolbox* has a host of useful Weibull utilities, including `wblpdf` and `wblcdf`. As an example, you might use `wblrand` to generate realistic winds to force a model, or use `wblstat` to estimate the mean and variance of a three-parameter Weibull.

Similar to our cumulative graphical test of normal distributions, you might use `wblplot` to evaluate how closely your data matches a theoretical distribution.

2.6.5 Log-normal transformations

Sometimes, whether you realize it or not, there is an intrinsic logarithmic character to your measurement, in the sense that the size of your measurement error may tend to depend in a complicated way on the size of your measurement. A transformation that is frequently used is the *log-normal transformation*. Many sample distributions from nature are said to have this characteristic: taking the log (natural or common) of the data points transforms the data into a normal population. But keep the following in mind:

$$y_i = \log(x_i) \tag{2.29}$$
$$\bar{y} = \text{mean}(y_i) \tag{2.30}$$
$$s_y^2 = \text{var}(y_i) \tag{2.31}$$

as usual, but:

$$\bar{x}_G = \sqrt[N]{\prod_{i=1}^{N}(x_i)} \tag{2.32}$$

$$s_x^2 = \sqrt[N-1]{2\prod_{i=1}^{N}\left(\frac{x_i}{\bar{x}_G}\right)} \tag{2.33}$$

where these last two are known as the *geometric mean* and the *geometric variance* of x, respectively (hence the subscript G). Here, the symbol $\prod_{i=1}^{N}(x_i)$ means the product of the series, i.e.

$$\prod_{i=1}^{N}(x_i) = x_1 \times x_2 \times x_3 \times \cdots \times x_{N-1} \times x_N \tag{2.34}$$

And keep in mind that:

$$\bar{x}_G = e^{\bar{y}} \tag{2.35}$$
$$s_x^2 \neq s_y^2 \tag{2.36}$$

which makes sense when you realize that the two variables don't even have the same units. We want to get our data into normal form (if we can) for some very good reasons because:

- the normal distribution is mathematically simpler than most of its rival distributions, requiring only two parameters (the mean and standard deviation); and
- the central limit theorem implies that the mean of means follows a normal distribution.

That last statement can be expressed mathematically in the following manner. If you assume that your measurement (x) is actually composed of the true value plus numerous random errors you could express it as:

$$x = x_t + \varepsilon_1 + \varepsilon_2 + \varepsilon_3 + \cdots + \varepsilon_N \qquad (2.37)$$

where x_t is the true value of your measurement and ε_i is any number of small, random (independent) errors. Karl F. Gauss demonstrated that if N is large enough, then x will be approximately normally distributed.

How does this apply to the log-normal distribution? The central limit theorem also applies to log-normal distribution, but with a few added twists. To start with, it is $\log(x)$ that is normally distributed, not x. If you make the *assumption* that the errors in your measurements are *proportional* to the measurement itself, then you could rewrite the above equation as:

$$x = x_t \times \varepsilon_1 \times \varepsilon_2 \times \varepsilon_3 \times \cdots \times \varepsilon_N \qquad (2.38)$$

where the variables have the same meaning as above. You can now easily see that the log transform of this equation is normally distributed for the same *Gaussian* reasons.

$$\log(x) = \log(x_t) + \log(\varepsilon_1) + \log(\varepsilon_2) + \log(\varepsilon_3) + \cdots + \log(\varepsilon_N) \qquad (2.39)$$

But why should the measurement errors be proportional to x? It is because in some types of measurements (most in fact), *size* matters. It is much easier to measure the width of your keyboard to within a few millimeters than it is for you to measure the distance from your home to your office to the *same level of accuracy*. This is why we often report our measurement errors relative to the measurement itself. When we do, we call it the *coefficient of variation* (or relative standard deviation) and it is given by:

$$s_v = \frac{s}{\bar{x}} \qquad (2.40)$$

The interesting twist is this. In log-normally distributed data, the error is already normalized to the mean. If you were to make a plot of the coefficient of variation versus the mean values they are derived from, you should get a horizontal line across the plot. Deviations from such a line give you an insight into where your data is *not* behaving in a log-normal fashion (typically at either the high or low end of the x-axis).

2.7 The central limit theorem

Experience will eventually tell you that although the normal distribution is a powerful and central concept in statistics, it is not all that often found in its pure, unadulterated state. Nature has a way of throwing curves at us (see the next chapter!). But don't despair, because we have a secret weapon, the *central limit theorem*. This theorem states that no matter what the underlying parent population is, the mean of the means drawn from the parent population will be normally distributed. And the average of the sample means will equal the population mean in the case when the number of sample means goes to ∞, or:

$$\text{mean}(\bar{x}_i) = \mu \tag{2.41}$$

where $i = 1, \ldots, N$ is the number of means and its variance is given by:

$$s_{\bar{X}}^2 = \frac{\sigma^2}{N} \tag{2.42}$$

The standard deviation of this is called the *standard error of the estimate of the mean* and is sometimes written as:

$$s_e = \sqrt{\frac{\sigma^2}{N}} \tag{2.43}$$

What this really means is that by subsampling and averaging data, we can ultimately flog our data into a form that allows us safely to assume a normal distribution. That is, the central limit theorem allows us to design statistical tests based on normal distributions, even when we know that the population we are sampling is not normally distributed.

Now there's an important nomenclature point we need to make here. Many students confuse the standard error of the estimate of the mean with the standard deviation. The two are *not* the same, and the standard error of the estimate of the mean is smaller than the standard deviation by a factor of $\frac{1}{\sqrt{N}}$. Think of it this way: making repeat measurements will progressively reduce the size of the standard error of the estimate of the mean, but the standard deviation will remain the same.

An example of where the central limit theorem might come into play is if you're measuring the activity of some radioisotope in seawater, say for example ^{228}Th. Assume that you have 1000 measurements over the course of a few days. Although counting statistics associated with an individual measurement will follow the Poisson distribution, averaging small groups of measurements, e.g. groups of 25, leads to the PDF of means (in this case 40 of them) being normally distributed. You can then apply the usual statistical tests to this data.

Another example might include conductivity/temperature/pressure (CTD) data acquired in the mixed layer at sea. In this thought experiment, the CTD device is hanging off the side of the ship aquiring data in an otherwise homogeneous environment. The instrument may be reporting conductivity, pressure, and temperature data at a rate of 25 times a second. Although the sensor data may be normally distributed about some central (presumably true environmental) value, since you're computing *salinity* from these properties using a nonlinear equation, the individual salinity measurements will *not* be normally distributed. If, however, you were to block-average the computed salinity values in 1 second increments, you would find that those averages would be normally distributed.

Now you need to be a little careful when you apply the central limit theorem. Bear in mind that if you're averaging groups of measurements that are *evolving* over time, you should not expect to see a normal distribution, because the means will have a temporal trend in them. An example might be if you were averaging radioactive counting experiments that progressed over a significant fraction of the half-life of the radioisotope. In other words, you need to be assured that you are sampling a time-independent process. We will return to this issue in a later chapter on time series.

2.8 Covariance and correlation

Here we are talking about a very important property, one which will show up many, many times in data analysis, and one which is key to many techniques. The variance of a variable (we'll call it x) or what you might loosely term "the scatter" relative to its mean, is like the standard deviation (see Section 2.2.2), and is given by:

$$s^2 = \frac{1}{(N-1)} \sum_{i=1}^{N} (x_i - \bar{x})^2 = \frac{SS}{(N-1)} \qquad (2.44)$$

That is, it is the mean square deviation of your data points relative to the average value. Also, we've introduced a kind of short-hand notation SS which stands for the *sum of squared deviations* for variable x. Now think about how you might go about computing this value. You would first calclulate the mean value $\bar{x}$ in the usual way. This would involve going through all your values, adding them together, and then dividing by the number of data points, N. Next, you would total the square of differences of all your data points from the mean. This is actually inefficient, particularly if you are doing things with a calculator (what a quaint idea!). You can do it all in one fell swoop by calculating the sum of the data points and sum of their squares and using:

$$s^2 = \frac{N \sum_{i=1}^{N} x_i^2 - (\sum_{i=1}^{N} x_i)^2}{N(N-1)} \qquad (2.45)$$

Equation (2.45) is the computational form, because it is slightly more efficient to compute it that way than the first "definition form". We honestly prefer the first form (2.44), since it is more intuitive, and also because it is less susceptible to roundoff and truncation errors than the computational form (remember Section 2.1.5). We only mention the computational form because that tends to appear in older textbooks almost as often as the definition form. It would be useful if you could convince yourself that the two are equivalent by substituting in the definition of the mean into the definition form and doing the algebra. In fact, that may make an interesting problem . . .

Let's do something a little different with Equation (2.44). We'll write it a bit more explicitly as:

$$s^2 = \frac{\sum_{i=1}^{N} [(x_i - \bar{x})(x_i - \bar{x})]}{(N-1)} \qquad (2.46)$$

Now if you have two possibly related variables (let's identify them as x_j and x_k, and keep in mind they are both the same length because you've measured them at the same time), you can look at their *covariance*, which you can define as:

$$s_{jk}^2 = \frac{\sum_{i=1}^{N} \left[(x_{ij} - \overline{x_j})(x_{ik} - \overline{x_k}) \right]}{(N-1)} = \frac{SP_{jk}}{(N-1)} \tag{2.47}$$

This, is clearly analogous to the variance equation (2.44) where, rather than multiplying the deviation by itself, you're cross-multiplying the deviations. Put another way, the variance is a special case of the covariance where $j = k$, so that $s_j^2 = s_{jj}^2$. Here again we've introduced a kind of shorthand where SP_{jk} is the *sum of products of the deviations* for the variables x_j and x_k. Think about it this way. If x_j tends to be above its mean $\overline{x_j}$ at the same time that x_k is above its mean $\overline{x_k}$, then the product will be positive. Similarly if they're both below the sum will be a positive value. Conversely, if x_j tends to be above when x_k is below (or vice versa) the sum will be negative. Finally, if there's no correspondence (e.g. sometimes they are both above or below, and sometimes one is while the other isn't) then the sum will average out to zero.

Let's consider the *correlation coefficient*, r_{jk}, a statistical measure of whether two variables, e.g., x_j and x_k, are linearly related. It varies from -1 for perfectly negatively correlated, through 0 (for uncorrelated), to $+1$ for perfectly positively correlated. You normally see this in statistics texts as a gigantic, computational form, equation, which is not very illuminating but is impressive. However, using our shorthand notation, it can be rewritten more compactly as:

$$r_{jk} = \frac{s_{jk}^2}{s_j s_k} = \frac{SP_{jk}}{\sqrt{SS_j SS_k}} \tag{2.48}$$

where we see that the correlation coefficient is nothing more than the ratio of the covariance of data j and k to the product of their individual standard deviations. Note that we've sneaked in subscripts j and k for the sum of squares to identify them with a particular variable.

2.8.1 Analysis of variance (ANOVA)

The statistical machinery we've talked about so far has been useful for comparing data sets to parent population (or theoretical) values and for comparing two data sets to each other. But how do we compare two or more data sets that contain groups of observations (such as replicates)? Suppose we had M replicates of N samples of the same thing. We could create a null hypothesis that there was no difference between the means of the replicates and the alternate hypothesis that at least one sample mean was different. How would we test this? We do this with a technique known as analysis of variance (ANOVA).

Table 2.2 *Simple one-way ANOVA*

Source	Sum of squares	Deg. of freedom	Mean squares	F-test
Among samples	SS_A	$N-1$	$MS_A = \dfrac{SS_A}{N-1}$	$F = \dfrac{MS_A}{MS_W}$
Within replicates	SS_W	$N(M-1)$	$MS_W = \dfrac{SS_W}{N(M-1)}$	
Total	SS_T	$N \times M - 1$		

In order to apply this technique we need to calculate some sum of squares. We calculate the sum of squares among the samples (SS_A), within the replicates (SS_W), and for the total data set (SS_T).

$$SS_T = \sum_i^N \sum_j^M x_{ij}^2 - \frac{1}{N \times M} \left(\sum_i^N \sum_j^M x_{ij} \right)^2 \tag{2.49}$$

$$SS_A = \sum_i^N \left[\frac{1}{M} \left(\sum_j^M x_{ij} \right)^2 \right] - \frac{1}{N \times M} \left(\sum_i^N \sum_j^M x_{ij} \right)^2 \tag{2.50}$$

$$SS_W = SS_T - SS_A \tag{2.51}$$

We then arrange these results into a table that has entries as in Table 2.2.

This is just like the F-test we performed in an earlier section, $\nu_1 = N - 1$ and $\nu_2 = N(M-1)$ where $N \times M$ represent the total of all measurements. This type of analysis will turn up many times; for example you will be able to use it to test the statistical significance of adding an additional term to a polynomial fit (in 1-, 2-, or M-dimensional problems). If your data matrix is an $N \times M$ matrix $\mathbf{X}$ (N samples and M replicates) then the table above essentially examines the variance in a data set by looking at the variance computed by going down the columns of the data matrix.

But what if one's data set contained N samples and M different *treatments* of that data (not replicates as above)? To perform an ANOVA on this kind of data set we use a two-way ANOVA. In this case the ANOVA looks at the variance between samples by going down the columns as above, but also at the variance between treatments by going across the columns of the data matrix. The table created for a two-way ANOVA appears as in Table 2.3, where the first F-test tests the significance of differences between samples and the second F-test tests the significance between treatments. In this two-way ANOVA we have two additional sum of squares to compute, the sum of squares *among treatments* (SS_B) and the sum of squares *of errors* (SS_e). They are given by:

$$SS_B = \sum_j^M \frac{1}{N} \left(\sum_i^N x_{ij} \right)^2 - \frac{1}{N \times M} \left(\sum_j^M \sum_i^N x_{ij} \right)^2 \tag{2.52}$$

$$SS_e = SS_T - (SS_A + SS_B) \tag{2.53}$$

Table 2.3 *Two-way ANOVA*

Source of variation	Sum of squares	Degrees of freedom	Mean squares	F-tests
Among samples	SS_A	$N - 1$	MS_A	$\frac{MS_A}{MS_e}$
Among treatments	SS_B	$M - 1$	MS_B	$\frac{MS_B}{MS_e}$
Error	SS_e	$(N - 1)(M - 1)$	MS_e	
Total variation	SS_T	$N \times M - 1$		

For more about a two-way ANOVA refer to chapters 2, 5, and 6 of Davis (2002).

2.9 Basic non-parametric tests

When the underlying population is *distinctly* non-normal, or when our sample size is small, or when we simply do not know whether our samples are taken from a normal population, it is sometimes best to apply non-parametric tests. This is because in these cases (and others), the central limit theorem cannot be invoked to apply parametric tests (i.e. the Student's t-test, the F-test, χ^2-test, etc.) directly. The real beauty of these tests is that no assumption need be made about the parent population's shape. But as always in life, the catch is that the less you assume about the parent population, the less you can do with the data. To put it another way, you can think of non-parametric tests as being less efficient than parametric tests. Some of these non-parametric tests are included in the MATLAB *Statistics Toolbox*.

2.9.1 Spearman rank-order correlation coefficient

As the name implies, this non-parametric test is a way of computing a correlation coefficient (in this case the Spearman rank-order coefficient, r_S) by comparing the order (rank) of the data pairs. If you let R_i be the rank of x_i amongst the x values and S_i be the rank amongst the y_i, then the rank-ordered correlation coefficient is given by:

$$r_S = \frac{\sum_i \left[(R_i - \overline{R})(S_i - \overline{S}) \right]}{\left(\sum_i (R_i - \bar{R})^2 \right)^{1/2} \left(\sum_i (S_i - \bar{S})^2 \right)^{1/2}} \tag{2.54}$$

where r_S will range from $-1 \leq r_S \leq 1$ and you can test the significance of a *non-zero* coefficient by calculating the following:

$$t = r_S \sqrt{\frac{N - 2}{1 - r_S^2}} \tag{2.55}$$

which has a distribution that is *approximately* like the Student's t-distribution with $N - 2$ degrees of freedom.

2.9.2 Kendall's tau

This next test is even more non-parametric than the Spearman's r_S. Known as Kendall's tau (τ) this statistic considers all $\frac{1}{2}N(N-1)$ pairs of data points, say (x_i, y_i) and (x_{i+1}, y_{i+1}) and so forth. In other words it considers only the relative ordering of ranks between data point pairs. But in this case we don't need to rank the data, each data point will be higher than, the same as, or lower than its matched pair by virtue of their relative values. We now define the following terms:

Concordant: if the x's relative rank is the same as the y's, say x_1 less than x_2 and y_1 less than y_2.

Discordant: if the relative rankings of the x values is the opposite of the y values.

If there is a tie, i.e. the data pair cannot be judged either concordant or discordant, then they are referred to as "extra" pairs. They are specifically called extra-y pairs if the xs tie and extra-x pairs if the ys tie. Data points cannot be paired with themselves and the order in which they are paired does not matter, i.e. once they have been paired you cannot count them again as a new pair by switching the order. We calculate the statistic according to:

$$\tau = \frac{\text{Con} - \text{Dis}}{\sqrt{\text{Con} + \text{Dis} + X_y}\sqrt{\text{Con} + \text{Dis} + X_x}} \tag{2.56}$$

where Con refers to the number of concordant pairs, Dis the number of discordant pairs and X_y and X_x the number of the appropriate "extra" pairs. Oh yes, if the tie is in the xs and ys, then it is not counted at all.

Kendall, who worked out this mind-boggling nightmare in combinatorics, has demonstrated that τ has a value range of from -1 to $+1$ and is approximately normally distributed with zero mean and a variance given by:

$$\text{var}(\tau) = \frac{4N + 10}{9N(N-1)} \tag{2.57}$$

Be warned: this statistic's computational load grows as $\mathcal{O}(N^2)$ (that is, increases with the square of N) whereas Spearman's r_S grows only at the rate of $\mathcal{O}(N \log N)$ (that is, more slowly than N^2). You can be in for quite a bit of number crunching if you have a few thousand data points. The statistic exploits a "weaker" aspect of the data and can be more robust than r_S, but it does "throw away" information that Spearman's would use and so might not be as good at detecting correlations in the data that r_S might find.

2.9.3 Wilcoxon signed-rank test

This is another ranking test, but in this one you pay attention to the sign. From your data points x_i and y_i you calculate a new "variable" $z_i = y_i - x_i$ and assume that $z_i = \theta + \epsilon_i$ where $i = 1, \ldots, N$ and the null hypothesis (H_o) is that $\theta = 0$. We now form an array $|z_1|, \ldots, |z_N|$ and form R_i from the rankings of $|z_i|$s in the following manner:

$$\psi_i = \begin{cases} 1 & \text{if } z_i > 0 \\ 0 & \text{if } z_i < 0 \end{cases} \tag{2.58}$$

From here we go on to calculate:

$$T^+ = \sum_{i=1}^{N} R_i \psi_i \tag{2.59}$$

and you can reject H_o if $T^+ \geq t(\alpha, \nu)$. That is to say, we use the Student's t-test with a level of significance α and ν degrees of freedom. The assumption about z_i does not require that the xs and ys be independent. If there are any zs with a zero value they are discarded and N is redefined as the number of non-zero zs. Ties in the ranking are handled in the same fashion as the Mann–Whitney rank-sum test (Section 2.9.5 below).

2.9.4 Kruskal–Wallis ANOVA

This non-parametric test can be used as a substitute for a one-way ANOVA (see Section 2.8.1). Below, the variable r_{ij} represents the rank of the ith observation in the jth sample and R_j represents the average rank of the jth sample. The null hypothesis is that all j samples come from a population with the same distribution, the alternate hypothesis is that at least one sample is different (i.e. $H_o : \tau_1 = \tau_2 = \cdots = \tau_k$). That is, the data (x_{ij}) is calculated as:

$$x_{ij} = \mu + \tau_j + e_{ij} \quad \text{where } i = 1, \ldots, n_j \quad \text{and} \quad j = 1, \ldots, k \tag{2.60}$$

where μ is the unknown overall mean, τ_j is the unknown sample effect, and e_{ij} is the independent error of each observation. The overall total (N) and average ranking per sample are given by:

$$N = \sum_{j}^{k} n_j \quad \text{and} \quad R_j = \frac{1}{n_j} \sum_{i=1}^{n_j} r_{ij} \tag{2.61}$$

From the above ranking (R_j) we calculate the Kruskal–Wallis H statistic:

$$H = \left(\frac{12}{N(N+1)} \sum_{j=1}^{k} \frac{R_j^2}{n_j} \right) - 3(N+1) \tag{2.62}$$

Reject H_o at significance level α if $H > x$ in Table A.12 of Hollander and Wolfe (1999). Davis (2002) points out that this table is one of the few known compilations of critical values for this statistic and that it contains up to five samples of up to eight observations or less. But, if you don't happen to have a copy of Hollander and Wolfe handy, H has a distribution that is close to χ^2 with $j - 1$ degrees of freedom.

2.9.5 *Mann–Whitney rank-sum test*

This "rank-sum" non-parametric test can be used as a substitute for the Student's t-test of the equality of means of two samples. In this test you rank both sets of variables by lumping them together, the N x values and the M y values to give $N + M$ values to rank in one array. But when you calculate the statistic, only calculate it for the xs, leaving the ys out. If there are ties in the ranking, then all of the values are assigned the same rank: the average of the ranks the data would have had if their values had been slightly different.

$$T = \sum_{i=1}^{N} R(x_i) - \frac{N(N+1)}{2} \tag{2.63}$$

Use Table 2.22 in Davis (2002) to test T.

2.10 Problems

2.1. **Unstable algorithms and roundoff error:** You can go to our website to find an m-file named `goldmean.m`. It demonstrates how an unstable algorithm can interact with roundoff error in an unhealthy way. Download the m-file and save it in your current/working directory. Start up MATLAB and type

```
goldmean
```

This algorithm is based on an iterative formula for the powers of the so-called *Golden Mean*:

$$\Phi = \frac{\sqrt{5} - 1}{2} \tag{2.64}$$

It can be shown (we leave the algebra to you if you should doubt it) that:

$$\Phi^{n+1} = \Phi^{n-1} - \Phi^n \tag{2.65}$$

This m-file will calculate Φ both ways: the hard way (raising the first equation to each successive power) and then the easy way (by calculating and making use of the fact that Φ^0 is always 1, then applying the recursion). When it has finished it will print on the screen a table of four columns: the power, the hard way, the recursion way, and the relative error (difference divided by the "correct" answer). Notice that MATLAB divides all of the numbers by 10^4, what we see is that by the 16th iteration we are starting to get a deviation between the two, supposedly identical, numbers. You will see that by the 38th iteration the two different ways of calculating are diverging by quite a bit (on the order of 10%)! The results will be in the MATLAB variable R and will be placed in your MATLAB workspace, so be sure to have your m-file display this variable. Now make a plot of the iteration (power) versus the absolute value of the relative error: you can "see" how this kind of error can explode in certain types of calculations.

2.2. **The robustness of skewness and kurtosis**: If you have some data and you compute a non-zero skewness and a kurtosis that is not exactly 3, what does this mean, and how much can you trust this data? Let's do an experiment to see how robust these statistical indicators really are. Generate a normally distributed data set of 50 points using MATLAB's `randn` command and evaluate the skewness and kurtosis using either the `Statistics Toolbox` commands or your own calculations (try doing both if you're adventurous). Next try doing this experiment with 100 points, and see how these statistics change. Finally, embed the experiment in a loop that doubles the number of points (starting at 50 and going to 51 200 in factors of 2). Plot the skewness and kurtosis of these results as a function of sample size on a semilog scale (log of N) using the `semilogx` function. Now do this experiment a number of times, to see how the resultant curves "wag" at the low end. Which statistic settles down as a function of N first? What might you conclude about the requisite sample size before you begin to believe the numbers?

2.3. **Evaluating normalcy and rejecting data**: Download from our fabulous website the data file `NormQuery.mat`. It contains a row vector xnq which you need to evaluate for normalcy. Do your evaluation in each of the three ways suggested in the text:

- plotting a histogram, and overlaying the appropriate PDF using the computed mean and standard deviation;
- plotting the cumulative probability distribution; and
- using the Lilliefors test

What is the nature of the distribution? If you were to apply Chauvenet's criterion, how many points would you reject? How do the distribution tests change in response to this "filtering"? What happens if you do it again, and again, and again?

2.4. **Using MATLAB's Student's *t*-test**: You suspect that there are two distinct populations of NO_3 values in a nutrient data set, one between 2000 and 3000 meters and the other below 3000 m. In this problem you will use some of MATLAB's powerful data manipulation capabilities to analyze this problem. You formulate a null hypothesis that there is no difference:

$$H_o : \mu_1 = \mu_2 \tag{2.66}$$

and we can test this hypothesis regarding their means using MATLAB's `ttest2` function.

First download `no3.dat` into your current/working directory. This is a NO_3 data set from the North Atlantic. Now you can load the data into MATLAB after starting it up by issuing the following command:

```
load no3.dat
```

MATLAB will now have a variable in its workspace called no3. To make things simpler on us (and to protect ourselves from overwriting the original data) we will create a new variable equal to no3 by:

```
D=no3;
```

Don't forget the ";" or you'll get a face full of numbers. Let's look at these data first. Use the plot command:

```
plot(D(:,2),D(:,1),'+');
```

What have we done? We have told MATLAB to plot column 2 of D (the NO_3 values) on the *x*-axis and column 1 of D (the depth) on the *y*-axis and we told it to represent each data point with only a "+" sign, no connecting lines. But it doesn't look right. That's because it is "upside down". We can fix this by multiplying every element of column 1 by −1:

```
D(:,1)=-1*D(:,1);
```

Try that plot command again. Now we can use a very powerful feature of MATLAB to sub-select this data into two more variables:

```
Du=D(D(:,1)<=-2000 & D(:,1)>-3000,:);
Dl=D(D(:,1)<=-3000,:);
```

What have we done here? We've used MATLAB's self-referential capabilities and told it to set variable Du (for upper) to all D's where the first column of D was less than or equal to −2000 AND greater than −3000 meters. In the next line we've set Dl (for lower) to all D's where the first column of D is less than or equal to −3000. Try plotting these variables the same way you plotted D: you should get data points only in the specified depth ranges. Now we're ready to apply the t-test:

```
[H,sig,ci]=ttest2(Du(:,2),Dl(:,2),0.05,0)
```

Do the results surprise you? Calculate the mean and standard deviation for each group of NO_3 using:

```
Dmu=mean(Du(:,2))
Dsu=std(Du(:,2))
```

Do the same for Dl (call them Dml and Dsl). You'll see that the means are not that far apart, but one glance at the plots tells you something else might be going on. Make sure your m-file makes it clear what your decision was: reject or keep the null hypothesis and report the values for Dmu, Dsu, Dml, Dsl. Issue the command:

```
help ttest2
```

What are the odds that you got the results you did by chance assuming the null hypothesis is true? What was your probability of committing a type-I error?

2.5. **Approaching the central limit theorem**: Download the data file CTD1.mat from our website. The file contains three vectors for raw data; P (pressure), T (temperature), and P (pressure). Study the raw data PDFs by plotting a histogram of the T, C, and P vectors. Do they appear normal? What test would you apply to decide this? There is a fourth vector that is computed from T and C. Evaluate it for normalcy. By block averaging in groups of 25 readings, how do the block averages distribute? Test for normalcy. Download the data file CTD2.mat and do a similar exercise. What happens and why?

3

Least squares and regression techniques, goodness of fit and tests, and nonlinear least squares techniques

If you're trying to establish cause-and-effect relationships, do try to do
so with a properly designed experiment.

Robert Hooke

3.1 Statistical basis for regression

Often, in working with data, we must ask specific quantitative questions about how well
the data reflect some underlying model of processes in nature. The question may arise,
for example, as to whether some property (e.g. dissolved oxygen in seawater) changes
linearly with time or whether two properties are related to each other. The actual rate of
change (the slope of the relationship) may be of quantitative interest, or the initial value (the
intercept of the relationship) may be important. Moreover, in addition to obtaining the most
accurate estimate of one or more of these parameters, we need to know how precisely we
know the values, or, more specifically, the confidence interval (range of probable values)
of the parameters. Finally, we must decide if our linear model (or some other function) is
appropriate or robust by checking the goodness of fit.

3.1.1 The chi-squared (χ^2) defined (and goodness of fit)

How do you judge (aside from the eyeball test) the goodness of a fit? The most common
goal is to reduce the "distance" between the observations and the model. The standard
measure, which can be derived from the Gaussian nature of the underlying distributions, is
the *chi-squared*, which is defined as:

$$\chi^2 = \sum_{i=1}^{N} \frac{(\hat{y}_i - y_i)^2}{\sigma_i^2} \tag{3.1}$$

where $\hat{y}_i$ is the model (the fit) estimate, y_i is the actual observation, and σ_i in the denominator is the uncertainty in the individual measurement (y_i). The value of χ^2 is computed
by summing over all of your observations N. You could of course choose other measures
of distance, e.g. the absolute values $|\hat{y}_i - y_i|$. But χ^2 has the nice property that it results in

49

relatively simple, analytical forms, and it can be derived for a normal distribution from the method of Maximum Likelihood Estimation (a topic we will return to in later chapters). You may arrive at the choice of σ_i in a number of ways:

- it could be the size of the smallest graduation on your measuring stick or analytical balance;
- there is some fundamental physical limitation to your measurements, which you derive from basic principles;
- there is some internal statistic associated with the measurement itself (e.g. you may be using the mean of repeated measurements as your observation);
- you may read it in a manual, or from manufacturer's specifications, probably obtained from one or more of the above.

Note also that this σ_i may/can/will vary from measurement to measurement and hence represents some kind of weighting factor that you might use when incorporating data into a larger set. That is, you would not want to value a poorly made or inaccurate measurement as much as a more carefully made, precise measurement.

Regardless of how you arrived at your choice of σ_i, you would tend to think that the *root mean squared* (RMS) deviation normalized to measurement errors would tend to be close to 1 if things are working correctly. Here we define just such a measure, the *reduced chi-squared* (note the subscript ν, to distinguish it from its bigger brother), as:

$$\chi_\nu^2 = \frac{1}{\nu} \sum_{i=1}^{N} \frac{(\hat{y}_i - y_i)^2}{\sigma_i^2} = \frac{1}{\nu} \chi^2 \tag{3.2}$$

where $\nu = N - n$ is the degrees of freedom, and n is the number of coefficients or parameters used in the regression fit. For example, computing a mean gives $N - 1$ as the degrees of freedom, and a regression to a straight line gives $N - 2$.

If your χ_ν^2 is much larger than 1, say 10 or 100, it means that you are either doing a lousy job making measurements (i.e. something is wrong with your apparatus or technique), or you have been overly optimistic about your measurement uncertainties. Another possibility is that you may have the wrong model, that is, an inappropriate fitting function. For example, the data better fit a quadratic or exponential relationship rather than a straight line. If χ_ν^2 is too small, say 0.1 or 0.01, it may mean that you have been too pessimistic about measurement errors. It is very important to pay careful attention to estimation of your measurement errors.

We can be somewhat more quantitative about testing the goodness of fit of a model to the observations. Random errors can decrease the apparent goodness of fit of a model and increase the resulting χ_ν^2. We can use the incomplete gamma function (gammainc in MATLAB) (Press *et al.*, 2007) to estimate the probability, P_{χ^2}, that a χ^2 value as large as that found will occur because of random errors, even though the underlying model is a good fit to the data:

$$P_{\chi^2}\left(\frac{\nu}{2}, \frac{\chi^2}{2}\right) = \text{gammainc}\left(\frac{\chi^2}{2}, \frac{\nu}{2}\right) \tag{3.3}$$

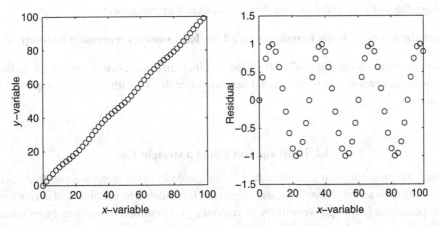

Figure 3.1 The data in the left panel appear to be a good fit to a straight line. But when you look at the data residuals (right panel) from the least squares fit, you see that the residuals actually have structure (in this case a sine wave)!

The lower the probability, the less likely that your model is indeed correct. The P_{χ^2} value for a good fit would be ~ 0.1 (remember *random* errors decrease the apparent goodness of fit), and the model could be acceptable with a P_{χ^2} as low as 0.01 or 0.001. Really incorrect models will have very small probabilities with $P_{\chi^2} \sim 10^{-16}$.

3.1.2 Look at your residuals

Finally, a good χ_ν^2 may not mean you have a good fit or model. It may be a conspiracy between the above factors (incorrect model, underestimated errors, etc.), or mean that you have overlooked something. Look at your data, and particularly look for patterns or structure in the residuals of the data. That is, plot up the differences between your data and the values predicted by your regression model. If the deviations between your data and your model fit show a characteristic large-scale structure, this is indicative of unresolved and unfit characteristics. Look, for example, at the two plots in Fig. 3.1.

The left-hand panel shows a data set that appears to follow a straight line rather well; yet when you subtract a least squares linear fit from the observations (panel on the right) you see a definite pattern in the residuals. The χ_ν^2 is just one of many diagnostics of model success and, like any single numeric indicator, does not tell the whole story.

If you do not have an independent determination of your measurement uncertainty (this, however, should be a most unusual and unphysical thing to happen), do not despair. You can use χ_ν^2 as a means of determining your measurement error, presupposing that you know you have a good "model" of your data, i.e. that you are fitting it to the correct curve. You simply adjust σ_i using a scaling factor so that χ_ν^2 equals 1.0.

Regardless of your situation, one thing is important to remember:

Minimization of χ^2 is the foundation of all the least squares regression techniques.

Hence the term "least squares". All of the routines discussed below are aimed at finding the model parameters (coefficients) that minimize the χ^2 with respect to a given data population.

3.2 Least squares fitting a straight line

Perhaps the most common data regression model (aside from the mean and standard deviation) is the fit to a straight line. It is also the easiest formulation to visualize and derive from basic principles (although we will try to convince you that there is an even more robust, succinct, general, and easy to understand approach in Section 3.3). By substituting into the definition of χ^2 the formula for a least squares regression $\hat{y} = a_1 + a_2 x$ (where a_1 is the intercept and a_2 is the slope), you have:

$$\chi^2 = \sum_{i=1}^{N} \frac{(a_1 + a_2 x_i - y_i)^2}{\sigma_i^2} \tag{3.4}$$

In this case x_i is the independent variable, and y_i is the dependent variable. For a so-called *Type I regression*, we will assume that x_i is known quite well (i.e. $\sigma_x \approx 0$). The name of the game then is to choose values for the coefficients a_1 and a_2 to minimize the value of χ^2.

What does that really mean? Let's consider a practical example. The data points in Fig. 3.2 are the annual mean, global upper ocean heat content anomalies as a function of time over the past several decades (Levitus *et al.*, 2005). A linear fit to the data (dashed line) suggests a long-term temperature increase in the upper ocean (0–300 meters), a trend that is consistent with global warming from excess carbon dioxide in the atmosphere (Gent *et al.*, 2006). If you look at Fig. 3.2, the least squares fit is equivalent to minimizing the sum of the squares of the vertical distances between the observations (y_i) and the regression estimated values ($a_1 + a_2 x_i$).

3.2.1 Doing things the hard way (the normal equations)

Now how do we choose the coefficients a_1 and a_2 to minimize χ^2? Well, you need to think back to your first-year calculus course. The extrema (maxima or minima) of a function are characterized by the first derivative going to zero. So if we differentiate χ^2 with respect to each of the coefficients and set the derivatives equal to zero, we have two equations in two unknowns. That is, we have:

$$\frac{\partial \chi^2}{\partial a_1} = \sum_{i=1}^{N} \frac{2(a_1 + a_2 x_i - y_i)}{\sigma_i^2} = 0 \tag{3.5}$$

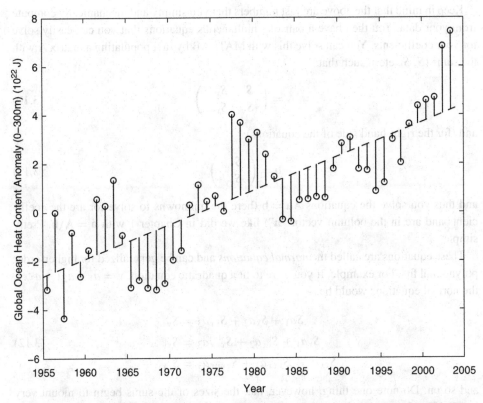

Figure 3.2 A plot of the annual mean heat content anomaly for the global upper ocean (10^{22} J) as a function of time (Levitus *et al.*, 2005). The basis of the least squares method is to minimize the sum of the squared model data misfit, for a Type I regression taken as the vertical distance between the data points and the fit line.

and

$$\frac{\partial \chi^2}{\partial a_2} = \sum_{i=1}^{N} \frac{2x_i \left(a_1 + a_2 x_i - y_i \right)}{\sigma_i^2} = 0 \tag{3.6}$$

This can be reduced to two equations in two unknowns (the coefficients),

$$S a_1 + S_x a_2 = S_y \tag{3.7}$$

$$S_x a_1 + S_{xx} a_2 = S_{xy} \tag{3.8}$$

where, for shorthand notation, we have used the notation that:

$$S = \sum_{i=1}^{N} \frac{1}{\sigma_i^2}; \quad S_x = \sum_{i=1}^{N} \frac{x_i}{\sigma_i^2}; \quad S_y = \sum_{i=1}^{N} \frac{y_i}{\sigma_i^2}; \quad S_{xx} = \sum_{i=1}^{N} \frac{x_i^2}{\sigma_i^2}; \quad S_{xy} = \sum_{i=1}^{N} \frac{x_i y_i}{\sigma_i^2}$$

$$\tag{3.9}$$

Keep in mind that the above are just numbers that you simply and mechanically compute from your data. You then have a pair of simultaneous equations that you can easily solve for your coefficients. You can solve this with MATLAB by first populating a matrix $\mathbf{A}$ with the sums (S, S_x etc.), such that:

$$\mathbf{A} = \begin{pmatrix} S & S_x \\ S_x & S_{xx} \end{pmatrix} \tag{3.10}$$

and (for the right-hand side of the equations):

$$\mathbf{b} = \begin{pmatrix} S_y \\ S_{xy} \end{pmatrix} \tag{3.11}$$

and thus you solve the equations $\mathbf{Aa} = \mathbf{b}$ (here the unknowns to solve for are the coefficients and are in the column vector "$\mathbf{a}$") like we did in Chapter 1 with $\mathbf{a} = \mathbf{A}\backslash\mathbf{b}$. Pretty simple.

These equations are called the *normal equations* and can be generalized for higher-order polynomial fits. For example, if you were to fit a quadratic equation, $y = a_1 + a_2 x + a_3 x^2$, the normal equations would be:

$$Sa_1 + S_x a_2 + S_{xx} a_3 = S_y$$
$$S_x a_1 + S_{xx} a_2 + S_{xxx} a_3 = S_{xy} \tag{3.12}$$
$$S_{xx} a_1 + S_{xxx} a_2 + S_{xxxx} a_3 = S_{xxy}$$

and so on. Do note one thing, however, that the sizes of the sums begin to mount very rapidly (for the quadratic, you are summing up the 4th power of x) and roundoff errors soon become a problem (see Chapter 2).

The solution to the normal equations for the straight-line regression can be easily obtained in MATLAB using the $\mathbf{a} = \mathbf{A}\backslash\mathbf{b}$ solution. Commonly, however, you'll find this solution solved by something called *Cramer's rule* (we won't go into this here; see Bevington and Robinson, 2003), which gives explicitly:

$$a_1 = \frac{S_y S_{xx} - S_{xy} S_x}{\Delta} \tag{3.13}$$

$$a_2 = \frac{S S_{xy} - S_x S_y}{\Delta} \tag{3.14}$$

where we have defined the denominator (actually the determinant of $\mathbf{A}$) as:

$$\Delta = S S_{xx} - S_x S_x \tag{3.15}$$

Remembering the concept of singular matrices, if $\mathbf{A}$ is singular, owing to inadequate data, then the determinant will be zero, and the solutions to Equations (3.13) and (3.14) will fail for obvious reasons. This is almost never a problem for straight-line fits but can be problematic for more complicated situations, where the normal equations become "almost singular" (we will talk about this more later).

3.2.2 Uncertainties in coefficients

But we also need to know the uncertainties in the coefficients. This is a relatively straight-forward thing to do, if a little tedious to calculate. If you assume there are no systematic errors in your measurements, then you can derive that the uncertainty in any parameter a with respect to the observations y is given by:

$$\sigma_a^2 = \sum_{i=1}^{N} \sigma_i^2 \left(\frac{\partial a}{\partial y_i} \right)^2 \qquad (3.16)$$

Here, we are interested in the parameters (a_1 or a_2) of the linear fit ($y = a_1 + a_2 x$) as functions of y, since that is what is uncertain. We differentiate Equations (3.13) and (3.14) with respect to the observations y_i (i.e. the contribution of each data point to the fit parameters) and substitute into the equation above to obtain:

$$\sigma_{a_1}^2 = \frac{S_{xx}}{\Delta} \qquad (3.17)$$

$$\sigma_{a_2}^2 = \frac{S}{\Delta} \qquad (3.18)$$

Note something quite profound in the structure of Equations (3.17) and (3.18): the size of the uncertainties in the coefficients depends not on the data you measured (the y_i values) but on where you made the measurements (the x_i values) and the uncertainties in the measurements (the σ values). No summation incorporating y_i is involved! This says something about experiment design; obviously maximizing Δ (the **A** determinant) in relation to S, S_x, and S_{xx} is a good thing.

Maximizing the determinant is a question of maximizing the difference between $S S_{xx}$ and $(S_x)^2$, which is equivalent to maximizing the spread (range of x values) of your measurements. The larger the range in your data, the better you know the slope and intercept (Fig. 3.3). That makes sense. Look again at the upper equation (3.17). The numerator (S_{xx}) says that the larger the 2nd moment of the x distribution (i.e. a measure of the squared distance between the centroid of your data and the $x = 0$ axis, where the intercept is defined) the larger your intercept error (Fig. 3.4). This also makes sense.

But where does the number of measurements come into this? You would expect that increasing the number of measurements (all other things being equal) would improve our knowledge of the coefficients. It only seems fair. Well, N does come into the coefficients, but through the determinant Δ. Consider a random distribution of measurements (i.e. various x_i) centered around zero (this argument works regardless of this stipulation, but it makes it easier to understand it this way). Note the terms in the equation that define the determinant. The second term is the square of S_x, which is the sum of x_i values. This sum will be close to zero, since about half of the x_i will be negative, and the other half positive. The first term, S_{xx}, which is the sum of the *squares* of the x_i, will always be positive,

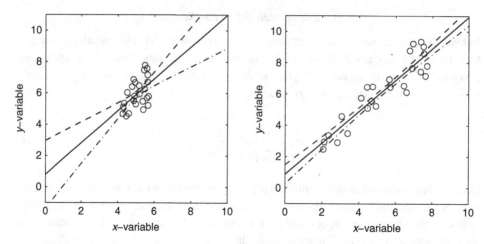

Figure 3.3 The error in the slope estimate depends on the *x* sampling locations; closely clumped samples in *x* will have a larger error. The confidence intervals for the expected *y*-value $\hat{y}$ (upper bound, dashed line; lower bound dot-dashed line) approach but do not meet the model fit line near the centroid of the data.

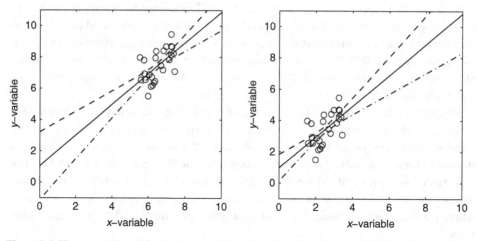

Figure 3.4 The error estimate for the intercept depends on how far the *x* sampling locations are from $x = 0$. The confidence intervals for the expected *y*-value $\hat{y}$ (upper bound, dashed line; lower bound dot-dashed line) approach but do not meet the model fit line near the centroid of the data.

and will continue to grow as the number of measurements increases. Thus Δ will always increase for increasing numbers of measurements. Figure 3.5 shows the uncertainties in the intercept (upper curve) and slope (lower curve) for a random group of measurements as we increase N (horizontal axis).

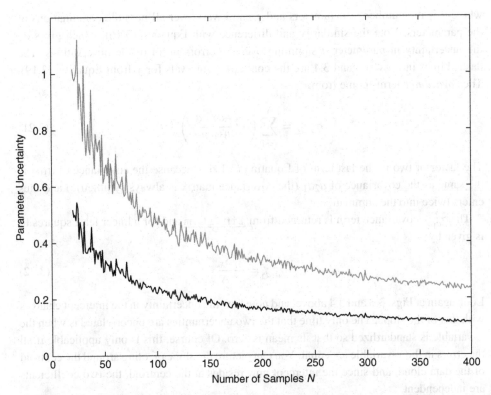

Figure 3.5 The uncertainty estimates (*y*-axis) for the intercept (upper, gray line) and slope (lower, solid line) parameters from a linear least squares fit decrease as the number of data points *N* increases (*x*-axis), shown here for randomly generated data.

3.2.3 Uncertainties in an estimated y-value

Now that you have determined the coefficients of your straight-line fit, and the uncertainties in those coefficients, you might want to go ahead and calculate some best fit value of $\hat{y}$ at some desired location $\hat{x}$. That's easy, $\hat{y} = a_1 + a_2\hat{x}$ does the trick. But how well do you know this new estimate? We can apply the error propagation formula shown in the last chapter, Equation (2.13). Keep in mind that the uncertainties in the intercept (a_1) and the slope (a_2) are *correlated* and thus we need a third term involving the covariance of the two uncertainties:

$$\sigma_{\hat{y}}^2 = \sigma_{a_1}^2 + \sigma_{a_2}^2 \hat{x}^2 + 2\sigma_{a_1 a_2}^2 \hat{x} \qquad (3.19)$$

The general form for calculating the uncertainty in $\hat{y}$ due to the uncertainties in the parameters a is:

$$\sigma_{\hat{y}}^2 = \sum_{j=1}^{n} \sum_{k=1}^{n} \sigma_{a_j a_k}^2 \left(\frac{\partial \hat{y}}{\partial a_j}\right) \left(\frac{\partial \hat{y}}{\partial a_k}\right) \qquad (3.20)$$

where n is the number of parameters and j and k vary over all possible combinations of the parameters. Note the similarity and difference with Equation (3.16), which gives σ_a the uncertainty in parameter a summing over the errors in all of the observations. The dashed lines in Figs. 3.3 and 3.4 are the confidence intervals for $\hat{y}$ from Equation (3.19). The *covariance* terms come from:

$$\sigma^2_{a_j a_k} = \sum_{i=1}^{N} \left(\sigma^2_{y_i} \frac{\partial a_j}{\partial y_i} \frac{\partial a_k}{\partial y_i} \right) \tag{3.21}$$

The factor of two in the last term of Equation (3.19) is because the covariance of $a_1 a_2$ is the same as the covariance of $a_2 a_1$ (the covariance matrix is always symmetric) and thus enters twice into the summation.

The $\sigma^2_{a_1 a_2}$ covariance term is returned from <u>linfit.m</u> and for a linear least squares fit is given by:

$$\sigma^2_{a_1 a_2} = -\frac{S_x}{\Delta} \tag{3.22}$$

Look again at Figs. 3.3 and 3.4 above, and note how the uncertainty in the intercept changes with that of the slope. The only time that the two uncertainties are uncorrelated is when the x-variable is standardized so that its mean is zero. Of course, this is only applicable if all the errors in the y-variable are equal. You can picture the slope pivoting around the centroid of the data cloud, and since the intercept now resides at the centroid, the two coefficients are independent.

If you do this, then $S_x = 0$, and Equations (3.13), (3.14), and (3.15) reduce to:

$$\begin{aligned} \Delta &= SS_{xx} \\ a_1 &= \frac{S_y}{S} \\ a_2 &= \frac{S_{xy}}{S_{xx}} \end{aligned} \tag{3.23}$$

And the uncertainties reduce to:

$$\begin{aligned} \sigma^2_{a_1} &= \frac{1}{S} \\ \sigma^2_{a_2} &= \frac{1}{S_{xx}} \end{aligned} \tag{3.24}$$

which makes life a little simpler. So all you need to do is to subtract the mean of x from your x-values before you do the regression and add it back in before you calculate your results. It pays to standardize! In the last chapter we standardized by subtracting the mean and then dividing by the standard deviation for the whole data set (Eq. (2.9)). It all depends on units; if x and y are in the same units simply subtracting the mean works well, but if they are in different units you may consider full standardization (Z-scores).

3.2.4 Example: ocean heat content

We return to the time series of ocean heat content (Fig. 3.2) to pull together the many threads of Type I linear regressions in a practical example. A basic question we might ask, given the data, is, "Has the heat content of the upper ocean increased, decreased, or stayed the same over the last 50 years?" To answer this question we want to examine the slope of a linear least squares fit and its uncertainty. The data are replotted in Fig. 3.6, but now with individual error bars for each observation, together with the linear model fit and confidence intervals.

You can easily do this yourself because we have a MATLAB routine called `linfit.m`, which you can download and use as a general straight-line fitting routine. It is very simple to use, requiring you to supply three (vector) arguments x, y, and sy (σ_{y_i}).

```
[a,sa,cov,r] = linfit(x,y,sy)
```

The routine returns the intercept and slope coefficients a_1 and a_2 (in the vector a), the uncertainties of those coefficients in sa, the covariance of the intercept on the slope in cov, and the linear correlation coefficient r (which we will discuss more in Chapter 4).

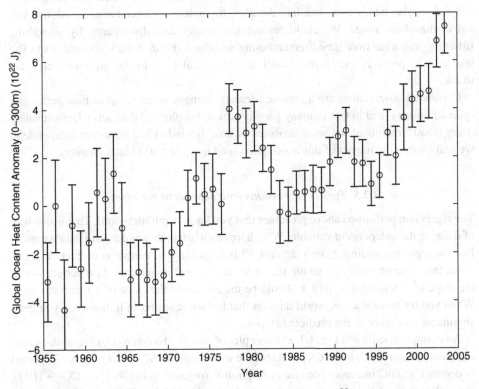

Figure 3.6 Data for the global upper ocean heat content anomaly (10^{22} J) replotted from Fig. 3.2 with error bars, a Type I least squares linear fit (solid line), and confidence intervals for the fit (dashed lines).

The slope of the linear fit $a_2 = 0.141 \pm 0.014$ $(10^{22}\,\mathrm{J\,y}^{-1})$ is much greater than the estimated error for a_2 (i.e. $0.141 \gg 0.014$), suggesting that a_2 is non-zero and the heat content is indeed growing over time. But a linear fit is not a terribly good model of the data in detail. The average σ_i is $1.3 \times 10^{22}\,\mathrm{J}$, somewhat larger for the first part of the record. A visual inspection of the residuals shows that many of the data are much farther than 1σ from the model curve. This impression agrees with the value of χ_ν^2 (2.01), which is uncomfortably greater than 1 (the "expected" value if our model–data misfit were simply due to random noise with a magnitude of σ). We can use Equation (3.3) to estimate the probability that the model fit is actually quite good but that random noise conspired to gives us such poor (large) χ_ν^2. The computed probability P_χ (1.1×10^{-5}) is quite low (remember we were looking for about 0.1), suggesting that our model is missing something.

The vertical lines in Fig. 3.2 correspond to the residuals from the fit, that is, the effective mismatch between the linear fit and the observations. There is considerable structure in the residuals, with definite correlations between the residuals from year to year (residuals from random noise would be expected to bounce around more from positive to negative values). Put another way, the model–data errors are not *independent* of each other. Of course one should not expect our simple linear model, which was designed to test for secular trends over multiple decades, to capture an interannual variability signal in the observations. We could resolve this model–data discrepancy by smoothing (filtering) the data over time (remembering to adjust the σ_i values accordingly). The lesson is to properly match the model and statistical tests to the question you want to ask.

Correlated observations are a common feature in many oceanographic time series and spatial transects (and in fact in many geophysical and ecological data sets). Unfortunately many (most) traditional statistical analyses assume that individual errors are independent; we will discuss the impact of this assumption (and its violation) in later chapters.

3.2.5 Type II regressions (two dependent variables)

The regression performed above presumes that you know x infinitely well. That is one way of defining the independent variable. What happens when both y and x have uncertainties? Do you regress y against x, or x against y? It turns out that neither is correct. You can prove this experimentally by taking the same data set and doing both. In a perfect world, the slope of y regressed against x should be the inverse of the slope of x regressed on y. When you try this for a real-world data set that has some scatter in it, however, you get a significant difference in the predicted slopes.

An example is shown in Fig. 3.7, a scatter plot of dissolved oxygen (O_2) versus dissolved inorganic carbon (DIC) in the deep water of the western South Atlantic. Oxygen is expected to decrease as DIC increases because of respiration (organic matter $+ O_2 \rightarrow CO_2 + H_2O$), but such simple relationships can be strongly distorted by mixing. The ratio of $\Delta O_2/\Delta DIC$ due to respiration is an important biogeochemical quantity. So how does one solve for the slope when there is uncertainty in both x and y? You use a so-called *Type II regression*.

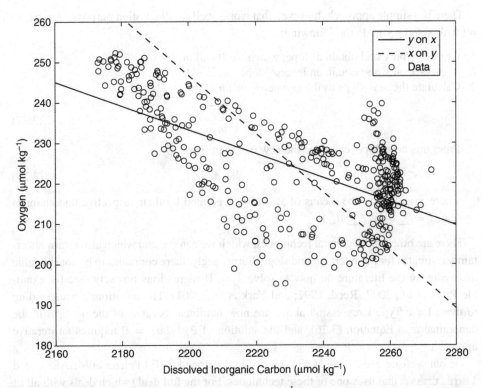

Figure 3.7 Scatter plot of dissolved oxygen (μmol kg^{-1}) versus dissolved inorganic carbon (μmol kg^{-1}) in the deep water of the western South Atlantic (Wanninkhof and Doney, 2005). Type I regressions (assume one independent variable and one dependent variable) can give very different slope estimates depending on whether y is regressed on x (solid line) or x is regressed on y (dashed line). For many real-world applications, both variables have error and a Type II regression should be used.

For a Type II regression, the process of minimizing the vertical distances between your y-data and the fit line is incorrect; you should be minimizing the perpendicular distance between your data points and the regression line (assuming a χ^2-like metric applies). This sounds simple, and it is conceptually:

$$\chi^2 = \sum_{i=1}^{N} \left(\frac{1}{\sigma_{xi}^2} (x_i - \hat{x}_i)^2 + \frac{1}{\sigma_{yi}^2} (y_i - \hat{y}_i)^2 \right) \tag{3.25}$$

where we now have an expected value (along the model line) for both x and y, $\hat{x}$ and $\hat{y}$. This can be rewritten for a straight line as:

$$\chi^2 = \sum_{i=1}^{N} \frac{(a_1 + a_2 x_i - y_i)^2}{\sigma_{yi}^2 + a_2^2 \sigma_{xi}^2} \tag{3.26}$$

Again we minimize χ^2, but the math becomes a little gnarly as we discuss below.

There is a simple approach, however, that works well for illustration purposes. For a data set with **x** and **y** you do the following:

1. Regress y on x and obtain a slope, which we'll call m_{yx};
2. Regress x on y and obtain an inverse slope, m_{xy};
3. Calculate the new slope as the *geometric mean* of the combined fits:

$$a_2 = \sqrt{\frac{m_{yx}}{m_{xy}}} \qquad (3.27)$$

4. From this new slope, calculate the new intercept from:

$$a_1 = \bar{y} - a_2 \bar{x} \qquad (3.28)$$

where $\bar{x}$ and $\bar{y}$ are the means of x_i and y_i weighted by their respective uncertainties (e.g. $\bar{x} = \sum \sigma_{xi} x_i / \sum \sigma_{xi}$).

There are other, more general techniques, which we can use and which also return uncertainty estimates for the intercept and slope. Surprisingly, there continues to be considerable discussion in the literature on how to solve Type II regressions properly (see for example Press *et al.*, 2007; Reed, 1992; and York *et al.*, 2004). The equations corresponding to those for a Type I regression above are now nonlinear because of the a_2^2 term in the denominator in Equation (3.26), and the solution of $\partial \chi^2 / \partial a_2 = 0$ requires an iterative approach.

On our website, we have a MATLAB m-file developed by Ed Peltzer at MBARI called `lsqfitma.m` that uses one of these techniques. For the full deal (which deals with all of the nasty math), we also have a more advanced least squares routine known as the *least squares cubic* (York, 1966; York *et al.*, 2004). However, this routine requires that you have error estimates for both x and y for each point; you can find this routine in `lsqcubic.m`.

As a final note, York *et al.* (1969) also describes what to do when σ_{xi} and σ_{yi} are correlated. Sounds unlikely? Actually it occurs frequently in radiometric age determinations when one plots one isotope ratio against another, where an isotope is in common between x and y (e.g. Kent *et al.*, 1990):

$$\frac{x_3}{x_1} = a_1 + a_2 \frac{x_2}{x_1} \qquad (3.29)$$

The presence of x_1 in the denominators of both x and y means that σ_{xi} and σ_{yi} for each data pair are correlated.

3.3 General linear least squares technique

3.3.1 Choose your model functions wisely

It is possible to set up the *normal equations* for any arbitrary set of *basis functions* (just as we did for the straight-line formula above). You can think of basis functions as building blocks for describing your data (as we discuss later, sometimes these functions have a mechanistic interpretation and sometimes they are just numerically convenient). The difficulty of using the normal equations is that the more complicated the functions, the more difficult the formulation, and the more the risk that the solution to the normal equations becomes numerically ill-behaved. This happens when the functions chosen are not very orthogonal (that is, they are more similar than not), or when the differing functions are not easily distinguished by the measurements actually made (e.g. the sampled range of conditions is too narrow). The latter happens more often than people realize. Finally, the choice of functions may not really do a good job of matching the actual observations.

Regardless of the underlying reason, when this happens the solutions end up being a delicate balance between very large numbers (recall for the normal equations we differenced rather large sums, e.g. $SS_{xx} - S_x S_x$). This may yield the minimum χ^2 for the actual measurements, but outside the range of the measurements, or sometimes in between data points if there are gaps, the solution may do rather strange things. A classic example (and often the worst offender) is a polynomial regression to an abruptly changing data set, which is otherwise very quiescent. Consider, for example, a step function in data, which you try desperately to fit with polynomials of increasing order (Fig. 3.8).

The upper left-hand panel in Fig. 3.8 shows the data, which is -0.5 for x less than 0, and $+0.5$ for x greater than 0 (with a little noise added on). The other panels show results of various order polynomial fits ranging from linear up to 10th order. Note that you have to go to very high order to begin to approximate the step function's sharpness, but the price you pay is tight little oscillations where you have data (kind of an induced "ringing") plus wildly aberrant behavior outside the data range. The latter may be particularly troublesome if you wish to try extrapolating your fits beyond your measurement range. Ringing will reappear in several more incarnations later in the book, particularly in Chapters 6 and 12 when we talk about time-series analysis and finite difference modeling.

3.3.2 *There is an easier way: the design matrix approach*

The solution by means of the normal equations can be unstable, and it is rather tedious to have to build the analytical forms for them when the functions are more complicated. Now we'll show you an easier, numerical way. By the way, the term *linear least squares* does not mean that the functions you are fitting are (necessarily) linear, but rather that the formulation is linear in the coefficients. Thus you could have a function to fit that looks like, for example,

$$y = a_1 + a_2 x^2 + a_3 \exp\left(\frac{x}{\pi}\right) + a_4 \sin(0.17x) \tag{3.30}$$

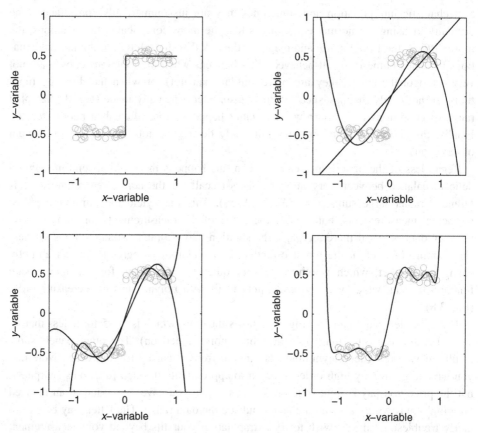

Figure 3.8 Discontinuities in data set such as the step function in the upper left panel can lead to all sorts of problems for polynomial fitting routines (upper right linear and cubic; lower left 4th and 5th order; lower right 10th order). Also note that once you get beyond the range of data either to the left or the right, the polynomial fits go haywire. Don't use polynomial fits for extrapolation!

which can be fit with linear least squares, as long as there are no unknown coefficients inside the parenthetically closed arguments to the nonlinear functions. For example, the following is definitely a nonlinear regression prospect:

$$y = a_1 \sin(a_2 x) + a_3 x \qquad (3.31)$$

The culprit is the a_2 term, which appears as part of the argument of a nonlinear function. We will deal with this kind of problem in Section 3.4.

Now about this supposed easy way. Well, think of constructing a series of simultaneous equations with one equation for each measurement (we'll use the nasty little equation we gave as an example in 3.30):

$$y_1 = a_1 + a_2 x_1^2 + a_3 \exp\left(\frac{x_1}{\pi}\right) + a_4 \sin\left(0.17 x_1\right)$$

$$y_2 = a_1 + a_2 x_2^2 + a_3 \exp\left(\frac{x_2}{\pi}\right) + a_4 \sin\left(0.17 x_2\right) \qquad (3.32)$$

$$\vdots$$

$$y_N = a_1 + a_2 x_N^2 + a_3 \exp\left(\frac{x_N}{\pi}\right) + a_4 \sin\left(0.17 x_N\right)$$

The above can be represented by a matrix called the *design matrix*, which would look like:

$$\mathbf{A} = \begin{pmatrix} 1 & x_1^2 & \exp\left(\frac{x_1}{\pi}\right) & \sin\left(0.17 x_1\right) \\ 1 & x_2^2 & \exp\left(\frac{x_2}{\pi}\right) & \sin\left(0.17 x_2\right) \\ \vdots & \vdots & \vdots & \vdots \\ 1 & x_N^2 & \exp\left(\frac{x_N}{\pi}\right) & \sin\left(0.17 x_N\right) \end{pmatrix} \qquad (3.33)$$

It would be a $N \times 4$ matrix (N measurements and four columns). One way to think about the problem is that the design matrix is just a set of basis functions that we are trying to fit to the data. Remember that all of the elements in this design matrix are simply numbers that you calculate from your x data vector. Then you would have a column vector consisting of your four unknown coefficients (which you want to solve for):

$$\mathbf{a} = \begin{pmatrix} a_1 \\ a_2 \\ a_3 \\ a_4 \end{pmatrix} \qquad (3.34)$$

and the column vector of your N observations:

$$\mathbf{y} = \begin{pmatrix} y_1 \\ y_2 \\ y_3 \\ \vdots \\ y_N \end{pmatrix} \qquad (3.35)$$

The matrix version of these equations now reduces to $\mathbf{Aa} = \mathbf{y}$. Looks simple, doesn't it? But what about the weighting factors (the σ values)? Well, like in the linear case, you just divide your x entries and your y entries by σ. That is to say, in this case we share the measurement uncertainty between the design matrix ($\mathbf{A}$) and the data array ($\mathbf{y}$). We won't repeat the matrices listed above, but for example the ith row of the design matrix of $\mathbf{Aa} = \mathbf{y}$ would look like:

$$\begin{pmatrix} \dfrac{1}{\sigma_i} & \dfrac{x_i^2}{\sigma_i} & \dfrac{\exp\left(\frac{x_i}{\pi}\right)}{\sigma_i} & \dfrac{\sin(0.17 x_i)}{\sigma_i} \end{pmatrix} \qquad (3.36)$$

and the ith element of the y-vector would also be divided by σ_i. Everything else proceeds the same as we discuss next.

3.3.3 Solving the design matrix equation with SVD

Now that we've shown you how to build a design matrix, etc., we'll show you how to solve for the parameters. You cannot uniquely solve this as a set of simultaneous equations because it is over-determined. That is, there are more equations than unknowns. Now since data are always imperfect, the data points will never agree on the true coefficients; the equations will always be inconsistent to some extent. This is equivalent to saying that not all the points lie exactly on the regression curve. But what you want to do is to minimize the square of the distance between the regression function and the observations. That is, you want to minimize the function $(\mathbf{Aa} - \mathbf{y})^2$. This is exactly what singular value decomposition does for you; remember what we said about the shortest vector in Chapter 1? So in MATLAB you enter the commands (after constructing the matrices, of course):

```
[U,S,V] = svd(A,0);                           % SVD of design matrix
W = diag(1./diag(S));                         % not checked for zero singular values
a = (V*W*U')*y;                               % your coefficients!
Covmat = V*W.^2*V';                           % compute covariance matrix
[n m] = size(A);                              % size of design matrix (row by column)
redchisqr = sum((A*a-y).^2)/(n-m);            % compute reduced chi squared
Covmat = redchisqr*Covmat;                    % estimate errors using reduced
                                              % chi squared
sa = sqrt(diag(Covmat));                      % uncertainties in coefficients
```

And those a and sa values are your answers. Note that the primes mean matrix transposes in MATLAB, and in the second line we have skipped an important step of checking for zeros in the singular value list before inverting. The second line is a bit tricky. It is equivalent to taking the diagonal of $\mathbf{S}$, inverting the individual elements of the resulting vector, and then reconstructing a square matrix with those new elements on the diagonal (and all of the off-diagonal elements equal to zero). Furthermore this code snippet assumes that you want an unweighted fit and/or you don't have the individual error estimates. The uncertainties in the coefficients are calculated from the χ_ν^2. The calculations look obscure, but they are efficient and powerful. You can use these six lines of code as an engine to a general linear least squares regression program. All you need to do is build the design matrix and data vector for the specific linear model you want to fit. Cool.

3.3.4 Multi-dimensional regressions

What if you want to fit your data to higher dimensions? For example, suppose you wanted to model the distribution of dissolved oxygen at some depth level in the North Atlantic. You might do this, for example, if you had observations in an area of the North Atlantic and you were interested in calculating the large-scale gradient (the rate and direction of change with distance). Supposing further that you believed that the distribution was best fit with a biquadratic function: that is, a function that is quadratic in both x (longitude) and y (latitude). Then your data, c, would be modeled after:

$$c = a_1 + a_2 x + a_3 x^2 + a_4 xy + a_5 y + a_6 y^2 \tag{3.37}$$

Be careful not to be confused here because we are now using "y" as an independent variable, rather than an observation, and we have introduced "c" as your dependent variable (observation). Actually, this whole thing sounds more complicated than it is, but it is mathematically the same as the general linear least squares for which you already have the equations. You calculate your design matrix just like before as:

$$
\mathbf{A} = \begin{pmatrix}
\dfrac{1}{\sigma_1} & \dfrac{x_1}{\sigma_1} & \dfrac{x_1^2}{\sigma_1} & \dfrac{x_1 y_1}{\sigma_1} & \dfrac{y_1}{\sigma_1} & \dfrac{y_1^2}{\sigma_1} \\[2mm]
\dfrac{1}{\sigma_2} & \dfrac{x_2}{\sigma_2} & \dfrac{x_2^2}{\sigma_2} & \dfrac{x_2 y_2}{\sigma_2} & \dfrac{y_2}{\sigma_2} & \dfrac{y_2^2}{\sigma_2} \\[2mm]
\vdots & \vdots & \vdots & \vdots & \vdots & \vdots \\[2mm]
\dfrac{1}{\sigma_N} & \dfrac{x_N}{\sigma_N} & \dfrac{x_N^2}{\sigma_N} & \dfrac{x_N y_N}{\sigma_N} & \dfrac{y_N}{\sigma_N} & \dfrac{y_N^2}{\sigma_N}
\end{pmatrix} \tag{3.38}
$$

The right-hand data column vector would be:

$$
\mathbf{c} = \begin{pmatrix}
\dfrac{c_1}{\sigma_1} \\[2mm]
\dfrac{c_2}{\sigma_2} \\[2mm]
\vdots \\[2mm]
\dfrac{c_N}{\sigma_N}
\end{pmatrix} \tag{3.39}
$$

and your coefficient matrix is given by:

$$
\mathbf{a} = \begin{pmatrix}
a_1 \\
a_2 \\
a_3 \\
a_4 \\
a_5 \\
a_6
\end{pmatrix} \tag{3.40}
$$

which you then solve with the MATLAB code we showed you above. Not too hard. You might want to look at an m-file called `surfit.m` on our website, which does an unweighted two-dimensional fit to an arbitrary order polynomial, to see how you might extend it to weighted fits (i.e. with different σ_i values for each data point).

3.3.5 *Transformably linear models*

There is one obvious case where the model equation is nonlinear in its coefficients, but you can transform your data to make the model linear. Consider the model equation:

$$
y = a_1 e^{a_2 x} \tag{3.41}
$$

If you take the logarithm of this equation, it reduces to:

$$
\log(y) = \log(a_1) + a_2 x \tag{3.42}
$$

So all you have to do is to take the log of your results (y_i), do a linear fit, and transform the first coefficient by taking the exp of the intercept. In this fashion, you can sometimes avoid doing nonlinear fits. Bear in mind, however, that because the data are weighted somewhat differently, the coefficients derived for Equations (3.41) and (3.42) will be similar but not identical. We refer you back to Chapter 2 for a more complete discussion on the log-normal distribution and some of its properties.

3.3.6 Non-coefficients

Also, beware the *non-coefficient* problem. You can sometimes introduce two model functions that are identical in behavior. That is, you can have two coefficients where you only really need one. The net result is that the normal equations become *exactly singular*. The SVD approach above, however, will give you an answer (after telling you there is a problem by giving you a zero singular value) just like it did in Chapter 1. Consider the following equation:

$$y = a_1 e^{(a_2 + a_3 x)} \tag{3.43}$$

which looks reasonable, until you realize that a_1 and e^{a_2} are essentially the same basis functions (both are constants multiplying the $e^{a_3 x}$ term, i.e. $y = a_1 e^{a_2} e^{a_3 x}$). You need to think carefully about the basis function you've built; the problem of non-coefficients happens even to the best of us.

3.4 Nonlinear least squares techniques

3.4.1 Iterative techniques

What happens when the model equations are nonlinear in the coefficients? You do the same thing: minimize χ^2. The difference is that you have to do it iteratively. There is no simple way of writing down and solving the problem as a set of linear analytic equations.

Most nonlinear fitting routines systematically search the coefficient space for a χ^2 minimum by repeatedly calculating the χ^2 for your model equation and data and then nudging the coefficients in differing directions until they reach a minimum in χ^2 (or as close to a minimum as you stipulate). The only difference between the techniques is the precise search mechanism, which can range from very crude to quite elaborate.

There are several basic approaches: grid search, gradient search, and expansion methods (e.g. see Bevington and Robinson, 2003, Chapter 8). The grid search and expansion methods work well when you are near the χ^2 minimum but are not very efficient or effective at moving large distances in coefficient space. The gradient search moves large distances well, since it travels down the path of steepest descent (in χ^2–**a** space), adjusting all the coefficients at once. But it tends to get trapped in long valleys and doesn't converge well near the global χ^2 minimum.

The more sophisticated routines tend to use initially a gradient search method, which adjusts the size of the nudge given to the parameters Δ**a** based on the corresponding size of $\Delta\chi^2$. Near the minimum these routines may switch to a grid search, and the last effort usually involves some kind of polynomial expansion (interpolation) of the χ^2 surface. The price is that you need to supply them with the means not only to evaluate your model function for your data (x_i) but also the gradient of χ^2 with respect to the parameters **a**. A popular method is the *Levenberg–Marquardt Method*.

Think of the algorithm like a marble rolling around on a surface of hills and valleys in the $\chi^2 - $ **a** hyperspace (a little like the arcade game "marble madness"). The ball will continue to roll downhill until it reaches the lowest point (sometimes oscillating around the minimum as it goes, depending on the character of the algorithm used). All of these routines require you to provide:

- an initial guess of values for your coefficients (from previous work, theory, or perhaps eyeball);
- the size of the initial incremental changes in the coefficients that you expect to make;
- the size of change in χ^2 that you would consider convergence;
- the name of the function (m-file) that computes your model results; ·
- sometimes the maximum number of iterations;
- sometimes the name of the function (m-file) that computes the model χ^2 gradient.

Keep one thing in mind. Nothing works better than a good initial first guess. You might get a little sloppy and tend to let the computer do the work. But the χ^2 surface in coefficient hyperspace may have more than one minimum, and your algorithm may get trapped in a *local minimum*, which is not nearly as good as a *global minimum* perhaps just over the next "hill" in hyperspace. As a corollary, you should always plot your results: the eyeball is not often fooled. You may not be able to distinguish precise fits, but you can tell if the fit is far from optimal. Sometimes it is worthwhile to do a grid search of initial starting values, or to approach the optimum from extremely different directions (e.g. from positive, then negative values).

If your problem is particularly plagued by *false minima*, one alternative method to consider is *simulated annealing* or the *Metropolis method* (Metropolis et al., 1953; see Press et al., 2007, and Section 10.4.1 for more discussion and example programs). In this and related techniques, the linear ("straight-down-the-hill") approach is somewhat abandoned for a more probabilistic view. In essence, the search routine starts by hopping randomly around parameter-space computing χ^2 and then iteratively homes in on the hoped for (but not rigorously provable) "global minimum". These searching techniques are practical for small to moderately sized problems. But be forewarned that the same caveats apply as for the grid and gradient searches.

For large data sets and/or complicated nonlinear functions with many coefficients, the iterative techniques are computationally expensive. You may find that you have no choice. Keep in mind that a sensible choice of model equations, based on a realistic physical, chemical, or biological model of the system you are fitting, will take you much further than any sophisticated mathematics or powerful computers. Common sense is a powerful ally.

3.4.2 Uncertainties in nonlinear coefficients

Computing the uncertainties in coefficients for nonlinear regressions is a challenging problem. It involves examining the shape of the χ^2 surface in coefficient hyperspace and estimating the boundaries of confidence intervals associated with elevation changes in χ^2. Many sophisticated routines return the covariance matrix (for the coefficients) on convergence from which the associated parameter uncertainties are computed. In addition to the excellent discussion by Bevington and Robinson (2003), the reader is referred to section 15.6 of *Numerical Recipes* for a discussion by Press *et al.* (2007).

3.4.3 Example: Exponential phytoplankton growth

To give you an idea of how this might work, we will first look at a "simple" nonlinear problem – exponential population growth. In our case the data are phytoplankton cell counts over time from a bottle incubation experiment (Fig. 3.9). If phytoplankton are provided with sufficient light and nutrients, growth can be modeled as a first-order process (don't worry too much about the differential equation – we will tackle these in detail in Chapter 8):

$$\frac{dP}{dt} = \mu P \tag{3.44}$$

where μ is the specific growth rate with units of 1/time. Integrating, we can find an equation for P with respect to time:

$$P = P_0 e^{\mu t} \tag{3.45}$$

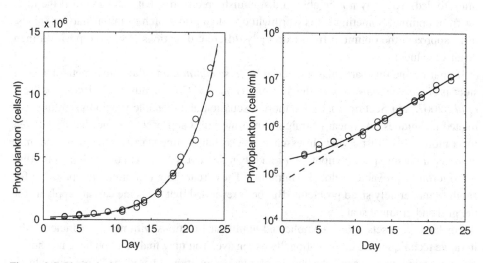

Figure 3.9 Phytoplankton concentration (cells/ml) as a function of time from a bottle incubation experiment. The right panel shows the same data but plotted as the base-10 logarithm of the cell concentration. The curves are nonlinear model fits to the data for an exponential population growth model with (solid) and without (dashed) a baseline concentration.

or a slightly more complicated one reflecting the fact there may be a pre-existing population prior to the beginning of the exponential growth:

$$P = a_1 + a_2 e^{\mu t} \qquad (3.46)$$

Populations growing exponentially are said to be in "log-phase" growth. Eventually either the nutrients run out or the bottle gets too crowded, and the population size will stabilize or crash.

So how do we fit our data to Equation (3.46)? The first thing you do is create a MATLAB m-file whose sole purpose is to compute your model function, in this case $P(t)$. You can call it anything you want, but here we will call it modfunc.m. It is a function that takes as its arguments the **x** data and vector **a** containing the coefficients and returns the model fitted data. The m-file in this example contains a whole two lines:

```
function out=modfunc(x,a)
out = a(1) + a(2)*exp(a(3)*x);
```

The function statement makes it a function call rather than just a simple script. You need to do this if you want it to return any values to MATLAB. One caveat about functions. In MATLAB, functions are only *local* in scope, so you must provide all the information you need as arguments of the function (in this case x and a); alternatively you can define some variables as *global*, in which case they can be used in the main routine and across functions. In our example, modfunc calculates all of the *y* values from the *x*s at once and is equivalent to the mathematical equation above except that we have added a parameter a(1) out front. Think of a(1) as the background population prior to the beginning of the experiment.

Next, you download from our website a nonlinear least squares fitting routine called nlleasqr.m and a helper routine called dfdp.m, which you also need. Both files ultimately came from The MathWorks website but have been fixed up a little to remove some bugs. The first file is the main engine. The second is a program that calculates the χ^2 gradient in coefficient space. It does this numerically rather than analytically. You can make your own analytic function (and call it some distinct name), which may do a better job, but it is not really necessary. Finally, there are the data phyto_growth.dat.

OK, now you have the m-files, you start up MATLAB and load in the data. You then plot it up, since you need to have some idea of your initial starting guesses for parameters. The left and right panels in Fig. 3.9 show the same data but plotted as P (left panel) and $\log_{10}(P)$ (right panel). For convenience, we divide the number of phytoplankton cells by 10^6 so that the model parameters are all close to 1. You eyeball the plot and guess that the baseline value (a(1)) should be about 0.1. The log-phase growth begins in earnest around day 10 when the cell concentration is about 0.2. You also guess that the specific growth rate is about 0.2 (a doubling time of roughly 5 days). So we set our initial parameter guess to be equal to ain:

```
ain=[0 0.2 0.2];
```

Note that the order is important and should be the same as you use in the m-file
`modfunc.m`.

Next, you type `help nlleasqr` to find out how to feed it information. Most of the
input parameters are optional, so don't be intimidated, and don't wear out your fingers.
We'll just try:

```
[f,a,kvg,iter,corp,covp,covr,stdresid,Z,r2]...
= nlleasqr(x,y,ain,'modfunc');
```

The stuff on the left is all the goodies you get back, but we'll only look at a few of them.
Your "answer" or coefficient vector is stored in a. First, let's look at what kind of job this
routine did for us. Let's plot it:

```
xf=[0:.1:25];              % dummy array to plot
yf=modfunc(xf,a);          % compute fit for the dummy array
plot(x,y,'*',xf,yf)        % plot both data and fit on same graph
```

Not bad. About the numbers: a is the output coefficients, kvg is a flag to say if convergence was achieved before the routine gave up, iter is the number of iterations, covp is
the covariance matrix for the coefficients (the square roots of the diagonal elements are the
uncertainties in the parameters), and r2 is the overall correlation coefficient squared. For
this example, we get:

```
coefficients =  [0.1372     0.0551     0.2306]
uncertainties = [0.1995     0.0211     0.0166]
r2 = 0.9833
```

Do we need the baseline parameter a(1)? From the parameter fit, it is not very well
defined by the data and is essentially zero (0.14±0.20). Let's redo the nonlinear fitting
exercise but fix a(1) = 0 (a simple rewrite of `modfunc.m` resulting in the dashed line
in Fig. 3.9). The growth rate parameter μ is similar (0.231 versus 0.222 day^{-1}) and the r^2
is nearly as good (0.9830). But an inspection of the curve fit shows that the model does a
poor job of matching the data early in the experiment. Whether or not to include parameters
depends on what we are asking the data; if your interest is in the bloom initiation (before
day 10), then yes, otherwise perhaps no.

3.4.4 Example: Gaussian on a constant background

Consider a somewhat more complicated example, a data set which you suspect to be a
Gaussian shaped peak embedded in a constant background, with some measurement noise:

$$y_f = a_1 + a_2 e^{\left(-\frac{(x-a_3)^2}{2a_4^2}\right)} \tag{3.47}$$

The measurements are represented by the dots in Fig. 3.10.

We follow the same steps, build an appropriate `modfunc.m`, choose our first guess for
the input parameters, and solve using `nlleasqr.m`:

```
function out=modfunc(x,a)
out = a(1) + a(2)*exp(-((x-a(3))/(2*a(4))).^2);
```

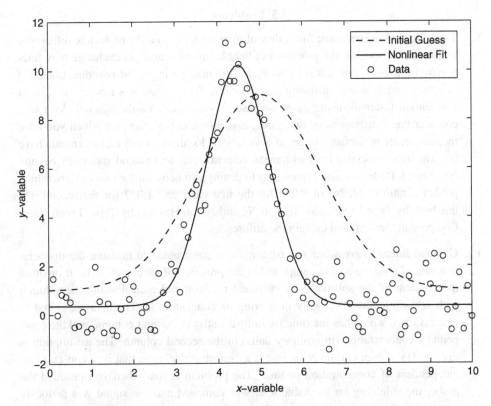

Figure 3.10 For some problems, such as the Gaussian peak in this data set, nonlinear least-squares fitting methods are required. The dashed line shows the curve for an initial parameter guess, and the solid line is the actual nonlinear fit to the data.

Note the "dot" before the "caret". This way you don't end up getting a matrix from squaring "x". You estimate the baseline value a(1) to be 1, the height of the Gaussian a(2) to be 8, and the center of the Gaussian a(3) to be at 5, and its width a(4) to be 2. So we set:

```
pin=[1 8 5 2]
```

and execute nlleasqr getting back:

```
coefficients  = [0.3557    9.8613    4.4815    1.0318]
uncertainties = [0.1226    0.3405    0.0281    0.0437]
r2 = 0.9148
```

Again, not bad. The nonlinear fit (solid line) is a significant improvement over the model curve with your initial parameter guesses (dashed line) (Fig. 3.10). Oh yes, the "real" values were [0.5 10.0 4.50 1.0]. The answer is roughly within errors of the real values, but will differ because of the noise that was added to the data. Your results may differ slightly when you do this experiment due to roundoff error.

3.5 Problems

3.1. **Type I and II regressions:** Since they ultimately depend on the molecular diffusivity of constituent species, the processes of bubble injection and gas exchange may fractionate in favor of one gas over another. This may go in one of two directions: (1) you may enrich slower diffusing gases because they escape less readily, or (2) you may enrich faster diffusing gases because they are more easily injected. As a test, consider the distributions of two gases, He.dat and Ne.dat, for which you have measurements in surface waters at a variety of locations. Both measurements have uncertainties associated both with measurement errors and natural variability beyond our control. Perform linear regressions to distinguish between the two models, which predict a ratio of He:Ne of 0.956 for the first case, and 1.027 for the second. Do this both by Type I regression (He on Ne and Ne on He) and by Type II regression. Compare the results, and explain the differences.

3.2. **General linear regression:** An experiment is constructed to measure the dissociation rate of a very unstable compound in the presence of ultraviolet light. It involves the creation of the substance by chemical reaction and then flashing the solution with UV while simultaneously measuring its concentration. The data are stored in photo.dat which has the time (in milliseconds) in the first column, and the compound's concentration (in arbitrary units) in the second column. The assumption is that the UV driven consumption rate is a "zeroth order" rate, that is, a constant rate independent of concentration or time. The problem is that when you installed the probe, the shielding on the cable lead was damaged and the signal was seriously contaminated with 60 hertz electrical noise from surrounding appliances. You must then do a general linear regression which accounts for this "contamination", using the model equation and of course, you know that $f = 60$. Also try the regression without the last term (i.e. just as a straight-line fit). Compare your results (the slopes) and explain.

3.3. **Nonlinear regression:** You need to separate two overlapping peaks on a sloping background from a chromatographic record. For the sake of simplicity, assume that the peaks are Gaussian, and that the background is changing linearly with time. The data are contained in chromo.dat (chromo(:,1) is the time in minutes, chromo(:,2) is the detector response) and you must use nonlinear regression. Give the best estimate elution times, peak widths, and the amount of material (area under the Gaussian curve). Also give uncertainties in your results (remember the propagation of errors). Devise a statistical test to determine if the background is really changing with time (your null hypothesis might be $H_o : a_2 = 0$). For reference, the area under a Gaussian curve:

$$\int_0^\infty a e^{-b^2 x^2} dx = a \frac{\sqrt{\pi}}{2b}$$

4

Principal component and factor analysis

'From a drop of water,' said the writer, 'a logician could infer the
possibility of an Atlantic or a Niagara without having seen or heard of
one or the other. So all life is a great chain, the nature of which is
known whenever we are shown a single link of it.'

Sir Arthur Conan Doyle

Suppose you're looking for patterns or relationships in your data. For example, you may be trying to quantify the presence and distribution of certain water masses in a hydrographic section, or you may be looking for evidence and patterns of nitrogen fixation or denitrification in some nutrient data. Perhaps you're trying to find the best way to account for interferences from other elements ("matrix effects") in your ICPMS data. You've gathered your data, maybe obtained from a cleverly designed experiment, or extracted from a hydrographic atlas or a collection of cruise data. The information you require lies within the relationships or correlations between the different properties or variables in your data set. But where (and how) do you look? If instinct leads you to look at the data covariance matrix, then your instinct is right! In this chapter we'll show you some techniques for extracting and analyzing this structure. We will start with some underlying basics that you'll need to understand these techniques, and we'll mention a few relatively intuitive approaches for analyzing data structure. However, our main objective is to show you two very powerful tools that allow you to probe your data set by analyzing the covariance matrix. These tools, *principal component analysis* and *factor analysis*, are in fact closely related. They have their origins in the social sciences and have been adopted and used in a wide range of disciplines. This success is a testimony to their utility, but also leads to some challenges. Each discipline has put its own stamp on terminology, so you may have to wade through some pretty confusing and inconsistent usage in the literature. We'll try not to compound the problem here, but in the thick of flying factors and dodging determinants, remember that the underlying principles are actually rather elegant and really quite simple.

4.1 Conceptual foundations

4.1.1 The data matrix and the covariance matrix

Analysis always starts with your *data matrix*, which you construct from your measurements in the following way: you make measurements of M different variables on N separate occasions (or *samples*), and arrange the results in a matrix that has M columns (one column for each variable) and N rows (one row for each sample). For example, you might have measured dissolved oxygen, temperature, and salinity on 1000 water samples during a cruise. Since you've also taken these samples from certain locations (specified by latitude and longitude) and different depths, you would have a resulting $N \times M$ data matrix $\mathbf{X}$ that is 1000 rows tall by 6 columns wide (your three location coordinates and your three observation values). From this you can construct the $M \times M$ (in our example 6×6) covariance matrix $\mathbf{C}$ by populating the matrix with the covariances between each of the variables using the relationship described in Equation (2.47):

$$s_{jk}^2 = \frac{\sum\limits_{i=1}^{N} \left[(x_{ij} - \overline{x_j})(x_{ik} - \overline{x_k}) \right]}{N - 1} \tag{4.1}$$

so that we have for our example

$$\mathbf{C} = \begin{pmatrix} s_{11}^2 & s_{12}^2 & s_{13}^2 & s_{14}^2 & s_{15}^2 & s_{16}^2 \\ s_{21}^2 & s_{22}^2 & s_{23}^2 & s_{24}^2 & s_{25}^2 & s_{26}^2 \\ s_{31}^2 & s_{32}^2 & s_{33}^2 & s_{34}^2 & s_{35}^2 & s_{36}^2 \\ s_{41}^2 & s_{42}^2 & s_{43}^2 & s_{44}^2 & s_{45}^2 & s_{46}^2 \\ s_{51}^2 & s_{52}^2 & s_{53}^2 & s_{54}^2 & s_{55}^2 & s_{56}^2 \\ s_{61}^2 & s_{62}^2 & s_{63}^2 & s_{64}^2 & s_{65}^2 & s_{66}^2 \end{pmatrix} \tag{4.2}$$

Note that the diagonal elements are simply the variances of the individual variables, and that the matrix must be symmetric about the diagonal (i.e. $s_{21}^2 = s_{12}^2$, etc.). This equivalence comes from the definition of covariance – see Equation (4.1) and check it out. The symmetric nature of the covariance matrix is a natural consequence of its definition, and also leads to some nice attributes that we'll exploit later on.

Suppose now we express our data matrix in its *deviate form* as $\mathbf{Y}$, that is, by subtracting off the mean of each variable (or column in your data matrix). Thus rather than using the "raw" data x_{ij}, we use $y_{ij} = x_{ij} - \overline{x_j}$. When you look at the definition of the covariance matrix (Equation (4.1)) then you realize that you can straightforwardly compute the covariance matrix (Equation (4.2))

$$\mathbf{C} = \mathbf{Y}'\mathbf{Y}/(N - 1) \tag{4.3}$$

which simplifies life, doesn't it? All you need to do to compute the covariance matrix is to express your data matrix in deviate form, then multiply it by its transpose, and divide by $N - 1$. Pretty convenient! But MATLAB makes it even easier for you with the cov function; using your data matrix $\mathbf{X}$ in its raw, non-deviate form:

```
C=cov(X);
```

gives you the covariance matrix.

Beyond this, your analysis rests on two important underlying principles. The first, *standardization and normalization*, is virtually a requirement for proceeding with analysis; importantly, it allows you to compare data measured in fundamentally different units, or over contrasting dynamic ranges. The second concept, that of *basis vectors* or *basis functions*, is an important mathematical principle that is really the key to understanding what factor analysis is all about. Let's discuss each of these further.

4.1.2 Standardization and normalization

Quite often our data sets are a mixture of measurements made on different scales and/or in different units. For instance, how can we compare or relate salinity in PSU to temperature in centigrade, to oxygen in micromoles per kilogram, to density in kilograms per cubic meter? How do you mix apples and oranges (other than in fruit punch)? Even when the units are the same, there can be difficulties. Consider combining horizontal separations, measured in distances ranging from 10^3 to 10^7 m, with water depths, ranging from 1 to 10^4 m. This approximately three order-of-magnitude disparity means that horizontal distance would totally dominate over depth in the covariance matrix, something you may not particularly want.

We use normalization or standardization to get around these sorts of difficulties and reduce the influence of one component with very large magnitudes compared with other data components that are small in magnitude. By *normalization* we mean the transformation of the variable vectors into vectors of unit length, as shown in:

$$\sqrt{\sum_i \hat{x}_{ij}^2} = 1 \tag{4.4}$$

By *standardization* we are referring to the transformation that changes the variable vector to have a mean of zero and a standard deviation of one, just like the Z-score transformation we talked about in Section 2.2.5:

$$z_{ij} = \frac{x_{ij} - \overline{x_j}}{s_j} \tag{4.5}$$

Typically these transformations are done column-wise on the data matrix, in effect making all variables have a zero mean and the same units – units of standard deviation. It's assumed that the data columns are close to normally distributed; this makes the transformation symmetric. An asymmetric transformation complicates the statistics and invalidates some of the underlying assumptions we made when we began the analysis. If some of your data violate this premise, you may still recover things by applying some kind of data transformation, or using the *central limit theorem* as discussed in Section 2.7.

To make life a little more bearable for you, we've created a couple of functions that you might find handy. The first is `colstd.m`, which standardizes a data matrix by columns, and the second is `rowstd.m`, which standardizes by row. We show you the first one, so you can see how it works:

```
function [Z, colmeans, colstds] = colstd(X)
colmeans = mean(X);      % compute means of each column
colstds  = std(X);       % compute stdev of each column
for i = 1:length(colmeans)     % and do each column separately
    Z(:,i) = (X(:,i)-colmeans(i))/colstds(i);
end
```

which returns a column-standardized version **Z** of the data matrix **X**, along with row vectors containing the column-wise means and standard deviations. Note that `rowstd.m` uses `colstd.m` and returns the row-standardized version of the data matrix along with column vectors of the row-wise means and standard deviations.

It is important to note one aspect of standardized data: the covariance matrix and the correlation matrix are one and the same. That is,

$$\mathbf{R} = \mathbf{Z}'\mathbf{Z}/(N - 1) \tag{4.6}$$

If you don't believe us, try making up a data set and use MATLAB's `corrcoef` and `cov` on the unstandardized and standardized data respectively; we think you'll see what we mean.

This brings us to another point. While it is usually a good idea to standardize your data, sometimes it's unnecessary work and more importantly may not be what you want to do. For example, if all of your data have been measured in the same units and over the same dynamic range, then it makes little difference in the end whether you work with the unstandardized covariance matrix or with the correlation coefficient matrix. Furthermore, standardization can have a significant effect on the covariance matrix structure, which you are trying to extract information about. For example, since each variable influences this structure in proportion to its variance, if you make all of the variables have a standard deviation (i.e. variance) of one, then they all have equal influence. We think you can imagine situations where you may want to know the true relative importance of each variable. This kind of nuance can become important when you perform factor analysis, and may color your results in unanticipated ways. It is therefore important to think very carefully about what you're doing to your data before feeding them to the mathematical meat grinder.

4.1.3 Linear independence and basis functions

Let's revisit our discussion of linear algebra in Chapter 1 for a moment. Remember the strategy of using matrix methods to solve systems of linear equations? Recall that the technique runs into trouble when one or more of the equations (and hence the row vectors associated with them) are not *linearly independent* of the others. That is, if you can "make up" one of the rows from some weighted combination of one or more of the other rows, that particular row has no uniqueness, and the matrix will be characterized as *rank deficient*. The way you can tell the number of linearly independent vectors is with the MATLAB rank function. By the way, this rule applies to either rows or columns.

If you think of a matrix as a bundle of vectors, you could in principle plot the vectors by using the individual matrix elements as the coordinates. What you will see is that the dimensionality of the space defined, or spanned by these vectors, is equivalent to the *rank* or number of linearly independent vectors in the matrix. This number will be less than or equal to the smaller dimension of the matrix (number of rows or number of columns). To see this, suppose you had the following data set, represented here, as a matrix:

$$\mathbf{X} = \begin{pmatrix} 0 & 2.5 & 5 \\ 6 & 3 & 0 \\ 4 & 6 & 8 \end{pmatrix} \tag{4.7}$$

This data set could be plotted as three column vectors as in Fig. 4.1. All three of these vectors turn out to be co-planar (they all lie in the same plane; note the dashed line connecting the ends of these vectors), and therefore lie within a two-dimensional subspace. Hence any one of these vectors can be created from the sum of the other two. Looking more carefully at the matrix defined in Equation (4.7), you'll notice that you can make the middle column

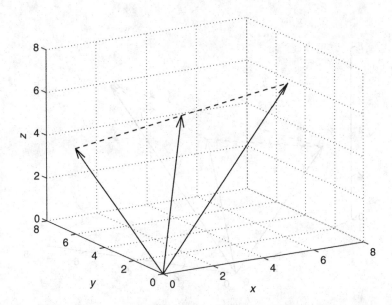

Figure 4.1 The three column vectors from the matrix **X** defined in Equation (4.7) are plotted by using the three matrix elements for each column as coordinates in *x y z* space. Note that all three vectors fall on the same plane.

by averaging the first and last columns. The fact that the three vectors fall on the same plane means that we need only two independent or *basis vectors* to define a space that contains all of the vectors in this matrix. Using the MATLAB `rank` function returns a value of 2 for this matrix.

But now consider a different matrix (set of vectors):

$$\mathbf{X} = \begin{pmatrix} 0 & 4 & 5 \\ 6 & 5 & 0 \\ 4 & 4 & 8 \end{pmatrix} \tag{4.8}$$

Plotting this group of vectors reveals a different set of circumstances, as seen in Fig. 4.2. In this case the three vectors are not co-planar (note the black dashed line that connects the tips of the arrows forms an open triangle rather than a straight line). No one of these vectors can be made up of a linear combination of the other two. Hence the data in Equation (4.8) span a three-dimensional space. This means that we need *three* basis vectors to fully define the space contained by these vectors. Using the MATLAB `rank` function returns a value of 3 for this matrix.

What do we mean when we talk about basis vectors "defining a space"? In the sense that any one of the vectors in Matrix (4.7) (Fig. 4.1) can be made up by some combination of two of the other vectors, we could choose any pair of non-parallel vectors that lie in this plane to do the job. It would be nice if they were *orthogonal* (i.e. at right angles to each other), but they don't have to be. What's more, even if we chose a pair of orthogonal basis

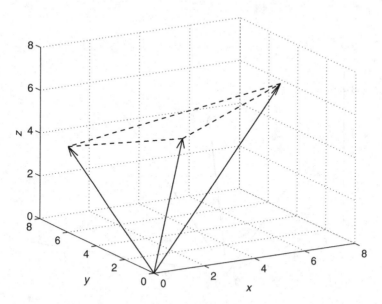

Figure 4.2 The three column vectors from the matrix defined in Equation (4.8) are plotted by using the three matrix elements for each column as coordinates in *x y z* space. Note that the three vectors do not fall on the same plane.

vectors, another pair obtained by rotating the original pair through some arbitrary angle would equally do the job. That is, there are an infinite number of basis vectors possible. So which ones do we choose? Well, that's the art and science of factor analysis, as you shall see.

4.2 Splitting and lumping

Before we jump into *principal component* and *factor analysis*, we'd like to get you in the mood first by looking at a few loosely related and commonly used techniques in this and the next section. As mentioned earlier, there are a number of statistical techniques used to discover structure or patterns within data built upon statistical methods that we have seen before. We won't be exploring them in detail, but you will find examples of these approaches in the literature, so we thought we'd mention them.

4.2.1 Discriminant analysis

Discriminant analysis is based on the assumption that your samples fall into natural groups that occupy separate regions within your data space. You use a priori knowledge about your samples and a transformation formula to maximize the difference or "data distance" between the group means and at the same time minimize the variance within each group. The basic idea is that you have independent knowledge that distinguishes two or more kinds of samples within your collection, and you are seeking a combination of *other* observables that allows you to discriminate between these groups in another collection of samples for which you don't have this independent knowledge. For example, you may have been able to identify two species of foraminifera based on morphology for some whole specimens that you had collected, but you wish to use trace compositional characteristics to distinguish other fragments that are too small to identify physically.

Let's consider the simplest possible scenario, namely two groups **A** and **B** with two measured variables x_1 and x_2 displayed in Fig. 4.3. Viewed from the perspective of either of the measured variables, the group populations overlap. The object is to find a composite coordinate or variable (corresponding to the diagonal line) made from some combination of the two variables that maximizes the separation of the two groups relative to their respective spreads. A sample's location along this line, and in particular its proximity to one archetype or the other, is a quantitative measure of its "**A**-ness" or "**B**-ness". To find the discriminant function (the diagonal line in Fig. 4.3) you need to solve the following equation:

$$\left[sp^2 \right] [\delta] = [D] \tag{4.9}$$

where the sp^2 are the pooled variances, the δ are the unknowns, and D is a measure of the separation of the groups. We should point out that we're using a notation with δ here rather than λ, which is typically used in discriminant analysis. (We didn't want you to confuse this with the *eigenvalue* λ used later in this chapter.) The value of D is given by:

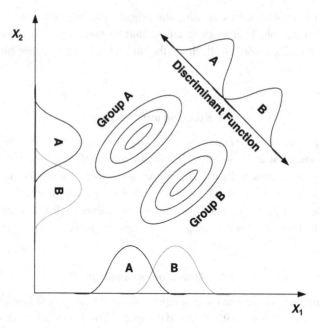

Figure 4.3 A plot of two bivariate groups showing overlap between the groups along both the X_1 and X_2 axes. The groups can be distinguished by projecting members onto the discriminant function. This figure was redrawn from Davis (2002).

$$D_j = \overline{A_j} - \overline{B_j} = \frac{1}{N_A} \sum_{i=1}^{N_A} A_{ij} - \frac{1}{N_B} \sum_{i=1}^{N_B} B_{ij} \qquad (4.10)$$

Here it is useful to keep in mind that we are talking about the ith observation of the jth variable for groups A and B. The D_j in Equation (4.10) is the difference between group means of the jth variable. Expanding Equation (4.9) we get:

$$\begin{pmatrix} sp_{11}^2 & sp_{12}^2 & sp_{13}^2 & \cdots & sp_{1M}^2 \\ sp_{21}^2 & \ddots & & & \\ sp_{31}^2 & & \ddots & & \\ \vdots & & & \ddots & \\ sp_{M1}^2 & \cdots & & & sp_{MM}^2 \end{pmatrix} \begin{pmatrix} \delta_1 \\ \delta_2 \\ \delta_3 \\ \vdots \\ \delta_M \end{pmatrix} = \begin{pmatrix} D_1 \\ D_2 \\ D_3 \\ \vdots \\ D_M \end{pmatrix} \qquad (4.11)$$

The sp_{jk}^2s are known as the pooled variances, calculated as:

$$sp_{jk}^2 = \frac{[SPA_{jk}] + [SPB_{jk}]}{N_A + N_B - 2} \qquad (4.12)$$

where the SPA_{jk} and SPB_{jk} refer to the sums of products for groups A and B and variables j and k. They are calculated in a similar fashion, so we show you only SPA_{jk}:

$$SPA_{jk} = \sum_{i=1}^{N_A} \left[\left(A_{ij} - \overline{A_j} \right) \left(A_{ik} - \overline{A_k} \right) \right] \tag{4.13}$$

Obviously when $j = k$ we are referring to the sum of the squares for that variable.

But you may wonder how you calculate the separation between groups when each variable potentially has different units, and/or may have different "spreads". Contemplation suggests that some kind of non-dimensionalized measure can be obtained by dividing by each variable's respective variance. The distance thus computed is known as the *Mahalanobis distance*.

$$D_\mathcal{M}^2 = \left[\overline{A} - \overline{B} \right]' \left[sp^2 \right]^{-1} \left[\overline{A} - \overline{B} \right] \tag{4.14}$$

Here the "division" by the multivariate equivalent of variance serves to standardize the distances.

If you now have a third group C which you believe contains elements from group A and group B, you can calculate a Mahalanobis distance between each sample in group C to the mid-points of group A and group B, and assign it to whichever group is closest. In addition to discriminating which group to put your samples into, you can also test the statistical significance of that assignment. The test is done as are all the statistical tests, and the distribution generally used is *Hotelling's T^2* test, not to be confused with Student's t-test seen earlier. More information about the use of this test can be found in many statistics texts, including Davis (2002).

4.2.2 *Cluster analysis*

If you are more interested in the relationships of various sample groups to one another, that is, to identify evolutionary or taxonomic relationships between sample types, you are likely to use one of the many types of *cluster analysis* used in environmental sciences. In contrast to discriminant analysis, these are regarded as unsupervised methods; that is, no a priori information regarding the makeup of the putative groups is required, and typically none is available. This has long been the specialty of taxonomists in following the lineage of the creatures they are studying, but molecular genomics has emerged as an area where these approaches are now flourishing.

Suppose you've measured M variables on each of N samples. The basic strategy is to use some criteria of relatedness to group your samples. This criteria may be some kind of Euclidean distance between individual samples in the M-space defined by your measurement parameters, given by

$$d_{ij} = \sqrt{\frac{1}{M} \sum_{k=1}^{M} \left(x_{ik} - x_{jk} \right)^2} \tag{4.15}$$

but might also include *association coefficients* consisting of binary "present/absent" elements, e.g. morphological features, nucleotides, or base pairs. There are a number of phylogenetic programs available to construct relationships (for example PHYLIP from the University of Washington) based on principles like *parsimony analysis*, which attempts to minimize the number of changes or substitutions required between "related" nodes. MATLAB's statistics toolbox contains several useful functions in this area. If you have this toolbox, type doc cluster, doc kmeans, and doc dendrogram for some examples and a starting point into their documentation.

Another approach, which we'll describe in a little more detail here, is to establish the degree of correlation between samples by computing a *similarity coefficient* (a kind of "quasi-correlation coefficient") row-wise in the data matrix:

$$\hat{q} = \frac{SP_{ij}}{\sqrt{SP_{ii}SP_{jj}}} \tag{4.16}$$

where we've used a notation like we introduced in Section 2.8, except that we have used a $\hat{q}$ (rather than the traditional $\hat{r}$) to emphasize the row orientation of the definition. That is, we're summing *across* variables (along each row in our data matrix) rather than column-wise:

$$SP_{ij} = \sum_{k=1}^{M} \left[(x_{ik} - \overline{x_i})(x_{jk} - \overline{x_j}) \right] \tag{4.17}$$

where there's a twist in our "averages" in that they're along rows:

$$\overline{x_i} = \frac{1}{M} \sum_{k=1}^{M} x_{ik} \tag{4.18}$$

Now, if you're thinking about this carefully, you'll realize that this kind of approach only makes sense if the variables (i.e. the columns in your data matrix) are in the same units and cover similar ranges, otherwise things can be really skewed toward one variable or another. Bear in mind, however, that despite the trappings of defining something that sounds like a legitimate statistical measure, this is not exactly a statistical process. It's a technique aimed at classification, not quantification. This similarity coefficient will vary between -1 for anticorrelated samples through 0 for uncorrelated, to 1 for perfectly correlated. Mechanically, you can compute these quasi-correlations by standardizing your data matrix row-wise (not column-wise, *cf.* Section 4.1.2), and then compute the transposed correlation matrix, which we'll call the *similarity matrix*.

$$\mathbf{Q} = \mathbf{ZZ}'/(M - 1) \tag{4.19}$$

Note the reversed order compared with Equation (4.6); it is the *major product* of the data matrix, which produces a $N \times N$ similarity matrix. Because you generally will have more samples (N) than variables (M), $\mathbf{Q}$ will usually be larger than the traditional correlation matrix defined by Equation (4.6). Here we've called it $\mathbf{Q}$ to distinguish it from the proper correlation matrix $\mathbf{R}$.

One can also think of clustering as a hierarchical linking of samples based on their "geometric" separation in some M-dimensional variable space. Once the criterion is established (and this is where most of the science is), the next step is to identify clusters of related samples, first in a pair-wise fashion. Pairs with a short Mahalanobis distance between them are selected for union. The measured variables for these pairs are averaged, and a new but smaller data matrix is created with the pair averages replacing those of their constituent samples. A new set of criteria are computed, and the next level of pairs are formed. This is graphically presented as a *dendrogram*, where horizontal lines linking the groups are drawn with ordinates matching the coefficients of the association (in this example, Mahalanobis distance). The procedure is iterated until the last pair of samples or groups is linked. An example of such a dendrogram is shown in Fig. 4.4.

The choices for construction of clustering criteria are strongly dependent on the scientific objectives and the field in which you're working, and the implementation of algorithms to efficiently accomplish this clustering, often in high-dimensional spaces, is a significant challenge, especially for large data sets. A particularly expedient algorithm for accomplishing clustering of larger data sets is the *k-means method* which appeared in the literature about 40 years ago. You need to be aware of one characteristic of this and other techniques. These techniques are iterative, and thus may converge on a sub-optimal solution (i.e. a local but not *global* optimum). If this is a serious pursuit for you, we suggest you consult a more

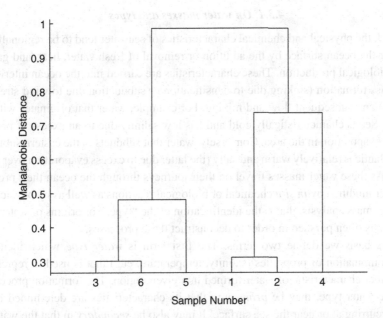

Figure 4.4 An example dendrogram of six hypothetical samples. The vertical axis is the "Mahalanobis distance". The smaller the distance, the more related the samples are to each other. In this example we see two groups of three samples that share a remote ($d_{ij} > 0.8$) relationship with each other.

complete discussion with practical applications such as Gordon (1999). Davis (2002) also has some useful suggestions.

4.3 Optimum multiparameter (OMP) analysis

Classical descriptive oceanography traditionally has involved itself with the documentation of the distribution of oceanic properties such as salinity, temperature, dissolved oxygen etc. and making inferences about the reasons for these distributions. This kind of approach, often termed *water mass analysis*, was practised largely in the early half of the twentieth century by many of the legends of oceanography, such as Sverdrup, Wüst, Defant, Worthington, and Reid. Much of what we know about the origins, pathways, and ultimate destinations of water in the oceans derives from these studies. Although this kind of approach was relegated to the back bench by more dynamics-oriented approaches in the latter half of the century, it underwent a more quantitative revival in the 1980s, largely driven by the work of Matthias Tomczak and others. The basic approach is best described in a paper by Tomczak and Large (1989), and we will explain it in a more consistent way here. But before plunging into the mathematical approach, we begin with little oceanographic background.

4.3.1 On water masses and types

In general, the physical and chemical characteristics of seawater tend to be regionally "set" at or near the ocean surface by the addition or removal of fresh water, heat, and gases, as well as biological production. These characteristics are carried into the ocean interior after water mass formation (sinking due to densification or subduction due to wind stress convergence) and subsequent flow and mixing. For example, water that originates within the Labrador Sea is characteristically cold and has low salinity due to an excess of precipitation over evaporation in the area. Conversely, water that subducts in the eastern subtropical North Atlantic is relatively warm and salty (the latter due to excess evaporation over precipitation). As these water masses travel on their journeys through the ocean their properties are further modified by *in situ* chemical or biological reactions as well as by physical mixing. Water mass analysis, that is the identification of the "types" or origins of water within the ocean, is often pursued in order to deconstruct these processes.

On this basis we define two terms. The first term is *water type* which identifies a unique combination of properties (salinity, temperature, etc.) that is used to represent an end-member characteristic of water formed in a given region. The formation process, and hence the water type, may be *primary* in that its characteristics are determined by processes occurring at or near the sea surface. It may also be *secondary* in that the water type is formed by the subsurface mixture of two or more primary water types, for example in a region where ocean currents or fronts bring waters into close proximity and mix them together. The second term is *water mass* which signifies a range of water properties found elsewhere in the interior of the ocean and which is formed as a variable mixture between two or more water types.

As will be discussed in Chapter 13, we can categorize most oceanic properties into two primary classes; *conservative tracers* (e.g. temperature and salinity), which are affected only by the physical processes of advection and mixing, and *non-conservative tracers* (e.g. oxygen and macronutrients), which are also affected by *in situ* biological or chemical processes. We ignore for now *radioactive* and *transient* tracers. Except in the unusual case of *double diffusion* (see Chapter 11) tracers tend to be affected equally by the physical transport mechanisms of advection and turbulent mixing.

We can show by algebraic manipulation of their advection–diffusion equations (Chapter 13) that two conservative properties must plot as linear functions of one another for water masses characterized by physical mixing between two end-member water types. For example, the two end-member water types would appear on a temperature versus salinity plot as two separated points with various mixtures of the two falling between these points on a straight line. At the next level of complexity, a water mass made of mixtures of three different end-members would populate a triangular space on a graph of these two conservative tracers. The vertices of this triangle would correspond to the end-member water types. One can easily see that as long as a water sample's properties fell within this triangular domain, it should be possible to describe the water mass as consisting of some fractional contribution from each of the three end-members. Finally, common sense would dictate that the sum of these fractional contributions, which are represented by the non-dimensionalized distance of a specific point within the triangle from each of the vertices ought to add up to unity.[1]

4.3.2 Classical optimum multiparameter analysis

When we survey the ocean by collecting hydrographic profiles, we quickly notice patterns in these properties. Large volumes of the ocean are characterized by systematic variations in conservative properties (especially temperature and salinity). There are often regions of linear variation between these properties, suggesting that these water masses are linear mixtures between these two characteristic water types. Examination of these property plots over larger ranges also reveals that there are breaks and bends in these relationships that point to distant end-members associated with key regions of water mass formation in the ocean (e.g. a low salinity core associated with Antarctic Intermediate Water).

Optimum multiparameter analysis (OMPA)[2] is an approach that utilizes conservative tracers and a choice of water types (end-member tracer properties) to deconstruct the relative abundances of those water types in the ocean. The underlying premise is that if the volume of water being studied comprises a mixture of M water types, then the measurement of $M-1$ conservative tracers is sufficient to uniquely deconstruct the relative amounts

---

[1] Ternary diagrams are also used in mineralogy where, for example, the composition of feldspars are represented as a function of potassium, sodium, and calcium (K-feldspar, albite, and anorthite) compositions.

[2] We are not sure why this acroynm became OMPA rather than OMA, but it is common usage now.

of these M water types at each point in space. The concept is quite simple. Consider the case for three water types and two tracers (temperatures and salinity). If we define the fractional contributions of the three water types as x_1, x_2, and x_3, and their characteristic temperature and salinity values as T_1, T_2, T_3, and S_1, S_2, S_3, we have three equations:

$$x_1 T_1 + x_2 T_2 + x_3 T_3 = T$$
$$x_1 S_1 + x_2 S_2 + x_3 S_3 = S \tag{4.20}$$
$$x_1 + x_2 + x_3 = 1$$

where T and S are the corresponding temperature and salinity of the water sample in question. Note that we have gained an additional constraint by logically insisting that the fractional contributions must sum to 1. Implicit in this (and this must somehow be enforced) is that each fractional contribution be positive, which means $0 \le x_i \le 1$. Now this, of course, is something you're basically familiar with. We can represent Equations (4.20) as a set of M simultaneous equations with a design matrix $\mathbf{A}$, a vector of unknown fractional contributions $\mathbf{x}$, and the water samples actual values $\mathbf{b}$ with:

$$\mathbf{Ax} = \mathbf{b} \tag{4.21}$$

where the first $M - 1$ rows of the design matrix are made of the $M - 1$ tracer conservation equations and the Mth row is the constraint that they add up to 1. This kind of problem is easily solved.

But other than temperature and salinity, what other conservative tracers do you have in your arsenal? The noble gases might be a good choice, if the information that they bear is significantly independent of temperature. Strangely, you might use oxygen or macronutrients (e.g. nitrate, silicate, or phosphate) if you are dealing with a situation where physical processes (advection and mixing) are sufficiently strong that biological processes are relatively unimportant, but in most cases you'll be dealing with the larger picture and you need something more immune to biogeochemical change. Wally Broecker introduced two such tracers, affectionately known as NO and PO. They are based on the fact that in the sub-surface ocean the nutrients nitrate and phosphate are introduced into the water column by oxidation of organic material in stoichiometric proportions to oxygen consumption. These so-called *Redfield ratios*[3] mean that you can to a close approximation correct for *in situ* production of these nutrients by using the dissolved oxygen concentration changes. These ersatz conservative tracers are defined as

$$NO = 9NO_3 + O_2 \tag{4.22}$$
$$PO = 135PO_4 + O_2 \tag{4.23}$$

where all properties are expressed in the same units, namely $\mu\mathrm{mol\,kg^{-1}}$. Also, for example in the Atlantic, dissolved silicate may be regarded as sufficiently inert that it can be used as

---

[3] Named after Alfred Redfield, who demonstrated through work in the 1930s to 1950s a remarkable uniformity in the N, P, C, and O_2 stoichiometry of the deep ocean.

a quasi-conservative tracer as well. Thus we can point to four or perhaps five conservative tracers that could be used to separate out water types.

If you get the feeling that Equations (4.20) and (4.21) sound too good to be true, then you are absolutely right! The challenge is that in the real world, data are not perfect. Measurement errors inevitably creep in, and it may be that your choice of end-member water types is less than perfect. Even if your choice was absolutely on target, nature may not be playing fair; conditions in the "source regions" of your water types may fluctuate or change over time. Moreover, there may be unaccounted-for water types that you have not chosen to represent in your analysis. Thus the matrix Equation (4.21) must be replaced with a more realistic least squares minimization scheme:

$$\mathcal{D}^2 = (\mathbf{Ax} - \mathbf{b})' \, \mathbf{W'W} \, (\mathbf{Ax} - \mathbf{b}) \tag{4.24}$$

where $\mathcal{D}^2$, the sum of the squared weighted residuals, is minimized subject to positivity constraints on the elements of $\mathbf{x}$ (after all, a negative water mass contribution is not physically realistic). This problem is still readily solved in MATLAB, and we will get to that shortly. However, there is the little matter of the weighting matrix $\mathbf{W}$. This is where the oceanography must come in (more on this later).

Let's consider the fact that all of these properties we are using are generally in different units, and even when in the same units likely have different ranges of variation and analytical precisions. Thus it is necessary to *standardize* (see Section 4.1.2) both the end-member water type data in $\mathbf{A}$ and the observations in $\mathbf{b}$. This is done row-wise for the design matrix. Remember that the corresponding value in $\mathbf{b}$ must be standardized the same way, giving:

$$G_{ij} = (A_{ij} - \overline{A_i})/s_i \quad \text{and} \quad d_i = (b_i - \overline{A_i})/s_i \tag{4.25}$$

where

$$\overline{A_i} = \frac{1}{M} \sum_{j=1}^{M} A_{ij} \quad \text{and} \quad s_i = \sqrt{\frac{1}{M-1} \sum_{j=1}^{M} \left(A_{ij} - \overline{A_i}\right)^2} \tag{4.26}$$

which crudely puts each variable (be it temperature, salinity, or some composite pseudo-conservative nutrient-like tracer) on the same footing; that is with a mean of zero and a variance of order 1 (see Section 4.1.2). Note that the above applies to all but the last row in $\mathbf{A}$, the mass conservation statement where the sum of the contributions must be unity.

The next issue, however, is that we don't measure these tracers equally well. It would be inappropriate to weight, for example, *NO* with the same confidence level as temperature. Typically, temperature is much better measured in relation to its range of variation than is *NO*. Thus we must construct the weighting matrix to account for this. That is, $\mathbf{W}$ is used to adjust the contribution of each tracer equation (and the conservation equation) to the least squares constraint. The structure of $\mathbf{W}$ could be simple in that its diagonal elements would consist of squared values of the individual tracer's range (the span in property space occupied by the various water types) divided by its analytical uncertainty:

$$W_{ii} = \delta_i / \sigma_i \tag{4.27}$$

where δ_i is the range (maximum span between source water types) and σ_i is the *effective uncertainty* of property i. By "effective uncertainty" we mean a combination of analytical uncertainty and uncertainty in the source water type values. The latter is estimated based on how well we know the parameter values for the end-member water types. Thus the diagonal elements of $\mathbf{W}$ represent the particular tracer's ability to resolve water mass differences.

Now the alert reader may have noticed that we haven't discussed the weighting for the last (i.e. the Mth) row of the design matrix: the unity-sum constraint. The weighting of this equation represents the degree to which the sum of the water type contributions may not be exactly 1. In a perfect world, they would sum perfectly to one, but the presence of measurement errors, uncertainties in the end-member values, and the possible contribution of some unknown water type can change things. If the other tracer constraints are properly standardized, and for example you are willing to accept a 10% uncertainty on the unity-sum constraint, then a value of 10 for W_{MM} would be appropriate. If you were to insist on a tighter constraint, e.g., 1%, then a value of 100 would be chosen. Some practitioners choose to weight the Mth equation the same as temperature, which is often the most well-constrained parameter.

In practice, we don't solve Equation (4.24) but rather revert to a reduced form of the design matrix and data vector by multiplying each row in the standardized design matrix $\mathbf{G}$ and data vector $\mathbf{d}$ by the corresponding diagonal element in $\mathbf{W}$:

$$\tilde{G}_{ij} = G_{ij} W_{ii} \quad \text{and} \quad \tilde{d}_i = d_i W_{ii} \tag{4.28}$$

Thus the expression $\tilde{\mathbf{G}}\mathbf{x} - \tilde{\mathbf{d}}$ constitutes a vector of the *Mahalanobis distances* (see Section 4.2.1) that represent the non-dimensional distances in tracer-parameter space between the observed tracer values for a water sample and the closest construct based on an optimal mixture of the end-member water types.

We can then solve this problem for each water sample (corresponding to each vector $\tilde{\mathbf{d}}$) in positive constrained least squares sense using the MATLAB function lsqnonneg. This function executes an iterative least squares minimization of $\tilde{\mathbf{G}}\mathbf{x} - \tilde{\mathbf{d}}$. Assuming that we have standardized and row-weighted our design matrix and stored it in the MATLAB matrix GT, and our hydrographic data have been similarly normalized (on a tracer-by-tracer basis), weighted identically, and stored in a $N \times (M-1)$ matrix DT (where $M-1$ is the number of corresponding water properties, and N is the number of water samples), we can find the relative water type contributions with

```
X=zeros(N,M);     % storage for water type contributions
for k=1:N                              % for each sample
X(k,:) = lsqnonneg(GT,DT(k,:))';       % pos'v LSQ
end
```

We can boil down the whole process of optimum multiparameter analysis into five steps:

1. Identify your suite of $M - 1$ conservative tracers that you will be using;
2. Define the characteristics (values of the $M - 1$ tracers) for your M end-member water types;
3. Construct your design matrix $\mathbf{A}$ including the $M - 1$ tracers and sum-to-unity constraint;
4. Standardize and weight the design matrix and data vectors;
5. Solve for x for your data set subject to the least squares minimization of $\tilde{\mathbf{G}}x - \tilde{\mathbf{d}}$ and subject to positivity constraints.

Remember that the important oceanographic decisions are tied up in steps 1, 2, and 4. Your choices of what tracers you use, what end-member water types (and their character- istics), and as importantly how you weight the individual constraints are key to finding a useful and credible solution (see Tomczak and Large, 1989, for a discussion). We urge you to examine not only the output (the spatial distributions of the water type contributions) but also the spatial structure of the residuals of $\tilde{\mathbf{G}}x - \tilde{\mathbf{d}}$. This may inform you where the OMPA works well, and as importantly, where it does not.

4.3.3 An OMPA example

We conclude this discussion with an example. This is based on an OMPA performed by Ruth *et al.* (2000). Their motivation was to separate out different component contributions to water types in the South Atlantic to resolve a volcanic ^{3}He plume emanating from the rift valley of the Mid-Atlantic Ridge. In order to do this, they needed to reconstruct the relative contributions of various water masses throughout the region and hence their relative levels of "background" ^{3}He. We will just demonstrate the basic OMPA approach used to do this reconstruction.

Their first steps were to select four conservative (or quasi-conservative) tracers and five end-member water types, and assign their respective tracer characteristics (see Table 4.1).

Table 4.1 *OMPA water type characteristics used in the Ruth* et al. *(2000) paper. For a description of the water types, see text.*

Water type	Temperature (°C)	Salinity (PSU)	Silicate (μmol kg^{-1})	*NO* (μmol kg^{-1})
AAIW	3.95	34.18	15	506
UCDW	2.53	34.65	67	468
UNADW	4.00	34.97	18	422
TDDW	2.30	34.91	36	442
AABW	0.14	34.68	113	526
Uncertainties	0.16	0.016	1.9	8.8

This was done on the basis of oceanographic knowledge of the important water types in this region, and employing property plots (e.g. temperature versus salinity, salinity versus silicate) to identify the *proximal* end-member characteristics of those water types. By "proximal" we mean the tracer values of those water types just outside of the region of study rather than the original values. The reason for this is that these water types undergo significant modification via interaction with other water masses in their journey to the region. The water masses used in this analysis are:

- **AAIW** – Antarctic Intermediate Water is a low-salinity water type that is formed near the Antarctic polar front and spreads throughout the Southern Hemisphere at a depth of around 600–800 meters.
- **UCDW** – Upper Circumpolar Deep Water is a silicate-rich water type formed by mixing within the Antarctic Circumpolar Current and spreads northward at about 1000–1200 meters depth.
- **UNADW** – Upper North Atlantic Deep Water is formed as a mixture between the Norwegian-Greenland Sea overflow water and Labrador Sea Water. It flows southward along the deep western boundary of the Atlantic at depths of 1000–2000 meters.
- **TDDW** – Two-Degree Deep Water is formed by mixing of North Atlantic Deep Water and Antarctic Bottom Water along the western boundary of the Atlantic.
- **AABW** – Antarctic Bottom Water is the coldest, densest, and deepest of the water masses modeled here. It results from mixing of even colder, denser waters exported from the Weddell and Ross Seas with water in the Antarctic Circumpolar Current. It has the highest silicate levels of all the water masses considered here.

While we won't dwell on their choices, we do want to emphasize that selection of water type characteristics is a crucial step in OMPA and requires careful analysis of regional tracer distributions, knowledge of the general circulation of these water masses, and a thorough interpretation of property–property plots. As importantly, assessment of uncertainties in the properties is necessary in order to achieve appropriate weighting of the constraint equations. These uncertainties play a crucial role in determining the result achieved, and are derived from a combined knowledge of measurement accuracies as well as how well we can determine the end-member water type characteristics.

Ruth *et al.* (2000) used a variant of OMPA. They utilized additional information by separating the analysis into two parts (above and below 2000 m). They recognized that AAIW would not contribute significantly to water masses below 2000 m and that AABW would not play a role above that depth. Thus they split the problem so that they were solving it as two over-determined systems with four water masses and five constraints (four tracers plus unity-sum). Adding information in this manner can be helpful in separating water masses that may have some similarity in tracer properties.

The sections resulting from the OMPA are shown in Fig. 4.5. The sections differ slightly from the results reported by Ruth *et al.* (2000) because they did not completely document how they standardized their data, or what weight was used on their conservation constraint.

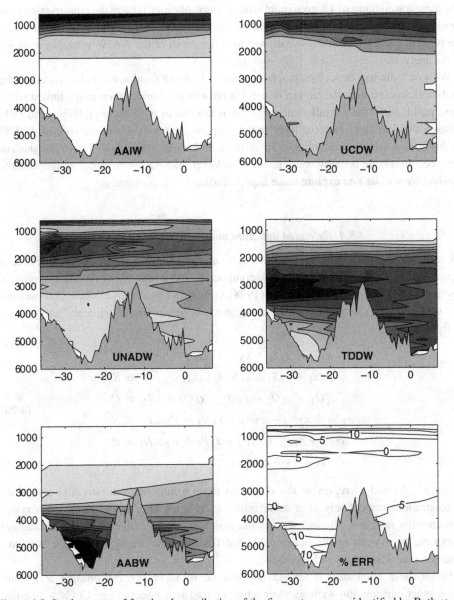

Figure 4.5 Section maps of fractional contribution of the five water masses identified by Ruth *et al.* (2000) for a hydrographic section along 19° S in the South Atlantic. The water mass "name" is printed at the bottom of each section (see Table 4.1), depth in meters on the vertical axis, and longitude along the bottom. The fractional contribution is grayscale coded from white (0%) to black (100%). The sixth panel is the percent error in the sum of the water masses.

We chose a weighting of 10, consistent with a "slop" of order 10% in the conservation constraint (i.e. that the water mass contributions sum to unity). Only depths below 600 meters are plotted, since water masses shallower than the depth of the AAIW were not included in the analysis.

We won't discuss the oceanographic character of these results, but rather focus on the analytical aspects; note the pattern of percent error in the conservation in the lowest right-hand panel. First, the overall scale of errors is consistent with our weighting term (10). Second, the deviations become large in the deep, western half where the core of AABW is, suggesting some compromise in its characteristics has been made. Third, the shallow error becomes large because water masses shallower than AAIW were not included in the OMPA. We will let you explore these aspects further in the problem set.

4.3.4 Extended optimum multiparameter analysis

We mention *extended optimum multiparameter analysis* for the sake of completeness, as it represents an interesting but potentially complicated extension of OMPA. This approach, discussed by Karstensen and Tomczak (1998), combines conservative tracers (temperature and salinity) with a non-conservative tracer (e.g. oxygen) by introducing an "aging term" as follows:

$$
\begin{aligned}
x_1 T_1 + x_2 T_2 + x_3 T_3 + x_4 T_4 &= T \\
x_1 S_1 + x_2 S_2 + x_3 S_3 + x_4 S_4 &= S \\
x_1 O_1 + x_2 O_2 + x_3 O_3 + x_4 O_4 - \alpha J_O &= O \\
x_1 N_1 + x_2 N_2 + x_3 N_3 + x_4 N_4 + R_N \alpha J_O &= N \\
x_1 P_1 + x_2 P_2 + x_3 P_3 + x_4 P_4 + R_P \alpha J_O &= P \\
x_1 + x_2 + x_3 + x_4 &= 1
\end{aligned}
\tag{4.29}
$$

where O, N, and P represent the dissolved oxygen, inorganic nitrate, and phosphate concentrations respectively, α is an effective water mass age, J_O is the *in situ* oxygen consumption rate, and R_N and R_P are the stoichiometric Redfield ratios of nitrate and phosphate to oxygen in the ocean (obtained from the literature), and where we have assumed four water mass end-members.

Given these equations, it should be possible to solve for the product αJ_O, and using other information such as radioactive or transient tracers (e.g. CFCs[4]), one can estimate α and thereby oxygen consumption rates. Further discussion of the assumptions and limitations of this approach is beyond the scope of this book, but we thought you needed to be aware that such an approach exists. The interested reader can also find additional MATLAB codes and information on OMPA activities at http://www.ldeo.columbia.edu/~jkarsten/omp_std/.

---

[4] Chlorofluorocarbons.

4.4 Principal component analysis (PCA)

Principal component analysis (PCA) may be regarded as a sophisticated mathematical procedure that attempts to recast the observations in terms of a new set of more "efficient" basis vectors. Doing this usually reveals something about the intrinsic structure of the variance within your data set and should give insight into the underlying processes. A great deal of the mystery surrounding PCA can be removed if we look at the rather simple recipe for forming principal components for any data set. For any $N \times M$ data set (N being the number of samples and M being the number of variables measured on each sample):

1. Create your $N \times M$ data matrix in deviate (and usually standardized) form.
2. Form the $M \times M$ covariance matrix from the data matrix (this requires pre-multiplication i.e. $\mathbf{X'X}$).
3. Extract the eigenvectors and eigenvalues from the covariance matrix.
4. The eigenvectors are the principal components and the eigenvalues are their magnitudes.

You might suspect that there must be more art to it than this simple, mechanical, four-step process, and there is, but the above recipe captures the fundamental essence of what we do when we do PCA. As you read on, you'll also realize that we're talking here about the most commonly used type of PCA called *R-mode analysis*, but more on that later … let's go through the mechanics first, and then we'll talk about the artistry.

4.4.1 Covariance matrices revisited

You need to construct your data matrix in at least the deviate form, i.e. with the mean values subtracted, and generally (but not always – see Section 4.1.2) standardizing by dividing through with the standard deviation. The decision as to how and why you do this is at the heart of your science and is specific to the particular problem you are working with. So we'll defer this to particular cases and simply refer to the data matrix as $\mathbf{X}$ from here on in. Also, we'll refer to the subsequent covariance matrix (or *correlation* matrix if the data matrix is standardized) as the "covariance matrix" and call it $\mathbf{R}$ from now on.

Earlier, in Section 4.1.1, we showed you how to create the covariance matrix as the *minor* product of the deviate form of the data matrix as $\mathbf{R} = \mathbf{X'X}/(N-1)$. This calculates the covariance column-wise *between variables*. In this case, if and when you do the standardization, you are subtracting the column-wise means and dividing by the column-wise standard deviations of the variables first. You are more likely to work with your data in this manner, that is in *R-mode*, as it implies that you are seeking relationships between your data variables to illuminate underlying processes.

However, if you are interested in the relationship *between samples*, which is what you do when you want to group samples, you could compute the covariance matrix from the *major* product of the data matrix as $\mathbf{Q} = \mathbf{XX'}/(M-1)$ (Equation (4.19)). Here, you would standardize the matrix by subtracting the row-wise means. This is called *Q-mode* and is the less commonly pursued path, and to our minds the more difficult to conceptualize. You might use this approach if you were interested in the relationships between the characteristics of your samples. This approach is more closely related to cluster analysis, which we discussed earlier.

4.4.2 Eigenanalysis of matrices

You'll recall from Chapter 1 that it is possible to decompose any matrix into three component matrices with specific characteristics (remember singular value decomposition or SVD?). For any $N \times M$ data matrix X, there exist two orthonormal matrices that satisfy the following condition:

$$X = USV'$$ (4.30)

where S is an $M \times M$ square matrix whose diagonal elements are the M non-negative *singular values* of X and whose off-diagonal elements are zero. The orthonormal matrices are

U = an $N \times M$ column orthonormal matrix, containing the N *eigenvectors* of Q, and

V = an $M \times M$ column orthonormal matrix, containing the M *eigenvectors* of R.

The *basic structure* of the data matrix X is embodied in the M singular values of S and the first M eigenvectors of U and V.

There's another way of looking at the relationship expressed in Equation (4.30) using the covariance matrix R. This matrix will have M *eigenvalues* λ and the corresponding eigenvalue equation is:

$$RV = V\Lambda$$ (4.31)

where V contains the eigenvectors of R and Λ is a diagonal matrix containing the vector of eigenvalues λ:

$$\text{diag}(\Lambda) = \lambda$$ (4.32)

Since R is defined by the (*minor*) product of X, it makes sense that the eigenvalues of R should be related to the singular values of X. In fact they are, through:

$$S^2 = \Lambda$$ (4.33)

that is, the eigenvalues λ of the covariance matrix are equal to the squares of the corresponding singular values. Actually, the relationship between S and Λ in Equation (4.33) and the equivalence of the eigenvectors V_i in (4.30) and (4.31) are true only if the eigenvalues and singular values are ordered in the same way (e.g. from smallest to largest or vice versa). Because this analysis is closely related to the duality between R the covariance matrix and the columns of the data matrix X, we call this *R-mode* analysis.

By convention, the eigenvectors are all of unit length, that is for the jth column vector $\sum_i^M V_{ij}^2 = 1$. The positive square roots of the eigenvalues of the covariance matrix (that is, the singular values of the data matrix) give the magnitudes of the principal components.

If we multiply the eigenvectors times the square roots of the eigenvalues we obtain the *factor loadings* for each variable on each component. This is given by:

$$\mathbf{A^R} = \mathbf{VS} = \mathbf{V}\sqrt{\mathbf{\Lambda}} \tag{4.34}$$

That is, Equation (4.34) tells you how to "build" the new principal components from your old observables. It is this relationship that tells you something about the linkage between the two. The eigenvectors are a new set of axes ($\mathbf{V}$) or basis functions. Bear in mind that these new basis vectors are a mathematical construct mechanically derived from PCA and may have no intrinsic or fundamental physical meaning.

The projection of each data vector (matrix column) onto these new component axes is called its *principal component score*. It is given by the following:

$$\mathbf{S_P^R} = \mathbf{XV} \tag{4.35}$$

We've put in the superscript "R" to remind you that this is an R-mode operation score, not singular values, and a subscript "P" to denote *principal component*. The following equation gives a related quantity:

$$\mathbf{S_F^R} = \mathbf{XA^R} = \mathbf{XVS} \tag{4.36}$$

known as the *factor scores*. They are essentially the same thing except that the factor scores have been scaled by the magnitude of the singular values. Here the subscript "F" is to remind you that these are the factor scores.

4.4.3 A simple example with a geometric interpretation

We have talked about factor loadings and factor scores mathematically, but what do these things look like? Let's start with a concrete but simple example. Consider a data set of the dissolved inorganic nutrients nitrate and silicate taken over a range of depths and a number of years at a location – the Bermuda Atlantic Time Series (BATS) site – within the subtropical North Atlantic. The concentrations of these nutrients are very low in surface waters owing to uptake by photosynthesizing organisms, but are reintroduced into the deeper water column by various processes acting on falling particles. As a result, both nitrate and silicate increase downward in the water column, but the two play different roles in the life-process and are released by different mechanisms. Nitrogen forms a part of living material, so nitrate is released by bacterial degradation of organic material. Silicate forms a part of skeletons in some organisms, and is released by chemical dissolution. Thus, although there is a general correlation between these two properties, there are systematic differences. The data file BATSNSi.txt is available from our website. You can load it into MATLAB and plot it with

```
load BATSNSi.txt
X = BATSNSi;
plot(X(:,1),X(:,2),'.');
```

which should give a plot something like Fig. 4.6.

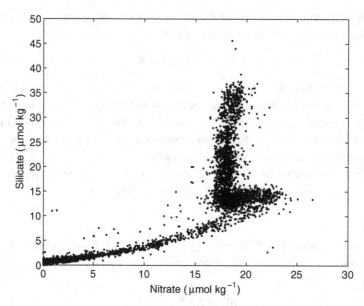

Figure 4.6 The relationship between dissolved inorganic nitrate and silicate from a 10 year period and all depths at the Bermuda Atlantic Time Series (BATS) site.

The question is whether we can use PCA to distinguish between the processes affecting these two properties. The answer is likely not, since we are only looking at a vertical profile that is a result of the horizontal displacement of water masses from a variety of remote locations. However, we proceed on the first steps of doing our PCA by computing the covariance matrix:

```
  R = cov(X)
R =
 74.3074    71.3253
 71.3253    97.3413
```

Notice that since the ranges of both properties (see Fig. 4.6) are similar, we've not bothered to standardize the data, otherwise the diagonal elements would be unitary. As expected, the covariance matrix is symmetric about the diagonal. Further, the total variance is the trace (i.e. the sum of the diagonal elements) of the covariance matrix, and is around 172. We can then compute the eigenvalues and eigenvectors for this covariance matrix using:

```
  [V,Lambda]=eig(R)
V =
  -0.7614     0.6483
   0.6483     0.7614
Lambda =
  13.5752          0
        0   158.0735
```

where V contains the eigenvectors as its columns and Lambda contains the eigenvalues on its diagonal. Now don't trust us! First check that V is truly an orthonormal matrix by multiplying it by its transpose. This means that the basis vectors defined by V are at right angles to one another and of unit length. Second, check that the trace (Lambda) is the same as trace (R). This means that the "new" representation embodies the same total variance. Convinced?

Now notice one other thing. If you think about the eigenvectors and their corresponding eigenvalues, they can be listed in any order. That is, much like the roots of a polynomial, there is no unique order in which they must be applied. Thus MATLAB can return eigenvalues/eigenvectors in any order, usually but not always in increasing eigenvalue size. It seems more logical to list eigenvalues in order of decreasing size. The reason for doing this will become evident once we get into factor analysis, but this awkwardness is easily remedied with following code snippet, which works regardless of the number of principal components you have:

```
[V, Lambda]=eig(R);        % obtain eigenvectors V, eigenvalues Lambda
[lambda, ilambda]=sort(diag(Lambda),'descend');
                           % sort in descending order with indices
Lambda=diag(lambda);       % reconstruct the eigenvalue matrix
V=V(:,ilambda);            % reorder the eigenvectors accordingly
```

This reorders the eigenvalues in order of decreasing size, and of necessity reorders the corresponding eigenvectors to match. Try this out to convince yourself that this works. You could then create a MATLAB function (call it eigsort.m) that you can use to obtain eigenvectors and eigenvalues sorted in descending order.

The eigenvectors, in combination with the eigenvalues, constitute a new set of basis vectors that more efficiently represent the covariance matrix, and hence the data. But what do we mean by "more efficiently"? First, look at the diagonal elements in the covariance matrix. The total variance, given by the trace of the covariance matrix $74.31 + 97.34 = 171.65$ is shared relatively evenly by the two variables: 43% for one and 57% for the other. The new principal factors, or eigenvalues (the diagonal terms in Lambda), share the variance quite differently: 92% for one and 8% for the other!

But is that really more efficient? We know you were taught from a very young age that sharing was a good thing, but think of it this way; we've dumped most of the variance onto the first basis variable, allowing that one variable to describe compactly most of the total variance. As importantly, the new basis variables are orthogonal, that is, *linearly independent*.

Let's look at this geometrically. First we plot the original covariance matrix vectors (column-wise) as arrows extending from the origin as in Fig. 4.7. The covariance matrix defines an ellipse in variable space that uniquely circumscribes the tips of these vectors (i.e. it is only possible to draw one ellipse centered on the origin that passes through the tips of the two arrows). Now here's the really neat thing: the two eigenvectors of the covariance matrix, when multiplied by their eigenvalues, correspond to semi-major and semi-minor

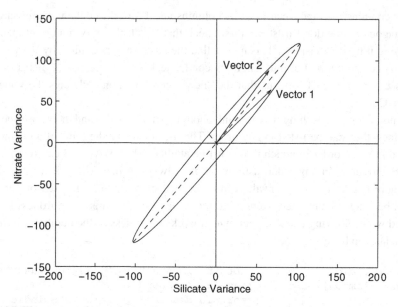

Figure 4.7 The vectors that make up the covariance matrix are plotted as vectors 1 and 2, and uniquely define an ellipse centered on the origin. The semi-major and semi-minor axes (dashed lines) of this ellipse are the principal factors.

axes of this very same ellipse. Instinctively, given the two new basis vectors, you could easily visualize their corresponding ellipse, but not so easily from the original covariance matrix vectors.

If, in Fig. 4.7, we were to project vector 1 (the first vector formed from the covariance matrix, whose original "coordinates" were 74.3 and 71.3) back onto the minor and major axes of the ellipse (the first and second eigenvectors), we would get the "more efficient" representation coordinates of -10.34, 102.48 respectively. Similarly for vector 2, we get 8.80 and 120.35. That is, most of the information comes from one of the principal components and this would be true of each individual sample as well. We call these more efficient coordinates the *principal component factors*. If one had to approximate their data with one "variable", the largest principal component score offers an obvious choice.

4.4.4 A more complicated PCA example

The previous example was fairly easy to think about, in that it contained only two variables with similar units and similar ranges. In fact, aside from the mathematical nicety of redistributing the variance "more efficiently" onto one of the principal components, we really didn't learn anything about the data than we didn't already know. Let's extend the analysis of the BATS data to include other properties. The file BATSA11T.txt contains a listing of depth, temperature, dissolved oxygen, nitrate, phosphate, and silicate. The profiles (plots vs. depth) of these properties are shown in Figure 4.8.

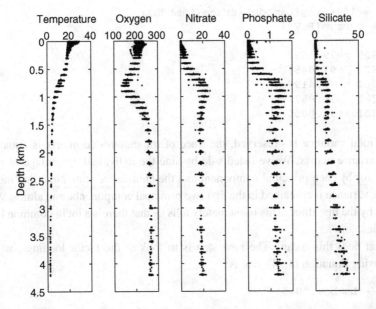

Figure 4.8 Profiles of temperature (in Celsius), dissolved oxygen, nitrate, phosphate, and silicate (all in μmol kg^{-1}) from the Bermuda Atlantic Time Series (BATS) site. Note the monotonic decrease in temperature, the broad minimum in oxygen, and the corresponding maxima in the other properties at about 0.8 km depth.

There's one other important thing to note. We need to standardize the data because there are different units (temperature is in Celsius degrees while the others are in μmol kg^{-1}) and because there are big differences in the ranges of values. There's an easy way to do this in MATLAB: just use `corrcoef(X)` instead of `cov(X)`. We now commence to do the PCA with

```
    load BATSAllT.txt
    X=BATSAll(:,2:6);      % don't include depth (1st column)
    R=corrcoef(X)
R =
    1.0000    -0.5287    -0.9348    -0.9429    -0.9072
   -0.5287     1.0000     0.2395     0.2900     0.5619
   -0.9348     0.2395     1.0000     0.9901     0.8363
   -0.9429     0.2900     0.9901     1.0000     0.8609
   -0.9072     0.5619     0.8363     0.8609     1.0000
    [V,Lambda]=eigsort(R);
```

Note that we left off a ";" so you could see the correlation matrix, whose diagonal elements are all unitary. Also, we've used our new function `eigsort.m` to provide the eigenvalues in decreasing order. We can see the percent of total variance associated with each principal component by entering the following:

```
    PoV = 100*diag(Lambda)/trace(Lambda);
    [PoV cumsum(PoV)]
ans =
    79.1828      79.1828
    17.8627      97.0455
     2.3744      99.4199
     0.4380      99.8579
     0.1421     100.0000
```

Since the total variance is preserved, the trace of the eigenvalue matrix is equal to that of the covariance matrix. We've listed side-by-side the individual percentage of variance accounted for by each principal component and the cumulative sum. Note that more than 97% of the variance is contained in the first two principal components, and almost 99.5% is subsumed by the first three. This immediately tells us that there's a lot in common between the variables.

But what does this mean? The next step is to look at the factor loadings, which we compute using Equation (4.34), that is:

```
    AR=V*(Lambda.^0.5)
AR =
    -0.9879      0.0045     -0.1014     -0.1153     -0.0207
     0.5267     -0.8448      0.0873     -0.0353     -0.0083
     0.9431      0.3147      0.0827     -0.0222     -0.0644
     0.9578      0.2634      0.0630     -0.0827      0.0495
     0.9481     -0.1052     -0.3000      0.0052     -0.0028
```

where each column corresponds to the factor loadings of a principal component on each of the original variables. Note that the "dot and caret" means we've taken the square root in "array form" (element by element). The factor loadings can be seen in Fig. 4.9. Also, remember that the sign of the eigenvectors is arbitrary, so that the direction can be arbitrarily reversed by changing the signs of all the components of an eigenvector (in a column of $\mathbf{V}$ or for factor loadings in $\mathbf{A^R}$).

Now let's think about what we're seeing here. The first principal component (PC) holds the lion's share (nearly 80%) of the variance, and is a reflection of the general tendency for all properties except temperature to increase with depth. Temperature is strongly anticorrelated with the other properties because it decreases with depth. Dissolved oxygen shows a slightly weaker loading on the first principal component relative to the other variables. You might argue that this makes sense, in that the dissolved oxygen does not monotonically increase with depth. Oxygen concentration changes because of two competing processes; increased physical dissolution from the atmosphere at lower temperatures results in higher oxygen content in deeper (colder) waters, while biological remineralization processes, which tend to increase nutrient concentrations, reduce the oxygen content. The second principal component, which accounts for about 18% of the total variance appears to reflect the latter process, showing anticorrelation between dissolved oxygen and macronutrients nitrate and phosphate. The loadings of temperature and silicate on the second PC are much

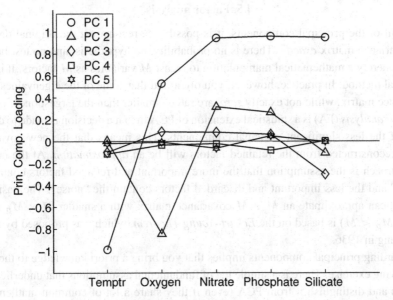

Figure 4.9 The factor loadings for the BATS data analysis example. Note that the principal components (Prin. Comp.) are numbered in decreasing order, with PC 1 being the largest. Temptr = temperature.

weaker. This is a reflection of the fact that the presence of silica is associated with dissolution of siliceous skeletal material rather than organic degradation. The third PC, which holds just 2.4% of the variance, may reflect the final process, dissolution of siliceous material, but this is just speculation on our part.

Now we finish with principal component analysis with one final demonstration. Remember that all of the information in the covariance matrix is contained within the new principal components. We've not lost any information here. In fact, we can easily reconstruct the original covariance matrix from the factor loading matrix with

$$\hat{\mathbf{R}} = \mathbf{A}^{\mathbf{R}} \mathbf{A}^{\mathbf{R}'} \tag{4.37}$$

where we've used $\hat{\mathbf{R}}$ to indicate that we're calculating something new here and $\mathbf{A}^{\mathbf{R}'}$ is $\mathbf{A}^{\mathbf{R}}$ transpose. Don't believe that it's that simple? Try this:

```
    Rhat=AR*AR'
Rhat =
    1.0000   -0.5287   -0.9348   -0.9429   -0.9072
   -0.5287    1.0000    0.2395    0.2900    0.5619
   -0.9348    0.2395    1.0000    0.9901    0.8363
   -0.9429    0.2900    0.9901    1.0000    0.8609
   -0.9072    0.5619    0.8363    0.8609    1.0000
```

Does that look familiar? Take a look at the original **R**!

4.5 Factor analysis

Given all of the principal components, it is possible to reconstruct the original data and the covariance matrix *exactly*. There is no probability, no hypothesis, and no test because PCA is merely a mathematical manipulation to recast M variables as M factors. It is not a statistical method. In practice, however, you often find that some of the eigenvalues of the covariance matrix, while not exactly *zero*, are rather smaller than the largest ones.

Factor analysis (FA) is a statistical extension of PCA in that a decision is made to discard some of the less significant principal components. This means that the new covariance matrix reconstructed from the retained factors will be an *approximation*. At the heart of this approach is the assumption that the more important and retained factors contain the "signal" and the less important and discarded factors contain the "noise". The basic idea that you can approximate an $M \times M$ covariance matrix with a smaller $M \times M_p$ matrix (where $M_p < M$) is based on the *Eckart–Young Theorem*, which was proposed by Eckart and Young in 1936.

Discarding principal components implies that you bring a priori knowledge to the problem solving exercise. There is a simple list of fundamental assumptions that underlie factor analysis and distinguish it from PCA (even if they share a lot of common mathematical machinery).

1. The correlations and covariances that exist among M variables are a result of M_p underlying, mutually uncorrelated factors. Usually M_p is substantially less than M.
2. Ideally M_p is known in advance. The number of factors, hidden in the data set, is one of the pieces of a priori knowledge that is brought to the table to solve the factor analysis problem. Traditional textbooks admonish that *you should not use factor analysis for fishing expeditions*. It may be pointed out that this is rather sanctimonious advice considering the way this technique has been used and abused.
3. The rank of a matrix and the number of eigenvectors are interrelated, and the eigenvalues are the square of the M_r (where $M_r < M$) non-zero singular values of the data matrix. The eigenvalues are ordered by the amount of variance accounted for.

Factor analysis starts with the basic principal component approach, but differs in two important ways. First of all, factor analysis is *always* done with *standardized* data. This implies that we want the individual variables to have equal weight in their influence on the underlying variance-covariance structure. In addition, this requirement is necessary for us to be able to convert the principal component vectors into factors. Second, the eigenvectors must be computed in such a way that they are *normalized*, i.e. of unit length or orthonormal.

4.5.1 Factor loadings matrix

As we stated above, we start factor analysis with principal component analysis, but we quickly diverge as we apply the a priori *knowledge* we bring to the problem. This knowledge may be of the form that we "know" how many factors there should be, or it may be more that our experience and intuition about the data guide us as to how many factors there should be. As before, we extract the M eigenvectors and eigenvalues from the covariance matrix. We then discard the $M - M_p$ eigenvectors with the smallest eigenvalues.

With PCA (Equation (4.34)), we can take the (unit length) eigenvectors and weight them with the square root of the corresponding eigenvalue:

$$\mathbf{A^R = VS = V\sqrt{\Lambda}} \tag{4.38}$$

Here $\mathbf{A^R}$ represents the matrix:

$$
\begin{array}{ccccc}
 & I & II & III & \cdots & M_p \\
X_1 & A_{11}^R & A_{12}^R & A_{13}^R & \cdots & A_{1M_p}^R \\
X_2 & A_{21}^R & A_{22}^R & A_{23}^R & \cdots & A_{2M_p}^R \\
X_3 & A_{31}^R & & \ddots & & \\
\vdots & & & & & \\
X_M & A_{M1}^R & A_{M2}^R & & & A_{MM_p}^R
\end{array}
\tag{4.39}
$$

where the M variables X run down the side and the M_p *factors* go across the top and the A^R represent the *loadings* of each variable on individual factors. When $M = M_p$ you have the same thing as PCA.

Returning to our BATS example discussed in Section 4.4.4, we might be inclined to retain the first two factors, because we're thinking that perhaps just a few processes are important in controlling nutrient and oxygen distributions in the water column. This is consistent with the fact that more than 97% of the variance is folded into the first two principal components. We thus select only the first two factor loadings with

```
Ar = AR(:,1:2);
```

which are tabulated in Table 4.2 and graphed in Fig. 4.9 as PC 1 and PC 2 (the circles and triangles respectively). Qualitatively, one might argue that the first factor corresponds to

Table 4.2 *Factor loadings for the first two factors from the BATS data. Note the significantly lower communality (Section 4.5.2) for silicate.*

Variable	Factor 1	Factor 2	Communalities
Temperature	−0.988	0.005	0.976
Oxygen	0.527	−0.845	0.991
Nitrate	0.943	0.315	0.989
Phosphate	0.958	0.263	0.987
Silicate	0.948	−0.105	0.910

the depth dependence of the properties: temperature decreases with depth while the other properties increase. That is, the deeper you go, the "older and colder" the water. The second factor resembles the organic degradation process: oxygen is anticorrelated with nitrate and phosphate (oxygen is consumed, while nitrate and phosphate are produced), and there's no significant effect on temperature or silicate.

4.5.2 Communalities

By calculating the communalities we can keep track of how much of the original variance that was contained in variable X_j is still being accounted for by the number of factors we have retained. As a consequence, when $M = M_p$, $h_j^2 = 1$, always, so long as the data was standardized first. The communalities are calculated in the following fashion:

$$h_j^2 = \sum_{k=1}^{M_p} \left(A_{jk}^R \right)^2 \tag{4.40}$$

You can think of this as summing the squares of the factor loadings horizontally across the factor loadings matrix.

Examination of the communalities for our BATS example, shown as the last column in Table 4.2, reveals something rather interesting. Recall that the first two factors account for 97% of the variance. The first two factors do even better for the first four variables (temperature, oxygen, nitrate, and phosphate). The bulk of the "missing variance" appears to be in one variable: silicate, which has a significantly lower communality. Looking back at the very first example (the data plotted in Fig. 4.6) we can see that despite a general correlation at lower values, silicate continues to increase with depth while nitrate remains constant. Phenomenologically this can be attributed to either dissolution from bottom sediments or the deep inflow of silicate-rich water from elsewhere. Regardless of this, we appear to have "thrown out" this trend by not keeping the third factor. If we expand our study to include a third factor, we get a more balanced set of communalities (see Table 4.3). This makes sense when we see that the third factor is dominated by silicate. It seems likely the third factor is associated with the deep increase in this property.

Table 4.3 *Factor loadings for the first three factors from the BATS data.*

Variable	Factor 1	Factor 2	Factor 3	Communalities
Temperature	−0.988	0.005	−0.101	0.986
Oxygen	0.527	−0.845	0.087	0.999
Nitrate	0.943	0.315	0.083	0.995
Phosphate	0.958	0.263	0.063	0.991
Silicate	0.948	−0.105	−0.300	1.000

Table 4.4 *Listing the percent of variance contained in your factors.*

Factor	Eigenvalue	% Total Var	Cumulative % Var
I	λ_1	.....	...
II	λ_2	...	...
III	λ_3	...	...
$\vdots$	$\vdots$		
M_p	λ_{M_p}	...	$\leq 100\%$

4.5.3 Number of factors

Above we make several references to the fact that if $M = M_p$ (i.e. the number of factors equal the number of variables) then factor analysis is no different than PCA with standardized variables. But of course, in factor analysis, you want $M_p << M$ and so the question remains: how do you decide which factors to keep? When doing factor analysis it helps to organize the results (A^R, h_j^2, etc.) like we've done in Table 4.3, that is:

$$
\begin{array}{c|ccccc|c}
 & I & II & III & \cdots & M_p & h_j^2 \\
\hline
X_1 & A_{11}^R & A_{12}^R & A_{13}^R & \cdots & A_{1M_p}^R & h_1^2 \\
X_2 & A_{21}^R & A_{22}^R & A_{23}^R & \cdots & & h_2^2 \\
X_3 & A_{31}^R & \ddots & & & & h_3^2 \\
\vdots & & & & & & \vdots \\
X_M & A_{M1}^R & A_{M2}^R & & & A_{MM_p}^R & h_M^2
\end{array}
\tag{4.41}
$$

And in a fashion analogous to what we did in our PCA example (Section 4.4.4) you can list the percent variance contributions as in Table 4.4.

At first the A_{jr}^R could be the entries from $\mathbf{V}$ (the raw eigenvectors), but later you can use the entries from $\mathbf{A^R}$ (the factor loadings) as the analysis continues and the number of factors decreases. In this manner, you can keep track of the number of factors you are dealing with, how much of the original total variance is being accounted for, where the variables are loading on the individual factors, and the communalities on each individual variable. This will help you see how your choices in the number of factors kept have affected these measures of performance.

As to the question of *how* you decide, unfortunately there is no hard and fast rule as to how many factors to keep. One rule of thumb is to keep all of the factors whose eigenvalue is greater than one, provided you started with standardized data. In the BATS data example we discussed in Section 4.4.4 we would likely have been inclined to keep only the first two principal components. Examination of the communalities and thinking about the *oceanography* suggests that three would be better. The lesson in this is that you cannot rely on a single criterion, and you must apply your scientific understanding to the choice of M_p.

If you get a lot of factors with eigenvalues greater than one, then you might have to face the likelihood that maybe the factor theory approach isn't applicable to your problem, at least in the way you have presented it. Typically the more successful factor analyses have been those where a "few" factors account for most of the variance. See the discussion of simple structure concepts in the next section.

4.5.4 Varimax rotation and simple structure concepts

After you have chosen the few (M_p) factors you wish to keep in your analysis, you can "improve" the fit of this reduced dimensionality coordinate system to your data by a technique known as *factor rotation*. Even though the number of factors may have reduced the dimensionality of your problem, the factors may not be easy to interpret. Factor rotation allows you to reorganize the *loadings* onto rotated factors. This is accomplished by maximizing the variance of the loadings on the factors. For each (kth) factor we can compute the variance of the loadings as:

$$s_k^2 = \frac{M \sum_{j=1}^{M} \left(\frac{A_{jk}^2}{h_j^2} \right)^2 - \left(\sum_{j=1}^{M} \frac{A_{jk}^2}{h_j^2} \right)^2}{M^2} \tag{4.42}$$

where M_p is the number of retained factors, M is the number of original variables, A_{jk} is the loading of variable j on factor k, and h_j^2 is the communality of the jth variable. Using this expression of the variance of the loading on the kth factor, one maximizes the following:

$$\mathcal{V} = \sum_{k=1}^{M_p} s_k^2 \tag{4.43}$$

This is an iterative process where you rotate two factors at a time, holding the others constant, until the increase in the overall variance $\mathcal{V}$ drops below a preset value. This is the heart of the Kaiser varimax orthogonal rotation. Think of it as trying (iteratively) to find "better" eigenvectors.

The various factor rotation methods have, as a guiding principle, the *simple structure concepts*. That is to say, the results, after rotation, should have become simple in their appearance. To put it another way, these simple structure concepts should be considered when trying to determine whether a given factor rotation has clarified the underlying structure of the data. Five simple structure precepts have been put forth by Thurstone (1935):

1. There should be at least one zero in each row of the factor loadings matrix.
2. There should be at least k zeros in each column of the factor matrix, where k is the number of factors extracted.

3. For every pair of factors, some variables (in the R-mode) should have high loadings on one and near-zero loadings on the other.
4. For every pair of factors, several variables should have small loadings on both factors.
5. For every pair of factors, only a few variables should have non-vanishing loadings on both.

These five rules embody what Davis (2002) was trying to get across when he stated that the effect of factor rotation was to push the loadings of variables on factors to either $+1$, -1, or zero. In the real world, such happy circumstances are rarely achieved with orthogonal factor rotation, such as the Kaiser varimax. A case in point is the BATS example we've been discussing. You might be tempted to undergo a factor rotation to better match processes and factors. Taking the reduced factor loading matrix (for three factors), we can execute a varimax rotation. We've included a simple implementation with varimax.m, which you can use in the following way:

```
   Ar = AR(:,1:3);              % select first three factors
   Arot = varimax(Ar)
Arot =
   -0.8987     0.3884      0.1668
    0.1487    -0.9824     -0.1066
    0.9827    -0.0770     -0.1540
    0.9708    -0.1261     -0.1798
    0.7336    -0.4016     -0.5481
```

The resulting factor loadings are much more difficult to interpret here, since the rotation actually tries to separate the loadings in a purely mechanical way. Compare this rotated factor loading with the original pattern in Table 4.3. The rotation partially shifted the oxygen and silicate variance onto the second factor and shifted most of the nitrate and phospate variance onto the first. Separating the oxygen and nitrate–phosphate variances doesn't make much oceanographic sense. In other words, although the concept of factor rotation seems appealing in the abstract, you must *really think* about what you're trying to do.

There are a number of other factor rotation schemes, and we will only make passing mention of these. But you may hear of these other rotations, particularly *oblique* factor rotation schemes that promise much better "separation" of variables. Oblique factor rotation schemes can usually achieve this $+1$, -1, 0 loading, but the algorithms for accomplishing such rotations are beyond the scope of this book and the interpretation of such factors is difficult to reconcile with the initial, mutually independent factors assumption made at the start of the factor analysis. Or as one of our professors, Dan Hawkins, used to say: "The analysis is telling us something we already know, but in terms we cannot understand."

4.6 Empirical orthogonal functions (EOFs)

You'll likely encounter *empirical orthogonal functions*, or EOFs for short, in physical oceanography, remote sensing, paleoceanography, meteorology, and climate-related research. That is, EOFs tend to appear in areas that involve a combination of spatial patterns and temporal trends in a variable. It will often be part of an attempt at the compression of space-time data in such a way as to use a small number of orthogonal spatial structures as functions of time and account for the bulk of the variance contained in your observations. If this sounds vaguely familiar, it is just a variant of factor analysis. It is often shrouded in subtly different nomenclature, but the mechanics are fundamentally the same. The motivation may range from a form of data volume reduction, more efficient data description, through probing for underlying processes. The strategy is to represent the data in terms of a reduced set of orthogonal functions or *modes*.

Unlike our prior approach of relating a number of observables (dissolved oxygen, nutrients, etc.) as a function of one another, and perhaps of a few spatial variables, EOF analysis typically attempts to cast a *single* variable in terms of space and time. The traditional approach (often referred to as *S-mode analysis*) is to construct the data matrix with one column for each spatial point, and one row for each instant in time. Thus, if you had a 10 year record at monthly intervals of sea surface temperature on a square grid of 30 by 50 locations, the data matrix ($N \times M$) would be 120 rows by 1500 columns, that is $N = 120$ and $M = 1500$. A particular row, when extracted into a two-dimensional space, would constitute a map of the property at one specific time, and any particular column would be a time-series of the property at a specific location. Beware that some practitioners use the opposite convection (rows = space and columns = time), but it really doesn't matter in the end if you are careful about what you are doing (Emery and Thompson, 1998). Such a data set would consequently be large; the example cited above contains 180 000 observations. For EOF analysis you may not always want to standardize the data matrix (for one variable it's already all in the same unit and you may want to examine spatial patterns of variance); but you should, as good practice, remove the column means, which are simply the time averages at each spatial point.

Now what do you do? You already know how to do this! The prescription for prosecuting an EOF analysis is as follows:

1. Create your $N \times M$ data matrix $\mathbf{X}$ (consisting of N time slices and M grid points) in column deviate ("time centered") form.
2. Form the $M \times M$ *spatial covariance matrix* from the data matrix with $\mathbf{R} = \mathbf{X}'\mathbf{X}/(N-1)$.
3. Extract the eigenvectors $\mathbf{V}$ and eigenvalues $\mathbf{\Lambda}$ from this covariance matrix, arranged in *decreasing* eigenvalue magnitude. These are your EOFs.
4. Select a small number of EOFs with the largest eigenvalues.
5. *Possibly* rotate these factors (EOFs) according to your scientific criteria.
6. Examine the spatial structure and temporal variation of these select EOFs.

Alternatively you can find the same eigenvectors $\mathbf{V}$ and the singular values $\mathbf{S}$ using singular value decomposition (Equation (4.30)). The SVD approach is often more computationally tractable for large data sets.

A word of caution here. You'll be starting with a sequence of two-dimensional space maps, one for each instant in time. This data can be organized in a variety ways.

Forming your data matrix $\mathbf{X}$ will involve "unwrapping" the two-dimensional maps into one-dimensional vectors that are stacked as rows, with each row corresponding to a time slice. Your results will not depend on how you do this unwrapping, but how you look at the results depends on you reconstructing the spatial maps the right way. We strongly suggest that you pay careful attention to this, and check your results by re-extracting some of the data fields from the rows of $\mathbf{X}$ and replotting them. The MATLAB `reshape` function can be used for these purposes, but we suggest you experiment with simple matrices to evaluate the effects.

If we have extracted the ordered eigenvectors from $\mathbf{R}$ using our handy `eigsort` function, the columns of $\mathbf{V}$ correspond to the EOFs, and the EOF loadings or amplitudes are $\mathbf{A^R} = \mathbf{V}\sqrt{\mathbf{\Lambda}}$. You can accomplish the final step by contouring the spatial structure of the first M_p columns of $\mathbf{V}$ to get the spatial structure of those EOFs, and then plotting their time variation, known as *principal component time series* or sometimes called *expansion coefficients*, as a function of time. You can compute the time variation for each EOF using:

$$\tau_i = \mathbf{X}V_i \tag{4.44}$$

where V_i is the ith column of $\mathbf{V}$, that is the ith EOF. Just as the EOFs are uncorrelated (orthogonal) in space, the expansion coefficients are uncorrelated in time. Compare the expression to $\mathbf{S_P^R} = \mathbf{XV}$ in (4.35). We could have done it that way "in bulk", but we would have got a lot of unnecessary extra time series (remember $\mathbf{V}$ is $M \times M$). Moreover, since the *rank* of $\mathbf{R}$ is less than or equal to the smallest dimension of $\mathbf{X}$, a lot of the eigenvalues of $\mathbf{R}$ will be effectively zero. One solution is to create a smaller matrix $\mathbf{V}_P$ ($M \times M_p$) by extracting the first M_p columns from $\mathbf{V}$.

The relative strengths of the EOFs can be assessed by the magnitudes of the corresponding eigenvalues. Since the objective is to create a "cleaner" version of the data, you would argue that the "major" EOFs represent the actual "signal", and the remainder represent the random "noise". The question arises as to how many EOFs we keep. Again, this is a *scientific* question, but there are a few criteria that might be applied:

1. There may be an obvious partitioning of variance, as evident by the eigenvalues, between the important and insignificant EOFs.
2. The time histories and spatial structures of the important EOFs will likely have simpler or more ordered behavior.
3. You may have expectations about the spatial or temporal structure, perhaps based on natural laws.

4.6.1 An example: seasonal subtropical sea surface temperatures

The file `SargassoSST.mat` contains an already constructed (S-mode) data matrix $\mathbf{X}$ of monthly mean sea surface temperatures from a 30×30 degree grid of locations in the western North Atlantic. The data were taken from a climatological atlas, and we show

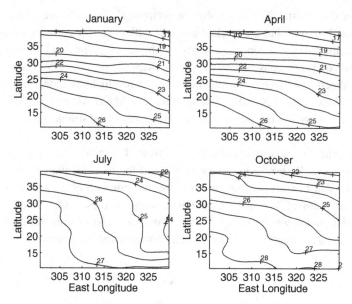

Figure 4.10 Select maps of monthly averaged sea surface temperaure in a small patch of the North Atlantic's Sargasso Sea showing the seasonal variation. The data file contains all 12 months.

in Fig. 4.10 the temperature distributions from four selected months in this series. We construct the covariance matrix and perform our EOF extraction using

```
t=1:length(X(:,1));                 % define month index
XX = X - ones(size(t'))*mean(X);    % remove column mean
R=cov(XX);                          % construct covariance matrix
[V,Lambda]=eigsort(R);              % extract sorted eigenvectors/values
AR=V*Lambda.^0.5;                   % compute EOF amplitudes
PoV=100*Lambda/sum(diag(Lambda));   % percent of variance
```

Note that the second step may take a little time, depending on your computer, because you're extracting eigenvectors from a 900 by 900 matrix. This process produces 900 eigenvectors (the columns of **V**) and 900 corresponding eigenvalues λ (the diagonal elements of $\mathbf{\Lambda}$). We could instead have accomplished the middle steps using SVD:

```
[U, S, V] = svd(XX,0);   % use SVD instead
Lambda = 1/(N-1)*diag(diag(S).^2); % compute eigenvalues
```

The eigenvalues from SVD are normalized by $1/(N-1)$ to match the unbiased covariance estimate returned by cov(XX).

Right off you should realize that the *rank* of **R** must be less than or equal to the smaller dimension of **X**, which in this case is 12. In fact, because it is computed from the *deviate form* of **X**, the rank of **R** is one less, 11, since you've used up a degree of freedom by subtracting the mean of each column. Try entering rank(R) and see what we mean. Therefore most of the eigenvalues will be effectively zero (they're not exactly zero due to roundoff error; i.e. somewhere around 10^{-14} or smaller), so we need to look at most at the first 11 only. You can get a look at their relative importance by typing

```
     [PoV(1:11)  cumsum(PoV(1:11))]
ans =
     97.2534    97.2534
      2.2635    99.5170
      0.2126    99.7296
      0.0726    99.8022
      0.0554    99.8576
      0.0402    99.8978
      0.0315    99.9293
      0.0227    99.9520
      0.0200    99.9721
      0.0164    99.9884
      0.0116   100.0000
```

Now you will immediately recognize that the bulk of the variance is contained in the very first EOF, which is the seasonal signal. That, of course, is to be expected, and typical for these kinds of records. You could, of course, filter the seasonal cycle out with some appropriate detrending filter, but you need to be careful not to introduce artifacts with that kind of activity. Leaving the seasonal trend in doesn't really cause us any problems here. The more interesting stuff is in the next few EOFs and you would be safe arguing that the last seven are certainly "in the noise". Looking at the spatial structure of these EOFs requires some MATLAB gymnastics, because the spatial maps are stretched out in the **X** matrix, and similarly in the covariance (**R**), eigenvector (**V**), and EOF amplitude ($\mathbf{A^R}$) matrices. We tried to organize the spatial nodes in a way that makes sense, so you can extract the first four EOF amplitudes from $\mathbf{A^R}$ and plot them with

```
for i = 1:4                            % do the first 4 EOFs
    subplot(2,2,i)                     % plot 4 to a page
    eof = reshape(AR(:,i),30,30)';     % orient correctly for
                                         plotting
    clabel(contour(xlon,ylat,eof));    % contour and label
    xlabel('East Longitude');          % label your axes
    ylabel('North Latitude')
end
```

which produces Fig. 4.11. We used the `reshape` function to wrap the column-oriented eigenvector amplitudes in $\mathbf{A^R}$ back into the original spatial map structure. We also included the two vectors `xlon` and `ylat` in `SargassoSST.mat` so that you can use them to locate the EOF maps in real space when you do this. The spatial patterns you see for the first two EOFs are typical. You tend to see a single sign, approximately monotonic, large-scale meridional (north–south) trend in the first EOF, and an asymmetric "dipole" pattern in the second. The third is more zonal (east–west) and has more structure.

The time variation associated with each EOF is contained in the principal component time series or "expansion coefficients", which we can calculate using Equation (4.44) and plot with

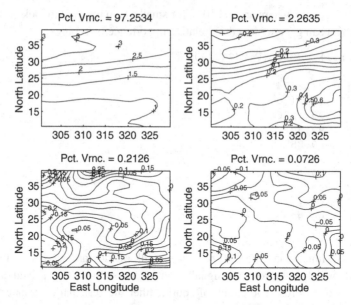

Figure 4.11 Maps showing the spatial structure of the amplitudes of the first (largest) EOFs for the sea surface temperature variations in the Sargasso Sea. The percent of variance (Pct. Vrnc.) accounted for each EOF is given above each map.

```
t=[1:12];        % time vector of months
for i=1:4
    subplot(2,2,i)      % plot 4 to a page
    pcts=X*V(:,i)/sqrt(Lambda(i,i)); % princ. comp. time
                                   series
    h= plot(t,pcts);    % plot vs time
    xlabel('Month')     % label your axes
    ylabel('Amplitude')
    title([ 'EOF ' sprintf('%d',i)])
end
```

which produces Fig. 4.12. When displaying EOFs, you have a choice of either mapping the (unit length) eigenvectors V_i or as we did in Fig. 4.11 the EOF amplitudes $\mathbf{A}_i^{\mathbf{R}} = V_i \sqrt{\lambda_i}$. If you chose to map $\mathbf{A}_i^{\mathbf{R}}$ then typically you normalize the principal component time series $\tau_i / \sqrt{\lambda_i}$ so that multiplying the spatial map by the time series recovers the full variability of the EOF.

You'll notice from Figs. 4.11 and 4.12 that the first two EOFs have fairly simple space- and time-structures. The time variation in the first EOF is a simple sinusoid, phase-lagged by a few months from the seasonal heating cycle. The second EOF time series is a less perfect sinusoid that is somewhat out of phase with the first, but you can reverse the sense of this by reversing the sign of the eigenvector, which is arbitrary. The third and fourth EOFs display much more spatial complexity than the first two, and appear much more

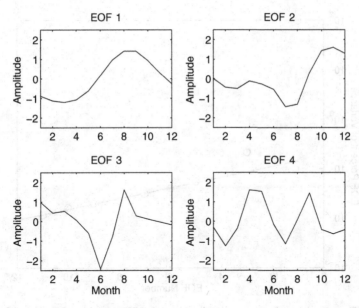

Figure 4.12 The normalized principal component time series (or expansion coefficients) for the first four EOFs in the Sargasso Sea SST example.

random in time as well. This, coupled with the fact that there's a roughly 10-fold decrease in the fraction of variance explained between the second and third EOFs, suggests that we should keep only the first two.

A more careful assessment of the variances (as revealed by the eigenvalues) shows yet another consideration. A semilog plot of the eigenvalues as a function of the EOF number (see Fig. 4.13) shows that beyond a certain EOF number, the variance decreases in a log-linear fashion, which is expected of pure noise. This shows that the third EOF may in fact have some significant "signal".

We have not rotated the EOFs in this example. There are circumstances where EOF rotation makes sense, and times when it does not. There's an idealized example in the problem set where rotation clearly makes sense. One caveat is that if EOFs are rotated, even though the EOFs remain spatially orthogonal (assuming you did an orthogonal rotation like *varimax*), the *expansion coefficients* or time series will no longer be orthogonal to one another. The question of whether EOFs should be rotated at all is a topic of discussion in the literature, just like factor rotation in general. Richman (1985) wrote an excellent review article on the subject, which you might consult.

4.6.2 Coupled fields

The concepts of EOF analysis can be extended to the comparison of the time evolution of two property distributions. For example, you may be interested in the co-evolution of sea

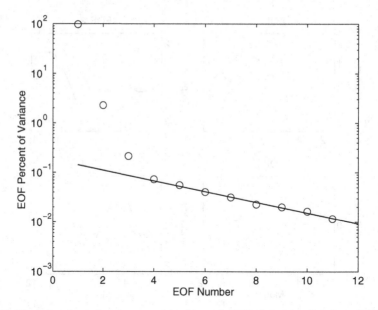

Figure 4.13 The percentage of variance contained in each EOF plotted as a function of EOF number for the Sargasso Sea SST example. Note that beyond EOF number 4, the variance decreases in a log-linear fashion, suggesting that these EOFs are pure "noise" rather than "signal".

surface temperature and heat flux, or the relationship between sea surface temperature and ocean color (chlorophyll). You can do this analysis by computing the cross-covariance R_{xy} between the fields by constructing the two data matrices (call them X and Y) in S-matrix form and building the cross-covariance with

$$R_{xy} = X'Y/(N - 1) \tag{4.45}$$

From the form of (4.45) it should be clear that the number of rows N in X and Y must be the same, but it's not necessary for the number of columns M_x and M_y to be the same. This means that you need to have sampled the two fields at the same times, or at least at an equal number of times separated by the same lag (for *lagged correlation*). This also means that the spatial maps don't have to be the same; you could be relating barometric pressure over the Pacific with sea surface temperature in the North Atlantic.

You can then use SVD to extract the singular vectors and singular values using

$$R_{xy} = USV' \tag{4.46}$$

where S contains the singular values, and U and V contain the *left patterns* and *right patterns* respectively for X and Y. We then compute the time histories with:

$$S_x^R = XU \tag{4.47}$$

and

$$\mathbf{S}_y^R = \mathbf{YV} \tag{4.48}$$

Analogous to the EOF analysis, we can compute the percent of covariance embodied in the ith pair of singular vectors (U_i and V_i) with:

$$\mathrm{PoV}_i = S_{ii}^2 / \mathrm{trace}(\mathbf{S}^2) \tag{4.49}$$

This approach (SVD of the covariance matrix between the two fields) is but one of a number of possible approaches. There's a useful review article published by Bretherton *et al.* (1992) that discusses and compares these various techniques.

4.6.3 *Some practical EOF issues*

There are a few things you need to keep in mind about EOF analysis. First, if you are going to delve into this field to any significant depth, we suggest you consult some of the many extensive texts on this topic. You'll most commonly hear about Preisendorfer's (1988) text. Second, be aware that the character of the EOFs may depend on the shape of the physical domain that you have selected. You may need to experiment with the size and shape of your spatial domain to see how *robust* are the extracted EOFs. Do they change radically? Sometimes EOF rotation helps in this case (see Richman, 1985). A third problem that you may have to deal with is missing data. The EOF and coupled fields approaches assume complete data sets. Sometimes in the real world, there are missing data points. All you can do in this case is to use some sensible interpolation scheme to fill in the missing points. If there's only a small percentage of missing points, you won't affect your results drastically.

Another issue you may encounter is the very size of the covariance matrix from which you are trying to extract eigenvectors/values. In the SST example we explored in Section 4.6.1 we had a 30×30 spatial grid that led to a 900×900 spatial covariance matrix. If you attempted to work with a 100×100 spatial grid, the resultant matrix would be $10\,000 \times 10\,000$. The lights would dim when you started that one! But don't despair. The number of operations for the covariance matrix PCA approach scales with the maximum of $\max[\mathcal{O}(M^3), \mathcal{O}(MN^2)]$ while the SVD approach scales with $\mathcal{O}(MN^2)$ (Emery and Thompson, 1998). Because there are often fewer time slices N than there are space grid points M, SVD is generally much faster.

Also this means that the *rank* of the data matrix $\mathbf{X}$ and hence the rank of the spatial covariance matrix $\mathbf{R}$ is no larger than the smaller dimension (N). This should suggest to you a suitable workaround: solve a smaller problem. We can define an alternate matrix:

$$\mathbf{Q} = \mathbf{XX}' \tag{4.50}$$

that may be considerable smaller ($N \times N$). Although it looks like the similarity matrix in Q-mode analysis, the $\mathbf{Q}$ matrix here is still column (time) standardized like the regular

EOF procedures, so it's really not a true "temporal correlation matrix". With a little matrix algebra we can construct an eigenvector/value equation analogous to Equation (4.31):

$$\mathbf{Q}\mathbf{V}^{\mathbf{q}} = \mathbf{V}^{\mathbf{q}}\mathbf{\Lambda}^{\mathbf{q}} \tag{4.51}$$

Here we've used the superscript **q** to signify that this a different dimension, **Q**-related eigenvector and eigenvalue problem. For our Sargasso Sea SST example **Q** will only be 12×12 rather than the **R** that is 900×900. The $\mathbf{\Lambda}^{\mathbf{q}}$ that you calculate here is related simply to the N largest elements of $\mathbf{\Lambda}$ that come out of the "full blown" eigenanalysis:

$$\lambda_i = 1/\lambda_i^q \tag{4.52}$$

where we've made some assumptions that things have been correctly ordered. In fact, if we left MATLAB to its own devices, it would report the Q-eigenvalues in some unpredictable order (likely ascending), so we would need to perform a descending sort to put them in decreasing order.

The above gives us the eigenvalues with much less computational work. We could make our selection of the most prominent EOFs based on eigenvalue size. Note, however, that this doesn't get us all the way there, since $\mathbf{V}^{\mathbf{q}}$ is *not* the same set of eigenvectors as the original **V** calculated in Equation (4.31). We can construct **V** from the data matrix, however, one column at a time with:

$$V_i = \mathbf{X}' V_i^q \sqrt{\lambda_i} \tag{4.53}$$

using the ith column of $\mathbf{V}^{\mathbf{q}}$ and the corresponding R-eigenvalue.

Finally, we should point out that EOF analysis yields only the "standing wave" components of the variance. If you're trying to deal with *propagating* time variation, e.g. for wave-like or transient changes, the covariance matrices will no longer be *real* valued, but will be *complex*. Extension to complex fields is not difficult, but beyond the scope of this book. We refer you to Preisendorfer (1988) for a more complete discussion.

4.7 Problems

4.1. **Playing with OMPA**. Develop an m-file to do optimum multiparameter analysis to reproduce the results shown in Fig. 4.5. The requisite data are contained in OMPAprob.mat, including the hydrographic data (**D**) (whose columns correspond to longitude, pressure, temperature, salinity, silicate, and NO respectively) and a bottom depths matrix. See Appendix A for hints on how to use MATLAB to create hydrographic sections. Enter the end-member water type characteristics from Table 4.1 and assign a weight of 10 to the unity-sum constraint. Once you have convinced yourself that you are replicating the water mass structures shown in the figure, plot sections showing the spatial structure of the corresponding residuals for all the tracers given by $\tilde{\mathbf{G}}x - \tilde{d}$ and explain what you think is happening.

4.2. **Using PCA**. The file <u>TraceMeast.mat</u> contains a selection of hypothetical trace metal measurements on sub-samples from a core. The belief is that the sediment sample contains a number of different minerals. Use PCA to evaluate how many minerals there might be in the core sample collection (the number of "significant components"), the relative abundance of these minerals (their respective eigenvalues), and the trace element composition of each component (factor loadings). Remember that the sign of the eigenvectors are arbitrary, so that you may need to reverse the signs of some factor scores. Why doesn't the rank of the correlation matrix reflect the number of significant components? Can you demonstrate that the factor loadings can be used to reconstruct the correlation matrix?

4.3. **An exercise in EOF rotation**. We have provided you with an artificial data set in the file <u>HypoGyre.mat</u>, consisting of a time series of maps of sea surface height in a hypothetical gyre. The data are organized in a three-dimensional matrix (longitude, latitude, and time) that you need to convert to S-matrix form, do EOF extraction and selection of primary EOFs. Plot the unrotated EOFs, along with their time variation. Perform *varimax* rotation, describe what happens and speculate as to why.

5

Sequence analysis I: uniform series, cross- and autocorrelation, and Fourier transforms

> Nothing so like as eggs; yet no one, on account of this appearing
> similarity, expects the same taste and relish in all of them.
>
> *David Hume*

5.1 Goals and examples of sequence analysis

Sequences of data, either in space or in time, appear all the time in ocean research. You may have a time series of measurements at a location (e.g. sediment trap data, or ocean surface temperature), a series of stations along a hydrographic section, or isotope measurements on a long sediment core. For the sake of simplicity (initially) we shall discuss only *regularly sampled data*; that is, samples taken at identical intervals in space or time. The analysis becomes more difficult and more complicated when we discuss irregularly spaced samples, but the principles are similar and best understood in terms of the simplest case first. What do we hope to achieve in the analysis of data sequences? There are as many reasons (or perhaps more) as there are data sequences. The next subsections outline briefly some of the major conceptual motivations.

5.1.1 Searching or testing for structure or periodicities

Within a single data set you might be looking or testing for changes in a system due to periodic forcing, for example the effect of seasonal changes on biological production, or the effect of lunar tides on shell-fish contamination. This may be extended to spatial regularity as well, in that you may be looking for evidence of large-scale Kelvin waves (rapidly propagating variations of the thermocline depth) on dissolved nutrients near convergence zones in the ocean. Structure is not restricted to periodicities. You may be interested in the *decorrelation timescale* or *distance*. For example, to what extent does the weekend weather depend on the previous Wednesday's, or how far apart should you space hydrographic stations (putting them too close together, i.e. within the decorrelation distance, results in redundant data)?

120

5.1.2 Correlation or correspondence between phenomena (and lags)

You may want to compare different data records, either from a data set collected at the same location or from data sets collected at different locations. For example, we could examine sea surface temperature and sediment trap flux at some great depth to establish relationships between two phenomena at that locale. Other examples might include comparing the El Niño–Southern Oscillation (ENSO) index and large-scale weather patterns, or oxygen isotopes in sediment cores and ice-core CO_2 concentrations. Lags between records may imply causality, or at least possible delay mechanisms associated with the interrelationships. Thinking for example about the sediment trap data, any lag between the oxygen isotope data in the sediment trap (as a proxy for sea surface temperature) and the actual sea surface temperature may yield the particle transit time from the surface to the trap. Bear in mind, however, that statistically significant correlation between data sequences does not prove a causal relationship; it may, in fact, point to *co-causal* relations (both sequences being driven by another, perhaps unobserved process).

5.1.3 Predictions (interpolation and extrapolation)

Of course, one of your goals might be to predict a certain value at a cherished location or time where data was not sampled, which becomes methodologically a regression exercise. The reasons for doing this may include regularizing your data when your record is incomplete owing to missing data points, or extending your data over longer periods. You may choose some method of regressing the series with a least squares algorithm, but you may also wish to map your data into frequency space to make your prediction. You may believe that there are predictable periodicities in your data that make your predictions more accurate in the frequency domain, or that it is more computationally efficient to perform the regression in this different domain.

5.1.4 Filtering and signal extraction

Another situation occurs when your signal is buried in a background of noise which you wish to filter out. There are techniques to digitally filter your data to remove unwanted noise, separate interfering signals, or in general improve the signal to noise ratio. For example, if you were measuring temperature and oxygen on a mooring, and there appeared to be a significant tidal signal causing temperature and corresponding oxygen variations on a regular basis, you could use a *notch filter* (one which does not pass a specific frequency). If there were further high-frequency variations associated with wave action (assuming you were interested in longer-term trends and variations), you could apply a *low-pass filter*. There are many types of digital filters, and we will talk about the most common ones in this and the next chapter.

5.1.5 *Power spectral analysis*

In many of the above applications, and indeed as a technique and end unto itself, you may wish to do power spectral analysis. This simply means that you want to see how the variance of the system (we will sometimes use the word "energy" in a very broad sense, since it can mean anything ranging from variability to power; but it is always variance) is distributed as a function of frequency or wavenumber. This kind of analysis yields clues as to the behavior of the system and the underlying physical laws operating. The shape of the spectrum tells you something about your measurement system and your experiment design as well. The study of sequential analysis also reveals the scope and fundamental limitations of the measurement process, we'll talk more about this in Chapter 6.

5.2 The ground rules: stationary processes, etc.

We will restrict our discussion initially to regularly (evenly) spaced data. A lot of data satisfy this requirement, although experience dictates that there is an equal amount that does not. The assumption of regularity makes the analysis much more straightforward mathematically, although we will get into unevenly spaced data at the end of the next chapter. An additional assumption, which you will often come across in sequential analysis, is that the series is *stationary*. This simply means that if you look at a sub-sequence of the record, it looks pretty much the same as some other sub-sequence. That is, there are no long-term trends in the data and means of different subsets are the same. We refer to this as *first-order* stationary. When data are examined, this isn't always the case but is usually easily fixed. All you need to do is to statistically test the data for the existence of a long-term trend, regress the trend and subtract it from your data prior to sequence analysis. This is also called *pre-whitening*, which we'll also run into in the next chapter. This makes sense because if your purpose is to examine the data for periodicities (sine and cosine waves, for example), then having an upward trend through your data means you might be sitting on the rising part of a very long sine wave. This would be a sine wave which you have grossly undersampled (you sampled only a small percentage of its period) so you don't want to include it in your analysis. How you subtract it off really doesn't matter, as long as it (a) makes physical sense, and (b) does a statistically good job. You already have some tools to do this from Chapter 3.

The concept of stationarity is actually much deeper than we described, and you can look for a more complete treatment in books like Jenkins and Watts (1968) if you need to. One of the underlying implications of stationarity is that the *statistics* of the data variation do not change with time (or location for spatial data). This may or may not be a particularly harsh restriction. If you can demonstrate that the means and the variances of any two sub-sequences are the same (think of F-test from Chapter 2) then you have a *second-order* stationary data set. It pays to think carefully about what is changing during the sampling period, not only from the viewpoint of the measurement process (e.g. are there calibration shifts?) but also in the fundamental physical nature of the phenomena you are measuring.

Now we realize that the world is not made of sines and cosines (well, maybe . . .), but it is a convenient set of *basis functions*. If you imagine a plot of a sine and cosine function with the same period and same starting point in time, then you are imagining two curves that are 90° ($\pi/2$ radians) out of phase with each other. Now this should immediately suggest to you (and it's true) that these functions are orthogonal and would therefore make a good basis function pair. We introduced the idea of basis functions back in Section 3.3.1 while discussing least squares regression. In that discussion we used a single function as a basis; why are we now suggesting you use two (or more)? Consider this: a time-varying signal can have a periodicity, an amplitude, and a phase. If we had a set of orthogonal functions whose linear combinations (think of a series of linear combinations) could account for multiple periodicities, amplitudes, and phases, then we would have a powerful tool indeed (see Section 5.6.1). We will use these combinations in the next two chapters to analyze time-varying signals found in nature.

In the remainder of this (and the next) chapter, we will talk in terms of time-series analysis. Keep in mind that this applies equally well to space-series analysis. Instead of time you'd use distance, and instead of frequency you'd use wavenumber (frequency is to period as wavenumber is to wavelength; they are inverses of one another). From time to time we will remind you of the correspondence, but not always.

5.3 Analysis in time and space

5.3.1 Autocovariance and autocorrelation

An obvious question to ask when you are collecting a time series or spatial transect is, "To what extent will the next measurement depend on the one that I just collected?" Or, for that matter, the one before? There is a simple statistical measure called the autocovariance, which is a measure of this dependence. The procedure is very easy, but somewhat laborious if you were to do it "by hand". Consider a time series of length N with equally spaced measurements, which we will call y. We will refer to each individual element as y_i. First, make a duplicate of the complete time series. Then, using the original and its copy, build a new series that is the lagged product of the two. You do this by multiplying every element in the original with every corresponding element in the copy, summing them up, dividing the sum by its length N, and subtracting from this the product of the means of the two series. Let's describe this lagged product series by considering first the "no lag" element given by:

$$C_{yy}(0) = \frac{1}{N} \sum_{i=1}^{N} (y_i - \bar{y})(y_i - \bar{y}) \tag{5.1}$$

$$= \frac{1}{N} \left[\sum (y_i)^2 - \sum (2y_i \bar{y}) + \sum \bar{y}^2 \right]$$

$$= \frac{1}{N} \sum y_i^2 - \frac{2\bar{y}}{N} \sum y_i + \bar{y}^2$$

$$= \frac{1}{N} \sum y_i^2 - 2\bar{y}^2 + \bar{y}^2$$

$$= \left(\frac{1}{N} \sum_{i=1}^{N} y_i^2 \right) - \bar{y}^2 \tag{5.2}$$

which you should immediately recognize as the variance (except we divide by N not $N - 1$), and we show it in both the definition (5.1) and computation (5.2) forms. We call $C_{yy}(0)$ the autocovariance of series y at lag 0, and it is the first element in the auto-covariance series (do not confuse this "C" with the concentration of some variable). In MATLAB, Equation (5.2) would look like:

```
cov0 = mean(y.^2) - mean(y)^2;
```

Note the dot before the first exponentiation; you are squaring the elements of "y", not creating a matrix. We called it `cov0` because MATLAB doesn't recognize a "*zeroth*" element in a vector or array. Now let's shift the second series to the left by one interval. Now since the series are of finite length, there will be one less element in the series overlap (picture one of the elements hanging out to the left in the second series, and one of the elements hanging out to the right in the first series, Fig. 5.1). You now compute a covariance at lag 1 as:

$$C_{yy}(1) = \left(\frac{1}{N-1} \sum_{i=2}^{N} y_i y_{i-1} \right) - \left(\bar{y}|_2^N \right) \left(\bar{y}|_1^{N-1} \right) \tag{5.3}$$

Notice that things are a little different now. For example, we divide by $(N - 1)$ because the series is shorter by 1. The summation is shorter too, since it has to start at $i = 2$. Also, the means are different now, since they are averaged over different intervals. We've used a somewhat unusual looking nomenclature to designate these different means, but it is analogous to limits of integration (see Equations (5.4) and (5.5)).

$$\bar{y}|_2^N = \frac{1}{N-1} \sum_{i=2}^{N} y_i \tag{5.4}$$

$$\bar{y}|_1^{N-1} = \frac{1}{N-1} \sum_{i=1}^{N-1} y_i \tag{5.5}$$

Again, pay careful attention to the summation limits. One average starts at $i = 1$, and the other starts at $i = 2$. The corresponding MATLAB statement would be:

```
cov(1) = mean(y(2:N).*y(1:N-1)) - mean(y(2:N))*mean(y(1:N-1));
```

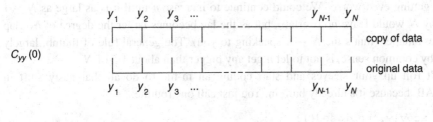

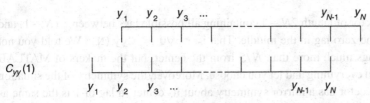

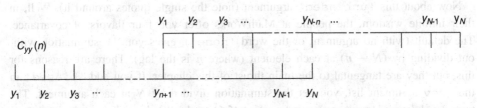

Figure 5.1 A cartoon showing a data series, its copy and how the individual elements of these data sequences line up with each other as one builds an autocovariance vector (N is the size of the data series, $N - n$ is the size of the overlap or window, and n is the size of the lag).

This is getting a little messy here. This is the second element in your autocovariance series. Now if we try two lags ... no, let's make that n lags, you have:

$$C_{yy}(n) = \frac{1}{N-n} \sum_{i=1+n}^{N} y_i y_{i-n} - \left(\bar{y}\big|_{1+n}^{N} \right) \left(\bar{y}\big|_1^{N-n} \right) \qquad (5.6)$$

The corresponding MATLAB code is:

```
cov(n)=mean(y(1+n:N).*y(1:N-n)) - mean(y(1+n:N))*mean(y(1:N-n));
```

where you have now only $(N - n)$ elements to compare, and a new definition for the means to subtract:

$$\bar{y}\big|_{1+n}^{N} = \frac{1}{N-n} \sum_{i=1+n}^{N} y_i \qquad (5.7)$$

$$\bar{y}\big|_1^{N-n} = \frac{1}{N-n} \sum_{i=1}^{N-n} y_i \qquad (5.8)$$

This is getting even worse! We could continue to increase n until it is as large as $N - 1$ (a lag by N would leave no overlap), but as the lag becomes larger the degree of overlap becomes smaller (that's the $N - n$ speaking to you). The general rule of thumb, largely driven by common sense, is not to let n get any bigger than about 1/5 of N.

Don't roll up your sleeves and start coding an m-file to do all that nasty stuff in MATLAB, because it's already built in. You just call one routine:

```
Cyy = xcov(y,'unbiased');
```

which returns a vector of length $2N - 1$ containing all possible lags between $-(N - 1)$ and $+(N - 1)$, with the zero lag in the middle. That is, `cov0 = Cyy(N)`. We told you not to believe those lags much more than $N/5$ from the center, but the makers of MATLAB decided to give you everything and let you choose. Moreover, the symmetry of the situation demands that the vector has a mirror symmetry about its center (a lag of 1 is the same as a lead of 1), so that `Cyy(N-1)=Cyy(N+1)`, `Cyy(N-2)=Cyy(N+2)`; but this only true of *auto*covariance.

Now about this `'unbiased'` argument (note the single quotes around it). Well, in their infinite wisdom, the people at MathWorks offer you four flavors of covariance. The default (with no argument or the word `'none'`), gives you the summation without dividing by $(N - n)$ at each element (where n is the lag). There are reasons for this, but they are tangental to the main thrust of this chapter. If you add `'biased'` in the `xcov` agrument list, you get the summation divided by N at each argument. The term `'unbiased'` gives you our Equation (5.6), and is *"unbiased"* because it corrects for the number of overlapping elements. Finally, if you put `'coeff'`, it normalizes the sequence so the zero lag element is 1.0, also known as the *autocorrelation*. What does the autocovariance tell you? Well, it speaks to you about the original question we posed at the beginning of this chapter. To what extent does a given measurement depend on its predecessors?

Let's look at an example. The tidal records collected in Woods Hole, MA, during May 2005 are plotted in Fig. 5.2. A plot of the autocovariance vs. lag time is called the autocovariance function (sometimes you will hear it called an autocovariogram). As you can see (and as most people know), there is a regular fluctuation in the tides at most locations, with an approximate 12 hour periodicity. Calculating the autocovariance (unbiased), and looking at the result as a function of lag, we see in Fig. 5.3 an undulating pattern rising and falling, shown on two scales (so you can see the detail). Note that the largest value of the covariance is at zero lag. This makes sense, because the series will never perfectly match up except at zero lag. The number decreases with increasing lag (left panel). Note that there is a kind of "decorrelation time"; the tidal level you see this hour is a good predictor of what you'll see next hour. This tails off crossing zero covariance many times as lag increases (right panel), so that by the time you get to a lag of 150 hours the autocovariance is less than or equal to one-half the zero lag value (not as good a predictor any more). The zero crossing is an important concept since it characterizes the length of time that must elapse before the time series becomes uncorrelated. This sequence is strongly periodic, so

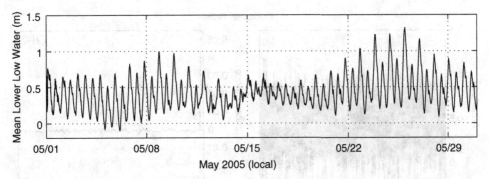

Figure 5.2 Tide data from Woods Hole, MA, in the month of May 2005. The tidal level is recorded every 6 minutes. Data obtained from NOAA's tides and currents web page (http://tidesandcurrents. noaa.gov/products.html).

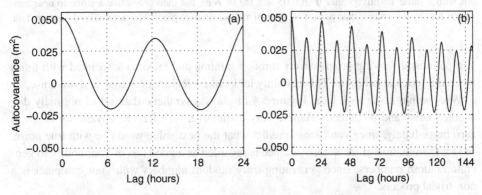

Figure 5.3 The autocovariance function (autocovariogram) of the tide data shown in Fig 5.2. (a) The autocovariance out to 24 hours; (b) the same autocovariance function out farther to 150 hours (or about $N/5$ of the data record).

that the autocovariance becomes significantly non-zero at longer lags (see Fig. 5.3a). Most sequences do not have such strong periodicity, and the autocovariance drops to near zero and stays there.

The first *zero crossing* is known as the decorrelation time. If we were looking at a space series (e.g. a hydrographic section of regularly spaced stations), the first zero crossing of the autocorrelation is a measure of the spatial decorrelation length. This could be a measure of the eddy mixing scale; stations closer than this distance will tend to be correlated, and hence not independent of one another. Now look at about 6 hours. There is a negative correspondence. If the tide is high now, it will likely be low about 6 hours from now, and vice versa. Notice that as you approach the 12 hour lag, the covariance rises to another maximum. Because there are longer-term variations in the tidal records, the match cannot be perfect, so the covariance never reaches the highest value seen at zero lag, but it does go through a maximum. This is the indicator of periodicity. Note that as you increase the lag,

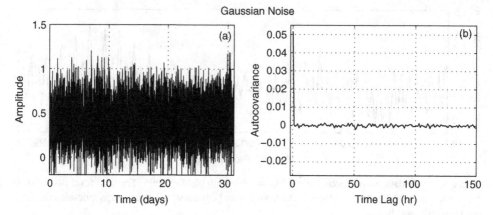

Figure 5.4 Some synthetic noise (a) and its autocovariance (b). Notice how the autocovariance starts out with a value slightly greater than 0.05 at a lag of zero, but then drops like a stone to near zero values for any lag greater than zero. This type of autocovariance plot is characteristic of random data.

you periodically swing progressively through minima and maxima associated with times when the series matches up. This certainly looks like a legitimate signal, and not noise.

What would noise look like? Figure 5.4 displays a synthetic data set of normally distributed noise and its autocovariance. Very distinctive, isn't it? The covariance drops to zero immediately, since you cannot predict what the next value would be with true noise. In fact examination of the autocovariance is regarded as a good test of your ability to generate random numbers, since generating truly random numbers with your computer is a non-trivial process.

You may have noticed that covariance has units, and those units determine the magnitude of the covariance. But suppose you wanted to compare several data series, measured in different units, and wanted the results to be on a comparable scale. The statistic we use in these cases is the correlation. Let us start with *autocorrelation*, which is simply defined as a normalized form of the autocovariance:

$$r_{yy}(n) = \frac{C_{yy}(n)}{s_y^2} \tag{5.9}$$

where $r_{yy}(n)$, sometimes written just r_n, is the correlation coefficient. Sensibly, its value starts at 1 for zero lag, and decreases thereafter, always remaining between -1 and $+1$. By the way, if you think back to Chapter 2, you could also get around the different scales by converting all of the data to standard scores. And you would get the exact same numbers as the correlation coefficients ... think about it. Have a look at Davis (2002; p. 247) for examples of various "idealized" autocorrelation plots.

The question remains, when you look at *autocorrelograms* how significant are the numbers? Instinct tells you that the autocorrelation of a random series should go to zero very quickly (like the example in Fig. 5.4). So for a given autocorrelation at a specific lag, we want to know: what is the probability that it could have been generated by a random series (i.e. that it is indistinguishable from zero)?

Remember our discussion of the normal distribution and hypothesis testing in Section 2.2.5. There we defined the normal distribution statistic (Z-score) as:

$$Z_i = \frac{x_i - \bar{x}}{s} \tag{5.10}$$

Additionally, the expected variance of autocorrelation for any random sequence of length N at some lag n is:

$$\sigma_n^2 = \frac{1}{N - n - 3} \tag{5.11}$$

The 3 in the denominator comes from a correction for the degrees of freedom used in computing the means of each of the two series used (even if they are the same) and in the choice of the starting point of the autocorrelation. So we take the ratio of observed correlation at lag n to σ_n and after a little simplification, we get:

$$Z_n = r_n \sqrt{N - n - 3} \tag{5.12}$$

This works out because $\bar{x} = 0$, i.e. the mean is zero. Now just like any statistical test, we can calculate the probability of occurrence for a value Z_n due to random processes, assuming a normal distribution.

So in a normal distribution if Z_n is 1.96 (or, since the Z-statistic is two-tailed, -1.96), then there is a probability of 0.05 that the null hypothesis is correct (that there is no difference between the series and a random series at that lag and therefore the autocorrelation coefficient is not significantly different from zero). Put another way, if

$$|r_n| > \frac{1.96}{\sqrt{N - n - 3}} \tag{5.13}$$

then it is likely ($>95\%$ confidence) non-random. This dependence, by the way, underlines the general importance of small relative lags. As the size of the lag increases, the uncertainty in the autocorrelation coefficient increases. This trend doesn't show in really long time series, but one ought to avoid doing this kind of analysis for time series shorter than 50–100 observations. Figure 5.5 shows the autocorrelation plots for both the tidal record and the random noise, along with the 95% confidence intervals ($Z_n = 1.96$). The tidal autocorrelogram has significant peaks all the way out to 144 hour lags (and beyond) and if you look very carefully at Fig. 5.5 you should be able to see that the critical r_n lines are diverging. Notice also that there are two critical r_n lines because anticorrelations can also be statistically significant.

5.3.2 Effective degrees of freedom

You may recall that in Chapter 2 we made a point about the assumption that the data must be independent, which is another way of saying that they must be uncorrelated. This

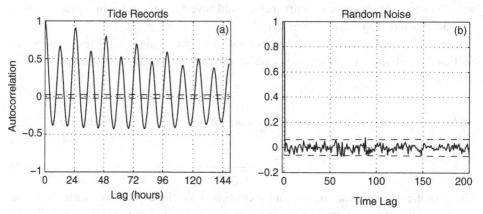

Figure 5.5 The autocorrelogram for (a) the Woods Hole tidal record data from Fig. 5.2 with the statistical insignificant region bounded by the dashed lines, i.e. if the autocovariogram is above the upper line or below the lower line, then the correlation coefficient is statistically significant. (b) The autocorrelogram of the data in Fig. 5.4 showing how most of the correlation coefficients are statistically insignificant as they stay within the two dashed lines. Note that in (a) and (b) the dashed lines are diverging as one goes to larger lags. This is due to the decreasing number of data points overlapping each other at these larger lags. In these autocorrelograms this effect is not that apparent because N is large, and we are showing only the first $N/5$ lags.

requirement is important not only for the statistical assumptions underlying the tests we were performing, but also for the correct calculation of the degrees of freedom. In Chapter 2 we maintained that you start with as many degrees of freedom as you have data points (N), then subtract one degree of freedom for each parameter that you had to calculate to perform the test (each mean, standard deviation, etc.). But wait a minute! If what we see in the autocorrelation of data is correct, it implies that data points closer together in time or space than the zero crossing are, to some extent, correlated. If you are testing some statistical theory about property distributions, this implies that the number of degrees of freedom may be less than the number of observations (or stations) minus the number of computed parameters.

Testing of regression parameters and other distribution statistics must take into account the *effective degrees of freedom*, which we designate ν^*. Emery and Thompson (1998) provide a straightforward way to use autocovariance to calculate ν^* based on the effective number of data points, N^*. Say you use N data points y to calculate a linear trend and intercept in your time observations. In order to test the statistical significance of this trend against a known or theoretical trend, you must use the effective degrees of freedom (in this case $\nu^* = N^* - 2$ for, say, a t-test). First you estimate the integral timescale of your data series, which is given by:

$$\mathcal{T} = \frac{1}{C_{yy}(0)} \sum_{k=0}^{m} \frac{\Delta t}{2} \left[C_{yy}(k) + C_{yy}(k+1) \right] \qquad (5.14)$$

where Δt represents your sampling interval and m is a somewhat arbitrarily chosen number of lags that is greater than the first zero crossing and a substantial fraction of N. Equation (5.14) is nothing more than the trapezoidal rule (more about this and other numerical integration techniques in Chapter 8) applied to the autocovariogram, normalized to the autocovariance at zero lag (i.e. the variance). The effective number of data points is given by:

$$N^* = \frac{N \Delta t}{T} \qquad (5.15)$$

remember to be careful to ensure that Δt and T are in the same units. Looking at Equations (5.14) and (5.15), and depending on how quickly your autocovariance function drops to zero, we think you can see that N^* can be significantly less than N. This can have a significant effect on statistical tests dependent upon the true degrees of freedom. For further reading and other references we direct the reader to Emery and Thompson (1998).

5.4 Cross-covariance and cross-correlation

Well, what you can do for one series, you can do for two. The *cross-covariance* (or *cross-correlation*) of two time series is a measure for discerning the relationship between them. For example, you might be interested in the influence of the sunspot index on the weather. There was a long period of anomalously cold weather, called the "Little Ice Age", several hundred years ago, which appears to be associated with a minimum in the sunspot index (the "Maunder Minimum"). This correspondence shows up in a number of other records, for example tree ring thickness and radiocarbon in tree rings and corals.

Of interest is not only the degree of correspondence between two records, but possible *time lags* between them. For example, there are records of atmospheric CO_2 concentrations in ice-core bubbles. There is a clear indication that atmospheric CO_2 levels were *lower* during glacial maxima. Were these lower concentrations a result of the glacial events, or did they cause them? If the CO_2 record *leads* the glacial advance/retreat, then this would imply, at least circumstantially, that there is a causal relation where the impact of CO_2 on the radiative balance of the Earth drives the ice advance and retreat. Further, the degree of the lag may be of interest. For example, the time lag between the CO_2 record and other glacial indicators indicates the response times of the various systems. Another example would be time lag between sea surface temperature and a temperature proxy indicator ($\delta^{18}O$) in foram tests in a sediment trap record. The lag would represent the settling time of particles, and hence the settling velocity.

The cross-correlation of two time series (x and y) at lag n with $N - n$ pairs overlapping is given in its definition form as:

$$r_n = \frac{N}{N-n-1} \frac{\sum \left[(x_i - \bar{x}|_{n+1}^N)(y_{i-n} - \bar{y}|_1^{N-n}) \right]}{\sqrt{\sum (x_i - \bar{x}|_{n+1}^N)^2 \sum (y_{i-n} - \bar{y}|_1^{N-n})^2}} \qquad (5.16)$$

which is just the same as the cross-covariance divided by the square root of the product of the variances of the individual series. Note that we calculate the variances only over the sections of data that actually overlap; remember our definition of how to calculate the mean in these cases, Equations (5.7) and (5.8). Also note that the time series x and y are on the same time grid, that is to say measured at the same (equally spaced) moments in time (or regridded to be so; see Chapter 7 for more about gridding).

Equation (5.16) should look familiar from our discussion of correlation coefficient in Chapter 2. The summations are assumed to be over an index i (not shown) varying from $n + 1$ to N. Here the zero-lag cross-correlation coefficient $r_{xy}(0)$ will in general not be 1, since the two different series will never perfectly match (the xy subscript refers to time series x and y). The cross-covariance (cross-correlation) vector, when plotted against its $-n$ to $+n$ lags, will be asymmetric about the cross-covariance (cross-correlation) at lag of zero ($C_{xy}(0)$).

Like the autocorrelation function, we can test the significance of the cross-correlation coefficient (remember the null hypothesis is that the series are uncorrelated) with the Z-statistic (Equation 5.10):

$$Z_{xy}(n) = r_{xy}\sqrt{N - n - 3}   \tag{5.17}$$

5.5 Convolution and implications for signal theory

The process of calculating the cross-covariance is closely related to a mathematical operation called *convolution*. This is mathematically represented for two functions in time $x(t)$ and $y(t)$ by:

$$(x * y)_j \equiv \sum_k y_{j-k}x_k \quad \text{or} \quad x * y \equiv \int x(\tau)y(t - \tau)d\tau   \tag{5.18}$$

where the right-hand equation is the continuum equivalent representation (an integral, after all, is just the continuum equivalent of a summation) of the left-hand, discrete, equation. In these equations, the "*" represents convolution and should not be confused with the MATLAB multiplication operator. It has a specific connotation in mathematics. The integral is done over τ, which is the time lag. If we were to standardize our data by subtracting the means and dividing by the standard deviations, i.e. make them "Z-scores", then the convolution of the two series becomes the cross-correlation function.

What does Equation (5.18) look like in MATLAB? The mathematical operation of convolution is mechanically the same as multiplying two polynomials. Let's take the two vectors $x_k, k = 1, \ldots, M$ and $y_i, i = 1, \ldots, N$ from the above equation and create a third vector z that is the convolution of x and y. The mathematical operation would be written as:

$$z_j = \sum_{k=1}^{j} x_k y_{j+1-k}   \tag{5.19}$$

summed over all k values that result in a valid subscript for x and y. The $+1$ in the y index makes allowance for the fact that there can be no $z(0)$ in MATLAB. This results in a vector $N + M - 1$ in length that incorporates the lags and leads of the shorter vector (typically x) over the other. If we were stubborn and wanted to do it "by hand" and if x and y have lengths M and N (let $N = M$ just to make the bookkeeping simpler), the MATLAB code would look like (in the code snippet below $*$ is just MATLAB multiplication):

```
z(1)   = x(1)*y(1);
z(2)   = x(1)*y(2) + x(2)*y(1);
z(3)   = x(1)*y(3) + x(2)*y(2) + x(3)*y(1);
...
z(N-1) = x(1)*y(N-1) + x(2)*y(N-2) + ... + x(N-1)*y(1);
z(N)   = x(1)*y(N) + x(2)*y(N-1) + ... + x(N)*y(1);
z(N+1) = x(2)*y(N) + x(3)*y(N-1) + ... + x(N)*y(2);
...
z(2*N-2) = x(N-1)*y(N) + x(N)*y(N-1);
z(2*N-1) = x(N)*y(N);
```

As you can see, there is quite a lot of "bookkeeping" and the same vector z could be obtained with the MATLAB command:

```
z=conv(x,y);
```

Convolution turns up in all sorts of places. One mathematical operation that you may already be familiar with, calculating a *running mean* or *running average*, is nothing more than the convolution of a boxcar window (zeros everywhere except for a band of ones the size of the desired averaging window) with the data, normalized by the length of the boxcar window. Say we want to look at the long-term, slowly varying part of the tidal record first seen in Fig. 5.2. One way we might choose to get at that kind of information, obscured by the twice daily tidal excursions, is to take a 2 day running mean (to average out the tides). But it is far easier to make the boxcar window (shown in Fig. 5.6b) and convolve it with the data. Keep in mind that the resultant vector is now $N + M - 1$ in length, and if you look at the above MATLAB snippet of code the beginning and end of it include only a few data points from the beginning and end of the arrays, so be sure to trim this part from the beginning and end to get a resultant vector the same length as the original, longer vector. Figure 5.6c shows the longer-term variability (of about a week) suggested by Fig. 5.6a but not immediately apparent. It also shows some small short-term variability as a result of the length of the window (two days are only four tidal cycles after all). In the next chapter (Chapter 6) we will discuss this and other windows and try to convince you that the boxcar, perhaps, isn't the best one to use for this purpose.

Convolution also has an interesting application to signal processing. Let's say you want to measure some kind of uncontaminated process, that is to say the true signal, with a device. We refer to this time series as $u(t)$. In this case, the device or instrument could be an actual measurement or a proxy method or process (e.g. sediments, ice cores). Every device will have some kind of response function, which we'll call $r(t)$, that will smear out

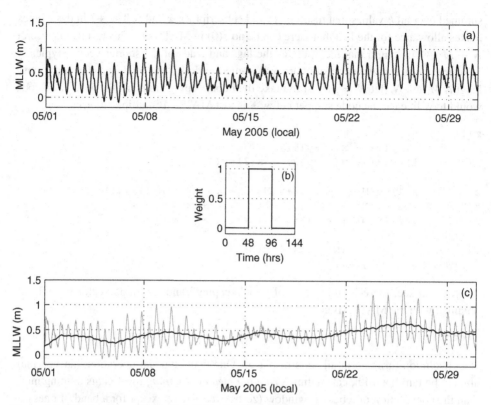

Figure 5.6 Tide data from Woods Hole, MA. (a) The same data shown in Fig. 5.2; (b) a boxcar window (vector) that is zero at the beginning and end and one in the middle 48 hours (c) subfigures (a) and (b) convolved and displayed as the dark line (original data plotted as a light gray line for reference). The values in the dark line are also divided by the length of the window they were convolved with.

your signal because the device has some finite response time. Then the output (signal $s(t)$) of your device will be the convolution of the response function with the signal, that is:

$$s(t) = r(t) * u(t) \tag{5.20}$$

What would the "ideal" instrument response function ($g(t)$) look like? Well, you'd want something that would put out exactly what came in, independent of previous or following measurements. That is, you'd want

$$s = g * u = u \tag{5.21}$$

where here the time argument is implicit. In the finite sum case, this kind of function would simply be a series that is 1 at zero lag, and zero everywhere else. In the continuum case, this corresponds to a *delta function* which is zero when it is not at exactly lag zero, and has an area of 1. This, you say, is impossible because if it is infinitely narrow, it must be infinitely tall to have unit area (area = height $\times$ width). Well, it can be done; don't worry about it. This idealized function is also called a *Green's function*, which simply means that when convolved with a function, it yields the same function back. Of course most instruments are not so perfect, there often being time lags or memory effects.

In real life, there is also always the additional problem of noise (random interference with your signal). We call this $n(t)$, and so the signal that finally makes it to your chart recorder (er, someone is showing their age) or computer ADC (analog to digital converter) would be:

$$c(t) = s(t) + n(t) = r(t) * u(t) + n(t) \tag{5.22}$$

Believe it or not, there are ways of filtering your data to extract the original, desired signal ($u(t)$). We'll get back to this later, but in general, *filtering* your data involves a similar process, whereby you design some kind of function (your filter) to alter the original data as:

$$p(t) = f(t) * c(t) \tag{5.23}$$

where $f(t)$ is your filter, and $p(t)$ is the final and (you hope) better product. You might do this to eliminate noise, to compensate for imperfect instrumentation, or to enhance some specific aspect of your data for further analysis (this is called deconvolution). MATLAB has a whole host of functions in its *Signal Processing Toolbox* that address these issues.

5.6 Fourier synthesis and the Fourier transform

5.6.1 Complete sets of orthonormal functions (basis sets)

There is a fundamental theorem in function theory which says that given a *complete* set of *orthonormal* functions, you can construct any other function. This is akin to saying that given enough bricks and mortar, you could build just about anything. There is an easy to understand analogy, that of component vectors. Think of a simple two-dimensional space (a graph) on which you place an arbitrary vector (from the origin to some point on the graph). You can decompose that vector into two "component" or "basis" vectors that sum up to that vector. For example, consider a vector from the origin to $x = 3$, $y = 4$. Then you could make that vector ($\mathbf{v}$) by a combination of two basis vectors of unit length in the x and y directions (i.e. parallel to the x and y axes) with:

$$\mathbf{v} = 3\hat{\mathbf{x}} + 4\hat{\mathbf{y}} \tag{5.24}$$

where $\hat{\mathbf{x}}$ and $\hat{\mathbf{y}}$ are the unit length *basis vectors*.

You can do this for any vector in that two-dimensional space (the graph). In fact, you have to do this implicitly when you give the location of the vector head with a co-ordinate pair. Well, the same is true of any arbitrary function, provided that the set of functions you use span the full dimensionality of your function space (just as the x and y unit vectors do in the two-dimensional graph space). It is also convenient if the functions are *orthogonal* and are *normal* (of unit size, just like the unit vectors). This can be expressed for a set of basis functions, which we'll call F, consisting of an infinite series of functions F_i such that:

$$\int_{-\infty}^{+\infty} F_i F_j \, dx = 0 \quad \text{and} \quad \int_{-\infty}^{+\infty} F_i F_i \, dx = 1 \tag{5.25}$$

where i is not equal to j. The first is simply a statement that two orthonormal vectors must have a dot product of zero (that is, they are at right angles); the functions must have a zero cross-correlation. The second statement is much like the unit length stipulation for a vector; the area under the function must sum to 1.

There are a number of groups of functions that satisfy this relationship, but the one we are interested in here is the sine/cosine series. It is convenient to use the transcendental representation of the pair:

$$e^{-2\pi i f t} = \cos(2\pi f t) - i \sin(2\pi f t) \tag{5.26}$$

Here i refers to the imaginary number $\sqrt{-1}$. We will also interchangeably use the frequency (f) and the *angular frequency* (ω), where:

$$\omega = 2\pi f \tag{5.27}$$

You can build any arbitrary function out of an infinite sum of these basis functions:

$$h(t) = \sum_{k=0}^{\infty} H_k e^{-2\pi i k f t} \quad \text{or} \quad h(t) = \int_{-\infty}^{+\infty} H(f) e^{-2\pi i f t} \, df \tag{5.28}$$

assuming you can determine the coefficients, H_k, which correspond to the lengths of the vector components in our two-dimensional example. The original function, $h(t)$, is a function of time, H_k are the coefficients for the discrete case, and $H(f)$ are the coefficient functions for the continuous case. In Equation (5.28) we call the left-hand equation the Fourier series of $h(t)$ and the right-hand equation the Fourier transform of $h(t)$.

An infinite Fourier series can fit any arbitrary data set. In practice we work with a finite subset of sines and cosines (called harmonics). Figure 5.7 displays an example of harmonics applied to the monthly multivariate ENSO index (MEI) from 1950 to 2005. In each subplot the actual MEI data are plotted as a thin black line, which we approximate with Fourier series of various lengths (harmonics). As you can see, the ability to mimic the behavior of the MEI increases with increasing series length, and the nature of the fit is that the "spikier" (high-frequency) elements are fit better by including more components of the Fourier series.

5.6.2 The Fourier transform and its properties

Equations (5.28) are equivalent to saying that we can "map" a function that is in the time domain into the frequency domain. This is the exact same thing as saying we can build any function in the time domain as an infinite series of sine/cosine pairs times a function

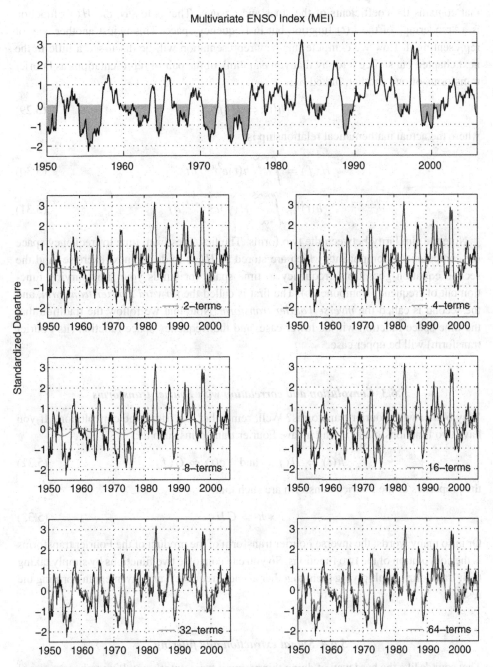

Figure 5.7 The monthly multivariate ENSO index (MEI), from 1950 through 2005, in standard deviation units is presented in its usual format in the top subplot. The remaining six subplots below are an example of how as one adds the harmonic terms of Equation (5.28) (2, 4, 8, 16, 32, and 64 terms) one comes closer and closer to matching the original data, both in terms of magnitude and spikiness (thicker gray line). Data are from NOAA-CIRES Climate Diagnostic Center (CDC), University of Colorado at Boulder (http://www.cdc.noaa.gov/people/klaus.wolter/MEI/).

that contains the coefficients of the sines and cosines. That is to say, the $H(f)$ function is a "rendering" of the $h(t)$ function but in frequency space. This is just another way of representing it. How we compute this new representation will be discussed a little in the next chapter, but suffice it to say that you can think of the pair of representations as Fourier transforms of one another, linked by:

$$h(t) \Leftrightarrow H(f) \tag{5.29}$$

where the actual mathematical relationship is:

$$H(f) = \int_{-\infty}^{+\infty} h(t)e^{2\pi i f t} dt \tag{5.30}$$

$$h(t) = \int_{-\infty}^{+\infty} H(f)e^{-2\pi i f t} df \tag{5.31}$$

Note the similarity between the two forms. The first goes from time to frequency space (or computes the components which are stored in $H(f)$ to do the next exercise) and the second equation goes from frequency to time space (or builds your time-varying function out of frequency components). The first is called the *forward Fourier transform*, and the second is called the *inverse Fourier transform*. Note that we follow the tradition that the time-space function will be lower case, and the frequency-space function (its Fourier transform) will be upper case.

5.6.3 Convolution and correlation with Fourier transforms

Why bother with Fourier transforms? Well, remember our friend the convolution? If you have two functions g and h which have Fourier transforms such as:

$$h(t) \Leftrightarrow H(f) \quad \text{and} \quad g(t) \Leftrightarrow G(f) \tag{5.32}$$

the properties of the Fourier transform are such that:

$$g * h \Leftrightarrow GH \tag{5.33}$$

Or in so many words, the inverse Fourier transform of the product of the Fourier transforms is the convolution of the two functions. So you can convolve two functions by simply taking their transforms, multiplying them together at each frequency, and inverse transforming the result.

5.6.4 Signal extraction and filtering

That sounds like the hard way of doing things, but it turns out it's not. Remember the signal theory problem where you want to *deconvolve* your signal to compensate for instrument response? Recall that the measured signal s is actually the convolution of your instrumental response function with the original signal:

$$s = r * u \qquad (5.34)$$

Now with your new-found power (Monsieur Fourier, we thank you!), you have:

$$s = r * u \Leftrightarrow S = R \cdot U \qquad (5.35)$$
$$u \Leftrightarrow U = S/R \qquad (5.36)$$

That's right. If you know your response function $r(t)$, which you might obtain from laboratory testing, from theory or from the user's manual, you can take its Fourier transform and divide it into the Fourier transform of your observations. Then you do an inverse transform of the result to get your uncontaminated signal. Equation (5.35) is called the *convolution theorem* and one of the many applications of this theorem (such as Equation (5.36)) is called *deconvolution*.

Sound too good to be true? In a way it is. While the process works in principle, you only know your response function to some limited accuracy, and the ratio S/R tends to be somewhat unstable because where your response function goes toward zero, the ratio becomes large. Small errors in R will lead to large problems with U. However, with a certain amount of caution, this kind of thing can be (and is routinely) done. In the next chapter, we will talk about a technique for dealing with the noise in your data too.

5.7 Problems

5.1. Download the data in `woods_hole_tides_2005_05.dat`. These data were used to create Fig. 5.6. Using these data and the Hamming window (use `doc hamming` or `help hamming` to learn how to generate a window in MATLAB) smooth the tidal data with this window and the convolution (`conv`) function. Here are some things you will need to know: the data were collected once every six minutes, this time series starts at midnight on 1 May 2005 and continues to 23:54 on 31 May 2005, the time series is continuous (there are no missing data points), and the columns of this data set are the following:

Station ID	Year	Month	Day	Hour	Minute	Predicted MLLW	Verified MLLW

Here, MLLW means mean lower low water.

(a) Plot the original data and the smoothed data on top of the original data (remember that the convolved product is longer than the original time series; it will be $N + m - 1$ elements, so you will have to trim some from the beginning and end). Hang onto the code you use for this; we'll be returning to the data in a problem in the next chapter (Chapter 6).

(b) Now, starting at midnight 1 May 2005, extract one data point every 24 hours from both time series and plot them together. Calculate the monthly means and standard deviations of these two time series. Which one best represents the true variability of the process (tidal excursions) being studied?

5.2. You may have heard that bioturbation in sediments has the same effect on the sediment column constituents as does convolution of the bioturbation response function with the undisturbed profile. But what *is* the bioturbation response function? Let's use Green's function, (Equation 5.21) to find out. To Fourier transform your data use the `X=fft(x);` command, and use `x=ifft(X);` to obtain the inverse Fourier transform.

(a) Download `bioturbseds.dat`. This file contains three columns: depth (in cm) and two ash content (in g ash per g sediment) columns, the first the undisturbed sediment column and the second the bioturbated column. Knowing the disturbed and undisturbed sediment profiles and the convolution theorem, derive the bioturbation response function as a function of depth. Using the nonlinear least square regression routine from Chapter 3 calculate the e-folding depth this response function has (how far in the sediment column one has to go to see a $1/e$ decrease in whatever property being examined). Remember the undisturbed profile is a spike function.

(b) Plot the disturbed and undisturbed sediment profiles and the bioturbation response function in three subplots next to each other. Given the results in part (a), what does Green's function tell you about response functions for idealized initial profiles?

6

Sequence analysis II: optimal filtering and spectral analysis

Nine Zulu Queens Ruled China.
John Derbyshire

This chapter is, in many ways, a continuation of Chapter 5. We will conclude the topic of filtering and discuss issues associated with the venerable field of spectral analysis, i.e. the fast Fourier transform (FFT), power spectral analysis, Nyquist frequency, and windowing. We will finish off this chapter by introducing you to non-uniform sampled spectral analysis (Lomb's method) and wavelets.

6.1 Optimal (and other) filtering

Filtering, whether we realize it or not, is ubiquitous. Most of the data records we deal with have been filtered, whether it is by the instrument's response function as in Equation (5.20) or by the fact that we have "windowed" the data with a boxcar window (i.e. by choosing not to window the data at all; Section 6.4.2). In the last chapter we left you with an appreciation of convolution and the Fourier transform. Here we further develop these concepts to show how all of that is related to windowing and filtering.

6.1.1 The Wiener (optimal) filter

Picture a situation where you have an uncontaminated, real-world signal, $u(t)$, which is being measured by an instrument with a response function $r(t)$. The output of this instrument will be the convolution of the two functions, which we will call $s(t)$, defined by:

$$s(t) = r(t) * u(t) \tag{6.1}$$

We've treated this part of the problem before in Section 5.6.4. Now, as is inevitable, noise $n(t)$ is introduced into the process, which must be added to $s(t)$, to give us what we finally measure:

$$c(t) = s(t) + n(t) \tag{6.2}$$

141

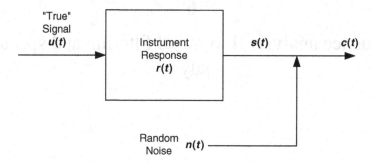

Figure 6.1 A schematic of the sources of signal that eventually become the thing that you measure.

This process can be viewed as in the schematic Fig. 6.1.

Now what we would like to do is to design an *optimal filter* that can be used to extract the original, uncontaminated signal from the mess that we actually saw in $c(t)$. Well, let's take stock of what we "know". We know $r(t)$ (presumably from reading the user's manual or running controlled experiments), and we've measured $c(t)$. If we somehow could guess at $n(t)$, we could solve the problem. It turns out (and we're jumping the gun a bit here) you can use *spectral analysis* to do just that. We will take advantage of the fact that "real" signals that come out of instruments are *bandwidth limited*; that is to say, they don't have much high-frequency power. This may be because of limitations in the instrument's electronics, response time of electrodes, ion transport in solution, etc. On the other hand, *noise* is generally not bandwidth limited, and has significant power at high frequencies. Quite often, the noise power is distributed relatively uniformly across the spectrum (noise that is uniformly distributed across all frequencies is referred to as "white noise"). Thus if you look at the spectrum of your data, which is a plot of the square of the Fourier transforms as a function of frequency, you would see something like Fig. 6.2.

Note that the labels (in Fig. 6.2) are deliberately upper case, since they are the Fourier transforms of the time domain signals. The heavy solid line is what you measure, and the long-dashed line is what you extrapolate from the high-frequency (noise) behavior. What is left is the S signal, the short-dashed line, which you infer by subtracting the extrapolated noise. This seems a very approximate process, but it turns out that the precision of your estimates need not be tremendous to make marked improvements in your determination of $u(t)$. The next step is to construct an *optimal filter* $\Phi(f)$, which in combination with the response function transform $R(f)$ gives you a best estimate of the uncontaminated signal. That is:

$$\hat{U}(f) = \frac{C(f)\Phi(f)}{R(f)} \tag{6.3}$$

We won't use the rigorous derivation for Φ but will argue that since $C(f)$ is the vector sum of $N(f)$ and $S(f)$,

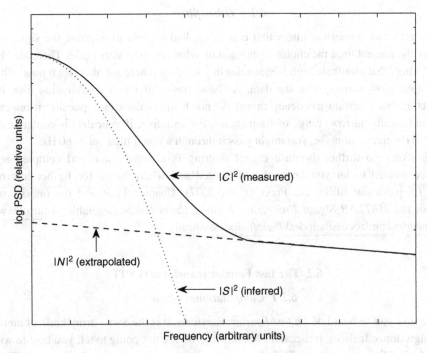

Figure 6.2 A schematic power spectrum of signal (S), noise (N), and what comes out of your instrument (C) vs. an arbitrary frequency axis. Figure redrawn with MATLAB from Press *et al.* (2007), Fig. 13.3.1. The log PSD axis is truncated for clarity of presentation.

$$|C(f)|^2 = |S(f)|^2 + |N(f)|^2 \qquad (6.4)$$

and since $S(f)$ and $N(f)$ are uncorrelated (true noise is uncorrelated with anything), then we argue that getting at $S(f)$ would require dividing $C(f)$ by some sum of $N(f)$ and $S(f)$. This amounts to having:

$$\Phi(f) = \frac{|S(f)|^2}{|S(f)|^2 + |N(f)|^2} \qquad (6.5)$$

where $|X|^2$ represents the modulus (magnitude) of the complex number X squared.[1]

Now this expression for $\Phi(f)$ is a kind of least squares minimization. This means that errors in our estimate of $\Phi(f)$ due to inaccuracies in our initial guesses of $S(f)$ and $N(f)$ are second order. Thus the crudeness whereby we estimated them is forgivable. Refining this estimate iteratively is almost always an activity of greatly diminishing returns, so the first stroke is the best.

---

[1] In complex number theory, the magnitude of a complex number is given by multiplying the complex number by its complex conjugate (e.g. if $X = 3 + 2i$ and $i = \sqrt{-1}$, then its complex conjugate is $X^* = 3 - 2i$, often written with the asterisk to denote the conjugate, and $|X|^2 = 13$).

6.1.2 Other filters

There are other numerical filters that can be applied to data to improve the signal and reduce the noise. Often the choice is dictated by what one considers noise. There are "low pass" filters that eliminate high frequencies in your data. There are also "high pass" filters that reject slow variations in the data. A "band pass" filter is a combination filter that rejects all but a certain frequency range. A "notch filter" does the opposite: it rejects a certain (usually narrow) range of frequencies. For example, if your data is contaminated with 60 Hz electrical noise, you might pass it through a notch filter set to 60 Hz.

We won't go further, because digital filtering is a rather large and complex area, but we wanted to let you know that it existed and where to go for further information. Of particular utility are Press *et al.* (2007), Chapter 13.3; and the tutorial section of the *MATLAB Signal Processing Toolbox Users Guide* (available at http://www. mathworks.com/access/helpdesk/help/helpdesk.shtml).

6.2 The fast Fourier transform (FFT)

6.2.1 Computational efficacy

We're sure you've heard of the *fast Fourier transform*. It is the Swiss army knife of numerical algorithms. It slices, it dices, it even minces. We are not going to tell you how to write the MATLAB code to do an FFT since there already exists a perfectly good MATLAB function `fft`. We will, however, tell you briefly how it is used, and point out why it is such a big deal.

You actually have a prescription for computing the Fourier transform of a function. It involves a convolution requiring of order N^2 multiplications (which mathematicians write as $\mathcal{O}(N^2)$). For a series of 1000 points, this amounts to about a million operations. For really long series, this could become computationally very expensive. The fast Fourier transform (or FFT) was discovered in the 1940s but did not become generally known until the 1960s. Rather than requiring $\mathcal{O}(N^2)$ operations, the FFT requires only $\mathcal{O}(N \log_2 N)$ operations. For a time series of a thousand points, this results in a saving in computer time of approximately a factor of 100. For a series of a million points, the saving is a factor of about 70 000. The latter would correspond to the difference between 4 seconds for your result (*if* that's how long your FFT took) and waiting 3 days! Of course, computers keep getting faster all the time, so this difference is only noticeable, nowadays, for very long series.

6.2.2 Limits and constraints

The one requirement for performing an FFT is that the length of your data array must be an integral power of two. This means that your data array must be of length 2, 4, 8, ..., 512, 1024, 2048, ... 65 536 ... 1 048 576 ... etc. That is, $N = 2^{\ell}$ where ℓ is an integer; we refer to this as "radix 2". This is a result of the underlying nature of the algorithm. You may think this terribly restrictive, and it is, but there are ways out of the problem. You can always pad the end of your data set with zeros (not a bad thing to do, so long as you remember to remove the mean first, i.e. pre-whiten), or throw out those few extra points that extend beyond the nearest boundary (not a fun thing to do). Well, if you have a million points, then a few thousand points here or there won't make a difference anyway. The computational savings may be sufficient to warrant that choice. If you have just a few thousand data points, you may also consider using the plain old, vanilla, slow Fourier transform.

MATLAB implements the FFT in an interesting way. You can invoke it simply with:

```
Y=fft(y);
```

where Y is a vector complex number ($Y \in \mathbb{C}$). Complex numbers are written $a \pm bi$ and are composed of real (a) and imaginary (b) parts. If the length of your data vector y is an integral power of two, then MATLAB uses FFT. If it is not, then MATLAB uses the slow way instead. If you want, you can insist that MATLAB use the FFT algorithm even when the length isn't an integral power of two by invoking the function as:

```
Y=fft(y,n);
```

where n is the integral power of two that is closest in size to your data vector length (N). If n exceeds your vector length, MATLAB pads your data vector y with zeros before doing the FFT. If n is smaller, it truncates y. For example, if your vector is 1000 elements long, you might try:

```
Y=fft(y,1024);
```

since 1024 is the closest radix 2 (2^{10}) number to 1000. The inverse transform is obtained from:

```
y=ifft(Y);
```

where you again have the option of specifying a length different from your input vector. Notice that MATLAB returns the same number of frequency estimates as you provide it with data points (N) or the radix 2 number of frequencies (n) that you forced it to use.

6.3 Power spectral analysis

Chances are that you are reading this chapter because you want to know how to create a power spectrum of a sequence of data that you have collected. And we're going to tell you how to do just that in this section, but first we are going to talk about the information contained in a power spectrum.

When we say *power* we are actually referring to a measure of variance. In the case of a power spectrum (more correctly, *power spectral density*), we are talking about the distribution of variance with respect to frequency contained in the signal analyzed. Forming a power spectrum, then, involves extracting the variance buried in one's data at specific frequencies (wavenumbers) or periods (wavelengths). It is another way of looking at the same information contained in the time domain, but in the frequency domain. The location of peaks and/or the overall shape of the spectrum can yield insights into the nature of the phenomenon you are studying.

Before we charge off into Fourier space, let's take a look at some time series and corresponding power spectra of known functions or signals. The spectrum of a fixed frequency, pure sine wave that goes on forever in the time domain would be a spike at that frequency (not shown); this would be referred to as "monochromatic". Figure 6.3 gives some

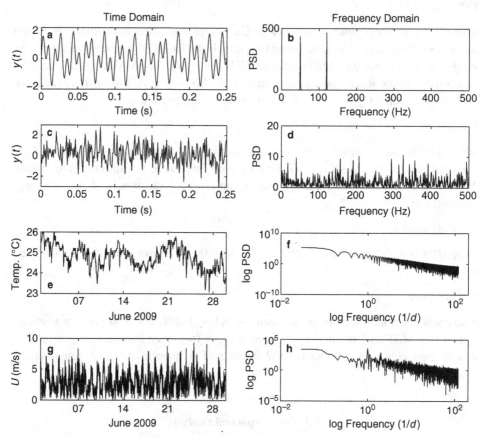

Figure 6.3 Examples of data and their spectra. PSD, power spectral density. (**a**) Two pure sine waves (50 Hz and 120 Hz) and (**b**) the spectrum (note the y-axis is not log transformed so the peaks at 50 Hz and 120 Hz really stand out); (**c**) random noise and (**d**) its spectrum (note there is approximately the same power at all frequencies, an example of a "white" spectrum); (**e**) water temperature data (°C); (**f**) temperature spectrum plotted on log–log axes; (**g**) wind speed data (m s^{-1}); and (**h**) wind speed spectrum also plotted on log–log axes. The water temperature and wind speed data were collected during June 2009 every six minutes by a NOAA National Ocean Service buoy located near Hilo, HI (19.730° N, 155.055° W), and reported via NOAA's National Data Buoy Center (http://www.ndbc.noaa.gov/).

examples of more complicated signals. Two sine waves with different frequencies result in two spikes in the power spectrum. The spectrum of random (Gaussian) noise would be a flat line (with some bumps and wiggles, but none significant). This is referred to as a "white" spectrum. A spectrum dominated by more variance in the low-frequency range is referred to as "red" (such as the water temperature record) and a spectrum dominated by higher-frequency variance is called "blue" (such as the wind speed record). Note the y-axis in Figs. 6.3f and 6.3h: the overall slope for the wind spectrum is flatter, hence it is "bluer" than the water spectrum.

6.3.1 Types of spectra

You need to be careful when you look at power spectra. There are a wide variety of ways used to compute the total power. Three common schema are:

1. The *sum squared amplitude*, given by

$$\sum_{j=0}^{N-1} |y_j|^2 \tag{6.6}$$

2. The *mean squared amplitude*, given by

$$\frac{1}{T} \int_0^T |y(t)|^2 dt \approx \frac{1}{N} \sum_{j=0}^{N-1} |y_j|^2; \quad \text{and} \tag{6.7}$$

3. The *time-integral squared amplitude*, given by

$$\int_0^T |y(t)|^2 dt \approx \Delta \sum_{j=0}^{N-1} |y_j|^2 \tag{6.8}$$

where y_j are the data in the time domain, N is the length of the series, and Δ is the sampling interval. We have pre-whitened (Section 6.3.2) the data so these are estimates of the variance (its mean is zero); check back with Chapter 2 if you're dubious. We use j here because, as John Derbyshire's little mnemonic (Derbyshire, 2003) at the beginning of this chapter reminds us,[2] we are now in the provenance of complex numbers (where $i = \sqrt{-1}$). We prefer to use the mean square amplitude (Equation (6.7)) and unless we state otherwise you should assume that is what we present in this chapter. These definitions are all written in the time domain form; it is more difficult to write similar equations in their frequency domain form. We give one approach below but leave it to the really inquisitive student to search out other definitions in Jenkins and Watts (1968) or Priestley (1981).

However you define the total power, something called *Parseval's theorem* must apply. It states that the total power computed in the time domain must equal the total power computed in the frequency domain:

---

[2] $\mathbb{N}$ natural, $\mathbb{Z}$ integer, $\mathbb{Q}$ rational, $\mathbb{R}$ real, $\mathbb{C}$ complex numbers.

$$\int_{-\infty}^{+\infty} |y(t)|^2 dt = \int_{-\infty}^{+\infty} |Y(f)|^2 df \tag{6.9}$$

This can be used as a check of your calculations. One can always calculate the total power in the time domain (the variance), and the total power you calculate in the frequency domain should equal the total power from the time domain.

More confusing yet are the *periodogram*, power vs. frequency, and *power spectral density* (PSD), power per frequency vs. frequency, which are what you actually plot when you plot a spectrum. The Fourier transform of a single array is symmetric just like the autocorrelation. For any given definition of the *total* power, you could define the periodogram or PSD to range over all positive and negative frequencies between $-f_c$ and $+f_c$ where f_c is the critical or Nyquist frequency, or over only the positive range (up to f_c). We'll tell you what MATLAB does in a moment.

We will learn more about the *Nyquist frequency* in Section 6.4, but for now we'll just assert that you cannot measure a frequency greater than one-half your sampling frequency.

6.3.2 Doing it with FFT

An important step in doing a spectrum, which is sometimes ignored, is to *pre-whiten* the spectrum. This is usually simply the task of removing any long-term trends and the mean from the data. Long-term trends represent more power at low frequencies (hence a "red" spectrum), and removing them "whitens" the spectrum. This can be done by fitting a linear line through your data and then subtracting this linear trend from the data. But quite often merely subtracting the mean of the data from each data point in the series will suffice. Another possibility is that the signal of interest is fundamentally bandwidth limited. For example, if you are looking for seasonal signals in a daily record of temperature, you are not interested in the high-frequency (or blue) end of the spectrum. So you low pass filter or smooth the data. Having done this, you proceed with the next few steps.

We show you this procedure to give you a rough idea of what bookkeeping is involved in the process, and so you will understand "what is going on under the hood". As an example, we will form the periodogram of the tidal data you saw in the previous chapter so that you can develop a feel for the mathematical operations. Feel free to play along with us at home.

First, let's read in the data set (data available at http://www.cambridge.org/glover). See problem 5.1 for a description of each column in woods_hole_tides_2005_05.dat. We'll create a vector of time t from this data in case you want to make a plot similar to Fig. 5.2 and create a time series y, which is the verified tide level from this station. In MATLAB, this is done with:

```
% Read in the tidal data from Woods Hole, MA

load woods_hole_tides_2005_05.dat;   % 7440 data points
TD=woods_hole_tides_2005_05;
```

```
% Create a time vector (data sampled every six minutes)

   t=datenum(TD(:,2),TD(:,3),TD(:,4),TD(:,5),TD(:,6),0);
   delta_t=6/60; % sampling interval in hours

% Put the verified tide observations in the "y-variable"

   y=TD(:,8);
   y=y-mean(y); % pre-whiten the data, y is the time series
```

We now do a $2^{13} = 8192$ point FFT; remember that FFTs prefer radix 2 numbers, and 8192 is the closest radix 2 number to 7440. Of course, we don't have 8192 data points, but MATLAB will pad the time series out with zeros. When MATLAB does the FFT, it will create values of the transform Y that range from $-f_c$ to $+f_c$ with 8192 distinct values. The manner in which the transforms are stored is that the series in Y starts at $f = 0$, goes to f_c at the $(n/2 + 1)$th element, and puts the negative frequencies (highest to lowest) in the latter half of the array. Hence for the 8192 point FFT, $Y(1)$ corresponds to $f = 0$ and is the mean, so $P_{yy}(1)$ should be zero for pre-whitened data. In $Y(2)$ is the coefficient for the lowest possible positive frequency; frequency then increases so that the f_c coefficient is stored in $Y(4097)$. The $-f_c$ coefficient is skipped (it is the same as for $+f_c$), and the coefficient for the next largest negative frequency is stored in $Y(4098)$ decreasing until $Y(8192)$ contains the coefficient for the lowest possible negative frequency. Remember, N is the total number of data points (7440) and n is the radix 2 size FFT (8192) that we asked MATLAB to perform. Next, we define the critical frequency (more about that later) f_c, which is the maximum observable frequency for the series. In our case f_c is $1/2\Delta t$, or $5\,\text{h}^{-1}$. We create a vector containing the frequencies associated with each element of the periodogram. Note that the frequencies are uniformly distributed from zero to the critical frequency.

Continuing on with the analysis, we perform the FFT.

```
n = 8192; % FFT length (radix 2 number!)
m = n/2; % number of distinct frequency bins
fc = 1/(2*deltat); % the critical frequency
f = fc * [0:m]/m ; % the frequency bins (4097).
Y = fft(y,n); % FFT of length 8192
Pyy = Y.*conj(Y) / n; % periodogram = |Y|^2
```

Note that in the sixth line above, we calculated the PSD by multiplying the Y elements by their *complex conjugates* (note the dot that indicates element-by-element or array multiplication). This is because the Fourier transform values are in general complex quantities – that is, they contain both real and imaginary components. The squared amplitude of a complex number is obtained by multiplying itself by its complex conjugate and is always real ($\in \mathbb{R}$). Further, we normalize the PSD by dividing by the series length (the length of the FFT, n) to make the total power the sum squared amplitude as in Equation (6.6), the simplest and most straightforward form.

We now must do a bit of bookkeeping while remembering how the frequencies are stored (see above). First, the real sequence `Pyy` is always symmetric about the zero frequency component. We throw away the upper half of the sequence (remember that m is the number of *distinct* frequencies) and multiply the power at the remaining frequencies by two, *except* for the zero frequency component `Pyy(1)`. This keeps the total power the same, and we can check this (remember Parseval's theorem, Equation (6.9)) simply by comparing the sum of the data squared and the power.

```
Pyy(m+2:n) = [ ] ;          % MATLAB trick for shortening a vector
Pyy(2:m+1) = 2*Pyy(2:m+1);  % compensate for missing negative freq.
[sum(y.^2) sum(Pyy)]        % print the total power side by side

ans =

380.8871   380.8871

loglog(f(2:m+1),Pyy(2:m+1)); % make the spectrum plot Fig. 6.4
```

Note that the expected frequency, corresponding to a period of 12 hours, comes through loud and clear with the largest power in Fig. 6.4. This is a raw spectrum, by which we

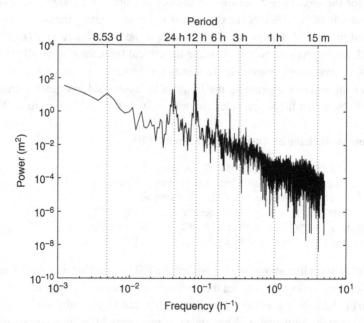

Figure 6.4 The periodogram of the May 2005 tide levels measured in Woods Hole, MA (shown in Fig. 5.2). We have made this plot a little bit more fancy than the commands in the text will produce, but the spectrum should be the same. As expected, the largest peak corresponds to a period of 12 hours. When the values to be displayed range over many decades it is conventional to plot spectra on a logarithmic plot to reveal as much detail as possible.

mean no smoothing or averaging was done to refine intrinsic features, although we will get to that presently. There are some other spikes in this spectrum as well as the 12 hour peak. You can easily devise a way of finding the frequencies and size of these maxima with MATLAB.

```
[Peaks IFreqs] = sort(Pyy,'descend');
```

By default MATLAB sorts in ascending order (smallest first); the second argument instructs it to sort in descending order (largest first). If you specify two outputs for sort, it also returns the indices of the sorted matrix, which you can use to display the frequencies. We now look at the five largest values, and their frequencies, with:

```
format short;
[Peaks(1:5) f(IFreqs(1:5))' 1./f(IFreqs(1:5))'] % top five peaks (power),
                                                 % frequencies (1/h),
ans                                              % and periods (hours)

180.0258    0.0806    12.4121
 35.5571    0.0012   819.2000
 22.9399    0.0415    24.0941
 20.9289    0.0391    25.6000
 17.3054    0.0781    12.8000
```

We straighten up the output here using the `format` command (four decimal places should be plenty). The units for power are meters squared (m^2); frequency, per hour (h^{-1}); and period, hours (h).

The biggest number is the peak in the spectrum and spot on the frequency we would expect from tide level data (if you've lived near the sea, you'll have noticed the tides aren't exactly 12 hours apart). Farther down the list there is another "12 hour" peak; this is very interesting because it is telling us that the 12 hour tide has power at both 12.4121 and 12.8000 hours, 23 mins and 16 s apart. Next, the second largest peak has a period of 819.2 hours (34.1333 days). The data set only has 31 days of data – but don't forget we padded the data with zeros. The amount of power at this frequency can be explained by the fact that subtracting the mean is not enough to pre-whiten the data properly; we leave it as an exercise for the student to remove a linear trend and observe the result. The next biggest peaks are near the frequency corresponding to a 24 hour period and are a factor of two (almost exactly) different from the two 12 hour peaks; they are *harmonics*. The power reported for these peaks is about the same (except the main peak). The areas under the peaks are perhaps more important than the peak magnitudes though, a topic we return to when we discuss the presence of noise in our data (Section 6.4).

Now we want to say a word about the zero frequency power estimation MATLAB appears to provide. The first component of the FFT is the *DC offset*; this, of course, is an electrical engineering term. It represents the power of very long period signals buried in the data. We were diligent and attempted to pre-whiten our data (i.e. remove the mean and render the data approximately stationary) and this value is as close to zero as makes no difference ($Pyy(1) = 1.1313 \times 10^{-29}$). Had we not pre-whitened the data, the "power" at zero frequency would be $Pyy(1) = 1244.7$, 32 orders of magnitude larger. This large number is an estimate of the power we would get at the zero frequency from an infinitely long time series based on our very short subsample. The mean we removed is a small number ($mean(y) = 0.4292$), but it makes a big difference because it approximately zeros out the power contribution from any long-term trend. This is why we don't attempt to plot the zero frequency power in the spectrum: we're supposed to have gotten rid of it (i.e. it is supposed to be near zero); besides, on a logarithmic scale the zero frequency power (large or small) would tend to distort the plot.

6.3.3 *Frequency binning and spectral error*

Well, now that you've done it the hard (and hopefully instructive) way, we'll show you an easier way. But first, we need to ask a general question about how well you know the PSD. Remember the basic philosophy; it is not enough to determine a quantity, you must also determine the uncertainty with which you know that quantity. Specifically, what is the variance in the PSD estimate as N (the length of the time series) goes to infinity? The answer seems terribly unfair: the variance in PSD remains the same, independent of the length of the time series, and it is 100% of the value. The reason for this is actually quite straightforward. Using the FFT technique described above, if we double the length of the time series, we halve the width of the corresponding frequency bins, and hence we double the number of frequency bins. All our efforts go into refining the frequency resolution. In other words, in the straightforward application of FFT, the number of estimated frequencies is the same as the number of data points ($n \approx N$), and you have used up all of your degrees of freedom.

We don't (and shouldn't) have to settle for 100% uncertainty in our PSD estimates. What we can do is average frequency bins to gain precision at the expense of resolution. This averaging has the effect of reducing the variance to one over the square root of the number of bins involved in the averaging (this assumes errors are uncorrelated), so that averaging 100 frequency bins reduces the error to 10%. An equivalent procedure (and one generally preferred for logistical reasons on longer data sets) is to break the time series up into M non-overlapping[3] segments of m distinct frequencies, computing the spectra for each of those segments, and averaging them together. This is exactly the same thing as doing one big spectrum and averaging down to m frequency bins with M frequencies in each bin. One more thing. When binning you get fewer distinct frequencies than the original FFT, of course, but this affects the lowest frequencies preferentially (i.e. there are fewer of them).

To make a better estimate of the power spectra with confidence limits (95%), we read in our May 2005 data file of tides, as before, and then do the following:

---

[3] Overlapping is allowed, and it is done all of the time to smooth the spectra further, but the reduction in the error is not as great. The variance reduction is itself reduced by the number of overlapping bins.

```
Fs=10;                                  % sampling frequency (10/h)
Hspec=spectrum.welch('Hann',256,0);     % the Welch spectrum object
Hpsd=psd(Hspec,y,'Fs',Fs,'ConfLevel',0.95); % the PSD object
Pyy=Hpsd.Data;                          % extract the PSD
Freq=Hpsd.Frequencies;                  % extract the frequencies
CI=Hpsd.ConfInterval;                   % extract the confidence intervals
loglog(Freq,Pyy);                       % make a log-log plot
hold;
loglog(Freq,CI(:,1),'--');              % lower 95% confidence level
loglog(Freq,CI(:,2),'--');              % upper 95% confidence level
```

The first statement establishes the sampling frequency in the units you want to use, in our case h^{-1}. The second statement creates the Welch spectral object using a Hann window (more about windowing in Section 6.4), divides the time-based sample array into segments 256 samples long, and sets the overlap between windows to zero. This statement controls the number of distinct frequency estimates ($m = 256/2 + 1$). If the original data array length (N) is not long enough for an integer number of 256 sample segments, the excess data are truncated (in this case $N = 7440$ is truncated to 7424, losing only 16 data points). Computing the spectrum in this fashion breaks the original series into segments of 256 elements each (because it can use FFT, there is an advantage to using sequences of radix 2 length). This produces multiple spectra with 29 estimates ($M = 7424/256 = 29$) at each frequency. Averaging these spectra improves the uncertainty in each estimate to a factor of $1/\sqrt{29}$. The third statement operates on the spectral object performing the mathematical operations necessary to form the PSD from the data in y and stores it in an object (Hpsd). In order to make some plots, we need to extract the PSD, frequencies, and confidence intervals from the Hpsd object.[4] Then we make a log–log plot as in Fig. 6.5. We do this for three reasons, first to expand the frequency axis at the lower end of the spectrum, second to reveal any power-law relationships in the variable analyzed (a straight line showing a $P_{yy} \propto f^{-4}$ relationship was added as an example to Fig. 6.5), and finally to create a plot with constant width confidence intervals. Note that psd is smart enough to know that even though you specified the sampling frequency (i.e. 1/sample_interval), the ultimate limit to frequency resolution is the *Nyquist frequency* (see Section 6.4), which is half the sampling frequency ($5\,h^{-1}$).

The above knot of code could have been simplified by merely issuing the third line of code without the PSD object as output, such as:

```
psd(Hspec,y,'Fs',Fs,'ConfLevel',0.95);
```

Then psd makes a plot similar to the kind of plot one gets using the semilogy command. If we had done this we could have skipped the last seven lines of code. But the plot would

---

[4] If you are not familiar with object oriented programming, just keep in mind that objects can contain several data variables and they can be accessed by combining the variable name with the object name separated by a period. In this case, the dot now separates the frequencies, confidence intervals, etc. from the object in which it is contained.

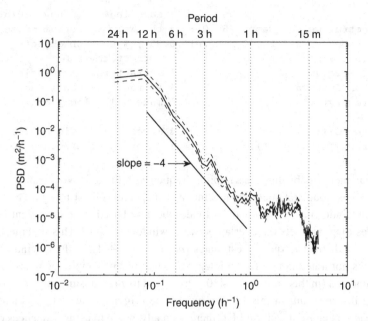

Figure 6.5 The power spectral density of the May 2005 tide levels measured at Woods Hole, MA, plotted by `loglog` from our example in the text. The solid line is the PSD and the dashed lines above it and below it bracket the ±95% confidence interval now averaged into 128 frequency bins (zero frequency is not plotted).

be a semilogarithmic y-axis plot, and the y-axis would be presented in units of decibels (dB, defined as $10 \log_{10}(P)$) ... which is great if you are an electrical engineer, but not so useful for most ocean science problems.

Figure 6.5 shows a solid line for the estimated PSD, and dashed lines above and below delimit the 95% confidence interval for the PSD estimate. The PSD is now given to you in units of power per frequency unit, which is why the units are given now as m^2/h^{-1}. Note earlier (Fig. 6.4) we were displaying power in periodogram form, where the power in each frequency bin is the total power in that bin. Now the power is further normalized on a per frequency basis, the power spectral density, and defined over the frequencies 0 to $+f_c$. Also note the trade-off between precision and resolution that is evident in this plot; the peaks, although smoother, are now wider and less distinct than the straight FFT-based estimate we did earlier. We have an exercise for the reader: try generating the time series sequence as we have done above using `spectrum.welch` with a variety of segment lengths, say, 32, 64, 128, 256, and 512. Observe what happens to both the width of the spectral peaks and the confidence intervals. Be warned that the y-axis will change from plot to plot. You can look at the numbers themselves too, since the `Hpsd` object that the MATLAB `psd` routine returns has both the PSD (as `Hpsd.data`) and the confidence range (as `Hpsd.ConfInterval`) within.

There are other spectral estimate functions provided by MATLAB. These, however, do not operate with (or on) spectral objects at the time of this writing. The first one forms the *cross power spectral density* of two time series (*x* and *y*):

```
[Pxy,F] = cpsd(x,y,window(n),noverlap,nfft,fs);
```

where Pxy is the cross power spectral density between the two time series, F is the frequency vector determined from fs, window(n) is the name of the window function and its width in samples (e.g. hann(128)), noverlap controls the number of samples by which the segments of data (size determined by window) are overlapped, nfft controls the number of FFT terms to compute, and fs is the sampling frequency. If you wish to use the default value and skip over one of the arguments (except data) just put an empty array in place of the argument (i.e. []). The other spectral estimate functions follow a similar format, *transfer functions*:

```
[Txy,F] = tfestimate(x,y,window(n),noverlap,nfft,fs);
```

where Txy is the frequency transfer function between the input x and output y of a model or system and the remaining arguments are as above. Finally the *magnitude squared coherence estimate* is given by:

```
[Cxy,F] = mscohere(x,y,window(n),noverlap,nfft,fs);
```

where Cxy is the magnitude squared coherence estimate between two signals x and y (a kind of covariance between two time series in frequency space). If you wish to use these functions we suggest you read the MATLAB documentation and see what Emery and Thompson (1998) have to say about their application.

We have one more question to settle when evaluating a spectrum: which peaks are the important ones? The largest (tallest) ones, right? Well, not necessarily, since the power spectra are plots of the distribution of the variance in the data when projected into frequency space. And since the actual points we calculate are really a frequency bin (width Δf) average, it makes sense to look at the total amount of variance each peak represents by integrating the area underneath each peak. So which plot is the most faithful to this purpose? In Fig. 6.6 we show the four most common ways of displaying a power spectrum. If we want to show in a glance the amount of variance beneath each peak, we need something where the expression:

$$\sigma^2 \left(f \pm \frac{\Delta f}{2} \right) = \int_{f-\frac{\Delta f}{2}}^{f+\frac{\Delta f}{2}} P_{yy}(f) df \qquad (6.10)$$

is true. This equation is nothing more than the equation one would write for integrating the area beneath the curve in Fig. 6.6a, and this kind of plot is known as a variance-conserving

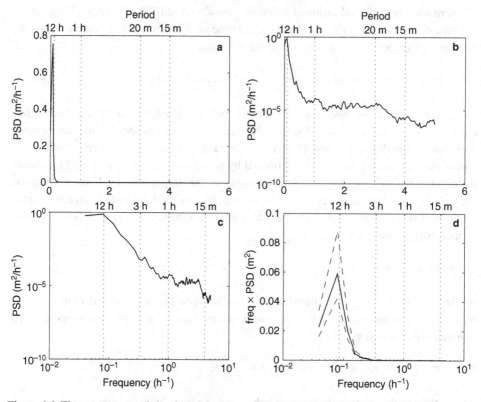

Figure 6.6 The power spectral density of the May 2005 tide levels measured at Woods Hole, MA, displayed four different ways. (a) The spectrum generated with the code shown in Section 6.3.3 appears on a linear–linear plot. The main peak is barely visible. (b) The same spectrum on a log–linear plot: the frequencies are distributed equally and some of the detail of the higher-frequency peaks are now apparent. (c) The same spectrum on a log–log plot similar to Fig. 6.5. Any power law dependencies are apparent. (d) The spectrum plotted as frequency × PSD vs. log(frequency) to display the variance-conserving form, i.e. equal areas beneath this plot map out equal amounts of variance.

plot of the power spectrum (Chatfield, 2004). But a linear–linear plot does not provide the eye with much detail as the larger peaks will effectively suppress all of the other peaks, important or not. Figures 6.6b and 6.6c cannot be integrated in an "equal area represents equal variance" fashion. What we need is a way of presenting spectra that is consistent with Equation (6.10) but also detail-rich as in Fig. 6.6c. This is accomplished by plotting the frequency times the PSD (on a linear scale) vs. the logarithm of the frequency, as in Fig. 6.6d. The integral of this plot is:

$$\sigma^2\left(f \pm \frac{\Delta f}{2}\right) = \int_{f-\frac{\Delta f}{2}}^{f+\frac{\Delta f}{2}} f P_{yy}(f) d[\log(f)] = \int_{f-\frac{\Delta f}{2}}^{f+\frac{\Delta f}{2}} P_{yy}(f) df \qquad (6.11)$$

since $d[log(f)] \equiv df/f$. Figure 6.6d is also a variance-conserving form for plotting power spectra (Emery and Thompson, 1998) and shows very clearly that the 12 hour peak is the most important (contains an overwhelming amount of the total variance) peak in the spectra. It is also the only subplot in which the confidence limits scale well enough to be seen (in Figs. 6.6a, b, and c the confidence limits plot so close to the PSD curve that they cannot be seen in plots this small). The peaks seen at $\sim$20 minutes and $\sim$2.5 hours in Figs. 6.6b and c, respectively, clearly contain very little of the overall variance and cannot be considered important in terms of the overall variability in the data.

6.4 Nyquist limits and data windowing

We've hinted that not all things are rosy in the Fourier garden. Don't blame Monsieur Fourier, because what you are about to learn about the limits of spectral techniques is a direct reflection of fundamental limitations in measurements. They are important to know because they impact how you design your experiments, or barring that measure of control, exactly what you can and cannot extract from a data set. Just don't shoot the messengers . . .

6.4.1 The Nyquist theorem and frequency folding or aliasing

Remember the "*critical frequency*" that we used in the previous section? We made the assertion that you cannot measure a frequency greater than half of your sampling frequency. This really amounts to the idea that you need to capture two points (at a minimum) in each cycle of a sine wave to identify it properly, one during its negative swing and one during its positive swing. Thus it makes no sense to talk about PSD outside the *Nyquist range*: that is, at frequencies greater than the Nyquist frequency. We will say it again, because it is so important: for any sampling interval (Δ) the Nyquist range is:

$$\frac{1}{n\Delta} \leq f \leq \frac{1}{2\Delta} \tag{6.12}$$

Never ever forget it! The lower limit of Equation (6.12) is more of a practical limit (it's not zero because that would imply an infinitely long time series; remember, you do want to graduate). If we have properly pre-whitened our data, then it makes no sense to estimate the zero frequency component anyway. Therefore, the lowest frequency you can get is controlled by the FFT length (n), which is influenced by the number of samples you start with (N), the amount of zero padding, and the size of the sampling interval (Δ) shown in Equation (6.12) for a one-sided spectrum. This explains why we were able to estimate periods greater than 31 days in Section 6.3.2; we padded our time series ($N = 7440$) out to the nearest integer power of two ($8192 \times 0.1 = 819.2\,h = 34.1\,d$). If you use frequency binning to reduce the uncertainty in your spectral estimates, then you increase the lowest frequency you can estimate because you have reduced the number of distinct frequency bins estimated ($(n/2)/M$ for no overlap) and the size of the sampling interval (Δ) remains the same.

That, unfortunately, is only half the story. This is not a question of "what you can't see can't hurt you." In fact, "what you can't see can hurt you!" Suppose for the moment that you are sampling something and that there is a high-frequency signal hitting your detector (measurement). Figure 6.7 shows this high-frequency signal (only because we made it; remember that you can't see it!). Now we superimpose on this signal the actual sampling interval that you have implemented, (the black dots on Figure 6.7a), which is too long to resolve the higher-frequency sine wave.

In the lower plot of Fig. 6.7, we show the actual values you measured (the black dots) and what you would infer to be the signal from your observations (imagine removing the thin gray sine wave and what you see is the thick line only). This looks like a much lower-frequency signal than what is really present. This phenomenon is known as *aliasing*. What happened here? Well, each sample you took on the fast sine wave was at a different part of the cycle. Since you could not sample at a rate that was an exact multiple of its (unknown to you) period, each time you took a sample it was systematically further along in the cycle, and you eventually "lapped yourself" repeatedly in a sinusoidal fashion.

Another way of looking at this is that the spectral energy above the Nyquist frequency has been reflected or folded down into the Nyquist range. Figure 6.8a shows a hypothetical "true" power spectrum. Below is the equivalent power spectrum when the function is sampled at discrete intervals that are characterized by a Nyquist frequency determined

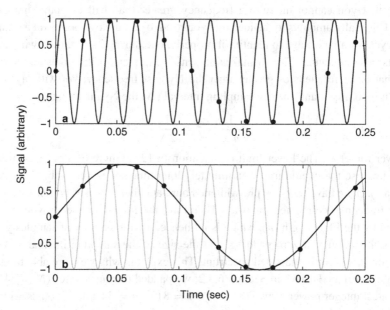

Figure 6.7 An example of aliasing. In (a) we show a high-frequency signal that is "hitting" your instrument (the thin line) and superimposed upon this we show when you actually sampled the signal (the black dots). In (b) we show the same high-frequency signal (thin gray line) with the perceived, lower-frequency, signal superimposed (the thick line).

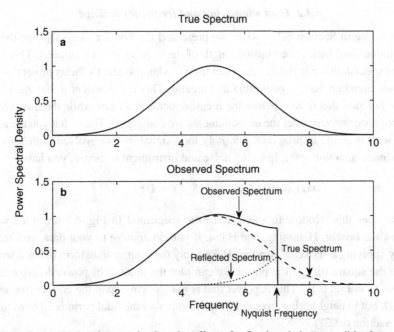

Figure 6.8 Aliasing higher frequencies has the effect of reflecting their "power" back onto lower frequencies (i.e. from above the Nyquist frequency). In (a) we see the actual true spectrum, but in (b) we see the effect of aliasing due to our actual sampling frequency. The dashed line is the true spectrum and the "reflected" portion is represented by the dotted line; their sum results in the solid line. The observed spectrum has more power at the higher frequencies just below the Nyquist frequency than it should.

by the sampling interval (Fig. 6.8b). The effect is to "reflect" the PSD above the Nyquist frequency down into the Nyquist range, inflating the PSD estimates greater than the true value. In Fig. 6.8b, note how the "real" spectrum (the dashed line) is reflected back (the dotted line) from the Nyquist limit (the vertical line). The observed spectrum is the sum of the "real" and the "reflected" spectra.

Now what can you do about this? Not a whole lot, it turns out. Aliasing is a *fundamental limitation*, not an artifact of the mathematics. It is a fact of life. But there are two things that you can do to minimize the impact. Since most signals in nature are fundamentally bandwidth limited, meaning you are generally not interested in the frequencies you have no hope of resolving, it pays to design your equipment to *analog filter* or smooth your signal over the higher frequencies before you sample. This has the desirable effect of blocking out those frequencies that not only cannot be measured, but which will alias and deceive you. Some devices inherently integrate and thus effectively analog filter for you (e.g. sediment traps) with no need to build in fancy electronics. The second thing you can do is to sample at a rate faster than any significant variability that may trouble you. This may work in some cases since most geophysical spectra tend to be "red", i.e. rich in low frequencies (Hasselmann, 1976).

6.4.2 Data windowing and frequency leakage

Did you notice in Section 6.3.1, when we presented the Fourier integrals, that the limits of integration (and hence the implied length of the time series) are *infinite*? This is not a realistic expectation for real data sets, even for a graduate student's thesis project! All time series measurements have a beginning and an end. This is equivalent to having a kind of *response function* that is zero before the measurements start, one while the measurements are in progress, and zero after the measurements are completed. Thus what you are actually dealing with is some, in principle, infinitely long signal $u(t)$ convolved with some finite length *window function* $w(t)$. Ignoring noise and instrument response, you have:

$$s(t) = u(t) * w(t) \Leftrightarrow S(f) = U(f) \cdot W(f) \tag{6.13}$$

But what does this window function do to the spectrum? In Fig. 6.9 are three window functions, the boxcar, Hamming, and Hann. If you do nothing to your data, you are inadvertently applying a "boxcar" window. If we apply the Fourier transform to these windows and take the square root of the power we can plot the amount of power *leakage* into the side-lobes as in Fig. 6.10. This is power that is not showing up at the correct frequency in your PSD, but in neighboring frequency bins (remember the tidal periods 23 mins and 16 s apart in Section 6.3.2?).

This should look familiar. Remember your first year physics lab, when you passed monochromatic light through a narrow slit, and got a diffraction pattern? This is the same thing. The "slit" is the window and the pattern is the "smattering" of power into adjacent

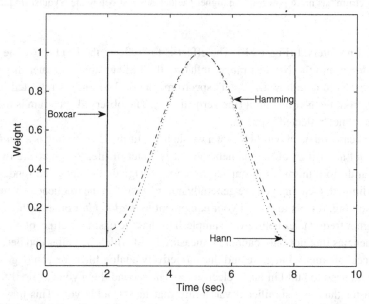

Figure 6.9 Three window functions in time space. The boxcar window is the implicit sampling window one gets by just collecting data without applying any windowing.

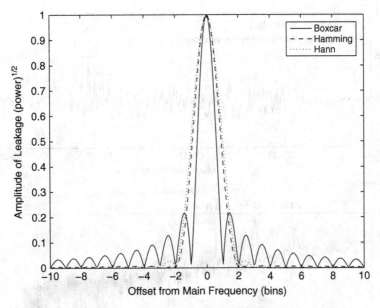

Figure 6.10 The power leakage of the three windows shown in Fig. 6.9 into the neighboring frequency bins. The boxcar window (the solid line) has the largest side-lobes (leakage).

frequency bins. What happens is that the finite length of the time series, coupled with the abrupt start and stop of the data, causes a leakage of power from the original, main frequency bin into adjacent ones. These little side-lobes of power leakage can be a real problem in resolving spectral peaks.

The amount of leakage has to do with the rapidity with which you open your data window and close it. The "default" window function that you use when you don't treat your data is the "boxcar" window that we described above. It is perhaps the worst window you could use. So it behoves you not only to pre-whiten your data, but also to use some other window. We use the tidal elevation data in Fig. 6.11 as an example. We show the data both unwindowed and windowed, and the resultant spectra. By multiplying the data by a window of the same size we dampen the beginning and end of the time series to zero. Yes, it looks as though we are recommending that you throw out hard earned data at the start and end of your time series, but the alternative is worse. Notice that the spectral peaks are much better defined in the results from the windowed data.

There are a variety of other windows, which have only slightly varying degrees of improvement. Their names, which you can drop at coffee hours or cocktail parties, include Bartlett, Blackman, Chebyshev, Gauss, Hanning, Kaiser, Taylor, Triangular, and Tukey. You probably don't really care which you use, but if you want to know more, look in Chapter 13 of Press *et al.* (2007). The Welch spectral object formed in Section 6.3.3 uses the Hann window by default.

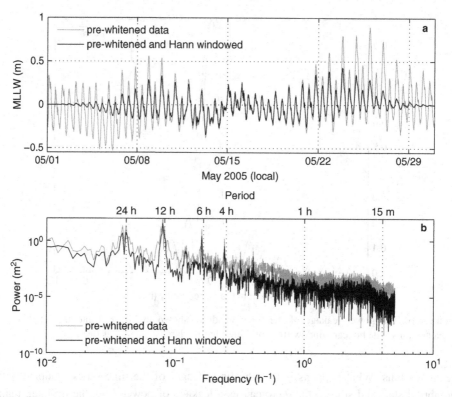

Figure 6.11 A comparison of unwindowed (boxcar) and windowed (Hann) May 2005 Woods Hole tidal data and spectra. (a) The time series of the tide elevation data, unwindowed (gray) and windowed (black). (b) The power spectral density plots of the data unwindowed and windowed. These spectra were created using the `fft` command as in Section 6.3.2.

6.5 Non-uniform time series

Everything we've done so far requires evenly spaced data. There is no fundamental reason why this must be so, just mathematical and computational convenience. The Fourier integrals presented in Section 5.6.2 may be taken over any arbitrary distribution of points in time, provided that the appropriate numerical integration scheme is applied and attention is paid to the real limits of accuracy of such an integration (especially related to the Nyquist limits). The latter becomes a serious, non-trivial matter because the Nyquist limit becomes a rather diffuse issue when points are *unevenly spaced* in time. Do you take the total time span of measurements divided by the number of points (i.e. the average sample interval) as the measure of Δ? Or do you take the closest (or farthest) spacing? We'll dwell on that a little later, but first we need to get down to pragmatic issues; how do we do the calculation if spacing is uneven? FFT seems definitely out, so what do we do now?

Some crude methods can be used. First, you may interpolate your data onto a regular grid. This is not too bad if the data are more-or-less evenly spaced and need to be tweaked into submission. If the data are more irregularly spaced, you may be destroying information by averaging over closely spaced data points. For example if some data points are very close together, you have some kind of high-frequency information (the pair have a much higher Nyquist frequency than the average of the series), and you throw that information out by gridding onto a coarser grid. Worse yet, you may create information by interpolating to a finer grid in some areas where your data are sparse. Depending on your interpolation scheme (e.g. splines) you may introduce all sorts of bizarre frequencies that have an unrealistic spectral signature.

Another situation arises when in fact you have a regular series, but some of the data are missing. This so-called missing data problem is quite common, and methods to deal with it range from simply putting zeros in where the pre-whitened data are missing (which is not as bad as it sounds), to clamping the missing points to actually measured values on either side of the hole. Interpolating the data segment can also be done. This may be forgivable if you are missing a very small percentage of your data, and the gaps cover one or perhaps two data points in length. Problems arise when the gaps become large or a significant portion of the record. Any patching technique introduces spurious frequencies, generally at the low-frequency end, just where you probably don't want them.

The more defendable approach is to use *Lomb's method*, which involves least squares fitting the time domain data points to sine/cosine series.

6.5.1 Lomb's method

Lomb (1976) devised a scheme of performing a general linear least squares regression of unevenly spaced data to a sine/cosine series of different frequencies. This is similar to the approach we used to solve problem 3.2. With Lomb's method you decide what frequencies to fit (usually a range of evenly spaced frequency values), build the normal equations, and solve them. That turns out to be a little tedious, so Lomb presented a simplification that permits you to solve for each coefficient independently. You first define the mean and variance, using the usual formulae:

$$\bar{y} \equiv \frac{1}{N} \sum_{i=1}^{N} y_i \tag{6.14}$$

$$\sigma^2 \equiv \frac{1}{N-1} \sum_{i=1}^{N} (y_i - \bar{y})^2 \tag{6.15}$$

Then you generate the Lomb normalized periodogram (P_N) for the ith angular frequency ($\omega_i = 2\pi f_i$) using:[5]

---

[5] There are no Fourier transforms in this method, consequently no complex numbers, and we are back to using i as an index.

$$P_{\mathcal{N}}(\omega_i) = \frac{1}{2\sigma^2} \left\{ \frac{\left[\sum_j (y_j - \bar{y}) \cos \omega_i (t_j - \tau_i)\right]^2}{\sum_j \cos^2 \omega_i (t_j - \tau_i)} + \frac{\left[\sum_j (y_j - \bar{y}) \sin \omega_i (t_j - \tau_i)\right]^2}{\sum_j \sin^2 \omega_i (t_j - \tau_i)} \right\}$$

$$(6.16)$$

where the sums over j are from 1 to N (the series length), $P_{\mathcal{N}}$ is the variance normalized power spectral density, and the constant τ_i is defined for a given angular frequency by:

$$\tau_i \equiv \frac{1}{2\omega_i} \arctan\left[\frac{\sum_j \sin 2\omega_i t_j}{\sum_j \cos 2\omega_i t_j}\right] \qquad (6.17)$$

This seems a rather tedious formulation but is easily coded up in MATLAB, which we give you as `lomb.m` for your computing pleasure. Now because we normalize Equation (6.16) by dividing it by σ^2, it turns out that the probability that the null hypothesis is true (i.e. that there is no difference between $P_{\mathcal{N}}(\omega_i)$ and one derived from a random sequence) is given by:

$$P(> z) = 1 - (1 - e^{-z})^M \qquad (6.18)$$

where z is the observed value ($P_{\mathcal{N}}(\omega_i)$), and M is the number of independent frequencies determined by the series. For an evenly spaced time series with unbinned statistics, the number of independent frequencies is N, the number of data points in the sequence (remember no Fourier transforms, no radix 2, and Lomb's is always one-sided). For *unevenly spaced* data, the answer is not so simple. But if the data are not particularly clumped, then $M \approx N$ is not a bad approximation. If the data are clumped in groups of n (e.g. groups of two), then the number of independent frequencies is approximately N/n. Knowing M very precisely is not really that important, for if you are thinking in terms of significance levels in the range of 0.1 to 0.05 (for the null hypothesis), a factor of five apart, then estimates differing in probability by a factor of two are not really of concern: would you feel significantly better about a 0.1% (99.9% confidence interval) than a 0.2% (99.8% confidence interval)?

The Lomb method has some nice features (aside from working!). First, it weights the spectral fit appropriately by the actual data points ($y_j - \bar{y}$). This is the natural character of least squares regressions. Second, it takes advantage of the fact that some of your data are clustered more tightly than the average sampling frequency ($1/\Delta$), where Δ is the average sample spacing obtained by dividing the total time span of the measurements by the number of points in the series. That is, if some of the points are separated by as little as δ, some small fraction of the average spacing, then you have made a partial determination of the power at this much higher frequency, i.e. $1/2\delta$, which is greater than the nominal Nyquist frequency ($1/2\Delta$).

As an example of *Lomb's method*, Fig. 6.12 displays a synthetic data set and resulting Lomb periodogram for integer frequencies from 1 to 200 Hz. The file that created this figure is `Lombtest.m` which generates a random series of 100 points between $t = 0$ and 1, computes some artificial data with noise with two dominant frequencies (50 and 130 Hz), performs interval analysis, and plots Lomb's periodogram and probability random. A uniform time series of 100 points in 1 second would have a sampling interval of 0.01 seconds, and hence a Nyquist frequency of 50 Hz. Because the data are scattered, there are enough points close enough together to pick out the higher frequencies. You can see this in Fig. 6.12b, which is a histogram of the $\log_{10}$(separation) of the points in the series. Although the bulk of the points are separated by a distance of 0.01 ($\log_{10}(\Delta t) = -2$), there are a number of points with finer spacing (to the left of -2). Note that both frequencies are picked out and that the lower of the two frequencies is exactly at the average Nyquist frequency, and the higher of the two frequencies (130 Hz) is above the average Nyquist frequency! It is the data spacings to the left of -2 that give you the high-frequency distribution. Furthermore, that is why there is less frequency folding (or aliasing) with this technique because you have some information at the higher frequencies.

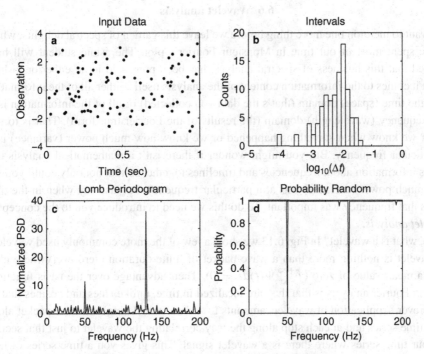

Figure 6.12 A demonstration of Lomb's method on unevenly spaced data in arbitrary units, as presented in (a). (b) A histogram of the $\log_{10}$ transform of the sampling intervals for these data. Lomb's method provides (c) the normalized PSD plot showing two significant peaks, the highest-frequency peak is above the average Nyquist frequency of this data. Note that the normalized PSD has no units and is not units^2. In (d) the probability that peaks this large would arise because of random noise is shown.

The test algorithm also returns the frequencies, values, and probabilities of the most probable peaks:

```
Frequency      Pn         Prob
----------------------------------
130.0000    35.9898     0.0000
 50.0000    11.3331     0.0024
118.0000     4.1685     0.9565
192.0000     4.0989     0.9654
 12.0000     3.5221     0.9976
```

Remember that `Prob` is the probability that there is no difference between your data at this frequency and random data.

The above points to an important tactic in experiment design. Although it is perhaps simpler to design your sampling around a regular grid, you may learn more by having an irregular sampling strategy. This applies to a grid in space as well as time.

6.6 Wavelet analysis

We want to mention one more thing before we leave this garden of spectral delights, where we've spent most of our time in Monsieur Fourier's plot. The astute student will have noticed that this business of spectral analysis has been presented in an either/or fashion when it comes to the information content of the analysis itself. Either all of the information is in the time (space) domain (that's the data you collected) or all of the information is in the frequency (wavelength) domain (the results of the Fourier transform). That is to say, either we know when something happened or we know how much power (variance) is at a particular frequency. But you might wonder if there isn't a mathematical analysis that yields information about frequencies and timelines together so that not only could you say how much power (variance) was at a particular frequency, but also say when in the time series that frequency was important. To do this we need to introduce you to the concept of *wavelet analysis*.

So, what is a wavelet? In Fig. 6.13 we show a few of the more commonly used wavelets. A wavelet is nothing more than a wave packet of finite duration (zero everywhere else) with a mean value of zero ($\int_{-\infty}^{+\infty} \Psi(t)dt = 0$). Their advantage over the basis functions used in Fourier analysis is that they are localized in time, unlike sines and cosines that go on forever. Application of wavelets amounts to sliding (translating) the wave packet along your time series and at each stop along the way correlating the wavelet to just that section of your time series where there is a wavelet signal. This gives you a time series of how well a wavelet of that period correlates at each stop. Now imagine that you could stretch or compress (scale) the width of the wavelet, thereby changing its inherent period, and do this analysis all over again. Now you have a new time series of how well the wavelet (with a longer or shorter period) correlates with your time series. Repeat this several times, stack these time series on top of each other in decreasing or increasing size of the inherent

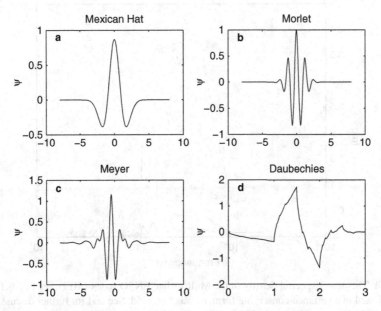

Figure 6.13 Four examples of wavelets: (a) Mexican Hat, (b) Morlet, (c) Meyer, and (d) Daubechies (pronounced "Dobe-eh-shee").

wavelet period, and you can see not only what periods (frequencies) in your time series have large power (variance) but also where in time those peaks occurred.

In order to help you appreciate wavelets (and provide a standard signal for comparison) let's look at a Fourier and a wavelet analysis of the Multivariate ENSO Index (MEI), which we first introduced in Fig. 5.7. Figure 6.14 was created using MATLAB code similar to the code used to create Fig. 6.6d. The interpretation of this figure is fairly straightforward. Most noticeable is the broad peak covering periodicities between 2 and 8 years. The higher-frequency peaks fall off following approximately a $f^{-2.5}$ power law (not shown) and are clearly not likely to be significant. The 95% confidence intervals are quite large; remember this is a variance-conserving plot. Because of these features of the power spectra overall and the broadness of the main peak, one cannot say for certain how often an El Niño will occur or how pronounced it will be when it does occur.

Can wavelet analysis tell us more? Figure 6.15 displays the MEI signal (Fig. 6.15a) and some views of the resulting wavelet analysis. Figure 6.15b shows the one-dimensional signal wavelet power spectrum (wavelets can be applied to two-dimensional, or higher, signals, but that is beyond the scope of this book). Here the wavelet power of the time series is shown as a gray-scale image ranging from approximately 10^{-5} (white) to 40 (black). Like the Fourier power spectrum, this power is a measure of the variance of the MEI. Also shown on this plot are dashed lines indicating where the power was large enough to be statistically significant at the 95% confidence level. This 95% confidence level is repeated in Figs. 6.15c and d. Furthermore, Fig. 6.15b has a thin black line indicating the

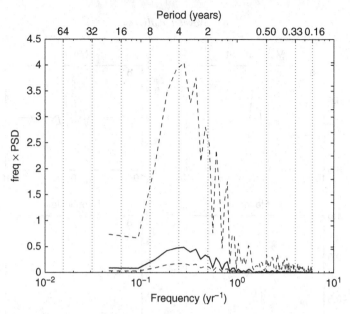

Figure 6.14 The power spectral density of the Multivariate ENSO Index (MEI), see Fig. 6.15a. This figure is plotted in a variance-conserving form, as was Fig. 6.6d. See text for further discussion.

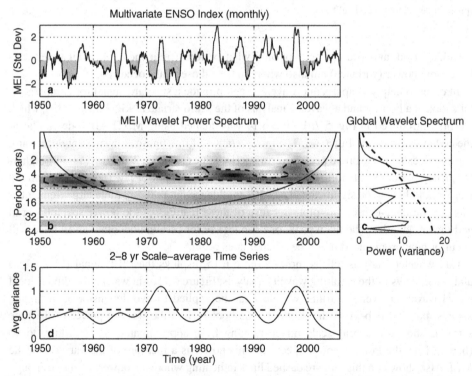

Figure 6.15 Wavelet analysis of the MEI time series generated with MATLAB code obtained from Christopher Torrence and Gilbert Compo at http://paos.colorado.edu/research/wavelets. See text for discussion of each subplot.

cone of influence. Below this line, *edge effects* (wavelets stretched so much they bump into the ends of the time series) become important enough to discount the power reported. Figure 6.15c shows a periodogram of averaged power created from the wavelet power spectrum by averaging sideways across Fig. 6.15b. Figure 6.15d shows a time series of the wavelet power created by averaging from top to bottom the periods between 2 and 8 years. In this subplot we see that there was significant variance (power) in the early 1970s, throughout most of the 1980s, and in the late 1990s corresponding to years when there were strong El Niño/La Nina events.

So, is wavelet analysis better than Fourier? Not necessarily. The amount of information in the time series was not changed by using a different analysis technique. Wavelet analysis is another way of displaying that information that allows one to observe variance in a time-frequency fashion. There is a lot more to wavelet analysis and we recommend that you read Torrence and Compo (1998), Strang and Nguyen (1997), or Mallat (1999) if you wish to learn more.

6.7 Problems

6.1. Download the data file `freqanal.dat`, which is a time series which runs from $t = 0$ to 10 seconds, in steps of 0.005 seconds. It contains three frequency components and some noise.

(a) Using the spectral objects we showed you in Section 6.3.3, do a power spectrum analysis, identify the three frequencies, and the relative power associated with each. Try different binning numbers (64, 128, 256) and identify what is happening. Hint: you can do this by sorting the output of the power spectral density object (`Hpsd`) in decreasing order, and taking the first several elements. For an example of using `sort` see Section 6.3.2.

(b) Now add to the data an additional frequency component of frequency 110 Hz. You do this with something like:

```
t=0:.005:10;
y=freqanal';
f=110;
y=y + 3*sin(2*pi*f*t);
```

and do the spectrum. Where does the new peak appear? Repeat the step again with $f = 160$; Where does this new peak appear? Now do it with $f = 146$. Explain your results.

6.2. Download the data files `tu.dat` and `fu.dat`. These are unevenly spaced samples in the time domain. Plot a histogram of the logarithm (base 10) of the sampling intervals. What is the average Nyquist frequency for the data set? Do a spectrum using Lomb's method by downloading the file `lomb.m` and identify the dominant frequencies (to the nearest hertz), and the probability of them being real (i.e. not random noise).

Hint: your maximum frequency is going to be some reasonable multiple (say a factor of 2–3) of your average Nyquist frequency. How do the dominant frequencies you identify from the spectral analysis compare to the average Nyquist frequency?

6.3. Download the tidal data `woods_hole_tides_2005_05.dat`. Demonstrate the low-frequency influence of unremoved linear trends in the data by repeating the exercise in Section 6.3.2 first without removing a linear trend and then with removing a linear trend (use `linfit.m`). Show the frequencies and magnitudes of the five largest PSD peaks and discuss the significance of removing the linear trend.

7

Gridding, objective mapping, and kriging

> To those devoid of imagination, a blank place on the map is a useless
> waste; to others, the most valuable part.
>
> *Aldo Leopold*

Most of you are familiar with topographic contour maps. Those squiggly lines represent locations on the map of equal elevation. Many of you have probably seen a similar mode of presentation for scientific data, contour plots with *isolines* of constant property values (e.g. isotherms and isopycnals). What many of you are probably not familiar with are the mathematics that lie behind the creation of those "maps" and their uses beyond visualization.

7.1 Contouring and gridding concepts

This chapter covers the question: "What do you do when your data are not on a regular grid?" This question comes up frequently with ocean field data, which are rarely sampled at exactly equal intervals of space or time. The grid dimensions could be latitude–longitude, like the familiar topographic map, or involve other dimensions such as time, depth, or even property values (e.g. temperature, oxygen, chlorophyll). Mathematical gridding is common in visualization because computers can only draw contour lines if they know where to draw them. Often, a contouring package will first grid your data using a default method, and this may be acceptable. But there is more to it than making pretty pictures.

You may have other good reasons for putting your data onto a uniform grid – for example to estimate values at places or times where you do not have measurements; to compare two or more data sets sampled in different locations; or to apply mathematical operations to your data, such as computing gradients or divergences, that are easier if your data are on a grid. In any case, you want a method that produces in some objective fashion the "best" estimates of the gridded values and a measure of the error in those estimates. The various techniques for doing this have many aspects in common, and later in the chapter we discuss the two most common approaches, objective mapping (sometimes called optimal interpolation) and kriging. But before we touch on these more advanced methods, we overview several more quick and dirty (and thus sub-optimal) approaches for gridding your data.

7.1.1 Triangulation

There is a straightforward way to contour irregularly spaced data, *Delaunay triangula-tion*. The individual data points are connected in a unique, optimal network of triangles (Fig. 7.1) that are as nearly equiangular as possible and with the longest side of each triangle as short as possible. To form the grid, we surround each of our irregularly spaced data points with an irregularly shaped polygon such that every spot inside the polygon is closer to our enclosed data point than to data points at the center of other poly-gons. These irregular polygons are known as *Thiessen* polygons. Straight lines drawn from only neighboring Thiessen data points create a Delaunay triangular network. Davis (2002) describes a sequential approach for building a Delaunay network. The locations of the contour lines along these triangulation lines are then computed by simple linear interpolation.

The resulting contours are straight lines within each grid triangle. In effect, we have turned our data into a set of flat triangular plates that can be used for computer visualization (Chapter 19). This approach is okay for producing contour maps but is difficult to use for derived products (gradients, etc.). Furthermore, if you were to sample at different locations you would get a different triangular grid. It is often better to put your data onto a regular, rectangular grid. A regular grid is easier for the computer to use but is more difficult for the user to generate. But the benefits to be gained from this extra trouble are large and lead into the optimal methods discussed in later sections.

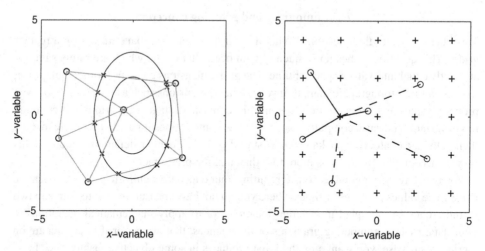

Figure 7.1 An example of two different gridding approaches. In the left panel, a triangular "grid" (gray lines) is built from the data points (o), and the locations of the contours (×) along the triangu-lation lines are calculated by linear interpolation. Choosing the optimal way to draw the connecting lines is a form of the Delaunay triangulation problem. In the right panel, a regular rectangular grid is built by nearest-neighbor method or interpolation from surrounding data points chosen either as N-nearest neighbors (solid lines) or from a larger region (solid and dashed lines).

7.1.2 Interpolation methods

Consider the case where you have measurements of some scalar Z in a two-dimensional x–y plane, and you want to estimate values $\hat{Z}_i$ on a grid. Most methods apply some variant of:

$$\hat{Z}_i = \sum_{k=1}^{N} w_{ik} Z_k \tag{7.1}$$

where the grid estimate is the sum of a set of weights (w_{ik}) times observations (Z_k). You have several options that differ in how you calculate the weights.

Nearest neighbor is a method that works in a way you might expect from the name. The grid value $\hat{Z}$ is estimated from the value of the nearest-neighbor data point. The distance d from a grid point i to the actual data points is given by:

$$d_{ik} = \sqrt{(x_i - x_k)^2 + (y_i - y_k)^2} \tag{7.2}$$

where the index i indicates grid number (in some sequential sense), and k identifies the actual data points. Then:

$$\hat{Z}_i = Z_k|_{\min(d_{ik})} \tag{7.3}$$

where $|_{\min(d_{ik})}$ simply means evaluated at the minimum distance d_{ik}. The nearest-neighbor method is of particular use for filling in minor gaps in data already on a regular grid or very nearly so. This method is fast, but has a tendency to generate "bull's-eyes" around the actual data points.

Inverse distance is a class of methods that use all of the data points within a specified distance from the grid point. One simple version weights the contribution of a data point by the inverse of the distance between the grid point and data point:

$$\hat{Z}_i = \frac{\sum_k Z_k/d_{ik}}{\sum_k 1/d_{ik}} \tag{7.4}$$

Sometimes this weight is raised to a power (2nd, 3rd, or even higher). As in Chapter 3, the data also can be weighted by the uncertainty in the individual data points σ_k^2:

$$\hat{Z}_i = \frac{\sum_k Z_k/(\sigma_k^2 d_{ik})}{\sum_k 1/(\sigma_k^2 d_{ik})} \tag{7.5}$$

Bilinear interpolation is commonly used for the reverse problem, estimating an unknown value $\hat{Z}(x_k, y_k)$ at a point k in space from two-dimensionally gridded data $Z_{i,j}$ (Fig. 7.2). It can also be used to transform data from one grid onto another. The approach is bilinear because it depends on separate linear interpolations in x and in y. The first step is to find the four grid points that surround (x_k, y_k). We then find the fractional distance in both x and y between the point to be estimated and one of the grid points, in this case $(x_{i,j}, y_{i,j})$:

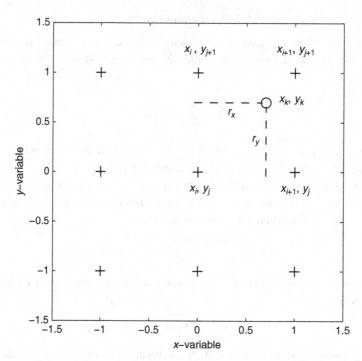

Figure 7.2 An example schematic of bilinear interpolation, which can be used to estimate data values at a point x_k, y_k from surrounding grid points weighted by the fractional distances r_x and r_y.

$$r_x = \frac{x_k - x_{i,j}}{x_{i+1,j} - x_{i,j}} \tag{7.6}$$

$$r_y = \frac{y_k - y_{i,j}}{y_{i,j+1} - y_{i,j}} \tag{7.7}$$

where i and j are the grid indices in x and y, respectively. We have assumed here that the boxes are rectangular for simplicity. The interpolated value would then be:

$$\hat{Z}_k = (1 - r_x)(1 - r_y)Z_{i,j} + r_x(1 - r_y)Z_{i+1,j} + (1 - r_x)r_y Z_{i,j+1} + r_x r_y Z_{i+1,j+1} \tag{7.8}$$

The fractional distances r_x and r_y vary from 0 to 1, and the coefficients in Equation (7.8) sum to 1.0, as they should do for an unbiased method. Bilinear interpolation can be extended into three dimensions, and there are logical extensions such as *bicubic interpolation*, which yields higher-order accuracy but suffers from the risk of over- or undershooting the target more frequently.

7.1.3 Splines

Like many of the topics we cover, splines could be a book unto themselves. But we would be remiss if we did not mention them here, at least briefly. Splines got their start as long flexible pieces of wood or metal. They were used to fit curvilinearly smooth shapes when the mathematics and/or the tools were not available to machine the shapes directly (e.g. hull shapes and the curvature of airplane wings).

Since then, a mathematical equivalent has grown up around their use, and they are extremely useful in fitting a smooth line or surface to irregularly spaced data points. They are also useful for interpolating between data points. They exist as piecewise polynomials constrained to have continuous derivatives at the joints between segments. By "piecewise" we mean if you don't know how/what to do for the entire data array, then fit pieces of it one at a time. Essentially then, splines are piecewise functions for connecting points in two or three dimensions. They are not analytical functions nor are they statistical models; they are purely empirical and devoid of any theoretical basis.

The most common spline (there are many of them) is the *cubic spline*. A cubic polynomial can pass through any four points at once. To make sure that it is continuously smooth, a cubic spline is fit to only two of the data points at a time. This allows for the use of the "other" information to maintain this smoothness. Consider a series of data points $(Z_{k-1}, Z_k, Z_{k+1} \ldots)$ as in Fig. 7.3. Cubic polynomials are fit to only two data points at a time $(Z_{k-1}-Z_k, Z_k-Z_{k+1}, \text{etc.})$. At each data point, the value and tangent (first derivative) for adjacent polynomials are required to be equal to those of the adjacent intervals. For the polynomial Z_k-Z_{k+1}, this means matching boundary conditions with $Z_{k-1}-Z_k$ at x_k and $Z_{k+1}-Z_{k+2}$ at x_{k+1}. Given these constraints, we can solve a series of simultaneous equations for the unknown cubic polynomial coefficients. Using MATLAB's spline toolbox, we can use a spline to interpolate values zz at a vector of locations xx from an original set of data $z(x)$:

```
zz = spline(x,z,xx)
```

See Davis (2002) and deBoor (1978) for more details.

There are a number of known problems with splines. Extrapolating beyond the edges of the data domain quite often yields wildly erratic results. This is because there is no information beyond the data domain to constrain the extrapolation, and splines are essentially higher-order polynomials that can grow to large values (positive or negative). Further, closely spaced data points can lead to large over- and undershoots of the true function as the methods attempts to squeeze a higher-order polynomial into a tight space. These problems also occur in 3D applications of splines. If your data are sparse and a smooth surface is what you are looking for, frequently a spline will give you a good, usable smooth fit to your data. For more noisy or closely spaced data, you might consider a *smoothing spline* (or *spline relaxation*), which like a least squares polynomial fit is no longer required to pass exactly through the data points. Smith and Wessel (1990) provide a very useful variant of smoothing splines, "continuous curvature splines in tension", which greatly alleviates the over- and undershoot problem and can be applied to irregularly spaced 2D data. Their method forms the basis for the gridding routines in the Generic Mapping Tool (GMT; http://gmt.soest.hawaii.edu/), which is used commonly in oceanography and geophysics.

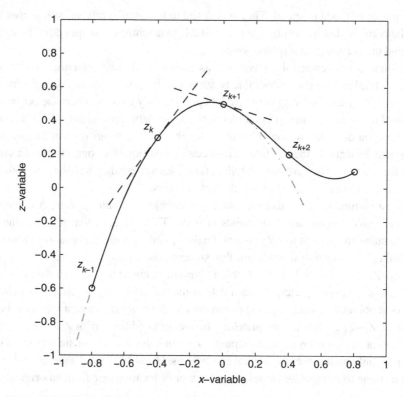

Figure 7.3 A cubic spline (solid line) consists of cubic polynomials (gray dot-dash line) fit piecewise between data point pairs. Because only two points are used at any one time, the additional information from the surrounding data points can be used to constrain the tangents (dashed lines) to be equal at the intersections of the piecewise polynomials.

7.1.4 Block and moving averages

Sometimes it is possible to put your data onto a regular grid through various averaging schemes. The simplest averaging techniques is the *block mean*. This method involves dividing your domain into a grid of sub-areas or "blocks" (Fig. 7.4). The block means, which can be thought of as grid point estimates centered in the middle of the blocks, are simply the average of all of the data points that fall within an individual block:

$$\hat{Z}_i = \frac{1}{N_i} \sum_{k=1}^{N} \delta_{ik} Z_k \tag{7.9}$$

The value of δ_{ik} is 1 when data point Z_k is within block i and 0 when it falls outside, and $N_i = \sum_k \delta_{ik}$ is the number of data points in block i.

An example of block averaging is shown in Fig. 7.5 for an AVHRR (Advanced Very High Resolution Radiometer) satellite image of sea surface temperature (SST) around Bermuda.

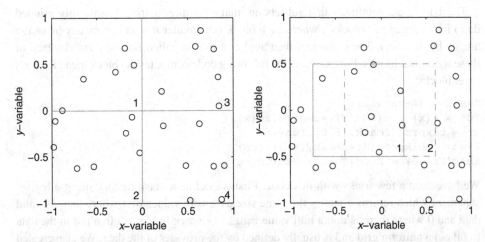

Figure 7.4 A schematic showing block and moving averaging applied to a two-dimensional domain. In the left panel, the domain is divided into four sub-areas or blocks, and block averages are computed as the means of the data inside each sub-area. In the right panel, a moving average is computed from overlapping blocks.

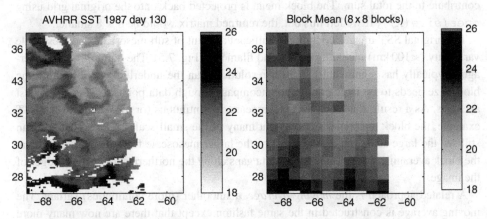

Figure 7.5 Example of applying a block mean to a 2D image of sea surface temperature (SST) from the AVHRR satellite. The original data are on the left (Glover *et al.*, 2002) and a block mean over 8 × 8 pixel blocks is on the right.

Because the satellite data are on a uniform 2D grid, we can take advantage of some powerful utilities in MATLAB's signal processing toolbox. The original image on the left is created with `imagesc(X,Y,Z)`, which displays individual pixels of the data scaled to a gray or color bar without any gridding or interpolation of the underlying data. To compute a block average, we could write a somewhat cumbersome code using nested for-loops to compute the sums in Equation (7.9); alternatively, we can use a more elegant (and versatile) m-file utility `blkproc` from the signal processing toolbox.

The `blkproc` command first subsets an image (or any matrix of uniformly gridded data) into a series of "blocks", where each block is a smaller $n \times m$ sub-matrix (n and m are set by the user). The command then applies a user supplied function `fun` to each of these individual blocks. For example, the following code computes the block mean for 8×8 pixel blocks:

```
Zmask = (Z~=znull);
fun = @(x) sum(sum(x))*ones(size(x));
Zn = blkproc(Zmask,[8 8],fun);
Zsum = blkproc(Z.*Zmask,[8 8],fun);
Zblk(Zn~=0) = Zsum(Zn~=0)./Zn(Zn~=0);
```

We have used a few tricks with this code. First we define a mask `Zmask` using a logical statement which returns a matrix the same size as Z with values of 1 where there is valid data and 0 where there is not; a null value `znull` or space holder is often put in the data to fill out a uniform grid and is usually defined by the provider of the data. We also created a *function handle*, `fun`, which is a quick way to create a function in MATLAB when you can get it to fit all on one line. Because there are missing elements in the data, we calculate the block mean by dividing the sum of the data in each block by the number of valid data points in the block. Remember that null values are set to zero using the mask so they do not contribute to the total sum. The block mean is projected back onto the original grid using `ones(size(x))`. If we left this out, the returned matrix would be 8 × 8 smaller.

The original SST images contain a significant amount of sub-mesoscale and mesoscale variability (<100 km) including eddies and filaments (Fig. 7.5). The output of block averaging typically has significantly coarser resolution than the underlying data because the block size needs to be large enough to encompass enough data points to generate robust averages. As a result, the block average appears discontinuous (or pixelated). In our SST example, the block averaging smooths out many of the small-scale features, but you can still see the large-scale trends and some of the larger mesoscale features. Notice too that the block averaging fills in some of the data gaps along the northern and southern edges of the image.

A related approach is the *moving average*, a smoother (more continuous) variant. The moving average is constructed in the same fashion except that there are now many more blocks, and the blocks are allowed to overlap. The greater the overlap, the higher resolution of the moving average but at a great computational cost and the price that the blocks are no longer independent. As with filtering, all points do not need to be weighted equally within the block (boxcar average). *Windowing* as a function of distance from the block centroid can improve estimates of higher-order derivatives and other properties of the moving average.

7.1.5 Trend surface analysis

Sometimes it is desirable or just convenient to have a function, or *trend surface*, that represents your data in terms of the coordinate system of your study area (e.g. in terms of longitude and latitude). A trend surface can be of any order and rank, meaning that the trend can be 1st order (a straight line, a flat plane) or 2nd order (quadratic curve, surface, or hyper-surface). The *order* refers to the highest power any independent variable is raised to, the *rank* refers to the dimensionality. You can set up the equations and solve them with either normal equations or the design matrix; in certain advanced cases you may need to apply the nonlinear fitting technique of Levenberg–Marquardt. Or, in most cases, you can use a handy little m-file called `surfit.m`, which uses the repetitive nature of higher and higher-order polynomials and the SVD solution to the normal equations to fit surfaces to your data of the form:

$$Z_{\text{trend}}(\vec{x}) = a_0 + a_1 x_1 + a_2 x_2 + a_3 x_1^2 + a_4 x_1 x_2 + a_5 x_2^2 + \cdots \qquad (7.10)$$

where $\vec{x} = (x_1, x_2, \ldots, x_n)$ is the coordinate vector in n dimensions.

Most grid generation schemes work best when $\bar{Z}(\vec{x}) = 0$; in order to accomplish this it is important to remove any large-scale trend from your data first:

$$Z_{\text{anom}}(\vec{x}) = Z(\vec{x}) - Z_{\text{trend}}(\vec{x}) \qquad (7.11)$$

Further analysis is then conducted on the anomaly field $Z_{\text{anom}}(\vec{x})$. Implicit is the assumption of a scale separation between the anomalies of interest and a low-frequency trend, but such scale separations are not always clear in real field data. At the very least you should remove a first-order, n-dimensional surface (n refers to the rank of your coordinate system) from your data before proceeding to run your grid generation routine. You can always add it back in to your grid estimation points because you now have an analytical equation for the trend surface. Alternatively you could remove a low-pass filtered version of your data field, or fit a higher-order, n-dimensional, polynomial surface. The higher order you go, the better your fit will be regardless of what you use as a goodness-of-fit parameter. But keep in mind the "better fit" may not be statistically significant, and you can use ANOVA to test for this.

7.2 Structure functions

In the sub-optimal approaches discussed above, we rather arbitrarily decided how to weight the observations (e.g. linear with distance) and did not generate any error estimates. More defendable statistical approaches can be developed, but only if we know in more detail the relationship between data points separated in space. This information is usually presented in the form of the structure function (*variogram*) or the closely related lagged autocovariance function.

7.2.1 Regionalized variables

Structure functions (and kriging) are part of geostatistics, an outgrowth of a school of thought largely credited to mining operations in France and South Africa. Structure functions are built upon *regionalized variables*, variables that are distributed in space with some continuous irregular pattern, but one that cannot be described simply by an analytical mathematical function. This space is not limited to real three-dimensional space and can be extended to include time, parameter space, property space, etc. Regionalized variables encapsulate two somewhat contradictory characteristics:

1. A local, erratic aspect that calls to mind a random variable; and
2. A regional-scale, structured aspect which can be captured by a functional representation.

Hence we are dealing with a naturally occurring property (variable) that has characteristics intermediate between a truly random variable and a completely deterministic variable.

Each observation can be thought of as combining a "true" regional-scale variable and a noise or error term:

$$Z(\vec{x}) = Z_{\text{true}}(\vec{x}) + \epsilon(\vec{x}) \tag{7.12}$$

Typically we will assume that the error term is independent across data points. In addition, field data can have what is known as a "drift" associated with it. These drifts are generally handled by subtracting a trend surface from the data (Section 7.1.5). Spatial fields are said to be *first-order stationary* when there is no trend or the trend has been removed; i.e. the mean of the field is the same in all sub-regions. For simplicity, for the rest of the chapter we will assume that the mean trend field has already been subtracted and drop the Z_{anom} notation. The more stringent constraint of *second-order stationarity*, when the variance of a data field is uniform from one sub-region to another, cannot be assured but should always be tested.

7.2.2 Experimental variograms

First remember the definition of the covariance at lag j for a uniformly sampled vector (like a time series):

$$C(j) = \frac{1}{N_j} \sum_{k}^{N_j} (Z_k - \bar{Z})(Z_{k+j} - \bar{Z}) \tag{7.13}$$

We could also define a spatial covariance as a function of distance d:

$$C^*(d) = \frac{1}{N(d)} \sum_{k}^{N(d)} [Z(\vec{x}_k) - \bar{Z}(\vec{x})][Z(\vec{x}_k + d) - \bar{Z}(\vec{x})] \tag{7.14}$$

where here we use an asterisk to indicate an experimental covariance computed from the data, and $N(d)$ is the number of data pairs separated by a distance d. Estimating the spatial mean $\bar{Z}(\vec{x})$ can be difficult, however, for irregularly spaced data. Semivariance $\gamma(d)$ is an analogous quantity but does not require the computation of the spatial mean. It quantifies the squared difference between data pairs as a function of their separation distance d:

$$\gamma^*(d) = \frac{1}{2N(d)} \sum_k \left[Z(\vec{x}_k) - Z(\vec{x}_k + d) \right]^2 \qquad (7.15)$$

The semivariance and covariance are related:

$$\gamma(d) = [C(0) - C(d)] \qquad (7.16)$$

The *experimental variogram*, given by the vector of semivariances, is generated by removing any large-scale trend or bias in the data using a trend surface $Z_{\text{trend}}(\vec{x})$, computing the lag or distance between all possible data pairs (or at least all the pairs within some plausible, computationally feasible lag), binning the data pairs by distance, and then calculating Equation (7.15) for each bin.

The idea behind the variogram and covariance is as follows. If there is indeed regional structure in the data field, and not just random noise, nearby data pairs should be more closely related and have a smaller semivariance (larger covariance) than distant data pairs. The variogram tends to start off small near zero lag, increases over some length scale, and then flattens off to some high background value (Fig. 7.6).

Outliers, data in the tails of your noisy, real-world data distribution that fall well outside Gaussian expectations, have a tendency to distort the results of the standard variogram. The *robust variogram* estimator (Cressie and Hawkins, 1980; Cressie, 1993) attempts

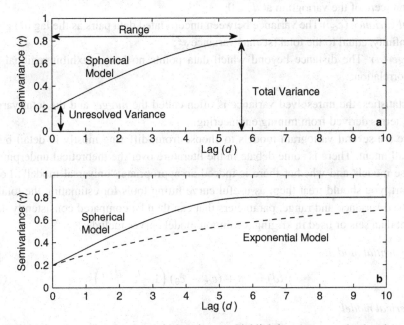

Figure 7.6 (a) A schematic of a theoretical variogram highlighting the unresolved and total variance and the range. (b) A schematic contrasting two theoretical variogram models. Note that the spherical (solid) and exponential (dashed) models have the same unresolved and total variance and range.

(mostly successfully) to make the experimentally determined variogram less sensitive to these outliers (hence, robust):

$$\gamma_{\text{robust}}^*(d) = \frac{\frac{1}{2}\left[\frac{1}{N(d)}\sum_k |Z(\vec{x}_k) - Z(\vec{x}_k + d)|^{1/2}\right]^4}{\left(0.457 + \frac{0.494}{N(d)}\right)} \qquad (7.17)$$

While looking somewhat overwhelming, upon inspection this is just Equation (7.15) modified. Note that we now use the absolute value of the difference between two data points separated by a distance d. The denominator is nothing more than a normalization to make γ_{robust}^* unbiased.

7.2.3 Variogram models

Model functions are often used to fit the shape of variograms and hopefully provide some insight on the spatial structure in the data field. These models are typically characterized by three parameters (Fig. 7.6):

Unresolved variance c_0: The sub-grid-scale variation or measurement error found as the intercept of the variogram at $d = 0$;
Total variance (c_∞) The variance between uncorrelated data pairs as the lag (d) goes to infinity, equal to the total (scalar) variance s^2;
Range (a): The distance beyond which data points no longer exhibit regional-scale correlation;

In geostatistics, the unresolved variance is often called the *nugget* and the total variance the *sill*, terms derived from mining engineering.

There are several variogram models to choose from, differing mostly in detail but not in overall intent. There is some debate in the literature over the theoretical underpinnings for these models and whether there is indeed an appropriate "universal model". For the most part, you should treat them as useful curve-fitting tools for estimating the total and unresolved variance and range, parameters that can then be compared consistently across different data sets or used in kriging. Common model curves include:

Exponential model:

$$\gamma(d) = c_0 + (c_\infty - c_0)\left(1 - e^{-d/a}\right) \qquad (7.18)$$

Spherical model:

$$\gamma(d) = \begin{cases} c_0 + (c_\infty - c_0)\left[\frac{3}{2}\frac{d}{a} - \frac{1}{2}\frac{d^3}{a^3}\right] & \text{for } d \leq a \\ c_\infty & \text{for } d > a \end{cases} \qquad (7.19)$$

Gaussian model:

$$\gamma(d) = c_0 + (c_\infty - c_0)\left(1 - e^{-d^2/a^2}\right) \qquad (7.20)$$

Linear model:

$$\gamma(d) = c_0 + bd \qquad (7.21)$$

The spherical model is similar in shape to the exponential but tends to approach the total variance value at a smaller lag for the same range. The Gaussian model displays parabolic behavior near the origin (unlike the previous two models which display linear behavior near the origin). The linear model is appropriate when semivariance continues to increase with distance with no indication of reaching a maximum value. In this case, there is no well-defined range, and the parameter of interest is the slope b that can be compared roughly with $(c_\infty - c_0)/a$ from the other models. The simplicity of the linear model leads to its common usage in kriging. But some caution is warranted when it is found in experimental variograms because it may reflect incomplete removal of a large-scale trend.

The simplest case to envision is when the semivariances and variogram model parameters are always the same, regardless of the direction being considered. But that is not always the case and it is often found that data display anisotropic behavior in their range. Nevertheless, if the data are second-order stationary, the total variance should be the same in all directions. If it is not, then this is a warning that not all the large-scale structure has been removed from the data. Knowledge of these anisotropies is necessary when designing an appropriate semivariogram model of your data prior to kriging.

7.3 Optimal estimation

Optimal interpolation and kriging are closely related methods to determine the *best linear unbiased estimate* (BLUE) for locations x_i where you have no measurements (including a uniform grid if you want). Almost as important, these statistical techniques produce error values for the spatial estimates. The methods are very flexible but require the user to bring to the problem a priori information, derived from the structure function (covariance function).

7.3.1 Optimal interpolation

Optimal interpolation, objective analysis, and simple kriging are mathematically equivalent (Bretherton *et al.*, 1976; Chilés and Delfiner, 1999) but often have quite different notation. The goal in any case is to find the weights w_k in Equation (7.1):

$$\hat{Z}_i = \sum_k w_k Z_k \qquad (7.22)$$

We can also compute the expected size of the error in the estimate; the estimated error variance $\hat{s}_e^2(i)$ is:

$$\hat{s}_e^2(i) = \langle [Z_i - \hat{Z}_i]^2 \rangle = \left\langle \left[Z_i - \sum_k w_k Z_k \right]^2 \right\rangle = \langle Z_i Z_i \rangle - 2 \sum_k w_k \langle Z_i Z_k \rangle$$

$$+ \sum_k \sum_m w_k w_m \langle Z_k Z_m \rangle \tag{7.23}$$

The expectation value $\langle X \rangle$ is a fancy way of saying the expected mean of X drawn from some large (and usually unmeasured) statistical population. The mean drift or trend is also assumed to be known and can be removed, i.e. $\bar{Z}(\vec{x}) = 0$. This turns out to be a critical point that differentiates simple kriging (optimal interpolation) from other forms of kriging (see below). Similar to a least squares problem, we solve for the unknown weights by minimizing the expected error with respect to the weights, by equating the derivative of the variance with respect to each weight to zero:

$$\partial \hat{s}_e^2(i) / \partial w_k = 0 \tag{7.24}$$

leading to:

$$\sum_m w_m \langle Z_k Z_m \rangle = \langle Z_k Z_i \rangle \quad k = 1, 2, \ldots, N \tag{7.25}$$

a set of N simultaneous, linear equations in w.

Of course we don't know the true value Z_i directly, so how can we compute the term containing Z_i in Equation (7.23) and (7.25)? For that matter, even for the observations we typically only have single measurements at any point in space while the formula calls for statistical averages. This is where the model covariance or semivariogram functions enter the problem. Since the spatial variables are all anomalies about a zero mean, the expectation terms (e.g. $\langle Z_k Z_m \rangle$) are equal to covariances (we could have kept a known, non-zero mean but it just makes the algebra a bit messier). The required data covariances, for example, can be estimated from the model covariance function by plugging in the distance d_{km} between the kth and mth points:

$$\langle Z_k Z_m \rangle = C(d_{km}) \tag{7.26}$$

Similarly, the covariance between an arbitrary unsampled location $\vec{x}_i$ and any data point can be derived using their separation distance d_{ik}. Notice that the actual data values Z_k do not enter the problem, only their locations and our model estimate of the covariance function.

In matrix form the original problem becomes:

$$\partial \hat{s}_e^2(i) / \partial w_k = 0 \text{ where } \hat{Z}_i = \mathbf{W}' \mathbf{Z}_k \tag{7.27}$$

where $\mathbf{Z}_k$ is the $N \times 1$ data vector. This translates into solving:

$$\mathbf{C}_{km} \mathbf{W} = \mathbf{C}_{ki} \tag{7.28}$$

C_{km} is the $N \times N$ data covariance matrix, C_{ki} the $N \times 1$ estimate–data covariance vector, and W the $N \times 1$ weight vector. The solution is:

$$W = (C_{km}^{-1} C_{ki}) \text{ and } \hat{Z}_i = (C_{km}^{-1} C_{ki})' Z_k \tag{7.29}$$

where C_{km}^{-1} is the matrix inverse of data covariance matrix C_{km}. Larger weights are assigned to data from nearby points that are expected to covary positively with the estimated value. The expected square error or confidence interval is:

$$\hat{s}_e^2(i) = C(0) - (C_{km}^{-1} C_{ki})' C_{ki} \tag{7.30}$$

The first term $C(0)$ accounts for the expected error if we had no observations and is equal to the covariance of the data as one approaches zero lag. That is, if we had no data, our best estimate would be no better than simply drawing a random point out of a population. The second term reduces the estimated error because of the information supplied by the nearby data points.

The equations above are in terms of the model covariance function. A variogram function could be used just as easily, using the corresponding matrices Γ of the model variogram function $\gamma(d)$:

$$\hat{Z}_i = (\Gamma_{km}^{-1} \Gamma_{ki})' Z_k \text{ and } \hat{s}_e^2(i) = (\Gamma_{km}^{-1} \Gamma_{ki})' \Gamma_{ki} \tag{7.31}$$

In some situations, the covariance function is not well defined but a variogram can still be constructed. This occurs, for example, when the variogram continues to increase with growing lag and does not settle down to a constant total variance level. In this case, a linear variogram model can be used for kriging.

7.3.2 Ordinary kriging

In *ordinary* or *punctual* kriging, we relax the assumption that we know the mean drift exactly and allow it to equal some unknown offset $\bar{Z}(\vec{x}) = Z_0$. In order to keep the estimate unbiased, an additional constraint equation needs to be added such that the weights always sum to one:

$$\sum_k w_k = 1 \tag{7.32}$$

A *Lagrange multiplier* λ (or slack variable) is also included in the simultaneous linear equations to ensure that this condition holds no matter the choice of w:

$$\sum_m w_m \langle Z_k Z_m \rangle + \lambda = \langle Z_k Z_i \rangle \quad k = 1, 2, \ldots, N \tag{7.33}$$

In matrix form this looks like:

$$\begin{bmatrix} C_{km} & 1_{N \times 1} \\ 1_{1 \times N} & 0 \end{bmatrix} \begin{bmatrix} W \\ \lambda \end{bmatrix} = \begin{bmatrix} C_{ki} \\ 1 \end{bmatrix} \tag{7.34}$$

$$\tilde{\mathbf{C}}_{km}\tilde{\mathbf{W}} = \tilde{\mathbf{C}}_{ki} \qquad (7.35)$$

Here we indicate the augmented matrices with a ˜. The solution and error estimate are found with the modified matrices in the same manner as in simple kriging.

7.3.3 Cokriging of multiple variables

The concept of *cokriging* is nothing more than a multivariate extension of the kriging technique. As we discussed earlier, a regionalized variable is one that is distributed in "space", where the meaning of space can be extended to include phenomena that are generally thought of as occurring in time. A regionalized phenomena also can be represented by several inter-correlated variables, for example nutrients, temperature, and salinity in the ocean. Then there may be some advantage to studying them simultaneously using what amounts to a coregionalized model.

With cokriging, we estimate values for all of the properties we wish to grid in a single calculation, instead of generating separate kriging estimates one property at a time. In addition, covariance information about the way properties are related to each other is used to improve the grid estimation and reduce the error associated with the grid estimates. This added information comes from the cross-semivariance, which is reminiscent of cross-covariance:

$$\gamma_{ZY}^*(d) = \frac{1}{2N(d)} \sum_k \left[Z(\vec{x}_k) - Z(\vec{x}_k + d) \right] \left[Y(\vec{x}_k) - Y(\vec{x}_k + d) \right] \qquad (7.36)$$

Here Z and Y are two different variables measured at the same locations; when $Z = Y$, Equation (7.36) reduces to the definition of the semivariogram, Equation (7.15). The cross-semivariance is symmetric in (Z, Y) and $(d, -d)$, which is not always the case in the covariance matrix formed from the data. Another interesting thing about the cross-semivariance is that it can take on negative values. The semivariance must, by definition, always be positive; the cross-semivariance can be negative because the value of one property may be increasing while the other in the pair is decreasing.

On the book website, we provide an example set of codes for 1D and 2D variogram analysis (`variof1.m`) and kriging/cokriging (`cokri.m`) in MATLAB courtesy of Denis Marcotte (Marcotte, 1991 and 1996). The variogram code utilizes 2D FFTs, greatly speeding up the computation when the data is already gridded (e.g. patchy satellite images like the ones we used for the block averages in Fig. 7.5).

7.4 Kriging examples with real data

7.4.1 Kriging sea surface temperatures

We will start with the simple, and somewhat artificial, example of estimating sea surface temperature (SST) at an unknown location from three surrounding data points (Fig. 7.7):

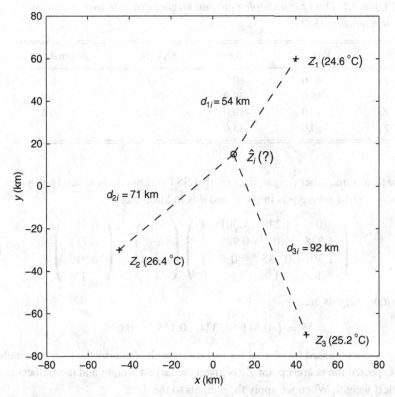

Figure 7.7 Layout of three control points and the grid point to be estimated in subsection 7.4.1. The distances (d) between control points are used to calculate the left-hand side of Equation (7.38), and the distances from control points to $\hat{Z}_i$ (dashed lines) are used to calculate the right-hand side.

$$\hat{Z}_i = w_1 Z_1 + w_2 Z_2 + w_3 Z_3 \qquad (7.37)$$

For three data points, the matrix equation for punctual kriging (7.34) using semivariances translates to:

$$
\begin{pmatrix}
\gamma(d_{11}) & \gamma(d_{12}) & \gamma(d_{13}) & 1 \\
\gamma(d_{21}) & \gamma(d_{22}) & \gamma(d_{23}) & 1 \\
\gamma(d_{31}) & \gamma(d_{32}) & \gamma(d_{33}) & 1 \\
1 & 1 & 1 & 0
\end{pmatrix}
\begin{pmatrix}
w_1 \\
w_2 \\
w_3 \\
\lambda
\end{pmatrix}
=
\begin{pmatrix}
\gamma(d_{1i}) \\
\gamma(d_{2i}) \\
\gamma(d_{3i}) \\
1
\end{pmatrix}
\qquad (7.38)
$$

and the error variance is:

$$\hat{s}_e^2(i) = w_1 \gamma(d_{1i}) + w_2 \gamma(d_{2i}) + w_3 \gamma(d_{3i}) + \lambda \qquad (7.39)$$

Now we will add in the actual SST data from Table 7.1. The first step is to calculate SST anomalies by removing the mean SST of 25.4 °C. The second step is to compute the distances between data pairs d_{km} and between the data and the unknown location d_{ki} using Equation (7.2). Then we need to apply a variogram model to the distances. Let's assume

Table 7.1 *Data for a simple punctual kriging example with sea surface temperature (SST)*

	X (km)	Y (km)	SST (°C)	Anomaly (°C)
Z_1	40.0	60.0	24.6	−0.8
Z_2	−45.0	−30.0	26.4	1.0
Z_3	45.0	−70.0	25.2	−0.2
$\hat{Z}_i$	10.0	15.0	?	

we know from some other (larger) data set that SST follows a linear semivariogram model with $\gamma(d) = 0.01d$ where γ is in $(°C)^2$ and d is in kilometers.

$$
\begin{pmatrix}
0 & 1.238 & 1.301 & 1 \\
1.238 & 0 & 0.985 & 1 \\
1.301 & 0.985 & 0 & 1 \\
1 & 1 & 1 & 0
\end{pmatrix}
\begin{pmatrix}
w_1 \\ w_2 \\ w_3 \\ \lambda
\end{pmatrix}
=
\begin{pmatrix}
0.541 \\ 0.711 \\ 0.919 \\ 1
\end{pmatrix}
\tag{7.40}
$$

The resulting weights are:

$$
\tilde{\mathbf{W}} = \begin{pmatrix} 0.511 & 0.334 & 0.155 & -0.075 \end{pmatrix} \tag{7.41}
$$

The weights are unbiased ($w_1 + w_2 + w_3 = 1$) as a result of adding the Lagrange multiplier. Also as expected the nearest point Z_1 is given the largest weight and the farthest point Z_3 the smallest weight. When we apply the weights to the data:

$$
\hat{Z}_i = 0.511 \times (-0.8) + 0.334 \times (1.0) + 0.155 \times (-0.2) + 25.4 = 25.294 \tag{7.42}
$$

where we have added back in the mean 25.4. The error squared is:

$$
\hat{s}_e^2(i) = 0.511 \times (0.541) + 0.334 \times (0.711) + 0.155 \times (0.919) + 1.000 \times (-0.075)
$$
$$
= 0.582 \tag{7.43}
$$

7.4.2 Kriging Argo float data

As a more sophisticated example application of kriging, let's consider Argo float temperature data from the North Atlantic. The Argo program (http://www.argo.ucsd.edu/) maintains a global array of about 3000 autonomous floats that profile upper ocean pressure, temperature, and salinity on an ∼10 day cycle. The number of temperature profiles and the widespread time/space coverage from Argo floats dwarfs that being collected from CTD casts from conventional oceanographic research ships, and autonomous floats are revolutionizing our understanding of ocean physical variability and climate.

Temperature data from Argo floats at 500 m depth in the North Atlantic are displayed in Fig. 7.8 for January through March 2005. Each observation is shown as a o, the size of which varies from small (cold) to large (warm). Because of the 10 day sampling schedule,

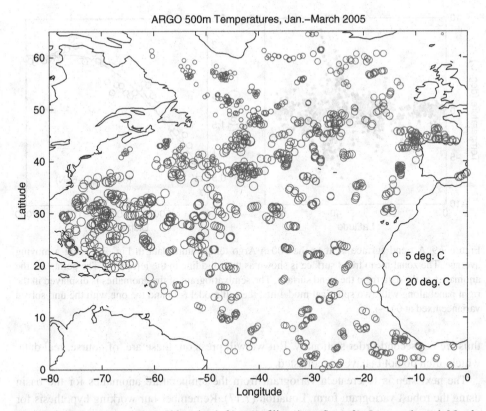

Figure 7.8 Temperature data at 500 m depth from profiling Argo floats for January through March 2005. The diameter of the data points (o) scales linearly with temperature, from cold (small circles) to warm (large circles). A legend with examples at 5 °C and 20 °C is included.

individual floats report data at multiple times and locations over the 3 month window. Thus you can discern float trajectories by connecting the overlapping or adjacent circles particularly in the Gulf Stream in the western Atlantic. The 500 m temperatures are warmer in the center of the subtropical gyre (20–40° N), where convergence of surface water causes a depression of the thermocline, and cooler both to the north in the subpolar gyre and to the south in the tropics where the thermocline is shallower.

The first step in kriging the Argo data is to remove the large-scale trend from the 500 m temperatures. The spatial structure is not easily represented by a low-order polynomial surface (we have tried). Instead, we use a 2D moving average to create a smoothed trend surface. The longitude-latitude averaging blocks are 8° × 8°, and the moving average is computed on a 0.5° × 0.5° grid. The warm waters and deep thermocline in the subtropics show up clearly in the zonal mean of the smoothed trend surface (left panel, Fig. 7.9). The anomalies from the trend surface, plotted in the same panel, are ±1–2 °C over much of the basin; but the anomalies can be quite a bit larger around 40–50° N where warm Gulf Stream waters abut cooler subpolar waters. An astute reader will notice that the anomalies

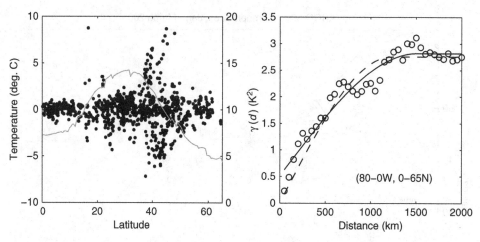

Figure 7.9 A trend surface is fit to the 500 m Argo temperature data in Fig. 7.8 using a moving average. The zonal mean trend surface is shown as the solid line in the left panel. The points are the anomalies of the data from the trend surface. The semivariogram of the anomalies is displayed in the right panel along with two spherical model fits, the full model (—) and the one with the unresolved variance fixed at 0 (- -).

thus are not second-order stationary. But we will press on; these are, of course, real data, which often do not behave as requested.

The next step is to create a variogram from the temperature anomalies for the basin using the robust variogram form, Equation (7.17). Remember our working hypothesis for a regionalized variable is that nearby sampling locations are more closely related than far away points. The variogram suggests that this is a reasonable assumption for the Argo temperature anomalies (Fig. 7.9). Variogram values are less than 1 $(°C)^2$ at short distances of 100–200 km, implying that the temperature variation between nearby data points is typically around $\sqrt{1}$ $(°C)$. The variogram increases with increasing distance, plateauing at a value slightly less than 3 $(°C)^2$ around a range of $\sim$1500 km lag. In this case this distance is not totally arbitrary and is roughly comparable to the size of the averaging boxes we used to create the large-scale trend surface; spatial variations on longer spatial scales were removed when we created the anomalies.

To be somewhat more quantitative, we use the nonlinear Levenberg–Marquardt routine from Chapter 3 to fit the spherical model (7.19) to the observed variogram values as a function of distance:

$c_0(°C^2)$	$c_\infty(°C^2)$	$a(km)$
0.512	2.82	1510
0	2.76	1230

$$(7.44)$$

We have done the fit twice with the full spherical model (top row) and a modified version where the unresolved variance c_0 (intercept at zero distance) is set to 0 (bottom row). The modified version does a better job at very short distances ($<$200 km), where the full model

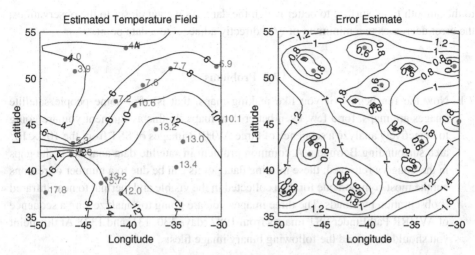

Figure 7.10 Kriging example for 500 m temperature data from Argo floats. The left panel shows the optimal estimated temperature field (°C) derived from the original float data, labeled at the individual sampling points. The right panel displays the error for the estimated temperature field.

appears to overestimate semivariance, but with the trade-off that the fit at larger distances is no longer as good and we lose an estimate of the unresolved variance. For the kriging example below we choose to use the modified $c_0 = 0$ version.

To illustrate kriging, we estimate temperature on a uniform grid from the float data for a single 10 day period over a sub-domain (50° W–30° W and 35° N–55° N) (Fig. 7.10). The kriging is done separately for each point on a $0.5° \times 0.5°$ grid using the temperature anomaly data. The trend surface is then added back to derive the estimated temperature field shown in the figure. The sub-domain near the Gulf Stream was chosen to highlight the large temperature gradients between cold subpolar waters to the north and northwest and warm subtropical waters to the south and southeast. Based on the observations, kriging has added to the optimal estimate regional structure that is not in the smoothed large-scale trend surface. Notice, for example, how the kriging routine bunches contour lines together in the southwest part of the grid to pass between the two nearby data points with values of 5.3 °C and 16.8 °C. The error field has smaller values $\hat{s}_e < 0.6$ °C near data points, increasing to $\hat{s}_e > 1.0$ °C in unsampled regions.

One way of thinking about how kriging works is to consider some end-member cases. If there are no observations closer than the variogram range a, the data are not helping to resolve the regional structure. The kriging routine will return a value for $\hat{Z}_i$ near zero (assuming you are working with anomalies where the mean and trend have been removed), and the optimal estimate will be close to the background trend surface. The corresponding error will be large, approaching $\hat{s}_e \Rightarrow \sqrt{c_\infty} = 1.66$ °C. For grid points closer than the range a to float observations, $\hat{Z}_i$ will be a weighted average of the surrounding data values, similar to our simple SST example above, and will act as positive or negative adjustments to the smooth trend surface to better match the data. As we get closer to the observations, the error decreases to a minimum of $\sqrt{c_0}$ directly adjacent to a data point.

7.5 Problems

7.1. Now the fun begins (if you like making maps, that is). For some people satellite images are maps, for a few, even better than maps. In this assignment you are going to objectively analyze a sequence of three AVHRR images of SST from the $10° \times 10°$ area surrounding Bermuda. A common problem in satellite data analysis is the gaps in the imagery retrieved; these missing data points can be due to a number of reasons but the most common one for data collected in the visible and thermal to near infrared is obscuration by clouds. The three images you are going to "analyze" are a sequence of AVHRR PathFinder SST images from 1987 (days 130, 131, and 132). At this point you should download the following binary image files:

- `A1987130h09da_sst.bin`
- `A1987131h09da_sst.bin`
- `A1987132h09da_sst.bin`

These are binary images and, consequently, will require a little manipulation so that you can krige them in MATLAB. We are providing you with several programs and a color map. The cokriging programs (below) were kindly donated to us by Denis Marcotte (Marcotte, 1991 and 1996). They are:

- `cokri.m`   the main cokriging program;
- `cokri2.m`   called by cokri.m;
- `trans.m`   called by cokri2.m;
- `means.m`   calculates row averages, used by cokri2.m;
- `variof1.m`   calculates variograms or covariograms;
- `surfit.m`   the trend surface fitting program;
- `excise.m`   removes any rows in a matrix containing NaNs;
- `byte2sst.m`   a program to convert binary satellite data into SSTs;
- `purp_map.mat`   a MATLAB MAT file with a color map useful for satellite data;
- `avhrr_init.m`   an initialization code to get you started with images in MATLAB.

(a) Run `avhrr_init.m`; when you get your plot on the screen, study it, notice the patterns of missing data; study the code, notice how the `imagesc` command is used and how NaNs are used to indicate missing data. Make an `.eps` file of the figure window.

(b) Compute a trend surface (use `surfit.m`) for each image (choose a low-order surface) and create a trend surface and residual image for each day. Save the trend and detrended tuples (groups of numbers) you'll need them later. Make a 3×3 plot of the original images (using `subplot`), their trend surfaces, and the resultant residual fields. Save this plot in a PostScript (`.eps`) file.

(c) Choose a grid for these images. These are "9 km" images and 10° yields an image that is 114 × 114 (12 996 potential grid points). We don't recommend that you create a grid with more than twice as many points as data points you start with, and in this case each image has more than half of its pixels with valid data. Use `variof1.m` to compute the 2D variograms of the three residual images. From these 2D variograms and covariograms extract 1D variograms and covariograms along the axis of the 2D variogram minimum trough and perpendicular to it. Note the angle of rotation.

Plot the empirical semivariances versus their lags. We suggest a 3 × 2 subplot arrangement; `hold` each of these subplots for later over plotting with the fitted thoretical variograms. Examine these scatter plots and decide on the following. Is there evidence for anisotropy? Are there any periodic structures (implying nonstationarity for the region analyzed)? If there are, decide how many of the empirical semivariances you will pass to your fitting routine (you don't necessarily have to use them all). Which theoretical variogram would fit well these empirical estimates? Is there any evidence for a nugget? If you decide to include the nugget effect in your theoretical variograms, be sure to use the reduced sill. Fit a 1D variogram model to each empirical semivariogram (using `nlleasqr.m`) and report the fit parameters (nugget, sill, and range) and their uncertainties (write these values out to a file). Plot these model variograms on top of the empirical points in the 3 × 2 plot and save this in a PostScript (.eps) file.

(d) Now cokrige. As you may recall from the text (Section 7.3.3), kriging is done to one variable (property) at a time, cokriging is done to p properties in one execution. It gains a lot in the exploitation of the covariance information between variables, that is to say from the cross-semivariance information.

The "clever" part about this multivariate approach is that by treating each image as just another property, we can cokrige all three images at once. Make another 3 × 3 plot of the original images, the objectively analyzed results in `x0s` (don't forget to add the trend surface back in), and the error fields. Discuss how the error fields (`s`) compare to distribution of missing data in the original images. Make this plot a PostScript (.eps) file.

8

Integration of ODEs and 0D (box) models

God does not care about our mathematical difficulties; He integrates
empirically.

Albert Einstein

Up to now we have been talking about data analysis, interpreting one's data with an underlying model in mind or exploring one's data looking for clues as to what model might
apply. Now we transition into modeling, building models to generate, for lack of a better word, *computa* (also known as *model output* or sometimes *synthetic data*). We will
start with simple models written in the form of ordinary differential equations (ODE);
these are characterized by a single, independent variable (e.g. time or space). Throughout most of this chapter we will use t as the independent variable because most of these
"zero-dimensional" problems involve the evolution of something (population(s), concen-
tration(s), etc.) in time. Ordinary differential equations can be written in general form as:

$$\frac{dC}{dt} = f(C, t, \text{constants}) \tag{8.1}$$

In this chapter, we show how to compute $C(t)$ from a single ODE or a set of coupled ODEs. We will present, explore, and motivate discussion about some important
applications, namely *box* or *population models* that are common in ecology, biological
oceanography, and geochemistry.

8.1 ODE categorization

The study of differential equations is aided by a system of categorization that, if nothing
else, makes them easier to talk about. Differential equations are broken down by type,
order, degree, homogeneity, linearity, and boundary conditions, the last of which is perhaps
the most important when it comes to numerically solving differential equations.

8.1.1 Type, order, and degree

Differential equations are broadly divided into two kinds: ordinary and partial. This chapter will focus on ordinary differential equations (ODE) and Chapter 12 will introduce a

key technique for numerically dealing with partial differential equations (i.e. finite difference methods). Ordinary differential equations occur when the dependent variable (for
example concentration C, number of items N, or generically y for the variable we are
modeling) is a function of a single independent variable (for example time t or a spatial
dimension x over which our model is varying). Partial differential equations (PDE) occur
when the dependent variable is a function of any combination of multiple independent
variables (x, y, z, t, etc.), but only one of these is allowed to vary at a time in the evaluation of a specific differential operator. In short, if the equations contain an ordinary d as
the differential operator, then the equation is an ordinary differential equation; if the differential operator is one of those funny "backward sixes", ∂, then it's a partial differential
equation.

The order of a differential equation is given by the highest-order derivative of the
dependent variable. For example:

$$\frac{d^3C}{dt^3} + \left(\frac{dC}{dt}\right)^2 - e^{-\lambda C} = 0 \tag{8.2}$$

is a third-order ODE.

The degree of a differential equation is given by the highest exponent of the highest-
order derivative of the dependent variable, once the equation has been cleared of fractions
and radicals in the dependent variable and derivatives. As an example:

$$\left(\frac{d^3C}{dt^3}\right)^2 + \left(\frac{d^2C}{dt^2}\right)^5 + \frac{C}{t^2+1} = e^t \tag{8.3}$$

is a third-order ODE of degree two.

8.1.2 Homogeneous and non-homogeneous equations

A homogeneous ODE is one in which every term contains the dependent variable or its
derivative. For example:

$$a_2\frac{d^2C}{dt^2} + a_1\frac{dC}{dt} + a_0C = 0 \tag{8.4}$$

is homogeneous if the as are constants or functions of t only. An equation of the form

$$a_2\frac{d^2C}{dt^2} + a_1\frac{dC}{dt} + a_0C = x^2 \tag{8.5}$$

is non-homogeneous (or inhomogeneous) because there is a term that does not depend
on the dependent variable (C). Note that replacing x^2 with a constant, or with x, in
Equation (8.5) also produces a non-homogeneous ODE.

A useful rule from calculus is that the sum of solutions to a differential equation is itself
a solution to the differential equation. We can then use the relationship between homogeneous and non-homogeneous differential equations when solving ODEs analytically. The

solution to an homogeneous differential equation is a general class of solutions. To find the specific solution for a non-homogeneous differential equation, solve the homogeneous equation first, then add the particular solution (usually based on boundary or initial conditions) for the non-homogeneous term.

8.1.3 Linear and nonlinear equations

The distinction between linear and nonlinear ODEs follows the same definition we use in describing other kinds of equations. Equations (8.4) and (8.5) are linear so long as the *a*s are constant or a function of *t* only. Some other examples are:

$$\frac{dN}{dt} = -\lambda N \tag{8.6}$$

$$F = -D\frac{dC}{dx} \tag{8.7}$$

both of which illustrate first-order, linear, homogeneous differential equations. You may recognize (8.6) as the radioactive decay equation and (8.7) as Fick's first law (a function of space, not time).

You may have noticed that if the sign on the right-hand side of Equation (8.6) were reversed the decay equation would become a growth equation. This change alone merely reverses the direction of the derivative. But if you were to add another term to slow this growth, such as:

$$\frac{dN}{dt} = \mu N - sN^2 \tag{8.8}$$

then this becomes a first-order, nonlinear differential equation. You will see this equation again shortly, as it forms the basis of carrying capacity population models.

8.1.4 Initial and boundary value problems

An *initial value* problem is a differential equation problem that has a starting point $C(t_0)$ given at an initial time (t_0), and is integrated forward in time, finding or predicting all $C(t_i)$ until some final time (or t_f). Typically, this type of problem involves a system evolving (changing) in time, although initial value problems do occur spatially as well.

A *boundary value* problem typically has multiple values of $C(x)$ (or $C(t)$) specified around the boundary of the modeling domain, and the trick is to solve for all positions within the domain simultaneously. A single ODE has two such points, the beginning and the end. Boundary value problems cannot be solved by picking one $C(x_0)$ (or $C(x_f)$) and integrating towards the other because this tends to be numerically unstable; there's no guarantee that everything will match when we get to the other point. Typically this type of problem is solved with iterative techniques or matrix inversions that converge on a stable answer.

We will discuss stability analysis in greater detail in a later chapter (Section 12.4). For the time being just think of stability as that characteristic of your model that prevents it from producing wildly varying results until the numbers are too large or too small for your computer to register. Generally stability, then, is a good thing.

8.2 Examples of population or box models (0D)

Zero-dimensional (0D) models are ubiqutous in ecology and geochemistry, from simple "teaching tools" to research grade investigations (Case, 1999). Frequently the "type" of box model comes down to the type of differential equations that describe the system you are modeling. The solution you come up with can vary from analytical to complicated numerical solutions. Because population models are important in any understanding of the biogeochemical processes that occur at the Earth's surface and because they provide good examples of the various types of equations that occur in 0D box models, we will use them in this section to illustrate some basic model "types". If the reader wishes to learn more about mathematical biology models than is covered in this text, there are a number of excellent books available, such as Murray (1993); Hastings (1997); and Fennel and Neumann (2004).

8.2.1 Exponential growth models

Here we discuss simple homogeneous ordinary differential equations of one dependent and one independent variable (usually time; the *zero*-D refers normally to spatial dimensions). Consider a simple birth–death population model (sometimes called an *exponential growth* model). We can model the number of individuals or concentration of biomass (N) as a simple, first-order, linear, deterministic ODE.

$$\frac{dN}{dt} = \mu N - \lambda N = (\mu - \lambda)\, N \tag{8.9}$$

where the population grows (has births) at a fixed rate (μ) and decreases (has deaths or is removed by sinking or other processes) at another fixed rate (λ). This ODE, of course, can be solved analytically as:

$$N = N_0 e^{(\mu-\lambda)t} \tag{8.10}$$

Population models based on analytical solutions like (8.10) assume that nothing limits growth. As demonstrated in Fig. 8.1, if $\mu > \lambda$, the population grows to infinity, and if $\mu < \lambda$, then the population will become extinct. The population is constant only when $\mu = \lambda$.

8.2.2 Carrying capacity models

The next level of complexity is to consider equations where something in the environment keeps the population from growing without bounds; this can be represented as a

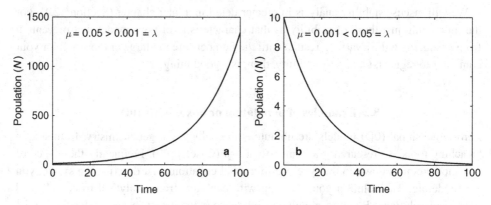

Figure 8.1 Solutions to Equation (8.9) with different parameters. (a) $\mu > \lambda$. The population grows without bound. (b) $\mu < \lambda$. The population decreases and eventually goes extinct. The other constant in this equation is the initial population ($N_0 = 10$). The units in this plot are arbitrary and are not shown.

carrying capacity of the environment. We introduce a limit to growth (other than death) to the equation(s):

$$\frac{dN}{dt} = f(N)N \qquad (8.11)$$

where

$$f(N) = \mu - sN \qquad (8.12)$$

and μ and s are positive constants. This is the simplest linear representation of an inhibitory effect, because as N gets large, $sN > \mu$, and $f(N)$ and dN/dt become negative. Recombining, we get:

$$\frac{dN}{dt} = (\mu - sN)N \qquad (8.13)$$

which is also known as the *logistic equation*. If we let μ be the net growth rate in unlimited conditions (absence of regulation) and s be the effect of an N-sized population on growth, then we can define $k = \mu/s$ as the carrying capacity. This gives:

$$\frac{dN}{dt} = \mu\left(1 - \frac{N}{k}\right)N \qquad (8.14)$$

As shown in Fig. 8.2, whenever $N > k$, the population is above the system carying capacity and the population decreases asymptotically towards k. But when $N < k$, the population grows asymptotically towards k; both behaviors are more realistic than unbounded exponential growth. Now this has, up to now, all been for a single population. How do the equations change for two (or more) components that interact (e.g. a predator–prey pair)? Again, population dynamics provides some good examples.

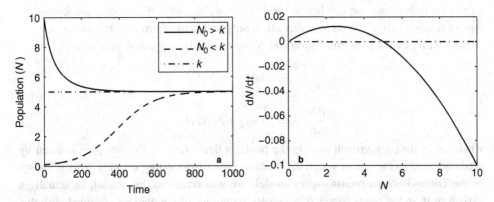

Figure 8.2 Results from a carrying capacity model, Equation (8.14) showing (a) solutions for initial conditions (N_0) both above and below the carrying capacity (k) and (b) the effect the size of the population (N) has on the rate of population size change (dN/dt). In (a), notice how the smaller initial population rises to its asymptote with a characteristic "S" shaped curve: this is the logistics or sigmoid curve. Model parameters here are $\mu = 0.01$ and $s = 0.002$, yielding a carrying capacity $k = 5$; as before the units are arbitrary.

8.2.3 Coupled linear first-order ODEs

Coupled ODEs arise when there is interaction between the dependent variables. That is when the right-hand side of the ODE for variable N_1 contains terms with other dependent variables. Consider competition, which comes into play when there are two or more populations competing for the same resource(s) (e.g. space, light, nutrients, funding . . .). First let's consider the logistic growth equation rewritten for two populations:

$$\frac{dN_1}{dt} = (\mu_1 - s_{11}N_1 - s_{12}N_2)N_1$$

$$\frac{dN_2}{dt} = (\mu_2 - s_{21}N_1 - s_{22}N_2)N_2$$

(8.15)

where s_{11} refers to competition within species (i.e. members of N_1 competing amongst themselves) and s_{12} to competition between species (N_1 vs. N_2). This can be simplified by assuming that within-species and between-species competition is basically the same for any given resource (i.e. $s_{11} = s_{12}$ and $s_{21} = s_{22}$).

$$\frac{dN_1}{dt} = [\mu_1 - s_1(N_1 + N_2)] N_1$$

$$\frac{dN_2}{dt} = [\mu_2 - s_2(N_1 + N_2)] N_2$$

(8.16)

where we now have two ordinary differential equations and each one is dependent upon both N_1 and N_2. Now all of this assumes that the two populations are in competition for a resource rather than one another. What do the equations look like when a *predator–prey* type of relation exists? Predation, after all, is only an intense form of competition.

Let's let N_1 be the prey population and N_2 be the predator population. Then:

$$\frac{dN_1}{dt} = (\mu_1 - b_1 N_2)N_1$$

$$\frac{dN_2}{dt} = (-s_2 + b_2 N_1)N_2$$

(8.17)

where μ_1 is the prey growth rate, s_2 the predator death rate, b_1 the rate prey is eaten by predators, and b_2 the rate at which the predator grows from eating the prey. This is known as the *Lotka–Volterra* predator–prey model. We will return to this model, or actually a variant of it, in the next chapter (Chapter 9). Examples of simulations obtained with this model are shown in Figs. 9.3 and 9.4.

8.2.4 Higher-order ODEs

It can be shown that any higher-order ODE can be rewritten as a set of coupled first-order ODEs. Take for example the following second-order, homogeneous, ODE (here we are using y_1 and y_2 as the dependent variables in this generic example):

$$\frac{d^2 y_1}{dt^2} - \mu(1 - y_1^2)\frac{dy_1}{dt} + y_1 = 0$$

(8.18)

If we let $dy_1/dt = y_2$, then

$$\frac{d^2 y_1}{dt^2} = \frac{dy_2}{dt}$$

(8.19)

By definition and through substitution:

$$\frac{dy_1}{dt} = y_2$$

$$\frac{dy_2}{dt} = \mu(1 - y_1^2)y_2 - y_1$$

(8.20)

we have converted the original Equation (8.18) into a pair (or system) of *coupled*, first-order ODEs (8.20). This example is known as the *van der Pol equation*, which describes an oscillator with nonlinear damping (van der Pol and van der Mark, 1927), although nowadays it gets a lot of attention as an example of *deterministic chaos* and has been used in fields as diverse as neurology and seismology, in addition to its oscillator work.

8.3 Analytical solutions

Before we dive into the numerical techniques that can be used for integrating ODEs, we would be remiss indeed if we failed to mention that many ODEs do have analytical solutions as well. We will discuss here three of the most common "tricks" used to find analytical solutions to ordinary differential equations. Our discussion will be limited to a brief description of the technique, the types of problems they can be used to solve, and some of our favorite textbooks (where the reader can go to learn more). You may wonder: what's the use in that? Sometimes knowing what something is called and where to look up more about it is half the battle.

Now some of you may be wondering why we discuss analytical solutions in a book about numerical techniques. Remember that analytical solutions can be very useful in testing numerical methods. Many times your problem can be simplified to an ODE that has an analytical solution. Even if the simplified form is not close enough to your real problem to be a useful solution, testing to make sure your numerical and analytical forms give the same results will give you confidence in your numerical solution of the more complex problem. Further, analytical solutions of simpler equations can be used to bracket your solution space, by which we simply mean you can place upper and lower bounds on your numerical solutions.

A word about finding analytical solutions. It can be intellectually challenging and it can also be very time-consuming. Most of the ODEs (that have an analytical solution) have been collected in extensive tables of integrals (Gradshteyn and Ryzhik, 2000). This is the ultimate trick in finding analytical solutions, using standard math tables. One particularly useful collection can be found in the *CRC Standard Mathematical Tables* (Zwillinger, 2002). There are no prizes for reinventing the wheel, and although it may not look like it, your ODE can often be rearranged to fit one of these standard forms if it has an analytical solution at all. So here is our advice — if you need (or want) an analytical solution to your differential equation, look in the standard integral tables first. Another approach, which we will not discuss in detail in this text, would be to use a symbolic math package symbolic Toolbox in (MATLAB, Mathematica, etc.).

What follows are our brief discussions of common ODE integration techniques: simple integration, variable replacement, and trignometric methods.

Simple integration

There is a class of ODEs, called *separable equations*, that are straightforward to integrate. Separable means one can group terms separately for the dependent and independent variables. Once separated, the individual terms can be integrated and if boundary conditions are available the constant(s) of integration can be derived through evaluation of the definite form of the integral.

Equation (8.21), familiar from radioactive decay, is such an equation:

$$\frac{dN}{dt} = -\lambda N \tag{8.21}$$

where λ here is the decay coefficient, because it can be algebraically rearranged and integrated. Most introductory calculus textbooks provide a solid description of this technique; one of our favorites is Thomas *et al.* (2009), although there are many others.

Variable substitution

Some ODEs can be analytically solved by substituting one variable for another, making the equation simpler and more straightforward to solve. Later, back substitution yields the actual solution you are after. Consider an oscillator with both a restoring term (pushes the dependent variable back to a set value) and a damping term (forces any oscillation to die out over time):

$$a\frac{d^2y}{dt^2} + by \pm c\left(\frac{dy}{dt}\right)^2 = 0 \tag{8.22}$$

where a is total mass of the system (e.g. particles, concentrations), b is the strength of the restoring term, and $c > 0$ is the drag on the system (the sign is chosen dependent upon the direction of motion of the system). While it may not look like it here, this is an example of a Bernoulli differential equation. Beware of Bernoulli-confusion: the Bernoulli family was large and prolific in seventeenth- through twentieth-century Switzerland. Differential equations carrying this name are the product of Jacob Bernoulli (1654–1705). The fluid dynamics equation relating pressure and fluid velocity was created by his nephew Daniel Bernoulli (1700–1782). Boas (2005) provides a good description of the variable substitution method.

Trigonometric methods

There is a large body of trigonometric substitutions that are used to solve ODEs. Suppose we have a second-order, linear, homogeneous equation of the form:

$$a_2\frac{d^2C}{dt^2} + a_1\frac{dC}{dt} + a_0C = 0 \tag{8.23}$$

The primary thrust to solving these types of ODEs is to rewrite the ODE as a characteristic equation. This equation will be in quadratic form and the main goal is to find its roots; the roots plus boundary conditions are used to write out the solution to the ODE. Strang (1986) and Thomas *et al.* (2009), in particular, have good explanations of how to apply this technique.

If the ODE is non-homogeneous, not all hope is lost. As we reminded you in Section 8.1.2, sums of solutions are solutions to differential equations. Here's how you can use this fact to exploit the relationship between homogeneous and non-homogeneous ODEs. First force your non-homogeneous ODE to be homogeneous by setting the single term without reference to the dependent variable to zero. Then find the general solution to the homogeneous equation as we have outlined above. Now all you need is one specific solution to your non-homogeneous ODE, add the two and you have a general solution to the non-homogeneous ODE. How do you find the specific solution? By inspection or by insight you might be able to guess at one, or you can use a technique known as parameter variation. If all of this sounds like a lot of work that also requires some luck, you are right, but if it's that complicated, why not take the numerical approach? That's probably why you started this chapter in the first place.

8.3.1 Taylor series expansions

Once upon a time a good friend of ours (David Musgrave) pulled a wet-behind-the-ears graduate student aside in the Alaskan wilderness and told him that the secret to navigating the numerical landscape was a mastery of Taylor series expansions. If you have not yet been exposed to this concept, there are a number of very good texts that cover this topic (Thomas *et al.*, 2009; and Boas, 2005). Here we briefly define this infinite series to refresh your memory.

In 1715 Brook Taylor published his mathematical *tour de force* (Taylor, 1715) containing his theorem for the approximation of a function with an infinite series summation of differentials of ever-increasing order. These are commonly referred to as *Taylor series expansions*. Expanding the evaluation of a function about a point t can be written as:

$$f(t+h) = f(t) + h\frac{df(t)}{dt} + \frac{1}{2!}h^2\frac{d^2 f(t)}{dt^2} + \frac{1}{3!}h^3\frac{d^3 f(t)}{dt^3} + \cdots = \sum_{k=0}^{\infty} \frac{h^k}{k!}\frac{d^k f(t)}{dt^k} \quad (8.24)$$

where h is some small arbitrary distance along the t axis (sometimes written as a Δt) and $k!$ is k-factorial ($k! = k \cdot (k-1) \cdot (k-2) \cdots 3 \cdot 2 \cdot 1$). Here

$$\frac{df(t)}{dt} = \frac{df}{dt}\bigg|_t \quad \text{and} \quad \frac{d^k f(t)}{dt^k} = \frac{d^k f}{dt^k}\bigg|_t \quad (8.25)$$

When $t = 0$ the expansion becomes known as a *Maclaurin series*.

To be of any practical use an infinite series must be truncated, which leaves a remainder, and we write this series as:

$$f(t+h) \simeq \sum_{k=0}^{\ell} \frac{h^k}{k!}\frac{d^k f(t)}{dt^k} + \mathcal{O}(h^{\ell+1}) \quad (8.26)$$

where $\mathcal{O}(h^{\ell+1})$ represents the after-truncation remainder and the magnitude of the error (of order $h^{\ell+1}$) introduced by the truncation. This suggests a way to estimate the magnitude of that error while approximating functions with a Taylor series expansion. The remainder, $\mathcal{O}(h^{\ell+1})$, is an infinite series, and the leading term is the largest part of the series and hence the largest piece of the error. Furthermore, one typically doesn't expand too many terms because as h becomes very small $h^{\ell+1}$ can become smaller than your machine accuracy (ϵ_m, Section 2.1.5). This series is used extensively in assessing error, and you will see it again.

8.4 Numerical integration techniques

When analytical solutions do not exist for your differential equation, numerical integration techniques can come to the rescue. The solution of a differential equation (its integration) takes on many forms, and we are going to discuss three of them for ODEs in this section (numerical techniques for partial differential equations are discussed in Chapter 12).

Although we do not expect you to be able to build your own integrator here (there are plenty of useful integrators in MATLAB), a basic understanding of the principles is valuable for using these techniques and recognizing which methods to apply to what problems, and what can go wrong sometimes.

8.4.1 Simple quadrature methods

We will start by looking at a simpler, but related topic – the numerical integration of a function. That is, we want to know the solution to $\int f(t)dt$ where you can evaluate the function $f(t)$ but don't have an analytical solution for the integral. If you remember back to your calculus you may recall being introduced to the concept that integration of a function is equivalent to finding the area underneath its curve. You may have been exposed to the idea that this action was approximated by drawing boxes underneath the curve, and as you made the width of these boxes smaller and smaller their area approached the true integral. This is essentially what quadrature means – finding the area underneath a curve when approximated by geometric constructs. If we look at Fig. 8.3 we see an example of this, where the area under the curve is approximated by the sum of the areas of the rectangles constructed under it. The equally spaced abscissa are given by $t_i = t_0 + ih$, and $f(t_i)$ is

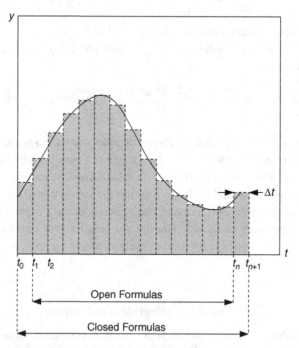

Figure 8.3 An example of quadrature from rectangles equally spaced along the abscissa. Notice how some parts of the rectangles extend above the curve being integrated while other parts seem to be "too short". The width of each rectangle is a constant $h = \Delta t$.

evaluated for all t_i. Sometimes it is not possible to evaluate $f(t)$ at the boundaries and in these cases open forms of the quadrature are used.

There are several forms of the closed formulas; we introduce two here that will be both useful and familiar, known as closed *Newton–Cotes* forms. The first is the *trapezoidal rule*, which is given by the following formula:

$$\int_{t_1}^{t_2} f(t)dt = h\left[\frac{1}{2}f(t_1) + \frac{1}{2}f(t_2)\right] + \mathcal{O}\left(h^3\frac{d^2 f}{dt^2}\right) \tag{8.27}$$

The reason for putting the 1/2 inside the brackets will become obvious when we examine the next closed Newton–Cotes quadrature. Another way of thinking about this expression is that we're multiplying h by a linear interpolation to the center of the interval (h). In other words, this is equivalent to the area of a rectangle cutting through the half-way (mean) point between $f(t_1)$ and $f(t_2)$. You may have noticed that there are no terms of even order of h in Equation (8.27); this is due to a symmetry of the formula that leads to a cancellation.

In the trapezoidal rule, the expression $\mathcal{O}(\ldots)$ refers to the *order of accuracy* of the approximation. In this case, the trapezoidal rule is accurate to the order of h^3 times the second derivative of the function being integrated. The final term in (8.27) is the error estimate of the integration approximation, and we see that quadrature becomes less accurate for larger steps (h) or for functions/regions with more curvature ($d^2 f/dt^2$). As you might have noticed (and it should be no big surprise) the trapezoidal rule involves a two-point formula and is therefore exact for polynomials of order one (i.e. linear).

A second closed form of Newton–Cotes formula, known as *Simpson's rule*, takes the form:

$$\int_{t_1}^{t_3} f(t)dt = h\left[\frac{1}{3}f(t_1) + \frac{4}{3}f(t_2) + \frac{1}{3}f(t_3)\right] + \mathcal{O}\left(h^5\frac{d^4 f}{dt^4}\right) \tag{8.28}$$

By using information from more points simultaneously, Simpson's rule does a better job of accounting for curvature. Another way of looking at this formula is to imagine one is taking a weighted average value (with the greatest weight on the middle value) of the function. Note that the sum of the fractional weights adds to 2, which is a result of the fact that we're taking the area over a distance of $2h$. Simpson's rule is a three-point formula and exact for polynomials of order two (i.e. quadratic).

Another advantage of closed forms is the ability to create "extended" versions of the form. We present here the *extended trapezoidal rule*:

$$\int_{t_1}^{t_n} f(t)dt = h\left[\frac{1}{2}f(t_1) + f(t_2) + \cdots + f(t_{n-1}) + \frac{1}{2}f(t_n)\right] + \mathcal{O}\left(nh^3\frac{d^2 f}{dt^2}\right) \tag{8.29}$$

where one adds up two $\frac{1}{2}f(t_2)$ terms, two $\frac{1}{2}f(t_3)$ terms, and so on and so forth until the $\frac{1}{2}f(t_n)$ term where there is only one, as with $\frac{1}{2}f(t_1)$ term. Equation (8.29) is the basis for most quadrature methods because it is quickly and infinitely expandable with minimal additional computation. Adding evaluation points between the integration limits (i.e.

Table 8.1 *Summary of MATLAB quadrature methods*

Function	Description	Supply m-file?
trapz	Numerical integration with Trapezoidal Rule	no
cumtrapz	Cumulative trapezoidal numerical integration	no
quad	Adaptive Simpson quadrature	no
quadl	Adaptive Lobatto quadrature	yes
quadgk	Adaptive Gauss–Kronrod quadrature	yes
dblquad	Numerically evaluate double integral	yes
triplequad	Numerically evaluate triple integral	yes

dividing Δt progressively by 2) requires only half of the function evaluations to be computed (the others having already been computed). The quadrature/numerical integration functions provided in MATLAB are summarized in Table 8.1. As with all things computer, we strongly recomend the student read the online help for any of these functions before using them.

8.4.2 Initial value methods

We have shown (Section 8.2.4) that higher-order ODEs can always be reduced to a system of coupled first-order ODEs. Just as important, while setting up an ODE problem, is the specification of boundary conditions that must be satisfied at specified points, but not necessarily in between these points. As we have already seen (Section 8.1.4), there are two broad categories:

Initial value problems: if $y = f(t)$ then $y_0 = f(t_0)$ is specified where t_0 is the initial condition, and all values of $y_i = f(t_i)$ are solved up to some final $y_f = f(t_f)$.

Boundary value problems: if $y = f(x)$, typically some $y_0 = f(x_0)$ and $y_f = f(x_f)$ are specified and must be satisfied by all $y_i = f(x_i)$ computed in between.

For the remainder of this chapter we are going to focus on the numerical solution of initial value problems. There are three broad categories of numerical integrators that can be applied to these kinds of problems. The first are the forward time stepping integrators; the most common and useful are the *Runge–Kutta* methods, a modification of *Euler*-style integrator. The second are the *extrapolation type integrators*, commonly referred to as *Bulirsch–Stoer* integrators, which try to extrapolate to a zero step size. Finally the third are the *predictor–corrector* methods that store the answers along the integration path and use that information to predict the next point.

Initial value problems are solved numerically using the differential equation and a specified time step to march the solution forward in time to a predetermined final time. We accomplish this by rewriting:

$$\frac{dy}{dt} = f(t, y) \qquad (8.30)$$

as a finite difference equation. The problem is more complicated than simply integrating a function (Section 8.4.1; quadrature) because now the right-hand side depends on our unknown (dependent) variable y. If this sounds like one of those time travel paradoxes, it isn't. We will get to finite difference equations when we develop techniques for partial differential equations in Chapter 12; for now just consider what the equation would look like if we replaced the dy/dt with $\Delta y/\Delta t$ and multiplied both sides by Δt:

$$\Delta y = f(t, y)\Delta t \qquad (8.31)$$

In this fashion we can predict changes in y with Δt steps forward, keeping in mind that Δt must be small enough to make a reasonably good approximation of the differential equation. The rest is bookkeeping.

If we write out the most straightforward implementation of this concept we get:

$$y^{n+1} = y^n + \Delta t f(t^n, y^n) + \mathcal{O}(\Delta t^2) \qquad (8.32)$$

This is known as *Euler's method* and $\mathcal{O}(\Delta t^2)$ is the truncation error (Section 8.3.1). We want to point out that the ns are not exponentiation, but rather refer to the time step (i.e. this is a formula for moving from time step n to time step $n + 1$). Although straightforward and simple, this method is not recommended because it is not very accurate compared with other methods of similar computational burden and it's not very stable. The reason for this is that we are using information from the derivative only at the beginning of the time step, that is t^n. Figure 8.4a shows, essentially, what's being done with Euler's method. Notice that the numerical integration rapidly diverges from the true curve in regions where the derivative dy/dt changes quickly.

As we have just said, the Euler method is a simple but not particularly accurate method. But suppose we took a "trial" step only half-way to the next time step and then used the derivative evaluation at the half-way point to predict the next full time step. Mathematically it looks like:

$$k_1 = \Delta t f(t^n, y^n) \qquad (8.33)$$

$$k_2 = \Delta t f\left(t^n + \frac{1}{2}\Delta t, y^n + \frac{1}{2}k_1\right) \qquad (8.34)$$

$$y^{n+1} = y^n + k_2 + \mathcal{O}(\Delta t^3) \qquad (8.35)$$

where $\mathcal{O}(\Delta t^3)$ refers to the truncation error of order Δt^3; that is to say, the method is second-order accurate. A graphical representation of these equations is shown in Fig. 8.4b.

This is known as the *Runge–Kutta method* of integration. When you use only one mid-point derivative evaluation it is said to be the Runge–Kutta 2,3 method (second-order accurate with a truncation error on the order of Δt^3). In MATLAB the function

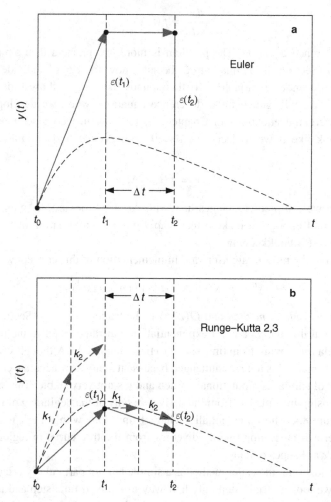

Figure 8.4 Graphical representations of numerical integration of a function $y = f(t)$, represented by the dashed line, via (a) Euler's method and (b) Runge–Kutta 2,3. In Euler's method, at each time step (t_i) the derivative (the gray arrow) is evaluated and extrapolated to the next time step. The arrows also trace out the actual path Euler's method follows while numerically integrating (truncated for clarity at time step t_2). The difference (or error) between Euler's method and the true curve is shown as the $\varepsilon(t)$ vertical lines; this method is first-order accurate, i.e. the truncation error is $\mathcal{O}(\Delta t^2)$. In the Runge–Kutta method the solid gray arrows represent the actual integration path this method follows (also truncated for clarity at the t_2 time step). The numbered, dashed gray arrows represent the evaluations of the differential used to calculate k_1 and k_2 (two estimations per time step). The mid-point derivative evaluation acts to increase the accuracy of the method to second order and the truncation error to $\mathcal{O}(\Delta t^3)$ (i.e. decrease the size of the $\varepsilon(t)$ mismatches).

that performs this algorithm is `ode23`. We can extend this method by making two mid-point derivative evaluations and an end-point evaluation, which leads to the following equations:

$$k_1 = \Delta t f(t^n, y^n) \tag{8.36}$$

$$k_2 = \Delta t f\left(t^n + \frac{1}{2}\Delta t, y^n + \frac{1}{2}k_1\right) \tag{8.37}$$

$$k_3 = \Delta t f\left(t^n + \frac{1}{2}\Delta t, y^n + \frac{1}{2}k_2\right) \tag{8.38}$$

$$k_4 = \Delta t f\left(t^n + \Delta t, y^n + k_3\right) \tag{8.39}$$

$$y^{n+1} = y^n + \frac{k_1}{6} + \frac{k_2}{3} + \frac{k_3}{3} + \frac{k_4}{6} + \mathcal{O}(\Delta t^5) \tag{8.40}$$

This is known as the Runge–Kutta 4,5 method (fourth-order accurate with truncation error $\mathcal{O}(\Delta t^5)$) and is implemented in MATLAB with `ode45`.

Now, a word about "higher-order" methods; higher order doesn't always mean higher accuracy. The 4th-order Runge–Kutta is usually superior to the 2nd-order, but beware of the size of the time steps and the number of them, because higher order *can* mean a greater number of smaller steps (to maintain stability) and that can add up to quite a bit of wall-clock time.

In MATLAB the Runge–Kutta routines all use a variable, continuously adjusted (adaptive) time step; you need only specify the first and last time step (and, under certain circumstances, you don't have to specify these either, see `help odefile` for more details on this advanced feature). MATLAB does give you the option of specifying the time steps at which you want solutions, but the underlying code will still use the adaptive time step method. There are versions of Runge–Kutta that are not adaptive time stepping, but you run the risk of getting really bogged down in many steps to keep it stable. The adaptive time stepping algorithm is controlled by a predetermined, desired accuracy (the variables `RelTol` and `AbsTol` in `ode23`, `ode45` and others). Basically this is a truncation error. The adaptive time stepping algorithm takes small steps where the function is changing rapidly and can take really big time steps where the function is smooth and not varying much. This ability can save you time measured in factors of tens to hundreds in compute speed and wall-clock time.

Press *et al.* (2007) give such good advice as to which ODE integrator to use that we are going to repeat it (paraphrased) here. In general, use Runge–Kutta 4,5:

- when you don't know any better;
- when you have an intransigent problem, i.e. the problem is not "smooth";
- when computational efficiency is not of paramount importance; and
- when adaptive time stepping is not a problem and may actually be a great boon.

We will provide a simple example of applying Runge–Kutta in Section 8.5.1.

8.4.3 Explicit vs. implicit methods

So far we have only used information or knowledge about the solution "from the past", as it were (t^n, t^{n-1}, etc.). This is what we mean by *explicit* methods, wherein we can algebraically calculate a new time step from the solutions we have calculated in earlier steps. But suppose, for a moment, you could use information "from the future", or at least from the time step you are trying to evaluate (t^{n+1}). This is what we mean by *implicit* methods. But wait a minute! we hear you cry, how can we use information about the very thing we are trying to evaluate in the first place? Try not to worry too much about it right now. It can be done and we'll show you some of the ways it is done later in this chapter and in other chapters.

Given that one can use information "from the future" (implicit methods) and that we live in a Universe where the second law of thermodynamics constantly reminds us that there is "no free lunch", you may wonder, "What's the catch?" Well, there is one, of sorts. It has to do with the trade-off between speed of execution and accuracy of results. Implicit methods allow us to take larger time steps than explicit methods and they are typically very stable. This can be a very useful feature, particularly when faced with integrating a system of stiff equations (Section 8.5.2). However, the gains in speed and stability are somewhat offset by a reduction in accuracy. We don't want to give the impression that implicit methods are inaccurate. There are higher-order implicit methods that are plenty accurate enough for global general circulation models (GCMs, Chapter 17), so they ought to be good enough for your applications too. If you need an implicit method, say, to make your model run in a reasonable amount of time, then by all means use one. Certainly we expect our readers to make their own, reasoned, choices.

8.5 A numerical example

Perhaps the best way of seeing how to use the MATLAB ODE solvers is to construct a simple model and work with it. We hope the application here will clarify how we use the tools, and incidentally, something about how the ocean works. Additionally, we will mention that in this example model we are going to use **x** as our unknown variable, i.e. the vector of phosphorus concentrations in time. We do this for two reasons: first, it is fairly standard in the modeling literature to use **x** for the thing you want to know; and second, we want to make the derivation of this model consistent with our usage of the same model in later chapters.

8.5.1 An example: a two-box global ocean phosphate model

We start with a two-box model of the global ocean made popular by Wally Broecker among others (Broecker and Peng, 1982), where we have two reservoirs (boxes) representing a surface layer and the deep ocean (see Fig. 8.5). In this model the global ocean cycle of phosphorus is relatively simple, with phosphorus added to the surface ocean by

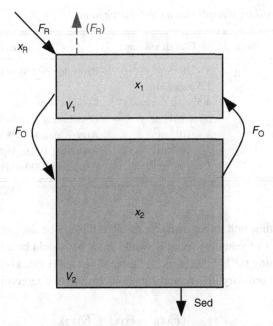

Figure 8.5 A two-box global ocean phosphate model showing water fluxes. In the implementation of this model in the text, we use the following parameters: box volumes $V_1 = 3 \times 10^{16}$ m³; $V_2 = 1 \times 10^{18}$ m³; river water flux $F_R = 3 \times 10^{13}$ m³ y⁻¹; and overturning water flux $F_O = 6 \times 10^{14}$ m³ y⁻¹.

river inflow and removed by sedimentation in the deep ocean. For the purposes of this model, phosphorus predominantly exists in the ocean in the form of inorganic phosphate, although it spends time within organic pools (ignored here for the moment). There is a net overturning of water in our ocean model, because in the real ocean, water sinks to depth in the polar regions and is balanced by an equivalent upwelling elsewhere. In fact, to balance the river input of water, there must be an equivalent amount of evaporation from the surface of the ocean (river flow must be balanced hydrologically by rainfall that finds its ultimate source in evaporation from the ocean surface). Since flowing water carries phosphate and evaporating water does not, our model river is a source of phosphate to the surface box.

To model the system, we need to put numbers on the river phosphate concentration, water fluxes, and box volumes (concentration changes occur as a result of material fluxes divided by volumes). The phosphate fluxes associated with overturning water and river inflow are given by the water fluxes times the concentrations of phosphate in the water. Table 8.2 gives the appropriate numbers for our toy model.

Note that we have striven to maintain a consistent system of units. This is vital in a model (or any ocean) calculation. We've chosen m³ for volume, m³ y⁻¹ for a water flux, mmol for phosphorus amount, and mmol m⁻³ for phosphorus concentration. The residence time of phosphorus in the oceans is probably of order tens of thousands of years, so we've

Table 8.2 *Global two-box phosphorus model parameters*

Name	Symbol	Flux or volume	Description
River water flux	F_R	3×10^{13} m^3 y^{-1}	River flux ~ 1 Sv $= 10^6$ m^3 s^{-1}
P conc. in river water	x_R	1.5 mmol m^{-3}	
River P flux		4.5×10^{13} mmol y^{-1}	Conc. $\times$ water flux
Overturning water flux	F_O	60×10^{13} m^3 y^{-1}	Overturning $\sim$20 Sv $= 20 \times 10^6$ m^3 s^{-1}
Surface area of oceans		3×10^{14} m^2	Approximately 70% of Earth's area
Surface box volume	V_1	3×10^{16} m^3	Assuming 100 m depth
Deep box Volume	V_2	1×10^{18} m^3	Assuming 3300 m depth

chosen years as our time unit rather than seconds. We'll likely be integrating our ODEs for thousands to millions of years, so seconds would work but would be a little inconvenient (try saying, "integrating for 10^{13} seconds ..." three times really quickly).

So let's set up the ordinary differential equations for the two reservoirs. The equations are:

$$\frac{dx_1}{dt} = \frac{(F_R x_R - F_O x_1 + F_O x_2)}{V_1} \tag{8.41}$$

$$\frac{dx_2}{dt} = \frac{(F_O x_1 - F_O x_2 - \text{Sed})}{V_2} \tag{8.42}$$

where x_1 and x_2 are the phosphorus concentrations in boxes 1 and 2 respectively at time t. Notice that the phosphorus mass budget is maintained because when we take an amount of phosphorus out of one box, we put it in the other; we say these two ODEs are coupled. In (8.42) Sed is the sedimentary phosphate flux in mmol y^{-1}. Inasmuch as we don't quite know what to do with this flux just yet, we'll arbitrarily set it to 1% of the downwelling flux of phosphorus from the surface box (not a very clever choice, but it's a start). Now the system of equations is given by:

$$\frac{dx_1}{dt} = \frac{(F_R x_R - F_O x_1 + F_O x_2)}{V_1} \tag{8.43}$$

$$\frac{dx_2}{dt} = \frac{(F_O x_1 - F_O x_2 - 0.01 F_O x_1)}{V_2} \tag{8.44}$$

We now need to code this in MATLAB in order to use the ODE solver. We write a MATLAB function, which we'll call phos2.m.

```
function dxdt=phos2(t,x)

V=[3 100] * 1e16;      % volume of reservoirs in m3
dxdt=zeros(2,1);       % Initialize output as a column vector
FO = 6e14;             % overturning water flux in m3 per year
FR = 3e13;             % river water flux in m3 per year
```

```
xR = 1.5;                    % river water P concentration in mmol per m3
Sed = 0.01*FO*x(1);          % sedimentary loss of P in deep box

dxdt(1) = (FR*xR - FO*x(1) + FO*x(2))/V(1);
dxdt(2) = (FO*x(1) - FO*x(2) - Sed)/V(2);
```

Notice how we've split out the terms for clarity and flexibility, so if we want to change the model in the future, it'll be easier (e.g. to change the water overturning flux we need to edit just one number in the fourth line instead of four numbers in the eighth and ninth lines). Note also that, as a `function`, this m-file returns a column vector of two values corresponding to the derivatives of the coupled differential equations. Now let's integrate the model.

Following our own advice we'll start with `ode45`, as it's generally the most accurate, efficient and stable of the algorithms. These routines are very easy to use. All we need to do is to specify the function name (here it's `phos2`), the time span for integration (let's say a mere 10^4 years), and the starting values for the variables (remember, it's an initial value problem). For the sake of simplicity, we'll start with zero values. Note also that the initial value and the output of the function must be column vectors.

But why 10^4 years? Well, to make things simple we decided to set the initial concentrations to zero, and let the river fill the ocean with phosphate. That being the case, we know from geochemistry that we will need to let the model run for at least the equivalent of one to several water overturning residence times before the model will show any detectable change. We can easily put bounds on the residence time of water in the deep model box with respect to overturning equal to $V_2/F_O \sim 1.67 \times 10^3$ years and the residence time of the whole model with respect to river inflow equal to $(V_1 + V_2)/x_R \sim 3.43 \times 10^4$ years. Based on these back of the envelope calculations, 10^4 years is a good compromise, long enough to allow the model to do something without committing to a very long integration. So we type the following:

```
x0=[0 0]';
[t,X]=ode45('phos2',1e4,x0);
plot(t,X)
```

Note that `ode45`, like the other integrators, returns two things: a column vector of the times, `t`, in the integration steps, and a multi-column matrix of the variable outputs, `X` (one column per variable, each row corresponds to a time step). Additionally, since it is a matrix of many rows by two columns, `X` is written capitalized. Also, we've chosen to integrate only for a shortish period of time, as an exploratory step. This is a prudent, standard approach.

In plotting the results, we see that the model has not reached equilibrium after such a short time (in Fig. 8.6a the concentrations are still increasing and have not leveled off yet). No real surprise there, so we run it again with a longer integration time span:

```
x0=[0 0]';
[t,X]=ode45('phos2',1e6,x0);
plot(t,X)
```

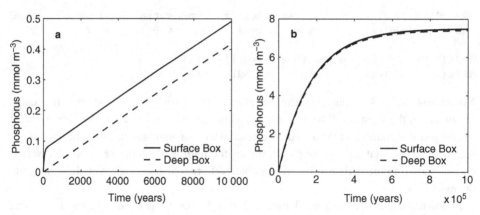

Figure 8.6 Results of our first two-box global phosphate model integration, (a) $t_f = 10^4$ y and (b) $t_f = 10^6$ y.

Figure 8.6b shows the results. Not surprisingly, given the simple structure, the model has the two boxes with rather similar concentrations. The surface box is slightly higher than the lower box. The sign of the difference is also not surprising, as the input is in the surface box, and the sink is in the lower box. The size of the difference makes sense too; the removal rate is about 1% per box turnover, which gives a timescale large relative to the overturning timescale ($V_2/F_O \sim 1670$ years) so that they should be very similar in concentration.

This, however, is not realistic as it disagrees with two things we know about the ocean. First, biological production occurs in the surface ocean, using up phosphate (plus other nutrients) and resulting in a particulate flux to the deep water. Most of this particulate flux is respired in the deep ocean but a small fraction of it ends up in the sediments. Second, as a consequence surface phosphorus concentrations in the real ocean are generally much lower than those in deep waters. Thus our model is not well formulated. Let's recast the equations in a slightly more realistic fashion.

$$\frac{dx_1}{dt} = \frac{(F_R x_R - F_O x_1 + F_O x_2 - \text{Production})}{V_1} \tag{8.45}$$

$$\frac{dx_2}{dt} = \frac{(F_O x_1 - F_O x_2 + 0.99 \times \text{Production})}{V_2} \tag{8.46}$$

$$\text{Production} = \frac{(x_1 V_1)}{\tau_x} \tag{8.47}$$

What we've done is assumed that phosphorus is removed from the surface box at a rate proportional to the amount of phosphorus in the box, $x_1 V_1$ (not a bad idea, actually, since biological production is largely nutrient limited). We've characterized this removal rate with a time-constant (τ_x) that might be visualized as the lifetime of phosphorus in the surface ocean due to biological uptake. In the deep box, this loss magically reappears (by oxidation of the falling particles), but it is not 100% efficient as some particles make it to the sediments and are lost from the system. In fact, this results in the system approaching steady state, i.e. $d\mathbf{x}/dt = 0$ ($\mathbf{x}$ is a two-element array representing the phosphorus concentration in both boxes), as the sedimentary loss must balance the river input. So we rewrite the m-file to add the biology (calling it `phos2b.m`):

```
function dxdt=phos2b(t,x)

V=[3 100] * 1e16;           % volume of reservoirs in m3
dxdt=zeros(2,1);            % Initialize output as a column vector
FO = 6e14;                  % overturning water flux in m3 per year
FR = 3e13;                  % river water flux in m3 per year
xR = 1.5;                   % river water P concentration in mmol per m3
Tau = 100;                  % 100 year residence time of P in the surface
Prodtvy = x(1)*V(1)/Tau;    % Production
ReminEff = 0.99;            % Remineralization efficiency

dxdt(1) = (FR*xR - FO*x(1) + FO*x(2) - Prodtvy)/V(1);      % surface box
dxdt(2) = (FO*x(1) - FO*x(2) + ReminEff*Prodtvy)/V(2);     % deep
```

and we integrate it with:

```
[t,X]=ode45('phos2b',3e6,x0);
```

to get Fig. 8.7 which shows the surface box has lower concentrations. Did you notice that the model took a little longer to run than the previous experiment (perhaps half again as long)? We increased the integration period by a factor of three (1×10^6 to 3×10^6 years) so

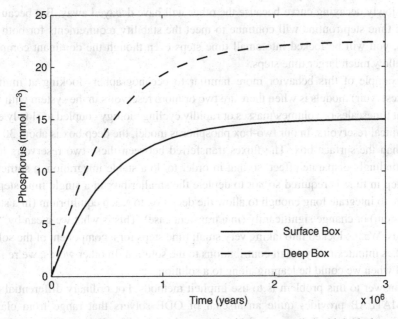

Figure 8.7 Our two-box global phosphorus model with more realistic biological removal of phosphorus from the surface box.

a half increase in execution time seems reasonable. Doing a whos shows that the number of time steps was larger, which fits (i.e. the computer took longer because it computed more time steps).

Now we suspect that the residence time of phosphate in the surface ocean is in fact a lot shorter than 100 years. Try running the experiment again, this time setting τ_x to 1 year. What you'll find when you do this is that ode45 takes forever to run the code. In fact, we had to kill the program with the task manager (Windows) or with kill -9 PID (UNIX). We finally got it to complete the integration by putting it on a really fast computer and letting it churn away for a long time. What happened? We didn't increase the computational load in the equations significantly. If you do a whos after running the code for a 3×10^6 year run, you find $\sim 3.7 \times 10^6$ time steps (vs. $\sim 110\,000$ for the 100 year time-constant case). It appears that the integrator has to choose smaller time steps for stability. Why ode45 needs to take so many more steps is the topic of the next section.

8.5.2 Stiff equations

A word needs to be said here about the problem of *stiff equations*. An ordinary differential equation (ODE) problem is said to be "stiff" when stability requirements force the solver to take a lot of small time steps; this happens when you have a system of coupled differential equations that have two or more very different timescales. For example, suppose your solution was the combination of (sum of) two exponential decay curves, one that decays rapidly and one that decays slowly. Except for the first few time steps, the solution is dominated by the slowly decaying curve because the other will have decayed away. But because the variable time step routine will continue to meet the stability requirements for both components, you will be locked into small time steps even though the dominant component would allow much larger time steps.

An example of this behavior more familiar to oceanographers looking at multi-box (multi-reservoir) models is when there are two or more reservoirs in the system with vastly different timescales or volumes/masses or rapidly cycling biology coupled to slowly evolving chemical reservoirs. In our two-box phosphorus model, the deep box is about 30 times larger than the surface box. The fluxes transferred between these two reservoirs have a correspondingly disparate effect, so that in order to do a stable integration, a sufficiently small step in time is required so not to deplete the smaller box in a single time step. But you have to integrate long enough to allow the deep box to reach equilibrium (in a steady-state system) or change significantly (in a transient case). This is what we mean by "stiff" equations. We get locked into taking very small time steps for a component of the solution that makes infinitesimally small contributions to the solution. In other words, we're forced to crawl when we could be leaping along to a solution.

The answer to this problem is to use implicit methods. For ordinary differential equations, MATLAB provides quite an arsenal of ODE solvers that range from classical applications of Runge–Kutta methods to rather sophisticated implicit, multi-step ODE integrators (see Table 8.3). How do we implement these approaches? We can use one of the stiff equation solvers in MATLAB in exactly the same way that we used `ode45`. This time we use `ode15s`, an implicit scheme that is somewhat less accurate, but far more efficient.

```
[t,X]=ode15s('phos2b',3e6,x0);
```

The `ode15s` scheme proceeds far faster than `ode45` and integrates over three million years in 65 steps. Of course, if you set τ_x back to 100 years, `ode15s` takes 63 steps to cover the same time span. This routine is only first-order accurate in Δt, so you might be tempted to artificially increase the resolution in time (as stability requires the adaptive time step to yield only a coarsely resolved system). Looking at the output of the integration above, we see that most of the action occurs in the first million years, so that we only need to integrate out to perhaps 1.2×10^6 years. Suppose we want to resolve at the century timescale. We can specify the reported time steps by providing a column vector of the time steps, e.g.

```
[t,X]=ode15s('phos2b',[0:100:1.2e6]',x0);
plot(t,X)
```

and `whos` confirms this by reporting t has 12 001 elements. But if you look at a plot of the results with 65 steps vs. 12 001 steps they appear to plot on top of each other. It turns out that for this problem the higher time resolution results differ from the coarse on the order of $0.01 \, \mathrm{mmol \, P \, m^{-3}}$ at 1.2×10^6 years. Small differences are expected in such a smoothly varying problem.

The question frequently arises, "Is MATLAB really using these extra time steps that I am specifiying?" The answer is both yes and no. Specifying a column vector of time steps does not change the way a MATLAB ODE integrator steps through the time interval. All of the ODE solvers in the MATLAB suite take time steps as required by their underlying method (Runge–Kutta, Bulirsch–Stoer, etc.). So, even though the ODE solver does not necessarily step to precisely the same time points you specified, the results provided at those time points are of the same order accuracy as the method being used. Specifying a lot of time steps has no effect on the efficiency of the method (because it is using its own internal time steps), but a great number of time steps can affect the amount of memory consumed.

8.5.3 Chaotic behavior

We should emphasize that although you may play by all the rules, e.g. choosing an efficient and accurate ODE solver for your system and perhaps dealing with stiff equations sensibly, the coupled equations may in fact exhibit a chaotic (seemingly random) behavior that is not a result of your numerical scheme. While chaos is a subject outside of the scope of this book, you may run into well-formulated biology models that exhibit this behavior (Roughgarden, 1998). This behaviour, in fact, may occur in natural systems. The difficult task is to convince yourself that what you're seeing is intrinsically chaotic behaviour and not a result of numerical instability. A number of approaches to rule out instability include the use of a fundamentally different ODE integration algorithm on the same system and changing (reducing) step size. If you find that the "statistical" behavior of the system remains independent of the scheme choice and step size, you are likely to be dealing with a truly chaotic system. By "statistical" we remind you that you need to examine the behavior over a number of "cycles" because different pathways or initial conditions will result in variations in individual realizations, and you need to compare the ensemble behavior of your solutions.

8.6 Other methods

You may think things are doomed if Runge–Kutta fails, but common sense tells you there must be a way. After all, these systems exist, and nature somehow finds a way to integrate them. In order to deal with stiff equations efficiently, some kind of implicit or semi-implicit scheme must be used. While we are big fans of Runge–Kutta for solving ordinary differential equations, as we indicated above there are times when other methods need to be used. So that you don't fall into the trap of using a mathematical tool as a "black box" we are going to describe in this section some of the more commonly used techniques to solve ODEs other than Runge–Kutta.

8.6.1 Predictor–corrector schemes

Predictor–corrector schemes are a subset of a class of multi-step, semi-implicit schemes that have some historical significance and have been used for efficient, precise solution of smoothly varying equations with complicated structure (i.e. many terms on the right-hand side). The most commonly used is the Adams–Bashforth–Moulton scheme. The approach is actually four basic steps done iteratively. As an example of the third-order version of this approach, the first step (the "P" or predictor step) is to make an estimate of the next step:

$$y^{n+1} = y^n + \frac{\Delta t}{12} \left(23 \left. \frac{dy}{dt} \right|^n - 16 \left. \frac{dy}{dt} \right|^{n-1} + 5 \left. \frac{dy}{dt} \right|^{n-2} \right) + \mathcal{O}(\Delta t^4) \qquad (8.48)$$

where in the above equation we've used a convenient shorthand for the time step index in the derivatives in the superscript (just checking to see if you are paying attention). In other words, we are estimating the next $(n + 1)$ value of y using the current position (n) and a Simpson's rule type evaluation using the current plus the two previous derivatives. The predictor is essentially a polynomial extrapolation.

The second step is the evaluation or "E" step in which you use the preliminary y^{n+1} estimate to evaluate the new derivative $\frac{dy}{dt}|^{n+1}$. You then take the corrector ("C") step to refine your estimate of y^{n+1} using:

$$y^{n+1} = y^n + \frac{\Delta t}{12} \left(5 \left. \frac{dy}{dt} \right|^{n+1} + 8 \left. \frac{dy}{dt} \right|^n - \left. \frac{dy}{dt} \right|^{n-1} \right) + \mathcal{O}(\Delta t^4) \qquad (8.49)$$

and finally take a fourth "E" step to re-evaluate the derivative at y^{n+1} to feed into the next predictor time-step calculation. Hence it's given the name "PECE". Of course, we do not have to keep track of all this bookkeeping as MATLAB provides us with `ode113`. An Adams–Bashforth–Moulton PECE method is useful for strict error tolerances and/or computationally difficult problems.

8.6.2 Bulirsch–Stoer extrapolation

The Bulirsch–Stoer algorithm for numerically integrating ordinary differential equations can achieve very high accuracy integrations with exceptional efficiency (i.e. with minimal computational effort). But, as with everything else in life, there are conditions. First of all, your problem (differential equation) must be very smoothly varying. Discontinuities or sudden jumps in value will give Bulirsch–Stoer problems. Second, there cannot be any singularities embedded within your limits of integration (any $\tan(t)$ that might approach $\pi/2$?). Third, as far as we can tell, MATLAB does not provide the Bulirsch–Stoer algorithm in its suite of ODE integrators. So this is a reality check; do you really need to use this algorithm or is there one listed in Table 8.3 that will suffice? Having said all of that, here's an introduction to the Bulirsch–Stoer method for those that need both high accuracy and high computational efficiency.

In Chapter 3 we discussed the concepts of interpolation and extrapolation. Now imagine one could take a mid-point integration scheme, like Runge–Kutta 2,3, and extrapolate the effect of crossing the Δt step not with two but with an infinite number of $\Delta t = 0$ steps. We know this sounds a little bit like Zeno's paradox, but this is essentially what the *Bulirsch–Stoer* method does. Now, of course, it doesn't really take an infinite number of $\Delta t = 0$ steps; it extrapolates to the effect of $\Delta t = 0$, with a relatively few ($\mathcal{O}(20)$ at the most), reasonable sized steps.

How is this done? Bulirsch–Stoer first estimates the integration by taking a big step (not at zero). It then refines the estimate by dividing that big step in two, then four, then eight, etc. Then assuming the error of these integrations are a polynomial function of step size, it builds higher-order estimates of the answer using using the polynomial recursion algorithm of Neville (Stoer and Bulirsch, 2002). As one builds up higher and higher orders of accuracy, the error (estimated from the order before) is checked; when it reaches a predetermined acceptable level, the algorithm stops. In reality, this usually happens long, long before it reaches the extrapolated infinite number of divisions (steps). If you feel you must have this algorithm to solve your problem, we recommend you check out the description (and code snippets) in Press *et al.* (2007).

8.6.3 Boundary value methods

When ODEs are required to satisfy boundary conditions in two different places along the independent variable axis, these problems are called "two-point boundary value problems". Boundary value methods involve iterative solutions from initial "guesses". Oceanographic examples might be the steady-state distribution of a pore water component when only the top and bottom concentrations (and the processes operating in between) are known. Similar problems arise in the water column under very special circumstances when lateral advection can be ignored. Earlier we said we would be concentrating on initial value problems, but there are problems and situations that cannot be solved by integrating forward from an initial condition. In this subsection we review some of the better-known boundary value methods and suggest the interested reader consult Press *et al.* (2007) to learn more.

The first type of boundary value method is known as the *shooting method*. At one boundary (call it the initial boundary) we specify values for all the dependent variables that are consistent with that boundary. We then integrate forward using initial value techniques to the other boundary (call it the final boundary) using a best-guess set of parameters for the differential equation. Quite often the values of these parameters are what is sought (e.g., diffusivities, advective velocities, and source and sink terms). Of course, our "shot" will have missed, but we will be able to calculate the discrepancies from the boundary conditions at the final boundary. We use this information to solve the multi-dimensional systems of equations and improve our initial parameters. Then we "shoot" again, and again, and again until our solution arrives at the final boundary within some predetermined accuracy and the routine terminates. In some respects, this resembles the nonlinear least squares approaches discussed in Chapter 3, and this is not accidental. One can imagine designing a similar χ^2 minimization strategy that seeks the optimal solution iteratively. Although conceptually straightforward, the primary difference would be the evaluation step (which would be an ODE initial value integration rather than chi-squared computation) so the iteration would be more computationally expensive, but manageable for simple systems. For more complex systems, a more efficient technique must be sought.

Another type of boundary value methods are known as *relaxation methods* wherein we make an initial guess at the solution of all the dependent variables in the domain. These guesses don't have to meet all of the boundary conditions, but the better our initial guess, the quicker we'll "relax" to the final solution. The relaxation involves modifying the solution of dependent variables in such a way that they come closer to agreeing with the differential equations and the boundary conditions. If this method suggests to you the solution of simultaneous equations, you are correct. If one can arrange one's space/time grid as a block diagonal, sparse matrix (see Fig. 2.7.1 in Press *et al.*, 2007) the inversion can be done quickly. Oddly enough, for the right kinds of problems (problems dealing with local subsets or problems involving one-dimensional approximations of more complex geometries), this approach is faster (more efficient) than the shooting method. We'll see one more "two-point boundary value" method in a later chapter when we discuss 1D models in sediment profiles (Chapter 14), the *tridiagonal algorithm*.

8.6.4 MATLAB ODE solvers

The ODE solvers in MATLAB fall, broadly, into two types: solvers for non-stiff and stiff problems. The plethora of ODE routines can be rather confusing, so in Table 8.3 we include

Table 8.3 *Summary of MATLAB ODE Solvers*

Solver	Problem type	Accuracy	Mathematical method	Recomended uses
ode45	non-stiff	medium	Explicit Runge–Kutta 4,5 method.	Try this method first. Often this is as far as you'll need to go.
ode23	non-stiff	low	Explicit Runge–Kutta 2,3 method.	Although it uses crude error tolerances, this method may be quicker than ode45. Also good for "moderately" stiff systems.
ode113	non-stiff	low to high	Variable-order Adams–Bashforth–Moulton predictor–corrector (PECE) method.	For strict error tolerances and/or computationally difficult problems. May give better results than ode45 at strict tolerances.
ode15s	stiff	low to medium	Based on numerical differentiation formulas (NDF), variable order.	If ode45 is slow (stiff systems). Try this one if ode45 fails. First-order accurate.
ode15i	stiff	low to medium	Fully implicit backward differentiation (BDF), variable order.	If your problem must be formulated in a fully implicit fashion. Initial conditions must be consistent. This one is a little different than the others.
ode23s	stiff	low	Modified Rosenbrock method of order 2.	For stiff systems when ode15s fails. Crude error tolerances, but higher-order accuracy.
ode23t	moderately stiff	low	The trapezoidal rule using a "free" interpolant.	If the problem is only moderately stiff and you need a solution without numerical damping.
ode23tb	stiff	low	An implicit Runge–Kutta method with first stage trapezoidal rule and second stage backward differentiation formula of order 2 (TR-BDF2).	If using crude error tolerances to solve stiff systems. May work when ode15s fails at crude tolerances.

some "words of wisdom" as to which ODE routine to use for which kind of problem. This table is inspired by and constructed from information found in the nearly 5000 pages of the *MATLAB 7 Function Reference* (MathWorks, 2009) and Shampine and Reichelt (1997). Our first piece of advice is to analyze the model. If it is a multi-box model, estimate response timescales based on fluxes and volumes. What would be the longest and shortest characteristic relaxation timescale of the system? When you do your first integration, do it for a short period (small compared with the final run) to get an idea of how effective the ODE solver is, and how long it might take for the real run (i.e. do you need to bring lunch, or a pup-tent?).

We advise you always to start with `ode45`; if it fails or seems to be taking a long time to integrate (even if you give it a small time interval to integrate over, as a test), then consider the following. Are you asking it to integrate solutions that are very small in comparison with the accuracy tolerances either provided to you by default or explicitly given by you? If that doesn't seem to be your problem, think about whether or not you may be integrating stiff equations with a method that wasn't meant to integrate stiff equations (like `ode45`). Some times the easiest thing to do is to try a different ODE solver, something like say `ode15s`. Don't make the mistake of trying to fix your stiff equation problem by switching to another non-stiff solver (like `ode23`). Be sure to look at Table 8.3 carefully, and if you have more questions, The MathWorks has provided extensive documentation about their ODE solvers on their website, and this documentation may be loaded onto your machine as well.

8.7 Problems

8.1. The Taylor series can be written in two equivalent forms. Typically most math books present the Taylor series in this form:

$$f(t) = f(a) + (t-a)\frac{df(a)}{dt} + \frac{1}{2!}(t-a)^2\frac{d^2 f(a)}{dt^2} + \cdots + \frac{1}{n!}(t-a)^n\frac{d^n f(a)}{dt^n} \quad (8.50)$$

and refer to this as the Taylor series expansion of $f(t)$ around point a where n represents the number of terms before truncation. Demonstrate with pencil and paper that this equation is equivalent to Equation (8.24).

8.2. Modify our biological, two-box model of the global phosphorus cycle to step systematically through the remineralization efficiency from 90% to 100%, making note of both the time it takes to reach equilibrium (t_{eq}) and the concentration difference (Δx) between the surface and deep box. Plot Δx vs. t_{eq} in a separate figure window. Do the same for time constants τ_x of 1, 5, 20, 50, and 100 years. Explain your results.

9

A model building tutorial

> Answering difficult questions is always easier than answering easy
> ones: you are not accountable for the inconsistencies. And asking
> simple questions is the hardest part of all.
>
> *Henry Stommel*

Until now we have concentrated on what may be loosely termed "data analysis methods". In some respects, this is a form of modeling in that we are attempting to interpret our data within the context of some intrinsic model of how our data should behave, whether it be assuming the data follow an underlying probability distribution, vary as a function of some other variables, or exhibit some periodic behavior as a function of time. We hope you are beginning to see that all of these methods share common mathematical and algorithmic roots, and we want you to realize that many of these tools will come in handy as we now embark on a more model-intensive course.

Before doing so, we want to outline the basic aspects of model design, implementation, and analysis. Selecting the most accurate and efficient algorithms and developing robust and usable MATLAB code is important, but most of your intellectual energies should be directed at the design and analysis steps. Moreover, although correct design is vital to any successful modeling effort, developing the tools to efficiently analyze model output is just as important. It is critical to assessing the mechanics of how a model is performing as well as ultimately understanding the underlying system dynamics and how well a model compares to observations.

To illustrate our points we will discuss the design, construction, and analysis of three example models. The models we will build are the simplest geometry possible, namely *box models*. We first show you how to build a simple ecosystem model, a predator–prey (phytoplankton–zooplankton or PZ) model. Next, we explore in more detail the simple two-box phosphorus model developed in Chapter 8. Finally, we extend the two-box model to a multi-box model of the world ocean. There are many more sophisticated representations of biogeochemical processes available, but our pedagogical intent here is to explore the simplest models that exhibit biogeochemically interesting behavior. Such models are by no means "state of the art", but they are sufficient to teach us some important lessons about modeling and modeling techniques. More sophisticated ocean models are presented in later chapters.

9.1 Motivation and philosophy

Conceiving, designing, and constructing a marine model is a tricky business; it requires a clear understanding of the physics, chemistry, and biology of the system you are modeling, plus an appreciation of the mathematical and numerical (algorithmic) tools at your disposal. Equally important is the process of using the model to reveal the underlying dynamics of the system.

The first step in creating a model is to identify your objectives. Why are you creating the model? A clear understanding of your motivation will inform many of the subsequent decisions (and inevitable compromises) that you will make along the way. We can think of a few overlapping reasons for modeling a system, but it is likely that your motivation may be a blend of two or more of these. If you think of others, let us know.

- **To replicate your observations** – Certain features in your data, or someone else's data, have caught your attention. Are they real, or an experimental artifact? Are the structures physically or biogeochemically possible? Can you construct a model that faithfully reproduces these features? What is the least complicated model that can generate the structures observed? If the model embodies all the processes that you believe important, if it is geometrically detailed enough to resolve everything, and you have done an exhaustive exploration of all reasonable parameter values, yet you still cannot replicate the observations, then you have obviously learned something. On the other hand, if you can simulate the observations with adequate fidelity, you may never know if you got the right picture for the wrong reasons.
- **To identify underlying processes** – You may need to ask basic questions about what causes something. For example, each summer in the subtropics, an oxygen maximum appears near the base of the euphotic (sunlit) zone. Is this due to photosynthesis or physical processes (see Chapter 15)? To answer this question, you might construct a model that includes the physical processes only, and ask whether the model can reproduce your observations. If not, you have answered your question. Selectively adding biogeochemical processes until the model emulates observations will allow you to assess which processes are important.
- **To quantify one or more processes** – In some ways, this is just an extension of the previous motivation, but is more "mission oriented". In essence, you would construct a model and then *tune* model parameters to match observations. In our example of the oxygen maximum, you may be asking what is the magnitude of gas exchange, net community production, or some physical process like vertical mixing. While some quantities, like the first two mentioned, may be of general interest, others like vertical mixing

are often highly *model specific* and may not be usable outside the context of the specific model. The reason is that your model cannot incorporate all possible physical and biogeochemical processes, and cannot span all space- and timescales. Thus things are left out, approximated, parameterized, or just plain ignored. For example, in order to successfully emulate seasonal mixed layer temperatures, vertical mixing in your one-dimensional biogeochemical model may be artificially large (or small) to compensate for horizontal heat fluxes (or more accurately, heat flux *divergences*) not included in your model formulation.

- **To extrapolate/interpolate in space or time** – In Chapter 3 we used mathematical functions to fit our data. These functions can then in turn be used to interpolate or extrapolate to a point in space or time of interest. A similar approach can be used with your model. If you can develop a model that incorporates all the important processes and has sufficient structural complexity that it can successfully emulate your observations, perhaps you can use this same model to interpolate or extrapolate to some location where you would like to have data, but don't. Bear in mind that many of the *caveats* described in Chapter 3 apply here. In addition, you have the related challenge of whether the same underlying processes remain unchanged in these inaccessible locations. If you're bold enough to want to use your model *prognostically* – that is, to predict the future – you face similar perils. Your model may indeed faithfully replicate current (and past) observations, but is it getting the right answer for the wrong reasons? Have you overlooked an important process? Will the underlying rules change with time? Here we repeat the standard Securities and Exchange Commission *caveat*: "Past performance is not necessarily a guide to future results."

A model should be no more complex than is necessary. Consider a famous paraphrase of Einstein: "Everything should be made as simple as possible, but not simpler." You should strive to construct the simplest possible model that will adequately reproduce the desired phenomena. Adding extraneous and unnecessary processes or geometric embellishments only serves to slow down model performance and confuse the picture.

It is not enough to construct the model and execute it in an accurate and efficient way; you must use it to enlighten yourself. For a simple model, this enlightenment may come from observing its behavior, or its response to changing boundary conditions and parameters. In more complicated models, you may need to invest considerable effort in characterizing or quantifying key aspects of the modeled system. Quantitative comparison of model properties to observations often leads to some *performance metric* or *figure of merit* that can be used to judge the model's fidelity and serve as a guide to optimizing its performance by adjusting poorly known parameterizations. An example of a figure of merit might be some weighted, root mean square deviation between model and observations. Decisions regarding model performance metrics involve deliberate choices about what features are key to assessing model fidelity, and what are not. These choices link directly back to your modeling objective. These *model analysis* activities are an important adjunct to model construction and execution, and are as important to successful modeling as well-designed numerical codes.

9.2 Scales

If you really think about it, the physical and biogeochemical processes operating in the global environment span an awe-inspiring range of scales. Processes such as molecular viscosity as well as biological and chemical reactions occur on molecular scales ($\sim 10^{-8}$ m), yet influence and are affected by processes occurring on macroscopic (mm to Mm) scales. For example, phytoplankton are influenced by molecular diffusive gradients of nutrients. These nutrients may in turn be provided by large-scale ocean currents influenced by planetary rotation. The same consideration applies in the time dimension. Chemical and biological reactions occur on timescales ranging from milliseconds to minutes, but may be influenced by events or processes occurring on diurnal, seasonal, annual, or even glacial–interglacial timescales. Thus we are dealing with a system that potentially can span 15 orders of magnitude in time as well as in two of three spatial dimensions (perhaps only 11 orders of magnitude in the vertical). Conceptually, let alone computationally, it is clearly impossible to construct a model that resolves and includes everything at all scales.

To state the obvious, a model must be big enough to encompass the system and processes of interest, but no larger. For example, although an upper ocean ecosystem in the subpolar North Atlantic may ultimately depend on a long-term nutrient supply that derives from upwelling deep water in the Southern Ocean and is carried northward by the upper limb of the global conveyor, it may be sufficient just to specify that some long-term supply arrives locally if you are only interested in modeling the spring bloom over one annual cycle. If, on the other hand, you want to model the same system's behavior over the last glacial–interglacial period, you are looking at a global model instead.

Given the outer bounds of your model, how fine a scale must it resolve? Your model will be some discrete representation of the continuum we call reality. That is, it will represent reality at a series of grid points, and anything that happens on scales smaller than the distance between those grid points will not be *resolved* and thus must be assumed (parameterized). Here there are practical computational limitations. Suppose you construct a model of the wind driven upper layer of the North Atlantic subtropical gyre. It's not important at the moment to ask what that means, but suffice it to say that it might have a horizontal extent of about 5000 km, and a vertical depth of about 1000 m. For the purposes of our study, we might desire a horizontal resolution (the spacing between the model's horizontal grid points) of 50 km, and a vertical resolution of 10 m. Why you might make such choices is something we discuss in later chapters, but we wanted to point out that such a model will thus have $100 \times 100 \times 100 = 1\,000\,000$ grid points. Doubling the resolution (i.e. halving the grid spacing) in all three directions increases the number of grid points by a factor of eight!

The bottom line is that particularly with two- and three-dimensional systems, practicality dictates the *dynamic range* (the ratio of the largest to the smallest scales resolved) over which you can effectively model. Everything smaller than your model's grid point spacing (i.e. *sub-gridscale*) must be represented as a parameterized process. Anything larger than your model must be imposed as a boundary condition. A consequence of this dynamic range limitation is that depending on the nature of the process or system being studied, a model will be characterized by the region of space and time that it encompasses, and the two are roughly correlated. Figure 9.1 is a schematic of typical space-time domains for ocean processes (and models).

As you might expect, there is a general correlation between the spatial scales and timescales; larger-scale motions are usually associated with longer timescales and *vice versa*. Although your model may generally fall within one of these ranges, it may in practice overlap with adjacent scales. No doubt it will be affected by smaller-scale, unresolved processes that must be parameterized in some fashion, and it will be influenced by larger-scale processes that must be imposed as boundary conditions in your model domain. We briefly describe these space-time regions, what phenomena they encompass, and how they are typically modeled.

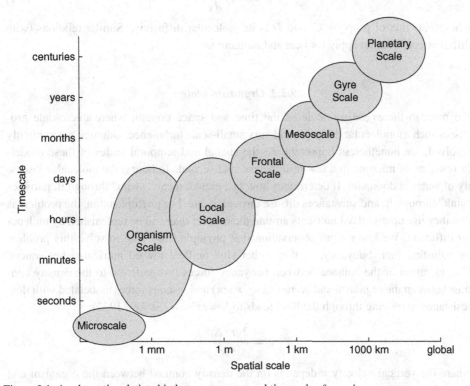

Figure 9.1 A schematic relationship between space- and timescales for various ocean processes.

9.2.1 Microscale

This domain encompasses the very smallest-scale motions, including molecular, that occur in nature. Molecular behavior, including thermal (Brownian) motion and the diffusion of chemical species is adequately described by statistical physics and thermodynamics. The details of chemical reactions and transport of ions across cell membranes are extremely complicated and must obey quantum-mechanical rules, and are the subject of sophisticated and compute-intensive research. Another example is the viscous film at the surface of the ocean, where surfactants and biological exudates play a role in controlling the thickness and persistence of this film. It is an important boundary layer for the exchange of gases between the ocean and the atmosphere because gas transport across this layer must occur by molecular diffusion and may be influenced by chemical reactions. It is also at this scale that the cascade of turbulent motions, driven by the larger-scale circulation and wind stirring, must ultimately be dissipated by molecular viscosity.

It is unlikely that as a marine scientist you will be modeling these phenomena directly. Fortunately, as unresolved sub-gridscale processes, they can be effectively represented in larger-scale models by well-established formalisms, e.g. Fick's first law for molecular diffusion, which states that the flux of a property is proportional to its gradient:

$$F = -D\frac{dC}{dx} \qquad (9.1)$$

where F is flux of property C, and D is its molecular diffusivity. Similar relations (with different coefficients) apply for heat and momentum.

9.2.2 Organism scale

We refer to the *organism scale* as the time and space domain where microscale processes such as molecular diffusion and very small-scale turbulence, although not explicitly resolved, are nonetheless important. At the spatial and temporal scales of these models (a few tens of microns to a few millimeters, and seconds to hours) the molecular viscosity of water is dominant. It determines how fast plankton can "swim" through it, particles "sink" through it, and substances diffuse through it, etc. For phytoplankton, the problem is that they use up dissolved nutrients around them faster than can be replenished by molecular diffusion. We know from observation that phytoplankton tend to solve this problem by adjusting their buoyancy, so they either sink or float toward nutrient-richer waters. Consideration of the balance between buoyancy forces (proportional to the density contrast between the organism and surrounding water) and viscous forces associated with flow resistance to moving through the fluid leads to *Stokes' law* (Stokes, 1922):

$$v = \frac{2gr^2\Delta\rho}{9\mu} \qquad (9.2)$$

where the vertical velocity v depends on the density contrast between the organism and water $\Delta\rho = (\rho' - \rho)$, g is the acceleration due to gravity, r is the organism's radius, and μ

the *dynamic viscosity* of the water (with units $\mathrm{kg\,m^{-1}\,s^{-1}}$). Substituting "typical" values, one obtains a vertical motion of order $10\,\mu\mathrm{m\,s^{-1}}$ (about $1\,\mathrm{m\,d^{-1}}$) for a $50\,\mu\mathrm{m}$ diameter plankter with a 1% relative density contrast.

Stokes' law is an idealization that assumes a spherical body moving sufficiently slowly that the fluid motion around it is laminar (i.e. non-turbulent). For non-spherical bodies, the net result is a slower vertical motion. The non-turbulent flow assumption is satisfied for planktonic organisms significantly smaller than 0.1–1 mm in size. For larger species or particles with greater density contrast with faster sinking or ascending rates, the flow around the body becomes turbulent. In this instance flow resistance is affected, so that another relationship is found to hold:

$$v = \sqrt{\frac{16gr\,\Delta\rho}{3\rho}} \tag{9.3}$$

Modeling in this domain would be aimed at exploring the detailed interaction of individual organisms with their environment and with other organisms and particles (e.g. predator–prey interaction). Turbulence within the marine environment consists of a cascade of energy from the very largest scales ultimately down to the molecular scale. The smallest scales over which turbulence is important is typified by the *Kolmogorov length scale*, which is given by:

$$\eta = \left(\frac{v^3}{\varepsilon}\right)^{1/4} \tag{9.4}$$

where v is the *kinematic viscosity* (the molecular diffusivity of momentum equal to the dynamic viscosity divided by density leading to units of $\mathrm{m^2\,s^{-1}}$) and ε is the rate of turbulent dissipation of energy. For a typical upper ocean environment, $100\,\mu\mathrm{m} < \eta < 1\,\mathrm{mm}$. The character of the turbulent flow affects not only the availability of nutrients, but also where the organisms may accumulate in a complex flow field (e.g. Jumars *et al.*, 2009). As with the microscale, computational demands tend to be high for other than the simplest scenarios.

9.2.3 Local scale

By "local scale" we mean systems that are large enough that individual organisms are not resolved, but small enough that geographical trends over the domain are inconsequential. In this instance we are thinking of scales from meters to perhaps a few hundred meters at most. Timescales may range from hours to days or longer. A particularly important application would be modeling the surface mixing layer of the ocean, the boundary layer between the ocean and the atmosphere. This is a critical interface where the ocean and the atmosphere interact. Gas exchange, biological production, and the thermodynamic and physical interaction between the ocean and atmosphere occurs here. This is where water mass characteristics are "set".

At the scales typical of the mixed layer, phytoplankton face the problem that the source of light is from above and the source of nutrients is generally from below. The part of the water column where there is enough light for photosynthesis is called the *euphotic zone*. And the light intensity through the water column tends to follow an exponential decrease with depth:

$$I = I_0 e^{-\alpha_\lambda z} \tag{9.5}$$

where I_0 is the light intensity just below the surface and α_λ is the attenuation coefficient, which is a function of wavelength λ and turbidity. In coastal waters, where biological activity is high (*eutrophic* conditions), the presence of particles, organisms, and their exudates serve to absorb sunlight so that it doesn't penetrate deeply. In open ocean (*oligotrophic*) waters, the relative absence of organisms allows greater light penetration.

At these scales the physics of the mixed layer play an important role in defining the biological and chemical environment. The gain and loss of heat at the surface, the *in situ* warming due to the absorption of sunlight, the stirring of the mixed layer by wind, and the time evolution of all these processes affect the structure and depth of the mixed layer and consequently passage of nutrients, plankton, and other properties through this boundary layer. We will investigate one form of mixed layer modeling in Chapter 15.

Another feature that might be modeled on the local scale is *Langmuir circulation*. Looking out from a ship over the ocean surface you sometimes see streaks and bands (sometimes called *windrows*) associated with 10–100 m scale overturning called *Langmuir cells*. Vertical velocities on the ascending limbs of these cells can reach $0.1 \, \mathrm{m \, s^{-1}}$, and bring phytoplankton and debris to the surface, concentrating them in horizontally elongated bands at the downwelling limbs. Such areas serve as attractors to fish and other biota that can be modeled on the local scale. Local-scale structure is generated in the subsurface ocean as well by internal waves, which play an important role in vertical mixing in the stratified part of the water column.

9.2.4 Frontal scale

If you look at detailed regional maps of sea surface temperature you see that large-scale gradients in temperature are punctuated by relatively abrupt transitions or *fronts* on the 1–10 km scale associated with surface convergent flows. These occur as a natural consequence of the turbulent cascade of motions from the very large-scale circulation down to the microscale. For this reason, we refer to this domain as the *frontal scale*, but it may also be called the *sub-mesoscale*. The interaction between the physics, chemistry, and biology may be very important on these scales, because the timescales associated with the circulation (hours to days and longer) coincide with the biogeochemical response of organisms. Numerical models (e.g. Levy *et al.*, 2001) show that explicit inclusion of processes on the ~1 km scale substantially increases the biological production seen in the model. Other systems where modeling on this scale is likely to be important include estuaries, river plumes, and coastal regions.

9.2.5 Mesoscale

The cascade of turbulence from the largest-scale ocean circulation to the microscale fills a spectrum of motions that has a peak around length scales of 10–300 km, reflecting the influence of planetary rotation on ocean flow. These eddy features appear ubiquitously in the ocean and are particularly important for the transport of material and momentum. Eddies are an important part of large-scale ocean dynamics, and their presence or absence in an ocean physics model is a key feature determining the model's behavior. Models that do not resolve eddies (*non-eddy-resolving models*) characteristically parameterize the effects of these unresolved motions as "Fickian" diffusion with large coefficients. That is, the turbulent flux of some property would be modeled as a function of some large-scale gradient and some property-independent effective diffusion coefficient:

$$F_{\text{Turbulent}} = -\kappa_H \nabla_H C \tag{9.6}$$

where $\nabla_H C$ is the horizontal gradient in property C and κ_H is the horizontal turbulent diffusivity (Chapter 11), which might vary between 100 and 1000 m^2 s^{-1}, depending on the model and the location (compare this with Equation (9.1)). Models that resolve smaller scales will typically require smaller turbulent diffusivity coefficients to account for sub-gridscale motions.

9.2.6 Gyre scale

In general, wind driven circulation tends to form basin scale gyres that circulate cyclonically (counter-clockwise in the Northern Hemisphere) or anticyclonically. Rather than being symmetrically circular or oval in shape, they characteristically have intense western boundary currents on their edge. Although permanent features, these currents exhibit a fundamental dynamical instability that causes them to meander, forming loops that from time to time pinch off into large circulating eddies called "rings" that tend to move with the overall circulation. Such features can survive for months or longer before "spinning down" or being absorbed into the western boundary current. These features, typically larger than mesoscale eddies, are important mechanisms for material transfer. Capturing or emulating their behavior can be a challenging modeling task.

Gyre models typically encompass 1000 to 10 000 km scales, and span timescales ranging from months to decades. If they are truly *eddy resolving*, that is they resolve fluid motions sufficiently finely to spontaneously generate and propagate mesoscale eddies, they will have resolutions of order 10 km or less. Such models are understandably computationally demanding, as they must span a much wider range of scales than the non-resolving models. As a compromise, some models have high enough resolution to incorporate eddies but not spontaneously generate them; termed *eddy permitting*, these models can be operated by "artificially" injecting eddy motion into the circulation. On an even coarser scale, the effects of eddies on material transport can be modeled as a turbulent mixing process (see Equation (9.6)). We will explore a simple example in Chapter 16.

9.2.7 Planetary scale

The largest scales from ocean basins to the globe are relevant for a myriad of problems such as climate variability and climate change, ocean CO_2 storage, and the long-term controls on large-scale patterns of nutrients, oxygen and biological productivity. Modeling the global ocean has advantages over smaller-scale models, such as more realistic bathymetry and coast lines and no pesky model lateral boundary conditions (except perhaps rivers) but this realism also comes at a substantial cost. The timescale of the ocean overturning circulation is roughly 1000 years and the timescale for adjustment of the deep ocean can be many thousands of years because of the slow response time of vertical eddy diffusion. Planetary-scale simulations are typically carried out in a *general circulation model* (GCM, see Chapter 17). These models can be forced with climatological or idealized atmospheric characteristics, but quite often are run as *coupled models*, that is with a model global atmosphere as well. Because of the huge computational challenges associated with such efforts, these are typically carried out by large research teams with access to supercomputers.

9.3 A first example: the Lotka–Volterra model

Models have a few standard parts, some of which you are probably already familiar with. In this section we discuss a small group of elements that are almost universally contained in all marine models. As an example, we build the simple Lotka–Volterra predator–prey (phytoplankton–zooplankton or PZ) model. In light of the previous section (9.2) this simple model (a "box model") is applicable to mixed layer to basin-scale models, but only in a highly idealized and somewhat unrealistic world.

9.3.1 Assumptions

The *assumptions* that lie behind the model form an important foundation. These are the precepts one uses to build the model, at first in one's mind and then later in one's computer. We emphasize that these are arrived at a priori; they are preconceived notions one brings to the table – to be modified later as circumstances dictate. We advise you to enumerate your assumptions explicitly and to keep them firmly in view as you go about your model construction, execution, and, most importantly, interpretation.

In the case of the Lotka–Volterra predator–prey model, Volterra (1926) was trying to imagine a model that would help explain the time dependent, oscillatory behavior ecologists observed in the rise and fall of predator and prey populations. Lotka was a chemist who wrote about oscillating chemical reactions with similar equations (Lotka, 1920; 1925), hence the name. The assumptions behind the Lotka–Volterra model can be listed as:

1. The prey are not limited by food; they grow at a constant specific rate and their numbers increase dependent upon how many are present. That is to say, without any predators around the prey grow in an exponential fashion.

2. Without prey to feed on, the predators die off at a rate proportional to their numbers. As they are starving, the predator population decays in an exponential fashion.
3. The predators eat the prey, of course, and this represents a type of energy transfer. But the rate of loss experienced by the prey does not have to equal the rate of gain by the predators. This is an important assumption, as we shall see.

These assumptions may not sound like a very realistic model, and many have pointed this out over the years. But as an introductory, simple box model the Lotka–Volterra model provides a good starting point. Furthermore, studying this model should help inspire you to think of ways (both mathematically and logically) it can be improved. However, we urge you to consult the literature rather than trying to reinvent the wheel. Case (1999) is a good starting place.

9.3.2 State variables

By definition, *state variables* are those properties that you need to know to describe the state of a dynamical system. Combined with the geometry of the system and the underlying equations, the state variables should in principle provide you with enough information to predict the future behavior and condition of the system. In a physical circulation model, the state variables might include the distribution of temperature and salinity (and hence density) along with fluid velocities. In an ecosystem model, the state variables would be the populations (or as used here more properly population densities) of predators (e.g. Z for zooplankton) and prey (P, for phytoplankton). Because of this naming convention, such models are often referred to as "PZ" models. We will use a more generic nomenclature for these state variables, namely x_1 and x_2 for the population densities of prey and predator, respectively (see Fig. 9.2).

9.3.3 Dynamics

Model dynamics embody the relationships or interactions between the state variables. We've represented the two populations (and hence state variables) by boxes in the schematic shown in Fig. 9.2, and the fluxes of material into/from and between them as arrows. It is, of course, these arrows that define the dynamics of the system modeled. Such a diagram allows you to visualize how the model functions, and we recommend that you construct such a diagram whenever possible as you design your model. In addition to documenting the basic elements of your model, it serves as an important communication tool to others.

We consider each of the arrows in turn. The first is an arrow pointing upwards into the prey box. This represents the first assumption (above), that there is an exponential, unconstrained growth of the prey that is proportional to the population. The constant of proportionality (a growth rate) is p_1. The second arrow represents the loss of prey due to grazing, which should be proportional to the product of the prey and predator population

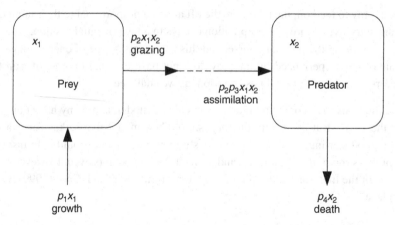

Figure 9.2 A schematic of our simple Lotka–Volterra predator–prey (PZ) model. In this model the state variables are prey (x_1) and predator (x_2). Note that the arrow from box x_1 to box x_2 has a dashed segment to represent the fact that $p_2 x_1 x_2$ does not necessarily need to equal $p_2 p_3 x_1 x_2$.

densities. This follows the simple logic that the rate of predation depends on the "collision rate" between predator and prey, and hence both the number of predators and the number of prey. The constant of proportionality (related to predation success) is p_2. The third, related arrow represents assimilation by the predator population. That is, predators may be sloppy or inefficient eaters, so not all the material coming from the prey makes it into the predator pool. The assimilation efficiency is p_3 and will be between 0 and 1. Finally, the fourth arrow is the death flux of predators (if we had cast the model in terms of biomass rather than number of individuals, p_3 would also account for the predator's background metabolism rate). In this model the only way the prey die is by being eaten by predators, while the predators themselves can just die (presumably some sort of net mortality rate due to old age, disease, being eaten by larger predators, smoking, or perhaps eating bad seafood). Collectively, the parameters p_i control the behavior of the model. You will likely set (using some a priori knowledge or personal prejudice) or vary these parameters depending on your modeling objectives.

Finally, the alert reader will see that this model doesn't really conserve mass. First, there is no limit to growth in the prey population. If we think of the system not in terms of population numbers, but rather in terms of nitrogen (a limiting nutrient), we ought not to be free to create more and more prey without consideration of what nitrogen may be available to "build" new cells. Second, the efficiency of conversion of prey to predator mass (p_3) will likely be less than 100%. We might regard that loss as releasing organic material (nutrients) back into the dissolved pool. Finally, the loss of predators to death will likely be manifested in cell lysis and settling of particulate material out of the system. Be that as it may, a more complete representation might include more of these effects, and perhaps more pools (state variables) like dissolved inorganic and organic nutrient and carbon pools to close the mass budgets.

9.3.4 Equations

At this point we can now write out equations that describe the model dynamics. These are based on the concept of conservation of mass among state variables, namely that any flux into one of the boxes in Fig. 9.2 results in an increase in the population density (the x_i state variable), and any flux out of the box leads to a decrease. We thus end up with two coupled ordinary differential equations (ODEs):

$$\frac{dx_1}{dt} = p_1 x_1 - p_2 x_1 x_2 \tag{9.7}$$

$$\frac{dx_2}{dt} = p_2 p_3 x_1 x_2 - p_4 x_2 \tag{9.8}$$

9.3.5 Boundary conditions

Because of the underlying scientific laws involved, a mathematical description of your model will almost always involve one or more differential equations. Your modeling efforts will by definition involve integrating these equations over time and/or space. To form a complete description of the system, that is to account for the "constants of integration", one needs additional information, called *boundary* or *initial conditions*. Paying careful attention to the choice of boundary or initial conditions is important, because the choice can sometimes have profound effects on the outcome of your experiment. Sometimes the impact will be how long it takes to reach a steady state, but sometimes it may play a critical role in how the system behaves. This will become evident in our example here. For now, we will "arbitrarily" choose starting values of our two state variables as $[x_1, x_2]_0 = [0.8, 0.171]$.

9.3.6 Computer code

As we did in the previous chapter, we will integrate these equations from initial conditions using a MATLAB Runge–Kutta integration routine. In order to do so, we construct a "kernel" function which we'll call LV_pz.m:

```
% LV_pz.m
% A function m-file for use with an ODE solver, a set of coupled
% ordinary differential equations for a simple ecosystem model where
% p(1) = phytoplankton growth rate
% p(2) = zooplankton grazing rate
% p(3) = zooplankton assimilation efficiency
% p(4) = zooplankton mortality
function dxdt = LV_pz(t,x)
    global p              % makes parameters accessible outside of func
    dxdt=zeros(2,1);      % reserve space for function output
    dxdt(1) = x(1)*(p(1)-p(2)*x(2));
    dxdt(2) = x(2)*(p(3)*p(2)*x(1)-p(4));
```

In the simplest implementation, you could "hard wire" the values of the parameters within this function, but changing the values would subsequently prove cumbersome. We choose to define the array of parameters as a `global` variable, because it provides a mechanism for us to change the values outside of the kernel function somewhere else in the model code. See Section 9.4.1 for a more complete discussion. We use this function by calling it from our model's "driver" program (`LV_pz_main.m`):

```
% LV_pz_main.m
% A simple 2-population Lotka-Volterra ecosystem model
%
global p          % define global parameter array (to match kernel)
% units of concentration (mmol/m^3) and fluxes (mmol/m^3/d)
p(1)=.1;     % phytoplankton growth rate (1/d)
p(2)=.4;     % zooplankton grazing rate (1/d)
p(3)=.2;     % zooplankton assimilation efficiency (dimensionless)
p(4)=0.05;   % zooplankton mortality (1/d)
T=[0:0.1:400]';  % use a fixed timescale to compare cases
n=length(T);
x0=[0.8 0.171]';  % initial conditions x1(t=0)=0.8, x2(t=0)=0.171
[T,X]=ode15s('LV_pz',T,x0);
```

where we link to the parameter array used in the the kernel function by defining p as global. For integration purposes we use `ode15s.m` to deal with any potential stiffness in the equations and at any rate produce speedy integrations. The reader, of course, could experiment with other integrators.

9.3.7 Exploring model output using time series and phase diagrams

Analyzing model output can lead to new insights into model performance, and perhaps into how the world works. The first and most obvious step is to plot the time evolution of the state variables over the course of the integration (the left panel of Fig. 9.3). This shows an interesting cyclical pattern that occurs despite the constancy of the system "forcing". (That is, we're not changing any of the parameters with time.) These oscillations are an internal dynamic of the system and are a typical manifestation of predator–prey relationships. A predatory population will grow in response to burgeoning prey populations, but will grow until the prey population begins to decrease because predation exceeds growth. The predator population will subsequently decrease through lack of food (death rates exceed income), and the prey population will begin to recover. This "delayed feedback" leads to oscillations, which in this case appear stable.

Plotting the two state variables against one another (Fig. 9.3, right panel) reveals the relatively stable relationship that has evolved, where the two populations endlessly circulate in "phase space". Such a diagram is also known as a *Lissajous figure*, commonly used in electrical engineering to diagnose the relationship between two time-varying signals. The shape and structure of such plots can tell us much about what is going on in a system. A closed loop, such as seen in Fig. 9.3, indicates that the model is in a stable configuration.

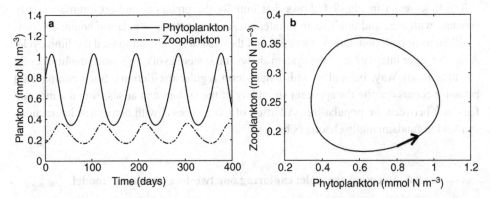

Figure 9.3 A plot of the results of a simple Lotka–Volterra predator–prey (PZ) model run with initial state variables of [0.8, 0.171]. (a) A 400 day time series where the solid line represents phytoplankton (the prey state variable x_1) and the dot-dashed line zooplankton (the predator state variable x_2). (b) A phase plane diagram of the two populations, i.e. the time series of P (state variable x_1) plotted vs. the time series of Z (state variable x_2), endlessly circling their limit cycle (the arrow provides the time direction). Clearly we have obtained some sort of stable oscillation in that each component repeats the same orbit, cycle after cycle.

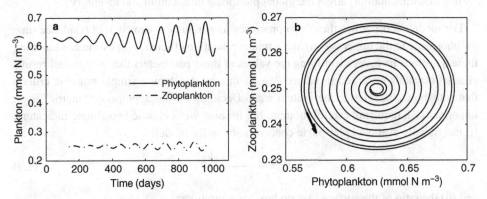

Figure 9.4 A plot of results from the same simple Lotka–Volterra predator–prey model used for Fig. 9.3 with two small differences: the initial condition vector is now [0.62, 0.25] and the time integration is 1100 days (roughly three years). (a) The time series plot of P and Z; notice how the oscillation amplitude of these populations appears to be growing with time. (b) The phase space diagram for this simulation showing the evolution of the two populations' limit cycle, which spirals outwards. Note the last orbit is much closer to the previous orbit than before suggesting the outward movement may be decreasing.

However, if we change the inital conditions of our model to $[x_1, x_2]_0 = [0.62, 0.25]$ we get very different results (Fig. 9.4). In essence, by choosing initial conditions that do not fall on the limit cycle in the state variable phase space shown in Fig. 9.3, we are now forcing the sytem to find its own way back. You can also now recognize that our "arbitrary" starting conditions for the first experiment were carefully selected.

The time series in Fig. 9.4 shows that initially the predators and prey remain nearly constant with time and don't seem to interact. As the predator population begins to grow, oscillations in both populations increase, and the system spirals out toward the limit cycle. A much longer integration of the system shows the system evolving toward the limit cycle not in a steady way, but rather with longer-term regular oscillations. Such complicated behavior occurs despite the apparent simplicity of the system and arises from the intrinsic feedback between the populations. Addition of more "pools" with associated interactions can lead to fundamentally chaotic behavior.

9.4 A second example: exploring our two-box phosphate model

As a second example, we revisit the two-box phosphate model that we introduced in Section 8.5.1 to demonstrate how we can explore parameter space. Our motivation for doing so is to ask the following basic questions of the model:

1. How important is biological productivity in shaping the distribution and inventory of phosphate in the global ocean?
2. How does the remineralization efficiency (or conversely the "lossiness" of the system due to sedimentation) affect the global phosphate distribution and inventory?

The model incorporates these two processes as parameters named τ (a residence time for phosphate in the upper ocean) and a remineralization efficiency. In essence, addressing these questions involves identifying the values of these parameters that give model results closest to what's observed in nature. We do this by developing simple numeric criteria that somehow characterize the system state. Deciding on the appropriate "metric" is an important task requiring careful thought, but for now we'll choose two simple indicators: (i) the globally averaged phosphate concentration in the model:

$$\bar{x} = \frac{x_i V_1 + x_2 V_2}{V_1 + V_2} \tag{9.9}$$

and (ii) the ratio of the surface to deep box concentrations:

$$\text{Ratio} = \frac{x_1}{x_2} \tag{9.10}$$

where x_1 and x_2 are the phosphate concentrations, and V_1 and V_2 are the volumes of the shallow and deep boxes, respectively.

Consulting a suitable oceanographic textbook, we learn that the globally averaged phosphate concentration in the ocean is approximately $2.5 \, \text{mmol m}^{-3}$ and average ratio of surface to deep concentration is roughly 0.05. Given these "diagnostics", we must decide over what range we could, realistically, vary our model parameters in order to emulate observations. (Allowing the model to explore unrealistic parameter values can lead you along the path to ruin – a reasonable looking model built on implausible physics, biology, etc.).

The critical biological parameter is the lifetime of phosphate in the surface box Tau (τ), where a very short time (e.g. 1 year or less) represents a strong biological forcing, and a long time (e.g. 1000 years) represents a very weak one. Biologically, the doubling timescale of phytoplankton in the oceans is just a few days, so one might argue that τ might be as short as 0.01 years. However, most of the nutrients are effectively recycled in the shallow ocean, and it's the export of carbon from the system that is important. Perhaps the better consideration would be to estimate the residence time of carbon in the oceanic biosphere by dividing the global export production (mostly sinking organic particles), which is something like 10 Pg carbon per year (1 Pg = 10^{15} g), into the shallow ocean biomass (phytoplankton + bacteria) which is of order 3 Pg. This yields a turnover residence time of 0.3 years. So perhaps a useful lower limit to our experiment would be τ = 0.1 years. The upper limit requires a different consideration. We would argue that τ is likely to be significantly shorter than the bathtub residence time of water within the shallow box, because if it weren't, there would not be much of a contrast in shallow and deep concentrations (that is, the ratio would be of order 1, not 0.05). The bathtub residence time is the shallow box's volume divided by the turnover flux:

$$T_W = \frac{V_1}{F_O} = \frac{3 \times 10^{16}\,\text{m}^3}{6 \times 10^{14}\,\text{m}^3\,\text{y}^{-1}} = 50\,\text{y} \qquad (9.11)$$

Thus a sensible upper bound to τ would be 10 years, but to make a point, we'll extend it to 100 years in our exploration of parameter space.

The remineralization efficiency parameter ReminEff can vary between 0 (which means 100% loss) and 1 (which means there is no sedimentation at all). This latter scenario is unacceptable, since it means that the phosphate concentration in the ocean would increase indefinitely. Biogeochemical lore says that only a few percent of the carbon that is fixed at the ocean surface by biological productivity actually makes it to the seafloor and into the sediments. A large fraction of the particles (and their constituents) are destroyed by bacterial oxidation and "remineralized" by dissolution. This carbon is a carrier for the phosphorus, which is actually tied up in the organic material that constitutes the particles falling down through the water column. Thus common sense would dictate that we would explore a range of loss rates between 1 and 10%, or conversely remineralization efficiencies between 0.90 and 0.99.

Finally, we might ask how we might vary these parameters, and at what resolution. Practicality asserts itself here. Doing a few experiments (say four) probably will not allow us to define a reasonable range with reasonable resolution: with four we could do all combinations of the extreme values for two parameters and no more. Realistically, we might be able to perform a few dozen (or perhaps a hundred) experiments, but certainly not thousands. Choosing a few hundred, and claiming that each parameter should be explored equally (an option, for one may be more important than another), we would then try 16 values for τ and 10 values of ReminEff, which would lead to 160 combinations.

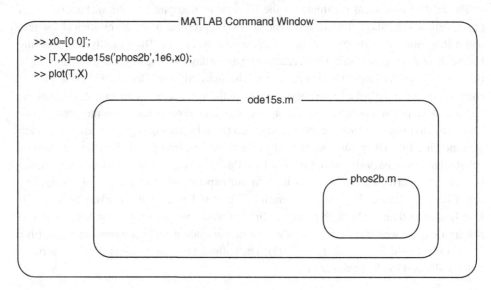

```
─────────────────────── MATLAB Command Window ───────────────────────
 >> x0=[0 0]';
 >> [T,X]=ode15s('phos2b',1e6,x0);
 >> plot(T,X)

           ──────────────────── ode15s.m ────────────────────



                                           ─── phos2b.m ───




```

Figure 9.5 The manual approach to running your model.

Your first instinct might be to change the values linearly, but this may not be the best strategy. In fact, you might suspect that there might be a logarithmic dependence on the parameters. For example you could try:

$$\text{Tau} = 0.100, 0.159, 0.251, 0.398, 0.631, 1.000, 1.59, \ldots, 100 \qquad (9.12)$$

$$\text{ReminEff} = 0.90, 0.91, 0.92, \ldots, 0.99 \qquad (9.13)$$

Note how the Tau values evenly divide the decades over the three orders of magnitude that Tau varies. We'll show you how to do that trick later. Finally, we need to use the technique described earlier (Section 8.5.2) about forcing the number of time steps over the integration to ensure that the number of steps and hence algorithmic accuracy is the same regardless of parameter values. A final consideration, however, would be to run the adaptive ODE integrator (ode15s) in the worst case scenario (or at the extreme values if you're not sure) to determine the shortest time step for stability requirements, just to ensure that your choice doesn't blow up.

9.4.1 Automating the search: global variables

Now we don't know about you, but we would find the prospects of sitting at the keyboard and sequentially modifying the parameters in phos2b.m, running the ODE integrator, and writing down the results for 160 times rather monotonous, if not impossible! You might at first think that you will be forced to do this, for reasons we'll explain below. Figure 9.5 illustrates how we were running the experiment manually in the first instance. That is, we

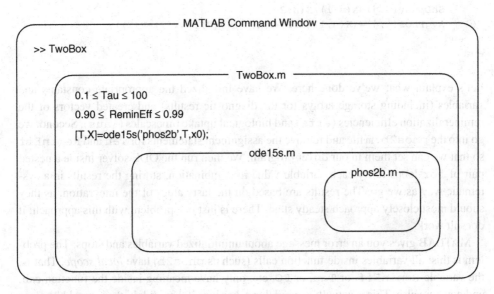

Figure 9.6 A not so clever way to automate your model.

were manually running the ODE integrator in the MATLAB command window. Now what you would like to do is to create a "driver program" that selectively changes the parameter values, and then runs the integration, while saving the results (see Fig. 9.6). An example of this might be a code like:

```
%   TwoBox.m
%       experiments with two-box global ocean phosphate model
%       by varying the life-time of phosphate in the surface and
%       the remineralization efficiency in the deep;
T = [0:100:1e6];    % fix timescale for comparability between experiments
n=length(T);        % pointer to last time step
x0=[0 0]';          % starting values for Phosphate
%
%       set up iteration parameters
%
effs=1-[1:10]/100;           % ReminEff = 99,98,97, ... ,90%
taus=10.^([-1:0.2:2]);       % Tau = .1 --> 100 y
AvgPhos=zeros(length(effs),length(taus));  % global average phosphate
ShDpRat=zeros(length(effs),length(taus));  % shallow-deep ratios
%       iterate through various parameters
for i=1:length(effs)
   ReminEff=effs(i);
   for j=1:length(taus)
      Tau=taus(j);
      [T,X]=ode15s('phos2b',T,x0);
```

```
      ShDpRat(i,j)=X(n,1)/X(n,2);
      AvgPhos(i,j)=(X(n,1)*V(1)+X(n,2)*V(2))/(V(1)+V(2));
   end
end
```

Let's explain what we've done here. We have initialized the appropriate constants and variables (including storage arrays for the diagnostic results) and created vectors of the remineralization efficiencies (effs) and biological uptake timescales (taus). Second, we go into the phos2b.m file and remove the assignment statements for Tau and ReminEff so that we can set them in our driver program. We then run the ODE solver inside a nested pair of for loops to try each variable value in combination, storing the results in a systematic way as we go. The results are based on the last values of the integration, as they should most closely approach steady state. There is just one problem with this approach; it doesn't work.

MATLAB gives you an error message about uninitialized variables and stops. The problem is this: all variables inside function calls (such as phos2b) have *local scope*. That is, the variable ReminEff, or Tau, or FO, etc. only have meaning inside the function call, and nowhere else. This is actually a good thing, because if they didn't, there would be mass confusion amongst different functions and subprograms. For example, you'd like to use a function such as svd without having to avoid using the variable xfil because it happens to be used inside svd. Another way of thinking about it is that the phos2b function is hidden inside the ODE integrator, so we cannot access it nor can we accidently change something we shouldn't. So how do we get around this?

We can make the relevant variables have a *global scope* by declaring them as global. When you think about it, you realize that this declaration must be made at the topmost level of the programming where it happens, and in each function that wants access to those global variables. Here is how the new version of the driver program will look:

```
%       TwoBox.m
global TAU         % P lifetime in surface box
global REMINEFF    % remineralization efficiency
global V           % box volumes
global FO          % overturning volume flux
global xR          % river water P concentration in mMol per m3
global FR          % river water flux
FO = 6e14;         % overturning water flux in m3 per year
FR = 3e13;         % river water flux in m3 per year
xR = 1.5;          % river water P concentration in mMol per m3
T = [0:100:1e6];   % use a fixed time scale for constant n
n=length(T);       % pointer to last time step
V=[3 100] * 1e16;  % volume of reservoirs in m3
x0=[0 0]';         % starting values for P
effs=1-[1:10]/100;      % ReminEff = 99,98,97, ... ,90%
taus=10.^([-1:0.2:2]);    % Tau = .1, 0.16, 0.25, 0.40, ... ,100 y
TotInv=zeros(length(effs),length(taus));    % storage for total inventory
```

```
ShDpRat=zeros(length(effs),length(taus));    % storage for shallow-deep
                                                 ratios
for i=1:length(effs)
   REMINEFF=effs(i);
   for j=1:length(taus)
      TAU=taus(j);
      [T,X]=ode15s('phos2g',T,x0);
      ShDpRat(i,j)=X(n,1)/X(n,2);
      AvgPhos(i,j)=(X(n,1)*V(1)+X(n,2)*V(2))/(V(1)+V(2));
   end
end
```

And here's the new version of the "kernel" program, now called phos2g.m to signify it being globally enabled.

```
function dxdt=phos2g(t,x)
global TAU                % P lifetime in surface box
global REMINEFF           % remineralization efficiency
global V                  % box volumes
global FO                 % overturning volume
global xR                 % river water P concentration in mMol per m3
global FR                 % river water flux
dXdt=zeros(2,1);          % Initialize output as a column vector
Prodtvy = x(1)*V(1)/TAU;  % Production
dxdt(1) = (FR*xR - FO*x(1) + FO*x(2) - Prodtvy)/V(1);     % surface box
dxdt(2) = (FO*x(1) - FO*x(2) + REMINEFF*Prodtvy)/V(2);    % deep box
```

Note that we've done two things here. We've declared all of the relevant parameters global, and moved their initialization up into the driver program. This frees us to change any of the critical parameters without major fiddling again. Second, note that the global variables are all capitalized. This is not required, but it makes good programming hygiene, since any time you come across an all-caps variable, you are reminded that it has global scope, and that you should be careful what you do with it. Not a bad idea. So now we are running the experiment much like Fig. 9.7. As a final note, you might try running the experiment either for shorter integrations or over a limited parameter range in order to gauge how long it will take for the full run.

9.4.2 Examining the results

We accumulated the results in two matrices corresponding to the ranges of the two parameters. The results are computed from the last (nth) values of the integration as they should be the closest to the steady-state values. The question as to whether the time span of the integration is long enough is not simple, but you can compare it to the bathtub water residence time of the deep water, which is:

$$T_{DW} = \frac{1 \times 10^{18}\,\mathrm{m}^3}{6 \times 10^{14}\,\mathrm{m}^3\,\mathrm{y}^{-1}} \sim 1700\,\mathrm{y} \tag{9.14}$$

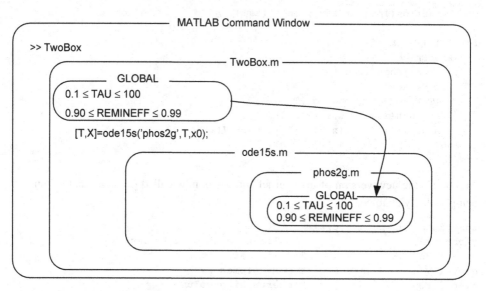

Figure 9.7 A much better way to automate your model.

An even more stringent test would be to compare the deep water phosphate fluxes against the riverine input flux.

$$T_{DW} = \frac{x(2)V(2)}{F_R x_R} \sim \frac{7 \times 1 \times 10^{18} \, \text{mmol}}{1.5 \times 3 \times 10^{13} \, \text{mmol} \, y^{-1}} \sim 150\,000 \, y \qquad (9.15)$$

It's not entirely clear which is the more definitive, but the above suggests that integrating for 10^6 years is probably safe. You could/should test this by rerunning your experiment with a different (half or twice) time span, and comparing results.

Okay, so you're convinced you have the "correct" numbers. How do you look at them? One way would be to print up the table of numbers, and see how they look. Since there's only a couple of hundred of them, this is tractable. A slightly more efficient way of looking at them is to do a contour plot of the diagnostics in parameter space. Be careful to keep track of x and y (column and row) space when you do it. The way we've constructed the nested for-loops makes taus the x-variable and effs the y-variable. Thus you get a labeled contour plot of the normalized inventories by running TwoBox.m and then entering

```
clabel(contour(taus,effs,AvgPhos))
set(gca,'Xscale','log');
```

The outermost function labels the contours while the innermost draws the contours. We've embellished things by converting the x-axis to a logarithmic scale to show the region that is really the most important.

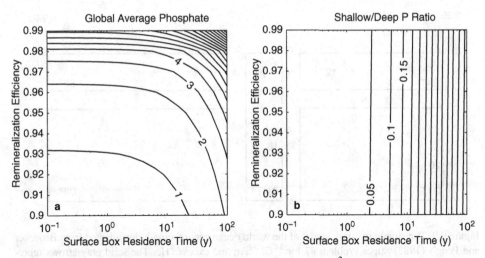

Figure 9.8 (a) Global average phosphate concentration (in mmol m^{-3}) and (b) ratio of shallow to deep phosphate as a function of productivity timescale and remineralization efficiency.

Remembering the target observations, namely that the global average phosphate concentration is about 2.5 mmol m^{-3} and the ratio of shallow to deep values is about 0.05, inspection of these plots gives $\tau \simeq 2$ years and `ReminEff` $\simeq 0.97$. These numbers seem reasonable and, given the crudeness of the model, perfectly acceptable.

What more can we learn from this figure? First, for relatively short biological uptake timescales the global average phosphate concentration is largely a function of remineralization efficiency (the contours become horizontal at low τ in the left panel). This is because the global average is largely dominated by the deep phosphate concentration due to the larger volume, and for rapid stripping out of river phosphate, the abyssal inventory will be related to the remineralization efficiency. The ratio of the shallow to deep concentrations, however, is strictly controlled by the biological residence timescale. Clearly, we've taken this model about as far as we dare, for we know the global ocean, and within it the nutrient cycles, are far more complicated than our little two-box world. Perhaps it's time to move on.

9.5 A third example: multi-box nutrient model of the world ocean

If we can model the world distribution of phosphorus, albeit crudely, with two boxes, we ought to be able to do a better job with more boxes. In this section we will look at a multi-box model for ocean phosphorus to examine the steady-state distribution of phosphorus as a result of a rudimentary approximation of the world ocean general circulation. In Fig. 9.9 we show the world ocean divided into five boxes: surface Atlantic (SAT), Antarctic (ANT), surface Indo-Pacific (SIP), deep Indo-Pacific (DIP), and deep Atlantic (DAT). As you can see, it is only a little bit more complicated than the two-box model we

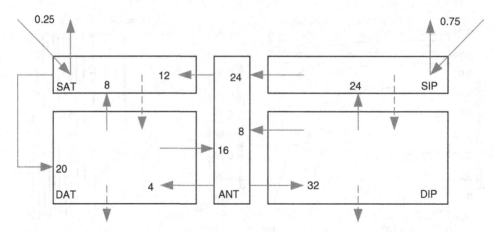

Figure 9.9 The simple five-box model of the world ocean general circulation adapted from Broecker and Peng's (1982) Super Problem #7 for ^{14}C, [39]Ar, and excess [3]He. The solid gray arrows represent the water flux between reservoirs and the dashed gray arrows represent the flux of phosphorus between surface and deep boxes or into the sediment.

used in Chapter 8. The solid gray arrows represent fluxes of water. The dashed gray arrows represent the flux of phosphorus due to biological conversion of phosphorus into sinking particles. We'll use an average river phosphorus concentration x_R of 1.3 mmol m^{-3} in this model.

9.5.1 A system of coupled reservoirs

As an example let us consider the following box model of the world distribution of oceanic phosphorus. We have been concerned with the distribution of phosphorus between the upper and lower oceans, but have a suspicion that the Antarctic waters act as some sort of mixing bowl/redistribution center and so we represent it with its own box. The volumes of the reservoirs (boxes) (in 10^{16} m^3) are as follows: $V_{SAT} = 3$, $V_{ANT} = 15$, $V_{SIP} = 9$, $V_{DIP} = 81$, and $V_{DAT} = 27$. The fluxes of water are in Sv (10^6 m^3 s^{-1}) and are shown at the end of each solid arrow in Fig. 9.9. The concentrations of phosphorus (x) are in mmol P m^{-3}.

9.5.2 The fluxes (between boxes and within)

We assume that 95% of the phosphorus in the surface waters is converted by biological activity into sinking detritus and that the remaining 5% stays in solution transported by the water flows. Of the phosphorus transported to the bottom by sinking particles all but 1% remineralizes back into solution; the remaining 1% is removed permanently into the sediments and this flux of phosphorus is balanced by the input of phosphorus from the river inputs (0.25 and 0.75 Sv in the surface Atlantic and Indo-Pacific respectively).

So the fluxes are given simply by the water flow (in Sv) times the concentration of the box it is leaving. For example, let's consider the surface Atlantic box (SAT):

$$\frac{dx_{SAT}}{dt} = \frac{[(8x_{DAT} + 12x_{ANT} + 0.25x_R) \times (1 - 0.95) - 20x_{SAT}]}{V_{SAT}} S_{sv} \qquad (9.16)$$

In Equation (9.16) the first term is the amount of phosphorus delivered from the deep Atlantic (DAT), the next term the amount of P from the Antarctic box (ANT), and the third term the amount of P delivered from the rivers. These terms all have to be multiplied by $(1 - 0.95)$ to represent the phosphorus lost as particles. Then we subtract off the phosphorus sent back (as dissolved P) to the deep Atlantic. All of that is then divided by the volume of the surface Atlantic to put it back into concentration units.

But wait, what are the units in Equation (9.16)? The flows are in terms of sverdrups, the concentrations in mmol P m^{-3}, and the volumes in 10^{16} m^3. A conversion factor S_{sv} is added to balance the units:

$$S_{sv} = \frac{10^6 \, m^3}{Sv \, s} \frac{3.16 \times 10^7 \, s}{y} \qquad (9.17)$$

where we have used the definition of a Sv (10^6 m^3 s^{-1}) and the number of seconds in a year (roughly 3.16×10^7), which will put everything on a mmol m^3 y^{-1} footing (this may not seem terribly important now, but see below under the Runge–Kutta solution).

9.5.3 The coupled equations (building them)

Now we can write the coupled equations that represent this box model:

$$\frac{dx_{SAT}}{dt} = \frac{[(8x_{DAT} + 12x_{ANT} + 0.25x_R) \times (1 - 0.95) - 20x_{SAT}]}{V_{SAT}} S_{sv}$$

$$\frac{dx_{DAT}}{dt} = \frac{[20x_{SAT} + 4x_{ANT} + 0.94 \times (8x_{DAT} + 12x_{ANT} + 0.25x_R) - (8 + 16)x_{DAT}]}{V_{DAT}} S_{sv}$$

$$\frac{dx_{SIP}}{dt} = \frac{[(24x_{DIP} + 0.75x_R) \times (1 - 0.95) - 24x_{SIP}]}{V_{SIP}} S_{sv} \qquad (9.18)$$

$$\frac{dx_{DIP}}{dt} = \frac{[32x_{ANT} + 0.94 \times (24x_{DIP} + 0.75x_R) - (8 + 24)x_{DIP}]}{V_{DIP}} S_{sv}$$

$$\frac{dx_{ANT}}{dt} = \frac{[16x_{DAT} + 8x_{DIP} + 24x_{SIP} - (4 + 12 + 32)x_{ANT}]}{V_{ANT}} S_{sv}$$

Notice that the biological export fluxes out of the surface boxes appear in the corresponding deep box (minus the 1% that is lost to the sediments). We have coded these equations up in MATLAB as an m-function for ODE45 (see `phos5box.m`). When coding them up we combined and simplified where possible so the equations in the m-file won't be exactly the same as these.

9.5.4 The Runge–Kutta solution

You can download the MATLAB file `phos5box.m` and use it in MATLAB with one of MATLAB's ODE integrators (our favorite is `ode45`, but it will work with the others such as their "stiff" ODE integrator `ode15s`). To make a run with this model and display the evolution of phosphorus to steady state use the following MATLABese:

```
[t,X]=ode45('phos',[0 1500],[1 1 1 1 1]); plot(t,X);
```

It's that simple. Or you can build a driver program of your own (a task we leave for the problem sets).

Now you may be wondering what all this unit business is about. If you look at the MATLAB statement that calls `ode45` you'll see that we have to give it start and stop times. We were careful to specify all of one constants in terms of years; if we had used seconds, we would have had to specify those bounds in seconds. That's not a very convenient unit of time to use if you have to integrate out to 1500 years! It makes no difference to MATLAB's `ode45`, it takes the same number of time steps whether you tell it to integrate from 0 to 1500 years or from 0 to 4.7336×10^{10} seconds. But it may make constructing figures from the results more readable. Figure 9.10 shows this model integrated out 2000 years.

The main point of this chapter is to introduce the parts of most biogeochemical models and to demonstrate via simple models how you would go about automating model

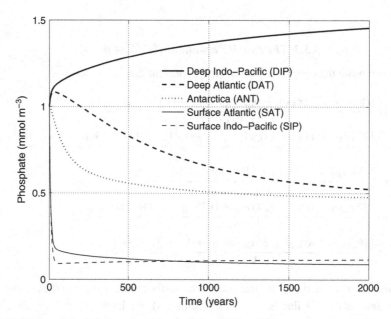

Figure 9.10 The evolution of phosphate concentrations in five different ocean "boxes". Some boxes reach their steady-state equilibrium faster than others, but we leave the estimation of the secular equilibrium time as an exercise for the readers.

integrations in order to systematically document model performance over a range of parameter values. In the process, you will have learned some new programming techniques (especially the use of global variables), and the concept of developing critical diagnostics of system response. You then plotted these diagnostics as a function of the parameters to illuminate model system behavior.

9.6 Problems

9.1. Employ the five-box model of the global distribution of phosphorus described in Section 9.5 to estimate the following:

   (a) Code up the model and a main (or driver) program in MATLAB. Prior to running the model, evaluate the model's "stiffness" and choose an integrator from Table 8.3. As a starting point, use a value of 10 years for the τ and 0.99 for the remineralization efficiency. Run the model out to steady state (run the model long enough to achieve steady state – effectively $dC/dt \approx 0$ or close enough to make no difference – but less than the life of the Universe) starting from an initial condition of zero everywhere, provide plots of the time evolution of the five model "state variables", and report the following quantities: approximately how long it took the model to achieve steady state, the steady-state concentrations of each box, the mean concentration of phosphorus in the global (model) ocean, and the ratio of the deep minus surface phosphorus concentration between the Pacific and Atlantic (i.e. Pacific over the Atlantic). This last quantity is Broecker's so-called "horizontal segregation", given as:

$$\frac{\left([P]_{Pacific}^{deep} - [P]_{Pacific}^{surf}\right)}{\left([P]_{Atlantic}^{deep} - [P]_{Atlantic}^{surf}\right)} \tag{9.19}$$

   (b) Do a sensitivity analysis of the diagnostics we identified in Problem 9.1(a) by making contour plots of the mean phosphorus concentration and horizontal segregation as a function of τ and remineralization efficiency (choose what you think is a good representative range of these parameters' values).

   (c) Broecker and Peng (1982) report that the global mean phosphorus concentration is approximately 2.3 mmol-P m^{-3} and the horizontal segregation is approximately 2. Is there a region in your parameter space where both of these diagnostic conditions are true? Use subplot to make a 2 by 2 plot of the surface Atlantic and Pacific and deep Atlantic and Pacific concentrations in your parameter space. What can you say about the way the model is constructed that would help explain your diagnostic and state variable results?

10

Model analysis and optimization

> Is it better to wear my purple and green Acapulco shirt, or
> nothing at all?
>
> *Hunter S. Thompson*

In the previous two chapters we familiarized you with the tools necessary to integrate ordinary differential equations (ODEs), the elements that go into a model, some basic approaches to analyzing model output, and our philosophy about model building. Now in this chapter, we show you some tricks for making your model better or, in modeling jargon, how to *optimize* your model. We begin with a discussion of the components of optimization and describe cost (objective) functions, contrast constrained and unconstrained optimization, as well as continuous vs. discrete problems, define local and global optimization, and discuss the important topic of convergence. Deriving error estimates of optimized model parameters is deferred for a later chapter (Chapter 18), where we address the complete inverse problem. We follow with descriptions of one-dimensional optimization techniques, detailing the algorithmic logic that guarantees that a minimum is found. From there we move to multi-dimensional searches and the techniques employed. We finish off this chapter with a twin numerical experiment demonstrating an unconstrained optimization applied to a variant of the simple PZ predator–prey ecosystem model developed earlier (Chapter 9).

So far, we have been doing (and talking about) *forward modeling*. The general flow of information in forward models is shown diagrammatically as:

$$\left.\begin{array}{l} \text{parameters} \\ \text{forcing} \\ \text{initial conditions} \end{array}\right\} \Rightarrow \text{model} \Rightarrow \text{``predictions''}$$

One generally starts with a model (a set of equations) and some values for the parameters in the model. Then one adds some forcing and puts the model into context with a set of initial conditions. Integrating (i.e. running) the model then produces a prediction that can be compared to observations. That comparison step is important and we do not want you to forget it. Frequently, the initial conditions are gleaned from a subset of the observations themselves.

There is also *inverse modeling*, of which model optimization is a variant. As we diagrammatically show below,

$$\text{data} \Rightarrow \text{model} \Rightarrow \text{estimates of} \begin{cases} \text{parameters} \\ \text{forcing} \\ \text{initial conditions} \end{cases}$$

the flow of information in inverse models is opposite that of forward models. Here one starts with the data and infers estimates of the parameter values, the forcing that produced the data, and the initial conditions from which the system evolved. Forward and inverse models are often used together, iteratively, to explore and expand our understanding of the systems under investigation.

10.1 Basic concepts

The topic of optimization is a book unto itself, in fact several books (Gill *et al.*, 1981; Nocedal and Wright, 1999; Venkataraman, 2002; Spall, 2003, to name just a few). What we want to do is give you a running start and point you at a number of excellent references to further your exploration of this fascinating field. It is important to emphasize at the beginning that there is no one perfect, all-purpose optimizer. The algorithms used change with the problems studied and the questions asked. To that end we adopt a pragmatic approach of first acquainting you with a number of the basic concepts and then presenting in the follow-on sections a user's approach to optimization methods:

- Methods that only require a cost function.
- Methods that require a cost function and optionally a cost function gradient.
- Methods that require both a cost function and its gradient.
- Methods that take a more global approach.

In order to make progress in ocean modeling, often we make a number of simplifications to the theoretical, ideal equations that describe ocean systems. Translation: we take complex, nonlinear, difficult to solve equations that would, in principle, completely describe our experiments, throw out the parts we can't solve, parameterize (sometimes very simply) the parts we know we can't live without, and apply techniques we learned about in school to solve the rest. This frequently leaves us with parameters that we cannot measure directly and have no first-principles way of estimating. What makes matters worse is that the data that we do collect to help constrain these parameters are often incomplete and subject to errors. Somehow we must take all of the information available to us and arrive at an answer that is, in some sense of the word, optimal. Optimization has been successfully applied to a number of oceanographic problems that range from finding the optimal value for bottom friction coefficients (Schöter and Wunsch, 1986) to determining the transport of nutrients and oxygen and bounds on export production in the world ocean (Ganachaud and Wunsch, 2002), to estimates of annual sea level cycle (Vinogradov *et al.*, 2008).

So, what *is* optimization? Well, put simply it is the process of finding the optimal parameters yielding model solutions that best fit some set of observations. The observations may be the distribution of nitrate in the water column, the amount of phytoplankton and zooplankton in your region of interest or the velocity ($\mathbf{u} = (u, v, w)$) of the water in your bathtub or the global ocean. This process has many steps but it generally follows a basic algorithm: start with an initial guess for the parameters ($\mathbf{p}_0 = (p_1, p_2, \ldots, p_M)$), evaluate a cost function (J) that is a measure of the model output–observations mismatch, adjust $\mathbf{p}$ with some technique, re-evaluate J and decide if you should repeat to find better (smaller) model–data misfit or stop and keep the final parameter set, $\mathbf{p}_f$. To help decide when to stop iterating, we must decide a priori on a *convergence criterion*. Generally speaking, one monitors the size of the cost function at each iteration and makes note how much it has changed since the previous iteration. In many optimization techniques, it will be decreasing, and the amount it decreases with each iteration will be smaller as one approaches convergence. Eventually the decrease will be so small that it will no longer be important in the context of the problem you are working on, and you can stop iterating. At this point, we say the optimization has converged.

MATLAB provides a very powerful optimization toolbox. But before we start talking about "trust regions" and "Hessians" we want to familiarize you with the concepts and techniques that form the basis of most modern optimization software. Furthermore, we want you to note that many numerical optimization textbooks treat the parameters they are seeking as unknowns and consequently refer to them as x. This is a common convention in many inverse methods texts as well. However, in this chapter and elsewhere we use the convention that parameters are parameters (represented by p) and the state variables that we are modeling are, after all, the unknowns (represented for data and/or model output as x). The model design matrix is $\mathbf{A}$, and the system of equations we are optimizing follow the $\mathbf{Ap} = \mathbf{x}$ syntax, which should appear familiar as a variant of the syntax introduced in Chapter 1 (i.e. $\mathbf{Ax} = \mathbf{b}$).

One more thing. In this chapter we stick to the following convention with regards to indices. For data and/or computa (model output), we use a lowercase i ($i = 1, 2, \ldots, N$). For the parameters of the model we use a lowercase j ($j = 1, 2, \ldots, M$). And for the iterates of the optimization we use a lowercase k ($k = 1, 2, \ldots$). You may have noticed that k has no final value, and that is because your convergence criterion is supposed to tell the optimization software when to stop. This is important, and we come back to this topic shortly. Frequently, software programmers provide a failsafe maximum number of iterations to prevent optimization runs from becoming stuck in an infinite oscillation around an uninteresting detail. Be sure to inspect the number of iterations. Did your optimization use all available iterations? If so, perhaps another method(s) should be used.

10.1.1 Components of optimization

There are several standard components to each optimization algorithm. In this section we discuss the ones our experience has shown are important to consider.

The cost function

The first thing we need in order to optimize is some sort of metric of the model skill; that is, how well does the model match the real data? The metric is known as the *cost function* or *objective function*. Often, and in this book, the cost function is represented by J written generically as:

$$J = \sum_{i=1}^{N} \omega_i^{-1} \left(\hat{x}_i - x_i \right)^2 \qquad (10.1)$$

where ω_i are the data variances and the inverses are the weights, $\hat{x}_i$ are the state variables from the model, x_i are independent observations made at the same coordinates, and N represents the number of observations. Does this equation look familiar? It should, as it is the heart of the χ^2 parameter we used in least squares fitting in Chapter 3. The more complete (and complex looking) cost function used in inverse methods is given in Equation (18.6).

We generally pick the objective function in such a way that we want it to be as small as possible. So, this becomes a minimization problem. You remember minimization problems from calculus, right?

$$\frac{\partial J}{\partial p_j} = 0 \qquad \text{for all } p_j \qquad (10.2)$$

This is read, "Find where the partial of J with respect to p_j equals zero, for all p_j." To solve this minimization problem you are going to need to know:

$$J(\mathbf{p}, \mathbf{x}) \qquad \text{and/or} \qquad \frac{\partial J}{\partial p_j}(\mathbf{p}, \mathbf{x}) \qquad (10.3)$$

where $\mathbf{x}$ are the independent variables (there can be, and frequently are, more than one, $i = 1, 2, \ldots, N$) and $\mathbf{p}$ are the parameters ($j = 1, 2, \ldots, M$) you are trying to optimize. If you can calculate both quantities in Equation (10.3), then you can use some very powerful techniques.

We've been throwing around the terms parameters and state variables pretty loosely. Often in numerical optimization there is a set of adjustable quantities (frequently called parameters) that directly influence the value of the cost function. For low-dimensional problems (1D or 2D) this allows the cost function to be plotted as a function of these parameters. And we will use this approach when discussing the merits of various algorithms. However, let us make it clear that in oceanographic problems one rarely has a direct graphical connection between parameters and cost function. Typically the parameters modify the behavior of a model, which has a demonstrable influence on the value of the cost function through model–observation discrepencies (think χ^2).

Before you reach into your MATLAB toolbox and pull out an optimization tool to solve your problem, you should ask yourself some questions. Is this a *constrained* or *unconstrained* problem? Am I doing a *global* or *local* optimization? The answers to these questions will, in part, guide you in your selection of a method.

Constrained vs. unconstrained

The difference between a constrained and an unconstrained optimization is relatively simple in concept. An unconstrained optimization is simply the process of finding the lowest cost function value (J) no matter where in parameter space the search may take you. Of course, this does not quite fit the bill for most optimization problems. Many times we know something about the parameters or state variables (or both) that prohibit such a free-wheeling approach. For example, to the cost function we could also add other things we know about the problem; we may know that some of the parameters must be positive or we may be able to calculate deviations of the parameters from a priori theoretical values. One of the most common examples of constrained optimization is find the lowest J while keeping the parameters positive. A specific example is that we may seek a solution where the turbulent diffusivity is positive (since a negative value is unphysical). Another example is to find the lowest J while ensuring all of the state variables are positive; after all, negative concentrations do not make much sense either.

Continuous vs. discrete

Another way of distinguishing between problems is whether the parameter values are continuous or discrete. In most problems we concern ourselves with in this book, the parameters are continuous. In the M-dimensional universe of all possible parameter values, our problems are infinitely divisible. However, there are certain problems where the parameters take on discrete values. Certain combinatorial problems involving discrete arrangement or ordering of components or cases/samples are known as discrete optimization problems. As a simple example of the latter, what is the best order of station occupation in a particular cruise? This is a variant of the Traveling Salesman problem, which we mention later on when we discuss simulated annealing.

Local vs. global optimization

Optimization techniques also divide along local and global lines. Local techniques usually start with the assumption that the optimum is "near" the initial guess, i.e. the search doesn't have too far to go. Local techniques further divide into methods that require only that you evaluate the cost function (J) and those that require evaluation of both the cost function and its derivative ($\partial J / \partial p_j$). In both cases these techniques demonstrate optimization methods that only move "downhill" on the cost function hypersurface. Figure 10.1 illustrates that a local technique would only find the global minimum if it started from point 1. Starting from point 2, it would find the local minimum, but would be unable to discover the dashed route that takes it over the hump in cost function to the global minimum. The role of $\partial J / \partial p_j$ is typically to accelerate finding a local minimum by reducing the number of J evaluations.

Global techniques, however, are heuristic and rely on methods for characterizing the *shape* of the cost function, J. Local techniques search only a part of the problem space, but global techniques search the complete (or at least more complete) parameter space. An example of a global optimization technique is *simulated annealing* such as the *Metropolis algorithm*, which we will talk about in Section 10.4.

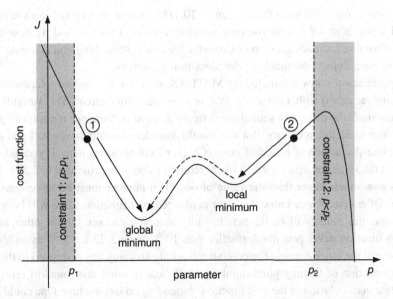

Figure 10.1 A plot of a one-dimensional cost function (J) vs. the single parameter (p). In this particular case, the value of the parameter is constrained to be $p_1 < p < p_2$ as indicated by the region between the gray areas. Note that p_1 and p_2 are just fixed values of p. Furthermore, we see that the minimum J obtained can depend on the initial guess of the parameter. Initial condition 1 arrives at the allowable global minimum, but initial condition 2 arrives at a local minimum. While there is no universal technique known that will guarantee that the optimization process will follow the dashed arrow, a lot of the work involved in optimization is to convince oneself that the global minimum has been reached.

Convergence

We mentioned convergence criterion at the beginning of this section. We know this sounds like a punchline from a cocktail party joke, but it is important to know when to stop. There are two important aspects of convergence that we wish to try to clarify for the reader: convergence rates and convergence tolerances.

It is a little confusing, but the convention in numerical optimization is to talk about convergence rates as being *linear*, *superlinear*, *quadratic*, etc. An optimization method is said to converge linearly if:

$$\frac{\|\mathbf{p}_{k+1} - \mathbf{p}_f\|}{\|\mathbf{p}_k - \mathbf{p}_f\|} \le c \quad (10.4)$$

where $\mathbf{p}_k$ and $\mathbf{p}_{k+1}$ are the parameter estimates from iteration k and $k + 1$, $\mathbf{p}_f$ are the final optimized parameters, and c is a constant with a value $0 \le c \le 1$. We write $\mathbf{p}$ because this could be a multi-dimensional problem in optimization, and so we take the norm ($\|\mathbf{p}\|$).[1] If the denominator of Equation (10.4) is squared the convergence rate is said

---

[1] This is the default `norm` in MATLAB, i.e. the Euclidean length of the vector $\mathbf{p}$ or the square root of the sum of the squares of its elements.

to be quadratic. An in-between case, given if (10.4) has a limit of zero as k goes to infinity, is called superlinear. Of course you may wonder, if we are trying to find $\mathbf{p}_f$ how can we use this if we don't already know $\mathbf{p}_f$? Convergence rate is something you compute from a known problem to test how quickly your algorithm converges.

All optimization software, including MATLAB, have a convergence tolerance with a variable (or variables) with a name like TOL or some variation thereof. This variable determines how small the change in something (usually J) must be from one iteration step to the next in order to declare a victory. But you should consider carefully how small you set this variable. Our discussion of roundoff error (Chapter 2) is pertinent here. If you ask MATLAB what the value for eps is, it will likely return a value of approximately 2.22×10^{-16}, which is how much larger than the value of one (1) a number must be to be considered different. Of course, if you know that your converged cost function value will be approximately one, this value will be the smallest tolerance you can set. On the other hand, if your cost function never gets much smaller than 10^{26}, using 2.22×10^{-16} as an absolute tolerance will not work because the computer has only so many bits dedicated to the internal representation of floating point numbers. Also bear in mind that roundoff errors may accumulate in calculation of the cost function, depending on its structure. One could define the tolerance as a ratio, but if the number you are converging to is close to zero this too can cause problems due to the singularity created by dividing by zero. Most software specifies a default tolerance around 10^{-6}, but you should carefully consider where your problem is taking your optimization in cost function space. A practical word of advice from Press *et al.* (2007) is that, as a good working start, you should set your convergence tolerance to $\sqrt{\epsilon_m} J_{\text{final}}$, where ϵ_m is the machine accuracy.

10.1.2 Linear vs. nonlinear programming

Before we begin our explanation of optimization routines of increasing complexity, we make a momentary digression into a discussion of linear vs. nonlinear programming. When you read about numerical optimization you will notice that many texts are divided along these lines. It can be quite confusing because texts (and MATLAB help files) throw around terms like *linear programming* and *sequential linear programming*, *quadratic programming* and *sequential quadratic programming*. To attempt to cut through this thicket of jargon run amuck, we offer the following guidance. Linear programming methods (sequential or otherwise) are applied to problems where the cost function, constraints, and model functions are all linear. A linear cost function, $J = \sum \alpha \hat{x}$, seems odd if you are used to χ^2 type problems with $J = \sum (\hat{x} - x)^2$. Nonlinear programming applies to all other kinds of optimization problems. Quadratic programming refers to a particular set of problems with a quadratic cost function and (generally) linear constraints. Technically, inverse methods that employ linear constraints with minimization of a cost function involving the square of model–data discrepancy are examples of what is known as quadratic programming, although they are referred to as linear inverse theory (see Chapter 18).

There is at least one linear programming algorithm that you are likely to find mentioned in your reading, the (linear) *simplex method*. We parenthetically add the linear to distinguish this algorithm from a nonlinear method that goes by the name of Nelder–Mead simplex, which we will introduce later. A "simplex" is a convex geometric shape. The simplex method is an iterative way of solving linear systems constrained by inequalities using this geometric approach. It works by following the edges of a simplex from one vertex to another until the maximum cost function is found. Typically the problem is stated in its familiar form:

$$\text{Maximize the function} \quad J \quad \text{subject to} \quad \mathbf{Ap} \le \mathbf{x} \quad \text{and} \quad \mathbf{p} \ge 0 \qquad (10.5)$$

where J is the cost function, $\mathbf{p}$ the parameters, $\mathbf{x}$ the data, and $\mathbf{A}$ the linear model under consideration. Note that the cost function must be linear in $\mathbf{p}$.

The inequalities in Equation (10.5) are inconvenient to program. The simplex method handles this problem in a practical fashion by adding *slack variables*, one for each inequality constraint. For example:

$$p_1 + 2p_2 \le 10 \qquad (10.6)$$

becomes:

$$p_1 + 2p_2 + p_3 = 10 \qquad (10.7)$$

The additional variable (p_3) is a non-negative slack variable whose cost is zero. For example, the cost function might be maximize $p_1 + p_2$, and the addition of the slack variable would change the cost function to $1 \times p_1 + 1 \times p_2 + 0 \times p_3$, which does not change the cost. The problem, however, does become a three-dimensional problem. The resulting linear system is $\tilde{\mathbf{A}}p = x$, where $\tilde{\mathbf{A}}$ is ($N \times \tilde{M}$), N is the number of constraint equations and $\tilde{M}$ is the number of total parameters (the sum of the initial parameters plus the slack variable parameters).

Figure 10.2 demonstrates the manner in which the simplex method works with a simple two-dimensional problem. We use this geometric explanation rather than an in-depth discussion of the rather elegant linear algebra because this method's strict linear construction has not found a lot of problems well suited for it in geophysics or geochemistry. Starting at $p_{\text{init}} = (0, 2)$ the method follows the upward gradient of the cost function until it hits one of the inequality constraints (at point A). It then follows the line described by $p_1 + 2p_2 = 10$ upward on the cost function gradient until it strikes another constraint (at point $p_{\text{final}} = (10, 0)$). At this point the only way the simplex method can continue up-gradient on the cost function surface is to violate one of the constraints, so it stops.

This algorithm is frequently applied to problems in operations research and financial management, where for example the $\mathbf{p}$ parameters might reflect the number of particular milling machines in a factory line and the goal is to determine the most cost efficient mix of machines for a factory given constraints such as total cost, floor space, number of operators, etc. A more earth science application might be heat or tracer transport, which is linear with velocity. The constraint that $\mathbf{p}$ must be greater than zero is obvious for the

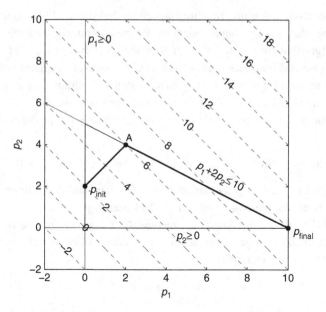

Figure 10.2 In this relatively simple, two-dimensional example of the simplex method we show the two primary constraints ($p_1 \geq 0$ and $p_2 \geq 0$) as well as a secondary constraint ($p_1 + 2p_2 \leq 10$). The solid black lines cordon off a triangular piece of parameter space as the feasible area. The cost function ($p_1 + p_2$) is contoured as dashed lines. The path the simplex method follows is shown as a heavy black line. See text for more details. This figure was inspired by similar figures in Strang (1986) and Press *et al.* (2007).

manufacturing example – you can't have a negative number of machines – but sometimes less so in geophysics and geochemistry.

10.2 Methods using only the cost function

In what follows we describe the basic techniques used in adjusting the parameter values. Knowledge of the pros and cons of these methods is important in making a decision of which one to use for your particular problem. A good choice can mean the difference between an optimization that converges in a reasonable amount of time and one that never converges.

Some optimization methods only require the user to provide a means to evaluate the cost function. Since many real world geophysical, geochemical and biogeochemical problems fall into this category we thought we would present them as a group to discuss their strengths and weaknesses. Since these methods have no access to gradient information about the cost function they are generally referred to as *line searches*. Once a direction has been chosen in M-dimensional parameter space, they use a method or combination of methods to find the minimum in the cost function in that direction. How these methods

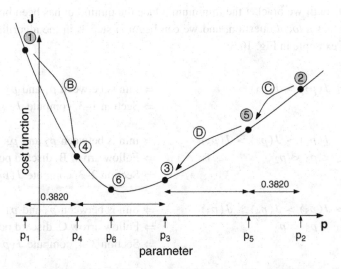

Figure 10.3 Here we illustrate the Golden Section search algorithm described in the text. As the Golden Section search narrows in on the minimum in the cost function, the remaining space to be searched is divided into $(3 - \sqrt{5})/2 \approx 0.3820$ and $1 - (3 - \sqrt{5})/2$ pieces. This figure is based on Press *et al.* (2007). The gray background numbers represent points discarded during the process of bracketing the minimum.

decide on the direction in parameter space to search is what sets them apart from each other. These are also known as *zero-order* methods.

Finally, we offer the following advice. First, optimization is an iterative process, therefore we instruct you to pay attention to the number of iterations used by each optimizer. Second, quite often the most computationally expensive step in an optimization technique is the repeated evaluation of the cost function. Some methods require many cost function evaluations for each iteration, so don't be fooled just by the number of iterations. There are other issues that are also important at times, such as amount of computer memory required to carry out certain computations.

10.2.1 Golden Section search

One of the most robust, if not terribly quick, algorithms for finding a minimum of a function in one dimension is the *Golden Section search*. This search technique is an excellent example of finding a minimum along a line, in other words a line search.

To start a line search, you need at least three points to bracket the minimum you are seeking. We start by picking points 1 and 2 with point 3 half-way between them. As we show in Fig. 10.3, points 1, 2, and 3 bracket the minimum and we know this to be true because it satisfies the mathematical relationship $J(p_1) > J(p_2) > J(p_3)$ and $p_1 < p_3 < p_2$ (the minimum would also have been bracketed if $J(p_2) > J(p_1) > J(p_3)$). If the above relationship does not hold, then we must repeat the process of choosing new sets of three points until we bracket the minimum. Once the minimum has been bracketed the problem becomes a *local* question, and we can begin at step A in the algorithm outlined below for the example in Fig. 10.3:

A $J(p_1) > J(p_2) > J(p_3)$ $\Rightarrow$ min is between p_1 and p_2
                                               $\Rightarrow$ Section 1–3, compute $J(p_4)$

$\Downarrow$

B $J(p_1) > J(p_2) > J(p_4) > J(p_3)$ $\Rightarrow$ min is between p_2 and p_4
    $p_1 < p_4 < p_3 < p_2$ $\Rightarrow$ Follow arrow B, discard point 1
                                               $\Rightarrow$ Section 2–3, compute $J(p_5)$

$\Downarrow$

C $J(p_2) > J(p_5) > J(p_4) > J(p_3)$ $\Rightarrow$ min is between p_5 and p_4
    $p_4 < p_3 < p_5 < p_2$ $\Rightarrow$ Follow arrow C, discard point 2
                                               $\Rightarrow$ Section 3–4, compute $J(p_6)$

$\Downarrow$

D $J(p_5) > J(p_4) > J(p_3) > J(p_6)$ $\Rightarrow$ min is between p_4 and p_3
    $p_4 < p_6 < p_3 < p_5$ $\Rightarrow$ Follow arrow D, discard point 5
                                               $\Rightarrow$ Section 4–6, compute $J(p_7)$ (not
                                               shown)

$\Downarrow$

$\vdots$

and so on. Of course the path of the algorithm and choice of which points to keep and which to discard would differ for different cost function curves. Although this method is not particularly fast, once the minimum is bracketed it is guaranteed to find the minumum and is useful for cost functions with difficult to locate minimums. One final note about this technique: in step A we divided the interval bracketed by points 1 and 2 in half to locate point 3. But thereafter we divide the larger segment at the $(3 - \sqrt{5})/2$ and $1 - (3 - \sqrt{5})/2$ point to find points 4, 5, 6, and so on. The fraction $(3 - \sqrt{5})/2$, known as the Golden Section and gives this particular line search technique its name, is more efficient than simply dividing the interval in half. This technique is repeated iteratively until the specified convergence criterion is met. The convergence of the algorithm is generally linear.

10.2.2 *Parabolic interpolation*

Perhaps you have noticed that as we iterate toward the minimum in the Golden Search technique the cost function about the minimum begins to look a lot like a parabola. This insight is the kernel of the parabolic interpolation technique. In this particular line search technique, we simply use three points along the cost function near the minimum and fit a parabola to those three points. It is then a simple matter of solving for the minimum of the fit parabola. The parabolic fit has the form:

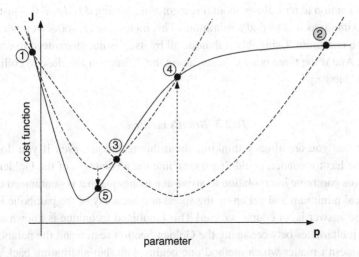

Figure 10.4 Parabolic interpolation of a cost function minimum. The solid line is the cost function and the dashed lines are parabolic approximations of the cost function constrained by the three points they pass through. Parabola 1–3–2 is used to find point 4, point 2 (grayed out) is abandoned and parabola 1–3–4 is used to find point 5, and so on. Once a search routine has gotten close enough to the minimum, approximating the locus of points immediately about the minimum as a parabola can lead to very rapid convergence.

$$\hat{J} = a_1 + a_2 p + a_3 p^2 \tag{10.8}$$

where $\hat{J}$ is the quadratic approximation to the cost function, p is the parameter you are varying, and (a_1, a_2, a_3) are the coefficients. Finding the minimum (or maximum) of a function is merely a matter of setting the first derivative to zero:

$$\frac{d\hat{J}}{dp} = a_2 + 2a_3 p = 0 \tag{10.9}$$

which gives us:

$$p = \frac{-a_2}{2a_3} \tag{10.10}$$

Because the cost function only approximates a parabola, we need to iterate. In the example in Fig. 10.4, start by computing the value of the cost function at points 1, 2, and 3. We then use Equation (10.10) to locate point 4 from points 1, 2, and 3. Point 2 is then discarded (grayed out) and the process repeated. We can then fit a parabola to points 1, 3, and 4 and use (10.10) to find point 5. Notice how much more rapidly this method is converging on the minimum of J than the earlier Golden Section method. Of course one could continue the process depicted in Fig. 10.4 and fit a parabola to points 1, 5, and 3 using the coefficients from that equation to locate a point 6. Once again, it would be the convergence criterion that determines when to stop iterating.

A word of caution to pass along about this algorithm. Setting $d\hat{J}/dp = 0$ is just as likely to find a maximum as it is to find a minimum. This method only works well when we are "close" to the minimum. Using this technique all by itself is not guaranteed to converge to a minimum. And if the three points chosen along the J function are close to collinear, this method will "blow up".

10.2.3 Brent's method

We are sure that you are already thinking about this technique, even if you don't know its name. One has to wonder, couldn't we combine the advantages of the Golden Section search with the parabolic interpolation to arrive at a technique that is guaranteed to find (at least) the local minimum and yet enjoy the speed and accuracy of the parabolic interpolation? And the answer is, of course we can! This combined technique is known as *Brent's method* and it alternates between using the Golden Section search and the parabolic interpolation. It doesn't matter which method one begins with, but alternating back and forth between these two techniques is sure to find the minimum either by virtue of the rapid convergence of the parabolic interpolation in the case of a smooth and well-behaved cost function or by virtue of the Golden Section's dogged, single minded pursuit of the minimum in the case of an uncooperative and poorly behaved cost function – the best of both worlds. Knowing when to switch is the trick.

In the parabolic interpolation step of Brent's method, the technique's housekeeping keeps track of the size of the step towards the minimum. On each iteration within the parabolic step the size of the step must be less than one-half the size of the step before. This requirement forces the quadratic steps to converge rapidly. Failure to meet this criterion forces the method to switch over to the Golden Section search. At this point the method redivides the bracketing interval as in Fig. 10.3 and then tries to take another parabolic step. If the parabolic step succeeds in meeting its criterion, the method continues by taking another parabolic step and so on until the convergence criterion is met.

10.2.4 Powell's method

For line searches in one-dimension the choice of direction is relatively limited; you search either to the right or the left and pick the direction that gives you the best decrease in J. But what do you do in higher-dimensional problems? Even a two-dimensional (i.e. two-parameter) search has an infinite number of directions one could search along; how do you decide which direction when you have no gradient (slope) information?

Before we answer the above question, we are going to describe and dismiss an obvious solution/answer to that question. Rather than decide which direction to search, why not let the coordinate system decide for you? In a two-dimensional case, one would start from $\mathbf{p}_o$ and perform a line search along the x-axis. Keep following that search direction so long as J decreases; when J starts to go back up, switch to the best decrease in the vertical direction. Keep repeating this process until your convergence criterion is met. The problem with this approach is that as the cost function becomes steeper and more elongated and at an angle to the coordinate system, this approach can take a very long time to converge because of all the steps back and forth. And if you can imagine this problem in three or more dimensions we think you can quickly see that this leads to far too many line searches and cost function evaluations.

The answer to the above question leads to what is known as zero-order methods. The most recommended of these is *Powell's method*. This method is a modification of the general idea of the *direction set methods* wherein we look for a better set of directions to perform our line searches than the coordinate system. Direction set methods share the characteristic that any new direction chosen will not undo any gains in reducing J by all of the previous directions used; these are known as *conjugate directions*. The modified Powell's method sequentially performs line search minimizations along the current set of directions (it starts with the coordinate system basis vectors) finding the next new direction by averaging all of the M-dimensional line search directions of the last iteration. The modification (from straightforward application of the above) is that it does *not* average the direction of the *largest* decrease in J in the last iteration. This helps prevent linear dependencies from building up in the direction set, which leads to degeneracy in the dimensional space explored (i.e. only a subset of all of the p dimensions are examined). Press *et al.* (2007) have an excellent version of the modified Powell's method.

10.2.5 Nelder–Mead simplex

In order to evaluate the performance of optimization algorithms that can search dimensions higher than one, we have programmed the *Rosenbrock function* into a select subsample of optimization routines. The Rosenbrock function is also known as the *banana function* because of its characteristic shape, especially when viewed as a contour plot (Fig. 10.5). The Rosenbrock function turns out to be a challenging target for numerical optimization, and is therefore often used to test candidate algorithms. We use it as the cost function:

$$J(p_1, p_2) = 100(p_2 - p_1^2)^2 + (1 - p_1)^2 \tag{10.11}$$

of the two parameters (p_1, p_2). This function is considered a strenuous test because of the large number of iterations and cost function evaluations usually required to find the function minimum (at $p_1 = 1.0$ and $p_2 = 1.0$, $J = 0.0$). First up is the Nelder–Mead simplex method.

The *Nelder–Mead simplex*, or sometimes the *downhill* simplex, is an algorithm that works its way downhill (down cost function gradient) without actually needing the cost function gradient to evaluate (Nelder and Mead, 1965). Like the linear simplex method (Section 10.1.2) it uses a geometric construction called a simplex, but the similarities between these two different algorithms end there. The general idea behind this algorithm is that you start with $M + 1$ initial guesses for $\mathbf{p}_o$ (instead of the more traditional single point). The Nelder–Mead then constructs a simplex and evaluates J at each of the vertices (in 2D the simplex would be a triangle). The algorithm reflects the vertex corresponding

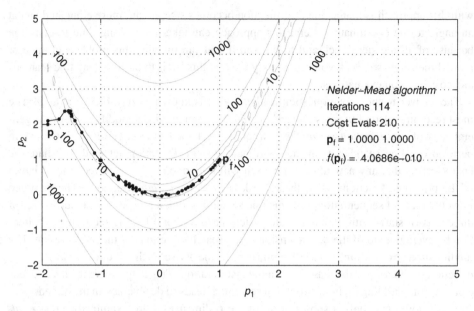

Figure 10.5 The Nelder–Mead simplex applied to the Rosenbrock function. Although the iteration and cost function evaluation counts are high, this method's persistent pursuit of the function minimum is generally successful, if not terribly fast. This figure was created using MATLAB's `fminsearch` routine.

to the largest J through the opposite face creating a new simplex. The value of $J(\mathbf{p}_{new})$ is evaluated and another reflection takes place. The reflections conserve the "volume" of the simplexes until there is an opportunity to grow through reflection and expansion. When that happens, typically the algorithm will contract the simplex in the direction perpendicular to the direction of expansion. If the algorithm finds plain reflection and reflection and expansion do not lead to a better (lower) J value it can contract the simplex along all directions and shrink the simplex volume. Slow and steady, this algorithm should not be overlooked, especially for problems where assumptions about the cost function behavior lead to wild extrapolations of parameters. The MATLAB optimization toolbox provides a version of this algorithm in the `fminsearch` function.

10.3 Methods adding the cost function gradient

Some methods have the ability to use information from the cost function gradient $(\partial J / \partial \mathbf{p})$ if it is available or make a fair approximation of it, if it is not available.

10.3.1 Steepest descent

An idea that is similar to the coordinate system choice of direction (mentioned above) is to search in the direction of *steepest descent*. This seems like an obvious choice; if we just go downhill always in the steepest direction, we're bound to arrive at the minimum. We apply this method to the Rosenbrock function as shown in Fig. 10.6 (notice this optimization was terminated because it exceeded the 250, user-defined, maximum number of cost function evaluations, long before it converged).

Starting from the initialization point $\mathbf{p}_o$ it performs a line search in the down gradient direction $-\nabla J(\mathbf{p})$, i.e. in the direction of $-\left[\partial J/\partial p_1, \partial J/\partial p_2, \ldots, \partial J/\partial p_M\right]$. One continues the search in this direction until one reaches a minimum in J along this direction. At this point one chooses a new line search direction down-gradient again. There are at least two problems with this approach. The first deals with too many line searches as it did in the coordinate choice method, although there are perhaps fewer searches. The reason there are still too many line searches is because at the minimum along a line search in any direction, the new steepest descent direction will be perpendicular to the last search. This is rarely, if ever, the most efficient direction in which to search next. Typically, in a short distance, a new minimum will be found and another perpendicular change in direction will ensue. We think you can see that in Fig. 10.6a, after an initial overshoot, this method quickly finds the valley of minimum and becomes stuck therein. In Fig. 10.6b we zoom in on the details of the steepest descent progress as it spends a lot of time zigzagging along the floor of the steep Rosenbrock function valley on its way to the minimum (the alternate directions are distorted in this graph in order to zoom in enough to show the zigzag pattern).

The second problem with this method involves the question of how you know which way is down-gradient. In the case where you *know* the functional form of the cost function, $-\nabla J(\mathbf{p})$ can be evaluated directly. Unfortunately this is rarely the case, and one must do a local search in many directions about each point $\mathbf{p}_k$, evaluating the cost function multiple times for each iteration k. In a two-dimensional problem one is only required to evaluate the cost function three times to get a fix on the local gradient, but in M-dimensions it is $M + 1$ evaluations. One of the things one strives for in numerical optimization is to evaluate the cost function as few times as possible, as this can quite often be the most computationally expensive operation of the optimization. So, it would seem, this technique will take too long. If, however, you insist on using this technique, use `optimset('HessUpdate','steepdesc')` when setting the options for some of the "medium-scale"[2] optimization routines provided in the MATLAB *Optimization Toolbox* (more about the Hessian below under conjugate gradients). Venkataraman (2002) also provides an example m-file for this algorithm.

10.3.2 Conjugate gradient

The *conjugate gradient* algorithm encompasses a number of ideas similar to the approach described in Powell's method. In particular, the idea of performing line searches only in conjugate gradient directions is critical in this method. This time we use the cost function

---

[2] The term *medium-scale* is a term The MathWorks uses to distinguish certain algorithms from *large-scale* algorithms.

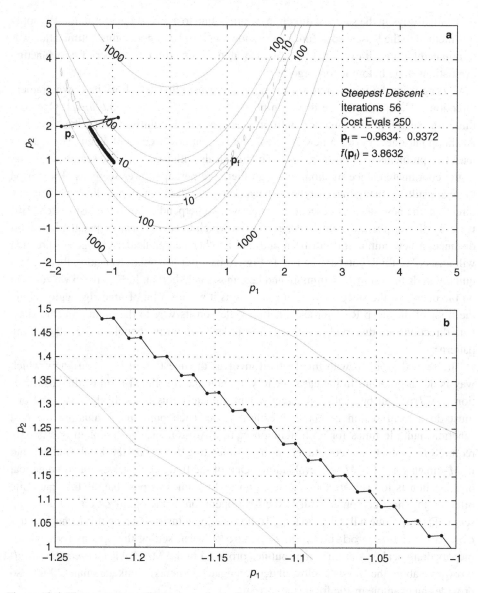

Figure 10.6 The steepest gradient approach applied to the Rosenbrock function. Terminated after 250 cost function evaluations, (a) shows the entire search domain and shows how little progress the algorithm had made after 56 iterations of the algorithm and (b) shows a zoomed-in view of part of the search domain where it can be clearly seen that this algorithm becomes trapped in this function making small progress as it zigs and zags its way along the bottom of a valley. The number of function calls and/or iterations can be increased, but we strongly advise against it owing to the even greater number of line searches that would have to be performed; there are better search algorithms. This figure was prepared with MATLAB's `fminunc` routine with `HessUpdate` option set to `steepdesc`.

gradient information, provided by the user. What is a *conjugate* gradient? Like conjugate angles in Euclidean geometry (their sum is always 360°) or complex conjugates (complex numbers whose imaginary parts are of opposite sign), conjugate gradients come in pairs that share a symmetric mathematical relationship. Following Press *et al.* (2007) we start with a quadratic approximation (a Taylor series truncated at its second order term) of the cost function (J):

$$J(\mathbf{p_o} + \delta\mathbf{p}) \approx J(\mathbf{p_o}) + \delta\mathbf{p}' \cdot \nabla J(\mathbf{p_o}) + \frac{1}{2}\delta\mathbf{p}' \cdot \mathbf{H}(\mathbf{p_o}) \cdot \delta\mathbf{p} \qquad (10.12)$$

where $\delta\mathbf{p}$ is a perturbation to the parameter vector about a base state $\mathbf{p_o}$, $\nabla J(\mathbf{p_o})$ is the gradient of the cost function at $\mathbf{p_o}$, and $\mathbf{H}$ is the *Hessian* matrix of the cost function, i.e.:

$$H_{ij} = \left. \frac{\partial^2 J}{\partial p_i \partial p_j} \right|_{\mathbf{P_o}} \qquad (10.13)$$

The Hessian of a function is a square, symmetric matrix of the multivariate second derivatives of a function. That is, it contains the curvature of the function for each variable. Note that near the minimum in J, the Hessian is also convex.

We now state a result (without proof) from linear algebra involving conjugate gradient and minimization of the above quadratic approximation (i.e. minimization of the cost function). Starting with two arbitrary vectors (directions) $\mathbf{r_o} = \mathbf{s_o}$, we build a set of conjugate gradient vectors, adding a new vector for each iteration step k:

$$\mathbf{r}_{k+1} = \mathbf{r}_k - \lambda_k \mathbf{H} \cdot \mathbf{s}_k \quad \text{and} \quad \mathbf{s}_{k+1} = \mathbf{r}_{k+1} + \gamma_k \mathbf{s}_k \qquad (10.14)$$

Here λ_k is a scaling factor applied to $\mathbf{H}$ minimizing $J(\mathbf{r}_{k+1})$, and γ_k is the size of the step taken from $\mathbf{p}_k$ to $\mathbf{p}_{k+1}$ in the $\mathbf{s}_k$ direction. Now we have a way to generate mutually conjugate directions from one optimization step to the next satisfying the orthogonality and conjugacy conditions:

$$\mathbf{r}_i \cdot \mathbf{r}_j = 0 \qquad \mathbf{s}_i' \cdot \mathbf{H} \cdot \mathbf{s}_j = 0 \qquad (i \neq j) \qquad (10.15)$$

All we need now is the Hessian ... which we typically do not have. But there is a mathematical operation we can employ to get around this problem. Start by setting $\mathbf{r}_k = -\nabla J(\mathbf{p}_k)$ (remember we have the gradient), and step to $\mathbf{p}_{k+1}$ in the $\mathbf{s}_k$ direction. Then set $\mathbf{r}_{k+1} = -\nabla J(\mathbf{p}_{k+1})$ and update $\mathbf{s}_{k+1}$ using the right-hand equation in (10.14). We are then set for another iteration, all without the use of $\mathbf{H}$.

The conjugate gradient methodology is used in many optimization algorithms to choose the direction to adjust parameters for each iteration step. Not only does it find directions that do not undo the gains in cost function minimization of the last iteration, it also accumulates information about all the previous conjugate gradient directions. Comparing this algorithm performance on finding the minimum of the Rosenbrock function (Fig. 10.7), we find it takes only 22 iterations or conjugate directions to find the minimum. The actual line search minimization was done with the Golden Section method alone, which may account for the high cost function evaluations. Furthermore no gradient information was given this

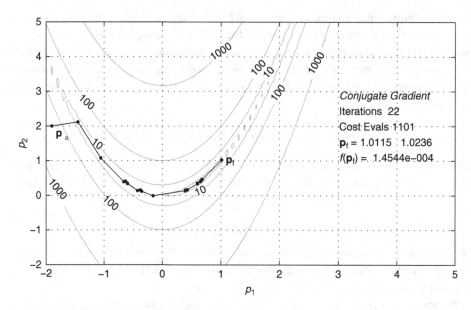

Figure 10.7 An example of the conjugate gradient search method in two dimensions applied to the Rosenbrock function. This figure was made with output from a modified conjugate gradient routine supplied in Venkataraman (2002).

particular run and this algorithm approximates the gradient with multiple cost function evaluations at each sub-step point.

10.3.3 Quasi-Newton methods

You may be familiar with *Newton's method* from introductory calculus, which uses the fact that the gradient of a univariate function $f(p)$ is zero at an extremum, and knowing that its second derivative is positive allows you to compute the location of a minimum. This can be done iteratively with

$$p_{k+1} = p_k - \frac{\partial f/\partial p_k}{\partial^2 f/\partial p_k^2} \tag{10.16}$$

which in essence assumes that a function can be approximated as a quadratic in the immediate vicinity of the extremum.

In optimization of mutlivariate functions, there is a class of methods, referred to as *quasi-Newton* or *secant methods* that are based on this concept. The second derivative of the function is the Hessian matrix, which must in general be estimated. Revisiting the Taylor series expansion of multivariate cost function (Equation (10.12)), we substitute an approximate Hessian $\hat{\mathbf{H}}(\mathbf{p_o})$ so that:

$$J(\mathbf{p_o} + \delta\mathbf{p}) \approx J(\mathbf{p_o}) + \delta\mathbf{p}' \cdot \nabla J(\mathbf{p_o}) + \frac{1}{2}\delta\mathbf{p}' \cdot \hat{\mathbf{H}}(\mathbf{p_o}) \cdot \delta\mathbf{p} \tag{10.17}$$

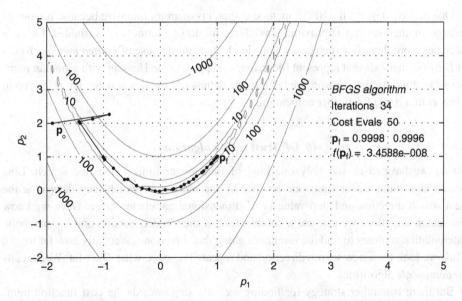

Figure 10.8 An example of the BFGS minimization method in two dimensions applied to the Rosenbrock function. This figure was made with MATLAB's fminunc with the LargeScale option set to off, but with gradobj to on.

Given this approximate Hessian, we can construct our Newton solver like Equation (10.16):

$$\mathbf{p}_{k+1} = \mathbf{p}_k - \hat{\mathbf{H}}^{-1} \cdot \nabla J(\mathbf{p}_k) \qquad (10.18)$$

Variants of the quasi-Newton methods use different methods to develop an estimate of the Hessian. Although more compute-intensive for each iteration step, it turns out that quasi-Newton methods are a generally faster alternative to conjugate gradient methods. For this reason, MATLAB optimizers tend to use them.

The most commonly used quasi-Newton algorithm is the *Broyden–Fletcher–Goldfarb–Shanno* or the *BFGS* method, which has been shown to be superior to its predecessor, the *Davidon–Fletcher–Powell* (or *DFP*) method, in terms of robustness against roundoff error and convergence issues. The method starts with an intial guess for $\hat{\mathbf{H}}$ and obtains successive approximations to the actual Hessian while constraining $\hat{\mathbf{H}}$ to be positive definite and symmetric, which it must be in the vicinity of optimum J. We won't delve into the details of the "machinery" here, but refer the interested reader to Press *et al.* (2007).

Figure 10.8 shows the results of the BFGS algorithm applied to the Rosenbrock function in "medium-scale" fashion. Like the steepest descent method, this optimization run has an initial overshoot but quickly recovers and finds the valley of the minimum. However, unlike steepest descent it moves right along the floor of this valley to the minimum in 34 iterations. Much like the conjugate gradient method it does not get trapped in the difficult to navigate, deeply incised valley. This method uses Brent's method to find the line search minimums, which greatly reduces the number of cost function evaluations.

One disadvantage of the BFGS method is that it is memory intensive because it requires storage of the Hessian approximations. Thus for large numbers of variables, memory becomes a problem. For such cases, you should consider the use of *limited memory BFGS* or *L-BFGS* methods that represent (describe) the approximate Hessian with a smaller number of vectors. Again, we won't delve into the details of how this is done, but refer you to Press *et al.* (2007) for further edification.

10.3.4　Trust region algorithms

So far we have discussed only one kind of function minimizer, the line search. Line search algorithms differ from one another by the manner in which they determine the next search direction and step value in M-dimensional parameter space. Once we know the search direction, we can use a combination of the Golden Section and the parabolic interpolation routines to find the minimum along that direction (essentially how far to go). Then we look for a new search direction and repeat. These are what The MathWorks calls *medium-scale* algorithms.

But there is another strategy for finding the next step towards the cost function minimum, *trust region* algorithms (also called *large-scale* algorithms). As before, we start with a cost function value for the current parameter estimate $\mathbf{p}_k$. In the parabolic interpolation method (Section 10.2.2) we take large strides toward the minimum in a single iteration by approximating the local shape of J in one dimension as a quadratic function and then setting the parameter values for the next iteration $\mathbf{p}_{k+1}$ equal to the minimum of the quadratic function. What if we followed the same idea in multi-dimensional parameter space?

Start by choosing the size of the domain over which we will fit a local model cost function J_{mod}; this *trust region* is an M-dimensional sphere or ellipsoid in parameter space about $\mathbf{p}_k$. The model function J_{mod} is typically a quadratic function built using the second-order Taylor expansion, Equation (10.12). Then set $\mathbf{p}_{k+1}$ equal to the minimum of J_{mod} within the trust region. Before we run off with the new parameter values, we need to test if the model approximation worked. The cost function evaluated at $\mathbf{p}_{k+1}$ should be less than the current cost function value, that is $J(\mathbf{p}_{k+1}) < J(\mathbf{p}_k)$. The size of the cost function increment $\Delta J = J(\mathbf{p}_k) - J(\mathbf{p}_{k+1})$ also should match expectations based on the shape of J_{mod}. If ΔJ is too small (or worse, negative) then our trust region is probably too large; that is, J_{mod} is not a good approximation to the local shape of the cost function. In this case, we would shrink the size of the cost region and start over. If ΔJ is about right, we would either repeat these steps using $\mathbf{p}_{k+1}$ as the center for a new trust region or expand the size of the trust region about $\mathbf{p}_k$ if, for example, the minimum in J_{mod} is at the edge of the trust domain (and the true minimum is likely outside the trust domain).

Many of the algorithms already discussed can be cast in terms of a trust region or large-scale version. For example, depending on how much we know or can approximate about the local cost function, we make an approximation of its Hessian $\hat{\mathbf{H}}$. If we set $\hat{\mathbf{H}}$ equal to zero, then we have the steepest descent algorithm again. But we can do better by assuming that $\hat{\mathbf{H}}$ is, at the very least, a symmetric $M \times M$ matrix as we did in the BGFS method. The

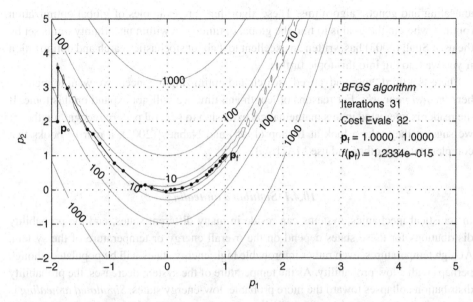

Figure 10.9 An example of the BFGS minimization method in two dimensions applied to the Rosenbrock function. This figure was made with MATLAB's `fminunc` with the `LargeScale` option set to on and `gradobj` to on.

initial choice of the trust region size is important; too small and it will miss opportunities to take large steps toward the minimum, too large and time will be wasted while the trust region is repeatedly shrunk down to a more appropriate size. Most modern trust region algorithms have an adaptive trust radius subroutine that adjusts the size of the radius based on past performance; for good performance the radius is gradually increased, while poor performance forces the algorithm to reduce the trust region size.

If we look at Fig. 10.9 we can compare the results to Fig. 10.8. Both of these searches for the minimum of the Rosenbrock function were performed with the BFGS algorithm. But in Fig. 10.9 we used the large-scale, or trust region, version. You'll note that the number of iterations is about the same and the number of cost function evaluations in Fig. 10.9 is only marginally smaller. What really stands out is the overwhelming (several orders of magnitude) accuracy in the final answer (deviation from the true minimum of $\mathcal{O}(10^{-15})$ vs. $\mathcal{O}(10^{-8})$). The only other algorithm we've shown you that comes as close is the persistent downhill searcher Nelder–Mead simplex (Fig. 10.5). One last thing, do not let the name "large-scale" throw you. These algorithms can be applied to ocean-basin-scale problems right down to petri-dish-sized experiments.

10.4 Stochastic algorithms

The methods of optimization that we outline below fall under the general title of *stochastic algorithms*. In particular, there are two that we want to bring to your attention, simulated annealing and genetic algorithms. These algorithms are examples of global optimization routines wherein the goal is to find the global optimum, to within uncertainty limits set by the user. Spall (2003) has written an excellent text about stochastic search and optimization if you want to dig into this topic further.

There is a third "more advanced" search and optimization concept we want to mention here, *neural networks*. For reasons of length and time we will not expand on this topic. If the reader is interested in this mathematical extension to general pattern recognition theory we suggest they take a look at Bishop (1995) and Nabney (2002), a pair of books that complement each other and use MATLAB.

10.4.1 Simulated annealing

In statistical mechanics, systems can reside in many different states, and the probability distributions for these states depend on the overall energy or temperature of the system. At high temperatures, even rather improbable high-energy states will be populated, though perhaps with a low probability. As the temperature of the system decreases, the probability distribution collapses toward the more probable, low-energy states. *Simulated annealing* is a mathematical analog where "states" are now vectors of different parameter values $\mathbf{p}$ and "temperature" is a measure of how much variability we allow in $J(\mathbf{p})$.

As you might guess, simulated annealing algorithms can be loosely compared to the physics of cooling. A system, any system, has a distribution of energies that can be expressed as probabilities in a statistical mechanical fashion. The probabilities can be related to their temperatures following a Boltzmann distribution. The transition from one energy state E_1 to a new energy state E_2 has a probability P:

$$P(E_1 \rightarrow E_2) = e^{-(E_2 - E_1)/(k_B T)} \tag{10.19}$$

where k_B is the Boltzmann constant, and T is the temperature. Metropolis *et al.* (1953) introduced the idea of this Boltzmann–Gibbs-like distribution (Equation (10.19)) to calculate the probability of moving from one energy state to another in an optimization context. If $E_2 < E_1$, e.g. $J(\mathbf{p}_{k+1}) < J(\mathbf{p}_k)$, then $P_{1,2}$ is greater than or equal to one, and so the algorithm sets it to one, i.e. the transition always takes place. When the relative energy is reversed there is some finite probability of the transition (uphill on a cost function hypersurface!). Given enough time at a constant temperature the system reaches thermodynamic equilibrium. Simulated annealing optimization creates a transition matrix for all of these possible transistions, so that at any "temperature" one can state the probability that the system is in "state" $\mathbf{p}_k$.

Now comes the annealing part. If we decrease the temperature slowly (this implies a schedule of some sort, usually an exponential decrease) then we can recalculate the transition matrix for points along the annealing schedule. Of course we're not really using temperature; instead some control parameter is used in place of thermodynamic temperature and that's what we slowly decrease. An example of a control parameter might be a scaling parameter used to multiplicatively increase or decrease an a priori vector of model parameter standard deviations. At each stop along the way the simulated annealing algorithm allows the system to go through some large, but not prohibitive number of recon- figurations (for discrete models) or iterations (for continuous models) and notes how many successfully reduce the cost function. When the control parameter decreases to the point that no new successful states are found, without waiting (iterating) a very long time, or some convergence criteria (there can be more than one) have been met, the algorithm stops and the parameter vector with the smallest J is the global minimum. Simulated annealing routines work solely with the cost function J; they do not require gradient information.

Simulated annealing has been used to provide solutions to the classical "Traveling Sales- man" problem: in what order should N cities be visited to minimize the distance traveled? But simulated annealing has also been used on many different oceanographic problems as well. Athias *et al.* (2000) used this approach to compare simulated annealing with other optimization routines (trust region and genetic algorithm approaches) to determine particle cycling parameters in a biogeochemical model. They found that the global optimization techniques in general worked better, with genetic algorithms (see next section) having a slight advantage. In an attempt to find an optimal combination of parameters for an ecosystem model applied at Ocean Station P, Matear (1995) applied simulated anneal- ing methods. Different configurations (simplifications) of the base ecosystem model were used, and up to 10 ecosystem parameters were successfully optimized. However, more data would be required to optimize the many more outstanding ecosystem model parameters. In a more engineering context, Alvarez *et al.* (2009) optimized autonomous underwater vehicle (AUV) hull shape parameters keeping the volume of the vehicle constant. They found that in "snorkel" mode, i.e. operations near the surface, the wave-induced drag was better compensated for by hull designs that deviated from the traditional torpedo shaped hull. Geoacoustic inversions have benefited from simulated annealing optimizations. For example, Fallat *et al.* (2004) used a hybrid adaptive simplex/simulated annealing routine to reduce inherent algorithmic noise, making these inversions more suitable for sensing environmental changes as a function of space and time.

10.4.2 Genetic algorithms

Genetic algorithms, as you might suspect, make mathematical analogies to natural evo- lutionary mechanisms observed in biology. In a Darwinian twist on normal optimization proceedure, the cost function is referred to as the *fitness function* and, of course, it is to be maximized, not minimized. The direction of the objective function can be switched by the simple multiplication of $-1 \times J(\mathbf{p}_k)$. Genetic algorithms are designed to mathematically mimic the biological processes of competition, reproduction, and mutation, the processes that are believed to drive evolution. The evolution of a single entity is not very informative, so genetic algorithms deal with populations of potential solutions allowing them to mate, compete, and develop new characteristics. If the algorithm is successful, what starts as a population of potential solutions (local minima) scattered across the complete, allowable parameter space will evolve, with iteration, clustering around one, global optimum. The analogy to biological systems is pushed even further by referring to the specific values of each $\mathbf{p}_k$ (remember there is a population of them now) as *chromosomes*. The values of these chromosomes are converted to binary bits (*genes*) to allow for efficient mathematical mixing of them when two potential solutions are chosen (at random) to become parents and reproduce, a process called *cross-over*. Encoding in bits (base-2) also allows for mutation by simple random bit flipping in the binary string; this operation is usually given a low probability to help preserve the healthy genes present in the population. Elitism, termination, replacement … the list of biological terms pressed into service is surprisingly a propos. These algorithms follow the following basic steps:

1. Initialize the population of potential solutions;
2. Select pairs for parenthood;
3. Allow them to reproduce and cross-over their genes;
4. Replace the parent population with the child population, with mutations and terminations;
5. Test the fitness function for convergence and either end or repeat at step 2.

Record *et al.* (2010) use genetic algorithms to parameterize a coupled biological–physical copepod dynamics model in Cape Cod Bay and find that optimized does not always mean correct. They also find that ensemble twin numerical models can reveal more about parameter uncertainties and importance. Canellas *et al.* (2010) use genetic algorithms to forecast and hindcast significant wave heights in Western Mediterranean Sea. Ward *et al.* (2010) compare variational adjoint (see Chapter 18) and genetic algorithm methods for optimizing parameters in a marine biogeochemical model applied to the Arabian Sea and the Equatorial Pacific. Like Matear (1995) they find model-data mismatches could be used to constrain up to 10 model parameters, but no large collections of parameters could be constrained with the data in hand. Huse and Fiksen (2010) use genetic algorithms to improve predator–prey interactions in ecosystem models, particularly with respect to fish–zooplankton interactions and zooplankton mortality. Results from their optimizations bolster their claim that zooplankton mortality needs to be tied more closely to environmental factors.

10.5 An ecosystem optimization example

We are going to apply a nonlinear, unconstrained optimization routine to a simple predator–prey ecosystem model similar to the Lotka–Volterra model we developed in Chapter 9. To follow along in this example experiment you will need access to MATLAB's optimization toolbox. We will use `fminunc`, the MATLAB unconstrained, nonlinear optimization routine. If you do not have access to this toolbox we have made some suggestions of alternate optimization software that can be substituted for the MATLAB function we will be using (Venkataraman, 2002; Nabney, 2002; Spall, 2003), although your results are likely to be slightly different.

We will perform a "twin" numerical experiment by first running the predator–prey model with fixed parameters to generate some model output, which we will use later as data to optimize a second model run after we add a little noise. The second model run will have adjustable parameters whose initial guess values will be far, but not too far, from the first model run in parameter space. Success (or skill) will be judged by how close the second set of parameters can be optimized to the original, fixed set of parameters

10.5.1 Predator–prey model (PZ)

In Section 8.2.2 we showed how one can limit the growth of an exponentially growing phytoplankton species by building in a carrying capacity. In this optimization experiment, we combine this concept with a Lotka–Volterra or predator–prey model similar to the one we constructed in Chapter 9. There are four parameters that describe the model's behavior:

$$\frac{dx_1}{dt} = p_1\left(1 - \frac{x_1}{p_2}\right)x_1 - p_3x_1x_2 \tag{10.20}$$

$$\frac{dx_2}{dt} = p_3x_1x_2 - p_4x_2 \tag{10.21}$$

where x_1 are the prey, x_2 are the predator, p_1 is the prey growth rate parameter, p_2 is the environmental carrying capacity of the prey, p_3 is predator grazing rate, and p_4 the predator mortality rate. We code these equations in a manner similar to the Lotka–Volterra model:

```
dxdt(1) = p(1)*(1. - x(1)/p(2))*x(1) - p(3)*x(1)*x(2);
dxdt(2) = p(3)*x(1)*x(2) - p(4)*x(2);
```

These are just the lines of code that are affected; the entire function is named below and is available online. This experiment will need one more subroutine, a cost function. The cost function is the sum of the square of the differences between the "data" we create and the optimizing model values:

$$J = \sum_{i=1}^{N}\left((x_{i,1}^{obs} - x_{i,1})^2 + (x_{i,2}^{obs} - x_{i,2})^2\right) \tag{10.22}$$

where $x_{i,1}$ and $x_{i,2}$ are the time-varying components of the model, x^{obs} are the "data", and the index i runs over each time step of the series. The cost function subsamples the time series the same way the initial model run does (see step 1 below). Remember to keep the time intervals the same for these kinds of comparisons. We code it as:

```
cost = sum((xobs(:,1)-X(:,1)).^2 + (xobs(:,2)-X(:,2)).^2)
```

In the next section we lay out the steps of the algorithm to perform this twin numerical experiment.

10.5.2 *An experiment using* fminunc

We have written some m-files that will run this twin numerical experiment for you. The underlined m-files in the list below are available on the website set up for this book, and if you wish to follow along on your computer (and we strongly suggest you do), download the codes list below. The steps in this experiment have the following sequence:

1. Run the model, once, with fixed parameters. Often it is simplest to create a separate driver program ecopz_main.m for this purpose. It uses the ODE solver of choice (ode15s) and the m-file with differential equations ecopz.m, but does not reference the optimizer or the cost function. At the end, this program subsamples the final time series and writes these values out to a file with MATLAB's save command.
2. Run the driver program ecopz_opt_main.m with access to the optimizer and the other m-files that we have created to perform the optimization. This includes an m-file to calculate the cost function ecopz_costfxn.m as well as the differential equations (same m-file as used in step 1).
3. The main program establishes global variables, reads in the "data" saved earlier, adds a small amount of normally distributed random noise to this data, sets the parameter initial guess (to different, but not too different, values from those used in step 1), and sets all of the other housekeeping variables the *same* as step 1 (time period, initial conditions of the predator–prey model, etc.), and calculates the initial value of the cost function.
4. The driver program makes a plot of the "data" and the initial run with the second set of parameters as shown in Fig. 10.10 and pauses after this plot. Hit any key to continue.
5. The driver program sets some of the optimization options with optimset, in particular LargeScale to off, and hands-off execution of the experiment to fminunc with the name of the cost function m-file and the initial set of parameters values as arguments.
6. With no gradient information fminunc performs a quasi-Newton line search of the cost function, and you can set up the program to print out the value of the cost function as the optimizer navigates its way to the minimum (note: the values will, occasionally, go up).
7. Each time the cost function is called, it first solves the set of ODEs using the initial values and parameters it was passed, and then calculates and returns the cost function value to fminunc.
8. Each time fminunc receives the cost function value it checks to see if it has decreased and if it has decreased enough to declare convergence. If not it adjusts the parameter values, dependent upon the method being used, and passes them back to the cost function m-file. This repeats until convergence or one of the maximum iteration flags is triggered.
9. When fminunc is finished, the answer is passed back to the main program, and one last plot is made as in Fig. 10.11. Various diagnostics can be set to print out at this point, controlled with the options variable.

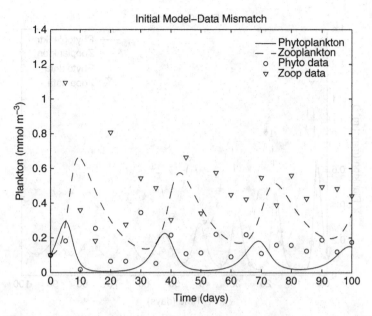

Figure 10.10 The model–data mismatch we find when we run our `ecopz_opt_main` program with our initial, best-guess for the parameters of the model $\mathbf{p}_o = (0.5, 2, 1.5, 0.1)$. Notice that not only are the magnitudes of both the phytoplankton (Phyto) and zooplankton (Zoop) too low, they are also largely out of phase with the "real" data from the twin experiment.

10.5.3 Epilogue

For this twin numerical experiment the "true" parameters are shown below on the left and the optimized values on the right. Since scaled, normally distributed noise is added to the "data" each time the optimization is made, one will get slightly different answers each time.

```
p=[1.0000    1.0000
   3.0000    2.9996
   2.0000    2.0001
   0.3000    0.3000]
```

The four decimal places are not called for, and we show them only to demonstrate that the optimization is not exact; rounding off to two significant digits would yield results indistinguishable from the "true" values. The final cost function value is 1.5×10^{-5}; that's a pretty good match. Further experimentation is left as an exercise for the reader to determine the random noise level at which the original parameters are difficult if not impossible to recover.

If this seems like a lot of trouble and you are tempted to find quasi-optimal values for your problem along the lines we discussed in the sensitivity analysis for the two-box ocean phosphorus model in Chapter 9, consider this. On one of the laptops we are writing this book, it took `fminunc` 51.828 seconds to perform a search of this problem's parameter space with 275 individual evaluations of the cost function. If we were to try the brute

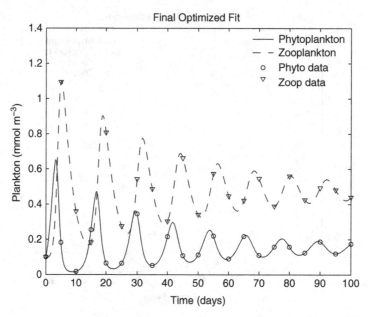

Figure 10.11 The optimized model–data mismatch yielded by `fminunc`, the unconstrained optimizer in the MATLAB optimization toolbox. Notice how much better the magnitude and phase match
between model and data is in comparison to Fig. 10.10.

force approach, with 10 values for each parameter, then we would need to evaluate the
cost function 10^4 times. Based on the above numbers we would estimate a parameter space
characterization run of approximately 15 hours. But, in truth, in order to locate optimal
values for the parameters precise enough to compete with this optimizer performance,
something more like 100 values for each parameter would be required. This would amount
to 100^4 individual evaluations of the cost function and would take the above mentioned,
put-upon laptop ~17 years. Certainly more powerful computers can do these calculations
more quickly, but the problem scales with both the size of the grid and the number of
parameters, so the extra amount of cpu power required to use brute force is not warranted.

10.6 Problems

10.1. Download the `ecopz` model codes mentioned in this chapter and adjust the scaling factor in a regular fashion until the `fminunc` optimizer can no longer recover
the original parameters. Make a plot of sum squared difference of the parameters
(similar to the cost function) versus the size of the scaling factor. Hint: the scaling
factor is multiplied with the `randn` values added to the `xobs` values in the main
program.

10.2. Set up a twin numerical experiment using the five-box phosphorus model from Chapter 9. Make the fluxes between the boxes the parameters to be adjusted. Is there a set of reasonable values that will permit a steady state solution ±5% higher than the answer obtained in Chapter 9? How about ±20% higher? Why or why not?

11

Advection–diffusion equations and turbulence

Big whorls have little whorls,
Which feed on their velocity;
And little whorls have lesser whorls,
And so on to viscosity
(in the molecular sense).

Lewis F. Richardson

11.1 Rationale

Why should we devote a whole chapter to advection–diffusion equations? It turns out that a significant part of oceanography involves the interpretation of the distributions of chemicals, particles, and organisms in the water column, and these distributions are the result of the combined effects of physical, biological, chemical, and geological processes. These processes are intimately intertwined in a way that requires us to "solve the physics" before we can begin to learn anything about the chemistry or biology. In fact, there is an active branch of marine geochemistry which involves the use of geochemical tracer distributions to make inferences about ocean circulation and mixing. That is, some people make a living from this kind of stuff. The processes involved play a fundamental role in the ocean's ability to exchange, sequester, and transport heat, carbon dioxide, and other biologically important properties. This in turn feeds back into the role of the ocean in regulating climate, global primary production, and all sorts of societally relevant things.

Put another way, you can think of the water column distribution of some property C, which might be a dissolved chemical concentration, concentration of suspended colloids or particles, abundance of some planktonic species, etc., as being affected by physical and biogeochemical processes:

$$\partial C(\mathbf{x}, t)/\partial t = P(C, \mathbf{x}, t) + J(C, \mathbf{x}, t) \tag{11.1}$$

where the space and time variations of the property of interest (C) are affected by some physical redistribution processes (the operator P) and some biogeochemical source/sink/transformation processes (J). For simplicity we've ignored the possibility that particles may sink, or that organisms may be motile.

We somehow must deconvolve the physical processes to understand the biogeochemical ones. In this chapter we discuss the general character of open ocean transport processes and the concept of turbulent diffusion. We do this because the nature of turbulent diffusion is not as clear cut and fundamental as is *molecular* diffusion. Yet it is an ugly necessity, for it involves a parameterization of processes that we have no hope of resolving directly, and that are so ubiquitous and important that we cannot ignore them. Our hope is to at least make you aware of the philosophical underpinnings of the concept, the strengths and weaknesses of the approach, and the caveats that need be kept in mind when dealing with turbulent diffusivities. At least that way you may have a better idea of why things appear the way they do.

11.2 The basic equation

Our starting point is really rather basic: we begin with the Daltonian concept of the conservation of mass. Also, we'll only do this in one dimension. If you are into self-inflicted pain you can do this in three dimensions or wait until Chapter 17 when we will do it for you. The conservation of mass for a tracer C in one dimension can be written as (Welty *et al.*, 2007):

$$\frac{\partial C}{\partial t} = -\frac{\partial}{\partial x}(uC) + \frac{\partial}{\partial x}\left(D\frac{\partial C}{\partial x}\right) + J \tag{11.2}$$

which looks more complicated than it actually is. Let's break down the terms one by one. The first term on the left, $\partial C/\partial t$, is simply the local time rate of change of concentration (mass or moles per unit volume) at a particular location. The second term reflects the contributions due to fluid flow or advection. The *advective flux* is just the velocity times the concentration, uC. The term $-\partial(uC)/\partial x$ is then the *advective flux divergence*. Tracer concentrations increase with time in locations where more mass is being brought in than is being removed by the flow field. Notice that the divergence term can change because of horizontal gradients in velocity u or concentration or both.

The third term is the molecular diffusive flux divergence (D is the molecular diffusivity, which for most substances is of order 10^{-9} m^2 s^{-1} in water). Similar to advection, tracer will accumulate (or deplete) if the amount of material diffusing into a location exceeds (or is less than) the amount diffusing out. In order for this to occur, the diffusive flux must change with distance. With the assumption that the molecular diffusivity is constant in space (a reasonable assumption in most systems) and neglecting advection the above equation can also be expressed in the possibly more familiar Fick's law form as:

$$\frac{\partial C}{\partial t} = D\frac{\partial^2 C}{\partial x^2} + J \tag{11.3}$$

but we'll stick with the other form for now. Finally, the last term J is there to represent possible source or sink processes, such as biological production/consumption or radioactive production/decay. Its form will remain unspecified here, but may be dependent on space, time, C, and possibly other variables. The form of Equation (11.2) is a quite general one. In lieu of C, we could equally have substituted energy, temperature, or momentum. Of course the corresponding meaning or value of D would be different, but the relationship would be the same.

Equation (11.2) is cast in terms of a fixed-space frame of reference, or an *Eulerian* (pronounced "oil-air-ian") frame of reference. The Eulerian derivative $\partial/\partial t$ is equivalent to sitting in a particular spot in the ocean and making measurements over time. Moorings and ship-based time series lend themselves to an Eulerian view of the ocean, as do many numerical models that are constructed on fixed geographic grids. An alternate approach, referred to as a *Lagrangian* frame of reference, follows individual fluid parcels. The Lagrangian derivative d/dt is the more natural framework for tracking the changing environment around a plankton cell or analyzing the data from a drifter or neutrally buoyant float.

We can translate between the two reference frames by writing the change in concentration dC as the sum of partial derivatives in space and time (Welty *et al.*, 2007):

$$dC = \frac{\partial C}{\partial t} dt + \frac{\partial C}{\partial x} dx \tag{11.4}$$

Dividing through by dt results in:

$$\frac{dC}{dt} = \frac{\partial C}{\partial t} + \frac{\partial C}{\partial x}\frac{dx}{dt} = \frac{\partial C}{\partial t} + u\frac{\partial C}{\partial x} \tag{11.5}$$

where we have used the fact that dx/dt following a fluid parcel is simply the flow or velocity u. Another way of saying this is that the local rate of change is the sum of changes due to the passage by of different water parcels and changes in the properties of the water parcels themselves.

A useful way to think of the difference between Eulerian and Lagrangian derivatives is to consider a weather analogy. Standing in your garden one evening you notice that the temperature is dropping. Is it due to local radiative cooling, or is a cold front coming in? Your friend in a hot air balloon passing overhead, since he or she moves with the air mass, feels dC/dt (here we're using C to mean the temperature), while you feel $\partial C/\partial t$ at your fixed point in space. Your friend might actually feel the temperature going up (maybe the sun hasn't set just yet), but because the cold air front is moving in, you feel a decrease.

Equation (11.2) is perfectly general and both mathematically and physically correct in all respects. If we could define the terms of this equation with adequate precision in both space and time, we could describe the evolution of C throughout the ocean. The trouble is, we can't.

11.3 Reynolds decomposition

Here's where things get a little complicated. So far, what we have said is fundamentally sound and quantifiable. The problem is that we cannot know, measure, or resolve all of the fluid motions that we know must influence chemical or biological properties in the ocean. Think of it this way: in order to completely describe the motion of the stuff that we have been talking about, we need to take into account movement on scales ranging from the molecular (order 10^{-8} m) to that of an ocean basin (order 10^7 m), for all of these motions can contribute to changes in, say, oxygen concentrations. This is clearly not possible, because we are talking about 15 orders of magnitude. However, it is possible to describe the molecular diffusion of a dye in a glass of water (at least ideally) without knowing the individual motions of individual molecules: there are fundamental statistical mechanics (or thermodynamical) laws that can be derived to describe molecular behavior in a statistical sense. The hope is that we can do something along those lines with turbulent motion.

We begin by thinking of the fluid motion as consisting of two parts: one that is the large-scale, mean flow, and the other that is some small-scale, randomly fluctuating component. This, we should warn you, is the fundamental, critical step in the development of a turbulence formalism to which you must pay very careful attention. It implies that there is some kind of *scale separation* that you can divide the motions into: large (mean) and small (random). It presupposes that the small-scale processes are both stationary and random, that there must be some particular timescale T over which averaging makes sense. This may not always be the case, so *caveat emptor!* This definition can be presented mathematically as:

$$u = \bar{u} + u'$$ (11.6)

where we have the mean flow $\bar{u}$ defined as:

$$\bar{u} = \frac{1}{T} \int_T u \, dt$$ (11.7)

which naturally requires that the velocity anomalies u' must be unbiased:

$$\int_T u' \, dt = 0$$ (11.8)

You can deduce this equation (11.8) by taking the time integral of Equation (11.6) and comparing it to Eq. (11.7).

This conceptual separation of velocity components is referred to as *Reynolds decomposition* and is a fundamental assumption underpinning the idea of *turbulent diffusivity*. Now the same thing could be argued about the concentration of material C in the fluid, i.e. there is some smoothly varying, mean concentration $\bar{C}$ coupled with some randomly varying component C':

$$C = \bar{C} + C'$$ (11.9)

where

$$\bar{C} = \frac{1}{T} \int_T C \, dt \quad \text{and} \quad \int_T C' \, dt = 0$$ (11.10)

Now if we substitute these definitions (11.6 and 11.9) back into Equation (11.2) we get

$$\frac{\partial}{\partial t}(\bar{C} + C') = -\frac{\partial}{\partial x}\left[(\bar{u} + u')(\bar{C} + C')\right] = -\frac{\partial}{\partial x}\left[\bar{u}\bar{C} + u'\bar{C} + \bar{u}C' + u'C'\right]$$ (11.11)

The alert and clever reader will notice that we're ignoring the molecular diffusion and J terms, but it won't change anything and we will put them back in a moment. Now if we average the above equation with respect to time over the same interval that we've defined the scale separation in our velocity and property variables, we have

$$\frac{\partial \bar{C}}{\partial t} + \frac{\partial}{\partial t}\left(\frac{1}{T}\int_T C'dt\right) = -\frac{\partial}{\partial x}\left[\bar{u}\bar{C} + \frac{\bar{C}}{T}\int_T u'dt + \frac{\bar{u}}{T}\int_T C'dt + \frac{1}{T}\int_T u'C'dt\right]$$

(11.12)

In doing this, we've taken advantage of the fact that you can reverse the order of integration and differentiation in continuous systems, and that the mean distributions are time-invariant over the averaging interval T so they can be pulled out of the integral. Note also that the integrals of the fluctuating components are by definition zero, except for the very last term. The last term, which involves the product $u'C'$, is non-zero because there is likely to be some correlation between velocity and concentration fluctuations, since the former likely causes the latter. We can therefore simplify the above equation to be:

$$\frac{\partial \bar{C}}{\partial t} = -\frac{\partial}{\partial x}(\bar{u}\bar{C}) - \frac{\partial}{\partial x}(\overline{u'C'})$$

(11.13)

where the first term on the right-hand side is the "macroscopic" advective flux divergence, and the second term is the divergence of the *Reynolds flux*. The "overbar" refers to time averaging as we have defined it for the velocity and concentration earlier. As a side note, if the property we were dealing with was momentum, then the equivalent term would be the *Reynolds stress*.

Where can we go from here? As we mentioned, the cross-correlation must be causal in nature since random displacements in the fluid acting on a macroscopic concentration gradient can result in apparent concentration anomalies. It can be argued that in a randomly moving fluid, there is a characteristic space scale of displacement, which we'll call ℓ'. This may be the mean vertical motion of a fluid parcel caused by breaking internal waves, or the horizontal movement caused by mesoscale eddies sweeping by. Now this random displacement coupled with a large-scale mean gradient in concentration will result in an apparent concentration fluctuation C' governed by:

$$C' = -\ell'\frac{\partial \bar{C}}{\partial x}$$

(11.14)

Note the negative sign. If the slope is negative then a positive displacement of the fluid results in a positive concentration anomaly, whereas if the concentration slope is positive then a positive displacement results in a negative concentration anomaly. Figure 11.1 shows this schematically. We suggest you look at a paper discussing the mixing length concept by Chris Garrett (1989). Thus we have:

$$\overline{u'C'} = -\overline{u'\ell'}\frac{\partial \bar{C}}{\partial x}$$

(11.15)

Since the macroscopic concentration gradient is time-invariant over the integration (averaging) timescale, we pulled it out of the averaging. Now Equation (11.15) represents an

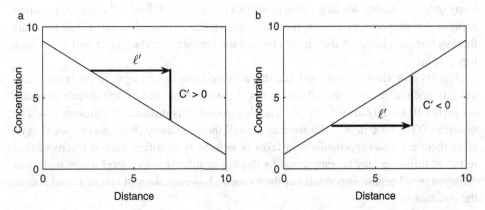

Figure 11.1 A schematic showing the effect of a positive going turbulent displacement ℓ' on a property distribution for (a) a negative property gradient, and (b) a positive property gradient. A similar diagram could be made for negative displacements.

important conceptual step, since we are now separating the causes of random concentration fluctuations into two components: one due to the large-scale distribution of C (the concentration-dependent part) and one due to the fluid motion (and hence not related to C).

Substitution of this term into Equation (11.13) gives the following equation:

$$\frac{\partial \bar{C}}{\partial t} = -\frac{\partial}{\partial x}(\bar{u}\bar{C}) + \frac{\partial}{\partial x}\left[(\overline{u'\ell'} + D)\frac{\partial \bar{C}}{\partial x}\right] + J \qquad (11.16)$$

where we have now come clean and put back in the molecular diffusion and J terms. Notice that we have bundled the $\overline{u'\ell'}$ term in with the molecular diffusion term since they are functionally similar in form.

Now what is the meaning of this $\overline{u'\ell'}$ term? It is a property of the fluid flow, but not of the fluid or the property that we studying. It is often called the *turbulent diffusivity coefficient* or interchangeably the *eddy diffusivity coefficient* since it appears in the equation like a diffusivity term, and typically represented as κ in equations. It is related to the cross-correlation between fluid displacement and the velocity. One would expect them to be more-or-less correlated in a turbulent fluid but not perfectly, since that only occurs for wave motion. Typically, the cross-correlation is around $1/4$ or so for most turbulent fluids (see, for example, Tennekes and Lumley, 1972).

So what is the size of this $\overline{u'\ell'}$ term? One might express this quantity as the product of the RMS (root mean square) displacement and the RMS velocity fluctuation, times the cross-correlation. In the thermocline, vertical motions associated with internal waves and tides of order of meters to tens of meters are not uncommon. This displacement occurs on timescales of a few hours, so associated velocities are of order $10^{-5}\,\mathrm{m\,s^{-1}}$ to $10^{-4}\,\mathrm{m\,s^{-1}}$. The product of these two yields a κ of order 10^{-5} to $10^{-3}\,\mathrm{m^2\,s^{-1}}$, which is many orders of magnitude larger than molecular diffusion (10^{-9} m^2 s^{-1}). In fact, measurement of this property, by various means, yields values of order 10^{-5} m^2 s^{-1} in the main thermocline, and larger in the mixed layer, in the benthic boundary layer and over rough topography.

How about horizontal motions? The disparity becomes even larger. In the open ocean, velocity fluctuations of order 0.01 to 0.1 m s^{-1} are common, and the space scale of eddies is of order 10 to 100 km (10^4–10^5 m). Thus horizontal eddy diffusion coefficients might be of order 100 to 1000 m^2 s^{-1}. So we may as well throw D away. Well, maybe not always, since there are cases (particularly in lakes or extremely stratified waters) where vertical turbulent diffusion may be suppressed by density stratification to a level where molecular diffusion may become important. For the moment, however, we will ignore D and rewrite the equation as:

$$\frac{\partial C}{\partial t} = -\frac{\partial}{\partial x}(uC) + \frac{\partial}{\partial x}\left(\kappa \frac{\partial C}{\partial x}\right) + J \qquad (11.17)$$

where we have dropped the overbars (the averaging symbols) with the understanding that we are dealing with the large-scale mean velocities and concentrations. Note the similarity to our "exact" relationship (Equation (11.2)), but with the molecular diffusivity D replaced by the numerically larger turbulent diffusivity κ.

There is one very important consequence of what we've just done. By throwing out the D, we're arguing that the turbulent diffusivity of a property in the ocean does not depend on the detailed characteristics of the species in question. That is, although the ultimate dispersion of material at the very smallest scales must be accomplished by molecular diffusion or small-scale dispersion of some kind, turbulent transport is dominated on all other scales by fluid motion, and hence is independent of a substance's molecular characteristics. So for example tritium will "diffuse" in the ocean in the same way that carbonate ion will, or viruses for that matter. This is probably a safe first-order assumption, although there are a couple of important exceptions to remember.

The first exception is the comparative turbulent diffusivities of momentum (Reynolds stress) and material (Reynolds flux). In addition to turbulent fluxes, momentum can be transmitted by wave motion (not just surface waves, but internal waves, Kelvin waves, etc.). As a result *eddy viscosity* tends to be significantly larger than eddy diffusivity. The second exception is the case of *double diffusion*, which occurs when small-scale vertical overturning in a thermohaline gradient of the appropriate sign results in separation of heat and salt anomalies. Because at a molecular level heat is about a hundred times more diffusive than salt and most other substances, heat is more readily extracted from an interpenetrating filament. The heat anomaly is "left behind" as the filament travels, with the counter-intuitive result that the slower diffusing species (the salt) is carried farther and has a higher turbulent diffusivity. This phenomenon, known as *salt fingering* or *double diffusion*, can be an important mixing mechanism in the subtropical main thermocline because both heat and salt play a role in determining the density of the water.

11.4 Stirring, straining, and mixing

Lest you be too comfortable and complacent with the basic formulation of turbulent mixing discussed above, we ought to mention the rather tangled problem of discerning mixing from straining and stirring. Consider, for a moment a patch of material (or dye) that has been "painted" on the water surface. If the patch transects the edge of some current or an eddy, some of the water, and hence the dye, will be advected further than the more slowly moving part of the patch. That is to say, the dye patch will be *strained* owing to a velocity shear over the dye patch. This serves to increase the separation between different parts of the dye patch, increasing its scale like diffusion would; but the scale will tend to grow linearly with time, rather than as $\sqrt{t}$, which you would expect from a diffusive process.

Thus some form of quasi-organized motion may act on a dye patch to streak it out, counter-intuitively increasing the property's spatial gradients and consequently its variance. One might refer to this process as *stirring*, whereby the originally simpler form of the property distribution is teased out by *straining*, and smaller-scale spatial gradients are enhanced. Beyond a certain point, smaller-scale turbulence takes over and these gradients are smoothed out. So in some sense, stirring and mixing work hand in hand to accomplish the same thing, but each in different ways. Stirring and straining serve to sharpen spatial gradients and increase the surface area over which smaller-scale mixing and diffusion can occur.

This may not be a great concern when modelling large-scale property distributions, since you will likely be looking at the long-term end stages of the processes, but during the development of a phytoplankton bloom associated with sub-mesoscale mixing processes, or larval dispersal, such effects will need to be considered. The take-home lesson from this is that when working with smaller-scale dye release experiments (e.g. like the NATRE experiment; see Ledwell *et al.*, 1998) the dye patch will evolve at different rates at different stages as it "feels" the various scales of motion. We suggest you consult Garrett (2006) for an excellent discussion of these issues.

11.5 The importance of being non

Virtually every property we deal with is measured in units, and often different units are used for the same property. Consider distance. You might be between 5 and 6 feet tall, carry a 3 inch long pocket knife, and live 4 miles from where you work on a planet that orbits approximately one astronomical unit from a sun that is 4.5 light years from its nearest neighboring star. And also consider time. It took you just a few seconds to read these sentences, but 20 minutes to drive to work, you've been struggling with the last problem set for several days, and you've been in graduate school for – well – years. It seems perfectly all right to speak in terms of different units in the same sentence; it's instinctive and people seem to understand you when you do it.

If you look at any advection–diffusion equation, you'll see it's riddled with units. Consider Equation (11.2). The first term on the left-hand side is in concentration units (mass per unit volume) and inverse time units (i.e. per unit time). The first term on the right-hand side, although it doesn't look it, is also in the same units. The property C is obviously in concentration units, the velocity is in units of length and inverse time, and the spatial derivative must be inverse length, so the lengths "cancel out" and we're left with concentration and inverse time units. The next term is a little more complicated but also works out. Diffusivity is measured in units of length-squared and inverse time (e.g. the molecular diffusivity of salt in seawater might be approximately $1 \times 10^{-9} \, \text{m}^2 \, \text{s}^{-1}$). The squared length units therefore cancel the inverse squared length units introduced by the compound spatial derivative. Throw in the C and you again have concentration units divided by time. The final term is easy, it's just the *in situ* consumption or production term, and must be in the same units, namely concentration divided by time.

Now there are three very important points about the above discussion that bear consideration. The first is that every equation must be *dimensionally homogeneous*. That is, you can't have one term that reduces to concentration units divided by time being equated to, added to or subtracted from another term that is in concentration units divided by length. That's an apples versus oranges comparison that simply doesn't make sense. This can be a very useful check when you're constructing or analyzing an advection–diffusion equation of any description. If you go through the exercise of *dimensional analysis* and the terms are dimensionally inconsistent, then you've made an error in constructing, remembering, or transcribing your equation. This is a very useful check on your model equation construction and we advise that you routinely do it. But we also want to point out that the converse isn't always true; the fact that an equation is dimensionally consistent doesn't guarantee correctness. Dimensional homogeneity is a necessary but not sufficient condition.

The second consideration is probably obvious to all of you, but bears mentioning anyway. Dimensional homogeneity requires that you use the same units for each property. While we routinely mix different, say, length dimensions in our day to day speech, you need to keep everything the same in equation-land. Philosophically, it makes perfect sense that any "law" expressed by an equation should not depend on an arbitrary decision we make on which units to express it in. That is, $E = mc^2$ should work equally well whether we choose to express it in mks (meters, kilograms, and seconds), cgs (centimeters, grams, and seconds), or in fhf (furlongs, hundredweights, and fortnights). Consider again Equation (11.2). It wouldn't make any sense if we used $\text{cm}^2 \, \text{s}^{-1}$ units for diffusivity but expressed spatial derivatives in inverse km units! This is a surprisingly common mistake among neophytes, and we urge you to make it a habit of working in consistent units (e.g. mks) and sticking to it religiously. You can avoid a lot of grief down the road that way.

Finally, there's a lot to be learned from doing dimensional analysis of the equations. The first is quite simply how individual terms in the equations depend on length- or timescales characteristic of particular system being studied. For example, in the case of Equation (11.2) examining the dimensional constitution of the diffusion term suggests that the timescale of diffusion is related to the square of the length scales. Conversely, the length scale associated with diffusive dispersion must depend on the square root of the elapsed time. By comparison, the advection term shows a linear time–distance relationship. So what we're saying here is that it is important to consider the dimensional forms of individual terms, which in turn represent physical or biogeochemical processes. There is much, much more that can be done with *dimensional analysis* techniques, including clever methods of data analysis as well as model interpretation, and we recommend a useful summary paper by Price (2003) for further reading.

So how do you *non-dimensionalize* an equation? It's actually quite simple, at least mechanically, but where it becomes more nuanced is in your choice of "scales". First, you go through the equation and perform a coordinate transformation for all the length scales. How you do this would be to argue that there's a specific length scale associated with your experiment – call it L – that you scale all your spatial measurements by. This might be the length of your model domain, the size of an ocean basin, the characteristic width of a boundary current, or something else. As you might guess, a specific situation may be characterized by more than one length scale, so you need to keep this in mind as you proceed, but you must use a single scale length for non-dimensionalizing your equation. Thus you would transform your space coordinate x (let's say it's in meters) as $x = \tilde{x}L$ where the new variable $\tilde{x}$ is now dimensionless. In the case where L is the length of your model domain or the scale of the ocean basin, you'd likely have $0 \leq \tilde{x} \leq 1$, but it doesn't have to be. Similarly, you would non-dimensionalize time so that $t = \tilde{t}T$ where T is some characteristic timescale. Also, you would scale the velocity according to some characteristic velocity U giving $u = \tilde{u}U$. Finally, you could non-dimensionalize your unknown, for example the concentration, by scaling to some initial value C_0. So in the case of Equation (11.2), we would have

$$\frac{C_0}{T}\frac{\partial \tilde{C}}{\partial \tilde{t}} = -\frac{UC_0}{L}\frac{\partial}{\partial \tilde{x}}(\tilde{u}\tilde{C}) + \frac{C_0}{L^2}\frac{\partial}{\partial \tilde{x}}\left(\kappa\frac{\partial \tilde{C}}{\partial \tilde{x}}\right) + J \tag{11.18}$$

which can be reduced by dividing by $(C_0)/T$ and assuming that κ is spatially invariant to

$$\frac{\partial \tilde{C}}{\partial \tilde{t}} = -\frac{UT}{L}\frac{\partial}{\partial \tilde{x}}(\tilde{u}\tilde{C}) + \frac{\kappa T}{L^2}\frac{\partial^2 \tilde{C}}{\partial \tilde{x}^2} + \frac{JT}{C_0} \tag{11.19}$$

Now we can group the advection and diffusion terms a little more:

$$\frac{\partial \tilde{C}}{\partial \tilde{t}} = -\frac{UT}{L}\left(\frac{\partial}{\partial \tilde{x}}(\tilde{u}\tilde{C}) - \frac{\kappa}{UL}\frac{\partial^2 \tilde{C}}{\partial \tilde{x}^2}\right) + \frac{JT}{C_0} \tag{11.20}$$

Notice the fact that within the parentheses the advection term can be compared to the diffusive term by a non-dimensional *number* κ/UL multiplying the latter term. You'll see the significance of this in the next section.

11.6 The numbers game

There's a logical extension to this dimensional analysis approach. Recognizing that the size of each term in the equation depends on the choice of units (e.g. whether you express length in cm or km), using consistent units means that this dependence won't matter in the end.

Moreover, it also means that you can divide the entire equation by the common dimensions (for example in the case of Equation (11.2) concentration times inverse time) to make all terms *dimensionless*, which means that these terms contain dimensionless *numbers* that represent "scale-independent" characteristics of the system being studied. These numbers provide important clues about how the system will behave, and how they are constructed informs us about what controls that overall behavior.

Before going on to talk about specific oceanographic results, we will digress to discuss various *numbers* that are of interest in oceanography. Numbers are, by definition, dimensionless quantities that somehow embody physically important characteristics of the systems being studied. More often than not, they indicate the relative importance of various processes, and appear as the ratio of either terms or timescales. The reason for describing your system in terms of these numbers is that quite often the character of model output or the behavior of the actual system tends to depend more on the ratio of terms than on their absolute value. So, for example, when faced with modeling the behavior of a system over a wide range of velocities and diffusivities, we may find that we need only perform experiments over a range of the *ratio* of two terms (the *Peclet number*) rather than doing every conceivable velocity and diffusivity combination. We will list some of the more important numbers, along with typical values and what they mean.

11.6.1 The Reynolds number

The Reynolds number, often referred to as *Re*, is a measure of the relative importance of inertial to viscous, that is momentum versus dissipation, terms in a fluid, and hence is characteristic of how turbulent the flow is. The higher the Reynolds number the more likely flow will be turbulent, and the lower the number the more likely the flow will be laminar. The Reynolds number is defined as:

$$Re \equiv \frac{uL}{\nu} \qquad (11.21)$$

where the denominator is the kinematic viscosity of the fluid, denoted by the greek letter "nu" (ν). This may also be regarded as the ratio of the (molecular) diffusive timescale to the advective timescale. For most "geophysical" fluids (the atmosphere, oceans, etc.) the Reynolds number is typically of the order of several thousand or greater. Thus most of these fluids are in a state of turbulent flow.

11.6.2 The Peclet number

The Peclet number (Pe) is a measure of the relative importance of advection to diffusion. The higher the Peclet number, the more important is advection. It is given by:

$$Pe \equiv \frac{uL}{\kappa} \tag{11.22}$$

and can be arrived at by non-dimensionalizing the advection–diffusion equation. This number may also be thought of as the ratio between the turbulent diffusive and the advective timescales. Remember that the longer the timescale, the less important the process. A typical open ocean is characterized by velocities of order $0.01\,\mathrm{m\,s^{-1}}$, lengths of order 3000 km (the size of ocean gyres), and turbulent diffusivities of order $1000\,\mathrm{m^2\,s^{-1}}$. This gives a Peclet number of order 20–30.

The trick, though, is in the seemingly arbitrary choice of the length scale L. Clearly, the bigger L becomes, the higher the Peclet number becomes. This is equivalent to saying that given enough time, advection *always* wins out over diffusion. This is because while the displacement of a particle increases linearly with time with advection, it only increases as the square root of time with diffusion. This can be seen by thinking about diffusion as a random walk experiment (which is what it mathematically is). It does boil down to this implicit ambiguity that the Peclet number (and hence the apparent relative role of advection and diffusion) depends on the spatial scale of the system being studied.

Radioactive tracers, with their built in decay constants, can define their own space scales. This can also be seen by non-dimensionalizing the advective–diffusive–decay equations. The characteristic length scale is the velocity divided by the decay constant $L_R = u/\lambda$, or quite simply the distance a fluid parcel would go before the tracer would be reduced to $1/e$ of its value by decay. Thus a *radiotracer Peclet number* would be defined as:

$$Pe_R = \frac{u^2}{\kappa \lambda} \tag{11.23}$$

Now for a given fluid flow, the length scale will be different for differing radiotracers, so that diffusion and mixing will be more important for one tracer than for another. Consider for example ^{7}Be, which has a half-life of 53.4 days, and thus has a decay probability of $1.51 \times 10^{-7}\,\mathrm{s^{-1}}$. For the subtropical North Atlantic, with velocities of order $0.01\,\mathrm{m\,s^{-1}}$, and horizontal turbulent diffusivities of order $1000\,\mathrm{m^2\,s^{-1}}$, this gives a Peclet number of order 0.7, which says that diffusion and mixing are as important as advection for this tracer. Consider the same situation, however, with tritium (half-life 12.45 years). The same calculation yields a Peclet number of order 50–60, which says that tritium is more affected by advection than diffusion. Now let's turn the problem around and say that if you were interested in studying the effects of diffusion, you'd be more interested in using ^{7}Be than tritium.

Similar arguments could be made for planktonic growth or mortality. For example, if the growth rate (doubling time) of a given species were short, then its effective Peclet number would differ from longer-lived substances or tracers. However, some caution must be exercised in that there is a characteristic scale-dependence associated with lateral turbulent diffusivities that will also come into play (see Section 11.8).

11.6.3 Some Richardson numbers

The ocean is, in general, stably stratified. That is, heavier water is overlain by lighter water. If the reverse were true, then the water column would be gravitationally unstable and vertical motions (convection) would result that would erase the condition. For turbulent displacement to occur vertically in a stratified water column, the fluid particles must overcome the vertical density (buoyancy) gradient. (We'll discuss this more in the next section.) Thus one would expect that the ability of the water column to resist this vertical turbulence will be related to the vertical density gradient. Now one model of the origin of the energy required to produce turbulent motions is the *vertical shear in the horizontal velocity*. That is, if the horizontal velocity is changing with depth, the different layers traveling at different speeds tend to "rub" against one another, and there must be an overall dissipation generated by the velocity gradient. This dissipation scales as the square of the velocity gradient, so that one defines a *gradient Richardson number* as:

$$Ri_G \equiv \frac{g \dfrac{\partial \rho}{\partial z}}{\rho_0 \left(\left| \dfrac{\partial \mathbf{u}}{\partial z} \right| \right)^2} \tag{11.24}$$

where g is the acceleration due to gravity, and $\mathbf{u}$ is the horizontal velocity (in any direction). In situations where Ri_G decreases much below 1, small-scale overturning readily occurs and turbulent diffusion becomes important, and can grow to a point where the system mixes vigorously. Laboratory experiments and analytical calculations indicate that a critical Ri_G of 0.25 is a good approximation for most systems.

Another form of the Richardson number is used when dealing with the stability of a well-mixed ("slab") layer overlying a sharp density gradient, such as the summer mixed layer (stirred effectively by wind stress) underlain by a seasonal thermocline created by summer heating. The *bulk Richardson number* is defined as

$$Ri_B \equiv \frac{g h \Delta \rho}{\rho_0 \left(|\Delta \mathbf{u}| \right)^2} \tag{11.25}$$

where $\Delta \rho$ is the density contrast across the transition layer below the slab layer, and $\Delta \mathbf{u}$ is the velocity contrast. The corresponding critical value below which entrainment of underlying waters into the slab occurs is 0.65, as again determined by laboratory experiments and analysis.

11.6.4 Two other numbers

Various other numbers crop up in different circumstances. Ones that you may hear of are the *Prandtl number* and the *Schmidt number*. The Prandtl number is the ratio of viscosity (i.e. momentum diffusion) to thermal diffusion:

$$Pr \equiv \frac{\nu}{\gamma} \tag{11.26}$$

Perhaps more appropriately for oceanography, the *turbulent Prandtl number* is the relationship between the eddy diffusivities for momentum (*eddy viscosity*) and heat:

$$Pr_t \equiv \frac{\kappa_M}{\kappa_H} \tag{11.27}$$

where the subscripts M and H correspond to momentum and heat respectively.

The Schmidt number appears in the characterization of air–sea gas exchange rates, and is the ratio of viscosity to molecular diffusion of the material of interest in the fluid:

$$Sc \equiv \frac{\nu}{D} \tag{11.28}$$

These numbers are often used to compare model or flux calculations between different situations or chemical species.

11.7 Vertical turbulent diffusion

The gradient Richardson number (Ri_G) tells us something about the vigor of vertical turbulent diffusion in a stratified fluid. Typical estimates of vertical turbulent diffusion in the ocean tend to vary inversely as the vertical density gradient. For example, consider the so called "mixed layer" of the ocean, which is generally isothermal (and hence uniform in density) and stirred effectively by wind stress. Vertical mixing rates are of order 10^{-3} to $10^{-2}\ \mathrm{m^2\,s^{-1}}$. In fact, some upper ocean models treat the mixed layer as perfectly mixed; such models are called *bulk mixed layer models*.

In the main oceanic thermocline, which in the subtropics is characterized by vertical density gradients of order $10^{-3}\ \mathrm{kg\,m^{-4}}$ (that is, a change in density of about $1\ \mathrm{kg\,m^{-3}}$ over a depth range of approximately $1\ \mathrm{km}$), turbulent diffusion is much less vigorous, and is typically seen to be of order $10^{-5}\ \mathrm{m^2\,s^{-1}}$ (see, for example, Ledwell *et al.*, 1998).

11.7.1 The Brunt–Väisälä frequency

Conceptually, we think of vertical turbulent diffusivity as being an inverse function of the vertical stratification, which can be expressed in terms of the *Brunt–Väisälä* frequency N. This is a measure, in inverse time units, of the fluid's density stratification or in other words its resistance to turbulence. The derivation of this concept is actually quite simple. Imagine displacing a fluid particle vertically in a stably stratified water column. If the particle were to be lifted a distance Δz, it would be denser than the surrounding lighter water and experience a downward force related to the difference in density between it and the surrounding water:

$$F = -g\Delta\rho = -g\Delta z \frac{\partial\rho}{\partial z} \tag{11.29}$$

which in turn would be a function of the distance of displacement and the density gradient. The equation applies for negative displacements as well, since the particle would be more buoyant than the surrounding water, and would want to bob back up like a cork. Now this restoring force, which is proportional to distance, should look familiar to those of you who have taken introductory physics: it is the spring equation. If we use the fact that $F = ma$ (a is the acceleration), then we can rewrite the equation (using the density ρ in place of the mass, and defining the origin of z so that it is zero at the equilibrium point and thus $\Delta z = z$) to have:

$$\rho a = -gz\frac{\partial \rho}{\partial z} \tag{11.30}$$

or

$$\frac{\partial^2 z}{\partial t^2} = -\frac{g}{\rho}\left(\frac{\partial \rho}{\partial z}\right)z \tag{11.31}$$

where the terms in front of the z on the right-hand side of the equation can be expressed as a constant, which we'll call N^2. The solution to this equation is simply:

$$z(t) = z_0 e^{iNt} \tag{11.32}$$

a simple harmonic oscillator with frequency N,

$$N = \sqrt{\frac{g}{\rho}\frac{\partial \rho}{\partial z}} \tag{11.33}$$

the Brunt–Väisälä frequency. That is, in the absence of dissipation, if you were to "pluck" a fluid particle above its equilibrium position, it would return to its position and then overshoot, oscillating back and forth with a frequency N.

This frequency is a measure of the water column stability, and is typically a few cycles per hour (10^{-3} Hz) in the main thermocline. Surveys by Sarmiento *et al.* (1976), Gargett (1984), and Gregg (1987) attempt to correlate apparent diffusivities with N and other parameters. The problem is extraordinarily complex, and not well resolved, but the following may be regarded as the typical ranges of diffusivities:

- ocean mixed layer, unstratified and strongly mixed: 10^{-3} to $10^{-1}\,\mathrm{m^2\,s^{-1}}$,
- main oceanic thermocline, stratified, weak mixing: $10^{-5}\,\mathrm{m^2\,s^{-1}}$,
- deep ocean far from boundaries and rough topography, weakly stratified: 10^{-5}–$10^{-4}\,\mathrm{m^2\,s^{-1}}$,
- benthic boundary layer, especially near boundaries: 10^{-4} to $10^{-3}\,\mathrm{m^2\,s^{-1}}$, and
- chemically, salinity, or thermally stratified lakes or estuaries: 10^{-9} to $10^{-6}\,\mathrm{m^2\,s^{-1}}$.

11.8 Horizontal turbulent diffusion

Horizontal motions are not generally inhibited by stratification. This is evident when you think of the aspect ratio of the oceans. Density contrasts which occur over a few hundred meters in depth are reflected in changes which occur over thousands of kilometers. The typical scales of motion associated with horizontal turbulence are the "eddy scale" which is of order 10–100 km. The kinds of estimates that you see for eddy diffusion range from a few $m^2 s^{-1}$ or even less to a few thousand $m^2 s^{-1}$, depending on location and scale.

The scale is important, because the kind of averaging implicit in the parameterization of the eddy diffusion, i.e. what you call $\bar{u}$ and what you call u', depends on the scales over which you are diffusing your material. The *spectrum* of horizontal velocities in the ocean is typically "red" (more energy at low frequencies), but also exhibits a local maximum at *mesoscales* (10–100 km). This can be seen in a classic paper (we recommend you read it, however old it is) by Okubo (1971). For a tracer patch shaped like a two-dimensional Gaussian and a constant turbulent diffusivity, you would expect the mean width of the patch to scale with the square root of time, or the variance (the square of the width) to vary linearly with time. Comparison with observed experiments, however, is not so simple, because the distributions quickly become distorted by motions and shearing on scales comparable to the size of the patch (remember, there is not really any actual scale separation; there is always a range of motions between the u' and the $\bar{u}$ ranges). The best you can do is to define some RMS width as $\sqrt{\sigma_{rC}^2}$ where:

$$\sigma_{rC}^2(t) \equiv \frac{\int_0^\infty r^2 C(t,r)dr}{\int_0^\infty C(t,r)dr} \qquad (11.34)$$

i.e. the second moment of the distribution (which obviously will not be symmetric or Gaussian). The argument is that in a kind of "central limit" sense, this will represent the mean spreading of the patch, and that it should behave as a Gaussian in some mean sense. However, because the spreading (with time) dye patch feels larger and larger motions as "turbulent", the diffusivity should increase with time.

Think of the dye patch spreading in the open ocean. Clearly, when the dye patch is small, only those turbulent motions small compared with the dye patch itself can play a role in its dispersal. However, as the patch grows, the range of turbulent motion accessible to the patch for dispersal grows, so that the amount of available energy put into dispersion grows with time and size. Thus there is an apparent increase in horizontal turbulent diffusivity with scale. Using the observed spreading, Okubo estimated an approximately $L^{1.1}$ dependence of the diffusivity on length scale.

Garrett (1983) suggested that the horizontal dispersal of a dye patch in a turbulent ocean proceeds through three stages. In the initial stage, where the patch is small compared with the mesoscale eddy field, dispersion occurs through a combination of small-scale horizontal swirling motion and vertical mixing. The magnitude of this process has been argued by Young *et al.* (1982), to be of order $(N^2/f^2)\kappa_z$, where f is the Coriolis parameter, and κ_z is the vertical turbulent mixing coefficient. Here the typical scale of the patch might be of order 10–100 m, and the area of the patch increases linearly with time, at a rate consistent with turbulent diffusivities of order $0.1 \, m^2 s^{-1}$.

Once the patch grows large enough to begin to feel the mesoscale velocity shear, that is once it reaches 1–10 kilometer scales, the tracer becomes distorted and streaked out by the strain field. Here the streak length, and hence the area painted by the dye patch, grows exponentially with time. Although this evolution cannot be effectively represented as a simple "Fickian diffusive" functionality, the characteristic mixing rate is of order a few $m^2\,s^{-1}$. This exponential growth continues until the patch size becomes larger than the typical mesoscale eddy size, so that eddy mixing becomes the dominant mechanism. The dispersive growth of the patch area now returns to a linear mode, with a rate in the range of 100–1000 $m^2\,s^{-1}$.

Your choice of turbulent diffusivities will therefore be dictated by the physical scale of the problem you are working with. This in turn may be linked to the timescales as well. For example, in coastal regions a relatively short-lived radionuclide (such as ^{223}Ra or ^{224}Ra) may feel only very small scales of lateral motion before it decays, compared with longer-lived isotopes (e.g. ^{228}Ra or ^{226}Ra). Thus the radium isotopes, because of their range of half-lives, their common chemical behavior, and their emission from sediments, could be used as tools for diagnosing the scale dependence of horizontal mixing in coastal regions.

Finally, you should be aware that because of the effects of rotation, horizontal turbulence exhibits an *anisotropy* so that mixing coefficients tend to be greater in the zonal than the meridional direction (see, for example, Sundermeyer and Price, 1998).

11.9 The effects of varying turbulent diffusivity

We could rewrite the general advection–diffusion equation (11.17) in the following way:

$$\frac{\partial C}{\partial t} = -u\frac{\partial C}{\partial x} + \frac{\partial \kappa}{\partial x}\frac{\partial C}{\partial x} + \kappa\frac{\partial^2 C}{\partial x^2} + J \tag{11.35}$$

where we have differentiated by parts the turbulent diffusion term. We have admitted the possibility that the diffusivity may vary in space because it is a property of the flow, not of the fluid.

The dimensions of $\partial \kappa / \partial x$ in the second term on the right-hand side are those of velocity. In fact, it behaves exactly like one. How important is it? Well, consider first horizontal effects, since those diffusivities are largest. Values as high as 2000 $m^2\,s^{-1}$ have been observed in the Gulf Stream area, and seem to taper off to a few hundred $m^2\,s^{-1}$ in the ocean interior. Assuming this takes place over a distance of 2000 km yields apparent "velocities" of approximately $(2000\,m^2\,s^{-1})/(2\times10^6\,m) = 0.001\ m\,s^{-1}$, or about an order of magnitude smaller than typical velocities in the upper thermocline. So the conclusion is that this is unlikely to be a problem for the upper thermocline, but may be important deeper down where velocities are small.

Now how about the vertical? Well, in the mixed layer, vertical diffusivities are of order $10^{-3}\ m^2\,s^{-1}$ or larger. We'll choose the lowest values for this calculation. Meanwhile, a few hundred meters down in the main thermocline, vertical diffusivities are of order $10^{-5}\ m^2\,s^{-1}$, so we calculate an effective "velocity" of order 10^{-6} to $10^{-5}\ m\,s^{-1}$. This doesn't sound like much, but when you realize that typical Ekman pumping rates (the rate at which water is pushed downward by wind stress convergence) are of order $50\,\mathrm{m\,y}^{-1}$ or about $10^{-6}\,\mathrm{m\,s}^{-1}$, this becomes really significant! Furthermore, abyssal upwelling velocities through the main thermocline (we'll get into that in Chapter 13) are 10 times lower. Thus we should be very careful when considering vertical balance models with turbulent diffusivities that change a lot with depth.

Now a footnote: the above considerations may be a little premature, since the physics associated with models that span large ranges in turbulent flows (which must drive these changes in turbulent diffusivities) is far from worked out. Our goal in pointing out these issues to you at this point is to make you aware of some of the pitfalls and foibles in such modeling. Two papers which discuss variable diffusivity effects are Armi and Haidvogel (1982) for horizontal gradients, and Jenkins (1980) for vertical.

11.10 Isopycnal coordinate systems

Let us first consider the advection–diffusion equation in three spatial dimensions. We have:

$$\frac{\partial C}{\partial t} = -\nabla \cdot (\mathbf{u}C) + \nabla \cdot (\kappa \nabla C) \tag{11.36}$$

where κ is now the three-dimensional *turbulent diffusivity tensor* and the operator ∇ is the three-dimensional *gradient operator*, given by:

$$\nabla = \hat{x}\frac{\partial}{\partial x} + \hat{y}\frac{\partial}{\partial y} + \hat{z}\frac{\partial}{\partial z} \tag{11.37}$$

where $\hat{x}$, $\hat{y}$, and $\hat{z}$ are unit-length vectors in the x, y, and z directions respectively.

Now what do we mean by a "three-dimensional diffusivity tensor"? Remember that there is typically a vast difference in strength of horizontal versus vertical mixing rates. In a "perfect world" one could represent the turbulent mixing coefficients as a kind of diagonal matrix multiplying the concentration gradient, so that κ would be represented by:

$$\kappa = \begin{pmatrix} \kappa_x & 0 & 0 \\ 0 & \kappa_y & 0 \\ 0 & 0 & \kappa_z \end{pmatrix} \tag{11.38}$$

where the two horizontal terms, κ_x and κ_y, will be much larger than the vertical term κ_z.

In the "real ocean", however, things are not so simple. Bear in mind that the distinction between "vertical" (buoyancy inhibited) and "horizontal" (non-buoyancy inhibited) mixing is really along the lines of *diapycnal* and *isopycnal* mixing. That is, there exist within the ocean surfaces of constant density, referred to as *isopycnals*, along which no buoyancy penalty is paid for fluid motion. These surfaces constitute a natural coordinate system for mixing. To the extent that these surfaces are approximately horizontal, the diffusivity tensor is more-or-less diagonal. We would account for this by populating the off-diagonal elements of the eddy-diffusivity tensor as:

$$\kappa = \begin{pmatrix} \kappa_{xx} & \kappa_{xy} & \kappa_{xz} \\ \kappa_{yx} & \kappa_{yy} & \kappa_{yz} \\ \kappa_{zx} & \kappa_{zy} & \kappa_{zz} \end{pmatrix} \tag{11.39}$$

where it is safe to argue that the tensor is symmetric, giving (for example) $\kappa_{xy} = \kappa_{yx}$.

However, the big problem is that since κ_x and κ_y are so very much larger than κ_z, just a little "bleed" from the larger terms can strongly influence the smaller one. How big an effect might this be? Consider that many isopycnal surfaces present within the main thermocline of the subtropics will outcrop (reach the sea surface) in the subpolar regions, and that even the deepest isopycnal horizons ultimately outcrop in the polar regions. In the former case, an isopycnal can rise as much as 500 m over a distance of 5000 km. This constitutes a slope of 10^{-4}, which, while seeming small, leads in a "Cartesian" x–y–z model to a significant projection of isopycnal mixing onto the vertical (approximately diapycnal) coordinate. Given the effective ratio of isopycnal to diapycnal diffusivities of order 10^8, the formulation of a model in purely Cartesian coordinates can exhibit an erroneously high diapycnal mixing rate. This could become extremely important in large-scale ocean models, climate simulations, or anywhere significant isopycnal slopes occur.

Early attempts to deal with this problem involved rotation of the eddy diffusivity tensor to accommodate isopycnal slopes, but more sophisticated modeling has been done using a true isopycnal coordinate system. A good example is MICOM (the Miami Isopycnal Coordinate Ocean Model). Such models have the added attraction of dealing well with thermodynamic (buoyancy modifying) processes. Bear in mind that such models have their own challenges, which we'll discuss in Chapter 17.

12

Finite difference techniques

It is better to take many small steps in the right direction than to make a
great leap forward only to stumble backward.

Ancient Chinese Proverb

Everything should be made as simple as possible, but not simpler.

Albert Einstein

12.1 Basic principles

Constructing numerical models of marine systems usually involves setting up a series
of partial differential equations, specifying boundary conditions and then "running the
model". Your purpose may be to establish the value of parameters (e.g. rates of reaction
or the magnitude of some property), estimate fluxes, or make some prediction about the
future state of the system. Although you can sometimes choose a physical problem that
is simple enough to be modeled with analytic solutions (an example would be Munk's
1966 "Abyssal recipes" model; Chapter 13), more often than not you will encounter situations where the processes or the geometry of the system are too complex to allow analytic
solutions.

Don't get us wrong; analytic solutions are nice. They can often provide you with a nice
conceptual, intuitive feel for how the system responds, especially in an asymptotic sense.
However, for realistic geometries, you will find that the few analytical solutions provided
in many books are infinite series solutions. Be very, very careful when dealing with those
series solutions. Pay particular attention to the assumptions made in deriving the solutions,
to the conditions under which they ought to be applied, and especially to convergence
issues. A classic example is a student we once knew who was attempting to compute the
rate of release of something from spherical grains as a function of time and temperature.
The diffusion equations appeared very simple, and the series solutions available in the text
he was using looked straightforward. He set the equations up and let them run on a personal
computer. It seemed to take a long time, so he let it run overnight on Friday. When he came
back after the weekend the computer was still calculating. On his request, one of us recast
the program into FORTRAN and put it on a much faster workstation, and then let him run it

there. After a week, he came back and complained that the solution still hadn't been arrived at. We asked him how many terms the computer had gone through in the series: "Oh, about 10–15 million," he said. "Well," we said, "have you considered roundoff error?" Each term in the series involved rather obtuse hypergeometric functions which in themselves were derived from infinite series and therefore somewhat difficult to compute. We won't go into the details of what assumptions he "violated" in setting up his solution, but basically there was no way that the series would have converged in his lifetime.

So what can you do in a situation like this? The answer is to approximate the continuum model equations with a discretized form, and run the model in discrete time steps. If it is a "transient" (time evolving) model, this is all you need to do. If it is a steady-state model, you would have to test for convergence; i.e. when the model solution ceases to change by some level of significance with each time step. Implicit in this strategy is the belief that the *finite difference equation* (FDE) asymptotically approaches the *partial differential equation* (PDE) as the grid scale becomes finer and finer (e.g. as Δx and Δt approach 0). Given the right attention to details, this is a rather simple process, at least conceptually. In practice, however, there are a number of subtle issues that need to be dealt with, and that is the topic of this chapter.

There are a host of different discretization schemes out there, each designed to be optimal for specific problems and goals. The bottom line advice we can give you, however, is to choose the simplest finite difference algorithm you can get away with and use it. The reason is that the simplest forms are more easily and thoroughly understood, especially regarding their limitations and weaknesses. Also, don't reinvent the wheel. You can rest assured that if you make up a new algorithm, someone has already thought of it. If there is no evidence of it in the literature, and it's not in common use, it's probably because it doesn't work! There are some very clever algorithms out there, and beyond some of the simple example applications we'll cover here, you will likely need to resort to them. But for learning purposes, we'll keep things as simple as possible . . . but not simpler!

12.2 The forward time, centered space (FTCS) algorithm

We'll be working with one-dimensional advection–diffusion models to make the points clear. Most but not all of the considerations map simply into higher dimensions. Starting with the continuum equation for time-dependent advection–diffusion, we have

$$\frac{\partial C}{\partial t} = -\frac{\partial}{\partial x}(uC) + \frac{\partial}{\partial x}\left(K\frac{\partial C}{\partial x}\right) \tag{12.1}$$

which in words simply means that the rate of change of concentration at a given point is equal to the sum of the advective flux divergence and the diffusive flux divergence. The advective flux is just the concentration times the velocity u, and the diffusive flux is the diffusivity K times the concentration gradient (i.e. Fick's first law). We use a K instead of a κ in this chapter because we don't write code with Greek keyboards. The reason for the negative sign in front of the advective flux divergence term is that concentration will only increase if the advective flux decreases downstream. In other words, it's a sign convention. You may be more familiar with the common (but less correct) form of this equation:

$$\frac{\partial C}{\partial t} = -u\frac{\partial C}{\partial x} + K\frac{\partial^2 C}{\partial x^2} \tag{12.2}$$

which is only valid for constant u and K.

12.2.1 The control volume approach

There are a number of ways of discretizing Equation (12.1), but we will derive a simple, commonly used way called *forward time, centered space* or FTCS. In fact, there are a number of ways of deriving this same finite difference scheme, but we will choose the most obvious and intuitive way. This is called the *control volume* approach, or also referred to as *fluid in a cell* or *particle in a cell*. The attraction of this approach is that if it is done carefully, it ensures that the algorithm actually conserves mass, which is a very desirable quality.

Consider a situation where you have a one-dimensional physical domain of finite extent, length L. You break it up into a number of discrete points, each separated by a length Δx. We will deal with one of the interior points, which we'll call i, and with the neighboring points at $i-1$ and $i+1$. We will draw a box centered on the ith point with dimensions Δx, and a facial area (at right angles to the line) of A. The faces of the box cut the line between the ith point and its neighbors exactly half-way. The amount of "stuff" in the box is simply the concentration at point i times the volume of the box. The rate of change of stuff is given by:

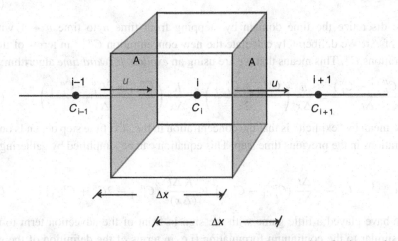

Figure 12.1 A schematic of the control volume approach. Consider a series of points each separated by Δx along a line, and an imaginary cell centered on the ith point with a length Δx and a cross-sectional area A.

$$\frac{\Delta \text{stuff}}{\Delta \text{time}} = (\text{stuff})_{\text{in}} - (\text{stuff})_{\text{out}} \tag{12.3}$$

Now the stuff coming into the box through one of its faces is equal to the sum of the material advected in plus the material diffused in. The advective flux is the velocity times the concentration times the area it's coming through, and the diffusive flux is the diffusion coefficient times the negative of the concentration gradient times the area it's coming through – that's just Fick's first law. Thus we can write Equation (12.3) in a semi-continuum form in the following way:

$$\frac{\partial (CV)}{\partial t} = \left[(uC)_{\text{in}} A - \left(K \frac{\partial C}{\partial x} \right)_{\text{in}} A \right] - \left[(uC)_{\text{out}} A - \left(K \frac{\partial C}{\partial x} \right)_{\text{out}} A \right] \tag{12.4}$$

but since the volume of the box is just the facial area A times Δx, we can divide the entire equation by the area so that it reduces to:

$$\Delta x \frac{\partial C}{\partial t} = (uC)_{\text{in}} - (uC)_{\text{out}} + \left(K \frac{\partial C}{\partial x} \right)_{\text{out}} - \left(K \frac{\partial C}{\partial x} \right)_{\text{in}} \tag{12.5}$$

From this point on, for simplicity we'll assume that the velocity and diffusivity are constant, which, of course, is not always true. If you need to, you could do this analysis with variable u and K.

The concentration at the faces of the box (ingoing and outgoing) can be estimated as the average between the two adjacent points, for example $C_{\text{in}} = (C_{i-1} + C_i)/2$. The gradients at each interface are estimated as the difference between the adjacent points divided by their separation, e.g. $(\partial C/\partial x)_{\text{in}} = [C_i - C_{i-1}]/\Delta x$. So we have:

$$\frac{\partial C}{\partial t} = \frac{u}{\Delta x} \left(\frac{[C_{i-1} + C_i] - [C_i + C_{i+1}]}{2} \right) + \frac{K}{\Delta x} \left(\frac{[C_{i+1} - C_i] - [C_i - C_{i-1}]}{\Delta x} \right)$$

$$\tag{12.6}$$

Now we discretize the time domain by stepping from time n to time $n + 1$ with an interval of Δt. We deliberately calculate the new concentration C^{n+1} in terms of the old concentrations C^n. This means that we are using an *explicit, forward time* algorithm:

$$\frac{(C_i^{n+1} - C_i^n)}{\Delta t} = \frac{u}{\Delta x} \left(\frac{C_{i-1}^n - C_{i+1}^n}{2} \right) + \frac{K}{\Delta x} \left(\frac{C_{i+1}^n - 2C_i^n + C_{i-1}^n}{\Delta x} \right) \tag{12.7}$$

What we mean by "explicit" is that the concentration in the next time step depends only on concentrations in the previous time step. This equation can be simplified by gathering time terms:

$$C_i^{n+1} = C_i^n - \frac{u \Delta t}{2 \Delta x} \left(C_{i+1}^n - C_{i-1}^n \right) + \frac{K \Delta t}{(\Delta x)^2} \left(C_{i-1}^n - 2C_i^n + C_{i+1}^n \right) \tag{12.8}$$

Note we have played a little game with the sign in front of the advection term to make it more similar to the continuum formulation (i.e. in terms of the definition of the gradient). We further simplify this equation by defining an advection number c and a diffusion number d such that:

$$C_i^{n+1} = C_i^n - \frac{c}{2}\left(C_{i+1}^n - C_{i-1}^n\right) + d\left(C_{i-1}^n + C_{i+1}^n - 2C_i^n\right) \qquad (12.9)$$

where

$$c = \frac{u\,\Delta t}{\Delta x} \qquad d = \frac{K\,\Delta t}{(\Delta x)^2} \qquad (12.10)$$

Both these numbers are dimensionless (check them out!), and the advection number is also known as the *Courant number*. Sorry about the potential confusion between the concentration C and the Courant number c. It's unfortunately traditional nomenclature and you'll have to be careful.

You can see that Equation (12.9) states a relation between the concentration at location i and time step $n + 1$ with the concentrations from the previous time step (n) at the ith and its neighboring points. We can then rewrite the equation for computational convenience as:

$$C_i^{n+1} = w_-C_{i-1}^n + w_0 C_i^n + w_+ C_{i+1}^n \qquad (12.11)$$

where

$$\begin{aligned} w_- &= \frac{c}{2} + d \\ w_0 &= 1 - 2d \\ w_+ &= -\frac{c}{2} + d \end{aligned} \qquad (12.12)$$

The reason we formulate Equation (12.11) this way will become evident when you see the MATLAB code, but generally you compute the weighting factors at the beginning and then perform the iterations with vector multiplications, which usually go much faster. Also, the algorithm – not accidentally, but because we have been very careful in using the control volume approach – passes a very important test; it is *conservative*. No, this is not a political statement, it is a statement that the algorithm conserves mass. You prove this by testing that the sum of the weights must be identically 1:

$$w_- + w_0 + w_+ = 1 \qquad (12.13)$$

meaning that the mass in the $n + 1$ step is the same as the prior time step. This is indeed a fine thing to have.

12.2.2 Truncation errors and formal precision

What is the effect of discretization? Well, it is a process of approximation. This process can be viewed as representing the continuum equations by a Taylor series expansion around the observation points. The value of a function at a location Δx away from a given location can be expressed as an infinite series:

$$f(x_0 + \Delta x) = f(x_0) + \Delta x\,\frac{\partial f}{\partial x}\bigg|_{x_0} + \frac{\Delta x^2}{2}\,\frac{\partial^2 f}{\partial x^2}\bigg|_{x_0} + \frac{\Delta x^3}{6}\,\frac{\partial^3 f}{\partial x^3}\bigg|_{x_0} + \cdots \qquad (12.14)$$

and a similar series expansion could be constructed for the function in time as well:

$$f\left(t_0 + \Delta t\right) = f\left(t_0\right) + \Delta t \left.\frac{\partial f}{\partial t}\right|_{t_0} + \frac{\Delta t^2}{2} \left.\frac{\partial^2 f}{\partial t^2}\right|_{t_0} + \frac{\Delta t^3}{6} \left.\frac{\partial^3 f}{\partial t^3}\right|_{t_0} + \cdots \qquad (12.15)$$

We could have derived the FTCS algorithm by approximating the continuum equation's terms with these Taylor series expansions (in both time and space) and arbitrarily truncating the series at specific terms. For example, we would arrive at the FTCS equation by ignoring terms higher than Δx^2 in space, and terms higher than Δt in time. Thus one could state that the FTCS algorithm is second-order accurate in space and first-order accurate in time. What does this "accuracy" really mean? Well, for example, if you divided your space domain into 100 pieces, then the errors introduced in your discretized calculation would be of order 1% in a "first-order accurate" algorithm, and of order 0.01% accurate in a "second-order accurate" algorithm. Similar things apply to the time domain (your time steps).

Thus you might be strongly tempted to say that a high-order accuracy is a good thing since you can get away with having a very coarse discretization. Yes, this is true, since you will find (in later discussions here) that for a one-dimensional system, the computation time goes up at least as the square of the fineness of your resolution, and generally as the cube of your resolution. For example, if you halve the size of your space step, you may have to quarter your time step, so to simulate an experiment for a given period of time, you have to do eight times the calculations. However, we should caution you that the subtleties introduced by going to higher-order equations can be a big problem. Consider, for example, the obvious step of making the FTCS second-order accurate in time; this simply requires that you replace the one-sided estimate of the time derivative with a centered time estimate:

$$\frac{\partial C}{\partial t} \approx \frac{C_i^{n+1} - C_i^n}{\Delta t} \Rightarrow \frac{\partial C}{\partial t} \approx \frac{\frac{1}{2}\left(C_i^{n+1} + C_i^n\right) - \frac{1}{2}\left(C_i^n + C_i^{n-1}\right)}{\Delta t} = \frac{C_i^{n+1} - C_i^{n-1}}{2\Delta t}$$
$$(12.16)$$

which looks a lot like the advection derivative term. This leads to the *centered time, centered space* (CTCS) algorithm, and it is second-order accurate in both space and time. You won't see much of this algorithm in use, because it has one minor problem: *it doesn't work!* It turns out to be unconditionally unstable (we'll learn about stability in Section 12.4). That is, it will "blow up" no matter what values of Δt and Δx you choose. So the moral of this story is that although truncation accuracy is a nice thing, it is not the only consideration.

12.3 An example: tritium and ^{3}He in a pipe

12.3.1 Constructing the pipe model

Now let's apply this algorithm to a simple one-dimensional model. As an example, we will simulate the penetration of a bomb-produced isotope *tritium* into the ocean. Picture a one-dimensional pipe with fluid flowing along it at a velocity u. In addition we will simulate turbulent mixing in the direction of flow. In this model we'll have surface water entering the pipe, and you might regard this pipe as a streamline that flows from the sea surface down into the thermocline around an ocean gyre. We choose reasonable values for the velocity ($0.01\,\mathrm{m\,s^{-1}}$) and turbulent diffusion ($1000\,\mathrm{m^2\,s^{-1}}$), considered typical for the eastern subtropical North Atlantic. The length of the tube is chosen to be $20\,000\,\mathrm{km}$ ($20\,000\,000\,\mathrm{m}$) so that fluid doesn't travel (or diffuse) beyond the pipe's length during the experiment. In this instance, the distance traveled at $0.01\,\mathrm{m\,s^{-1}}$ in 55 years is a little over $17\,000\,\mathrm{km}$. The diffusive length scale, approximated by $L \approx \sqrt{KT} = 1316\,\mathrm{km}$, is much shorter. The sum of the two is less than the domain size, but if they weren't, we would have to increase it to avoid boundary effects from the end of the pipe.

We discretize the problem with a distance increment Δx of $50\,\mathrm{km}$ ($= 50\,000\,\mathrm{m}$) and a time step Δt of around 2.5 days ($200\,000\,\mathrm{s}$). We remind you to be very careful to express all of your parameters in the same unit system. If you do something like put the decay constant in $(\mathrm{years})^{-1}$ and time step in weeks, diffusivity in $\mathrm{cm^2\,s^{-1}}$, and distance step in km, you will end up in a real mess!

For the demonstration, we'll run the model for a period of 55 years (from 1950 to 2005) using an idealized surface water history function depicted in Fig. 12.2. This curve is an estimate of how the surface water concentration of tritium – mostly produced by atmospheric nuclear weapons testing in the late 1950s and early 1960s – has varied in time for the subtropical North Atlantic. Tritium is usually reported in *tritium units* (TU), which is really an isotope ratio, being defined as 10^{18} times the atomic ratio of $^3\mathrm{H}/^1\mathrm{H}$. Prior to the bomb tests, there was a small background of natural tritium in seawater that is produced by cosmic ray interactions in the Earth's upper atmosphere. Tritium is radioactive and decays to $^3\mathrm{He}$ (a stable, inert noble gas isotope) with a half-life of 12.3 years. We assume that

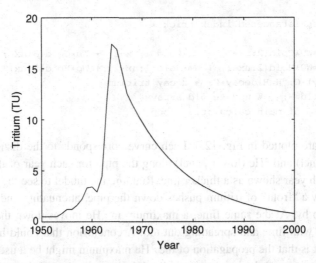

Figure 12.2 A schematic time history of bomb-produced tritium in subtropical North Atlantic surface water. TU stands for tritium units, and is 10^{18} times the atom ratio $^3\mathrm{H}/^1\mathrm{H}$.

excess ^{3}He is lost to the atmosphere in surface waters by gas exchange, but once below the surface (i.e. downstream in the tube) it can build up.

   Below is the MATLAB code for `dopipe1.m`. It takes less than a minute on a reasonably fast PC. `InitializePipe.m` defines constants, loads the surface tritium history, and initializes data arrays. `MakeWeights.m` calculates the finite difference weights. This needs to be done only once. Next is the main loop, which actually does the time stepping. Inside it is `PlotData.m` which executes once a year during the simulation, plotting the data so you can see the fields evolve. This is very important, since watching the evolution of the model not only tells you how the model is performing, but gives insight into the phenomena that you are studying. Note also that we've used a *staggered index* strategy to do the weighted calculation in one step. The calculation could have been done using a `for` loop to march through all the grid points along the pipe, but that would be substantially slower than doing a *vectorized* calculation using staggered indices. Look carefully at the code shown below to ensure you understand what we've done.

```
%      This is dopipe1
%      It models the flow of North Atlantic Surface Water
%      along a long pipe for 55 years (1950-2005)
%
InitializePipe      % Initialize tritium history and factors
MakeWeights         % Compute the weight factors
%
% the main calculation loop
%
for it=1:nt
   PlotData          % plot the curves annually
   C(1)=TSurf(i); H(1)=0;    %set incoming water to surface concentrations
%
% do time step (staggered index trick!)
%
   C(2:nx-1)=W0*Cold(2:nx-1) + Wm*Cold(1:nx-2) + Wp*Cold(3:nx);
   H(2:nx-1)=W0*Hold(2:nx-1) + Wm*Hold(1:nx-2) + Wp*Hold(3:nx);
   C=(1-decay)*C; H=H+decay*C; % decay tritium
   Cold=C; Hold=H;    % update old arrays
   end    % end of main calculation loop
```

   The results are plotted in Fig. 12.3. Each curve corresponds to the distribution of tritium (upper panel) and ^{3}He (lower panel) along the pipe for each year of the simulation, with every tenth year shown as a thicker line. Running the model to see the curves evolve shows you how a "front" of tritium pushes down the pipe, attenuating and smearing out as the years go by. At the same time, a maximum in ^{3}He moves down the pipe, generally increasing with time and spreading out too. A conclusion that could be drawn from this experiment is that the propagation of the ^{3}He maximum might be a useful diagnostic of the flow velocity, and the height or width might be indicative of the turbulent mixing rate.

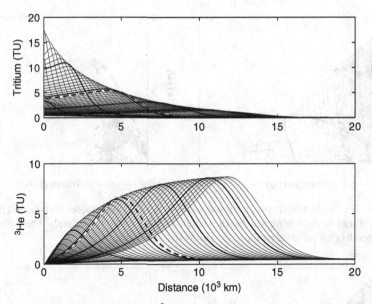

Figure 12.3 A simple transient tritium and ^{3}He pipe model. Each curve corresponds to the downstream distribution of these tracers for each year of the simulation. Every tenth year is plotted as a heavier curve, and 1981 is the heavy dashed line.

It would be nice to relate this highly idealized simulation to actual observations. However, nature is not so tidy as to provide us with a clearly marked stream tube along which we measure our tracers at known distances. Instead, we must rely on profiles of these isotopes made at locations within the subtropical ocean that may have intersected these hypothetical stream tubes at random locations. To deal with this, we look at the relationship between the two tracers. The m-file provided for this simulation also plots that as a function of time. Figure 12.4 shows the model simulations (left panel) and some observations taken in the early 1980s. For point of comparison, the heavy dashed line corresponds to model year 1981, when most of the data were taken.

Now there's good news and bad news here. The good news is that the shape of the model tritium–^{3}He relationship looks similar to the data. The hook-shape seems to show in both model and data. Moreover, the tritium values seem about right. We must be doing something right. The bad news is that the two – namely the dashed curve in the model simulation (left panel) and the observations (the dots in the right panel) in Fig. 12.4 – don't agree terribly well. The model predicts about twice as high a ^{3}He maximum as is observed. You can take our word for the fact that the data came from a wide range of locations within the North Atlantic subtropical gyre, so it looks like the model we've used, although based on very reasonable and accepted values of velocity and diffusivity, fails miserably. How might we redress this affront to our budding modeling careers? The first clue is that as the ^{3}He peak propagates down the pipe in Fig. 12.3, it spreads and attenuates because of diffusion. Perhaps the answer is to increase the diffusion!

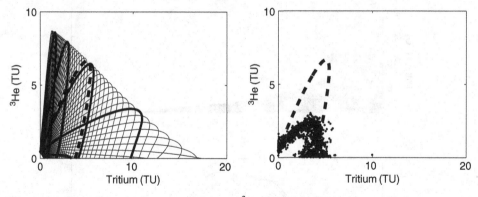

Figure 12.4 The relationship between tritium and ^{3}He from the pipe model (curves, left panel) and from observations in the subtropical North Atlantic in 1981 (dots, right panel). The dashed line in the model results (left panel) corresponds to the model year 1981.

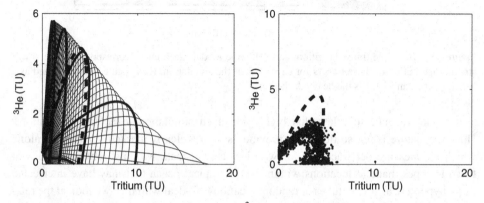

Figure 12.5 The relationship between tritium and ^{3}He from the pipe model (curves, left panel) run with a higher diffusivity ($5000\,m^2\,s^{-1}$) and from observations in the subtropical North Atlantic in 1981 (dots, right panel). The dashed line in the model results (left panel) corresponds to the model year 1981.

12.3.2 Experimenting with diffusion

Let's start by changing the diffusion coefficient K in `InitializePipe.m` to 5000. As a quick check, we realize that the diffusive length scale will be more than doubled (it's now close to 3000 km), so we should increase our domain a little, say to 25 000 km. So we run it again, and rather than displaying the along-pipe distributions, we look at the tritium–^{3}He relationships again in Fig. 12.5.

The model curve is now closer to our observations, so we're clearly heading in the right direction. The maximum value of ^{3}He observed in the model's 1981 simulation does seem to be a function of the diffusion. For 1981, the two model maxima differ by a factor of $7.2/4.5 = 1.6$. How does this scale with the diffusivity? That looks like a dependence that is a bit weaker than $\sqrt{K}$, and in fact around $\sqrt[3]{K}$, but can we determine it better? Well, let's

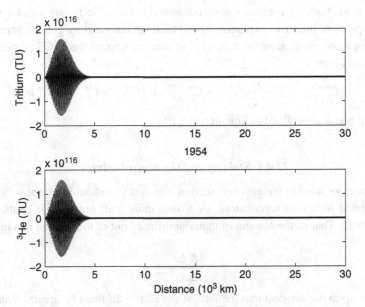

Figure 12.6 Instabilities can arise when more "stuff" is allowed to diffuse out of the control volume in one time step than was originally present in the control volume.

try increasing K to 8000. So we again modify our model program, and increase the model domain to 35 000 km to accommodate the extra spread.

Well, we only let it run for a few years, because clearly something is wrong. Thank goodness we plotted the data as we went along, because we might have blown the back off the computer if the numbers got any bigger (note the exponent on the plots in Fig. 12.6).

This kind of hedgehog plot is typical of some kind of oscillatory instability. Clearly our code is unstable. The clue? Well, look at the ws in your calculation. The w_0 term (look back at the equations; it is the weight assigned to the grid point itself for its next time step) is negative! Clearly this doesn't make sense. Going back to our control volume analysis, what is happening is that we are letting enough time elapse in our time stepping to allow more than the total amount of stuff in the control volume to diffuse out of the box. Thus it makes sense that the maximum time step, which somehow must be related to the spatial resolution, must be such that $d < \frac{1}{2}$. This comes from the demand that $w_0 > 0$. In the next section, we'll talk about how to analyze for this and other stability limitations in a more general way.

12.4 Stability analysis of finite difference schemes

The wild and unruly behavior witnessed in the previous section is obviously not desirable. We have exceeded the bounds of numerical etiquette, and we need to know what has gone wrong. It would be nice to have some idea of what reasonable parameter bounds are before we waste both the computer's and our valuable time. We'll start with the simplistic (and perhaps more intuitive) explanation, and then get onto a more general approach that reveals even more about numeric stability. But first, let's recall the FTCS algorithm given by Equation (12.9):

$$C_i^{n+1} = C_i^n - \frac{c}{2}\left(C_{i+1}^n - C_{i-1}^n\right) + d\left(C_{i+1}^n + C_{i-1}^n - 2C_i^n\right) \tag{12.17}$$

where the c and d are defined in Equation (12.10).

12.4.1 Stability and the control volume

The argument we used in the previous section was that you shouldn't be able, in the time step associated with your calculations, to diffuse more stuff out of your control volume than is in there. That is, the amount of material diffused out of the volume in a time step is a function of:

$$d = \frac{K\Delta t}{(\Delta x)^2} \leq \frac{1}{2} \tag{12.18}$$

This shows up in the demand that the weight $w_0 = 1 - 2d$ must be greater than or equal to zero. A similar constraint (which we did not break in our example) is that the advective component must be sufficiently small that we do not advect material more than one grid point during a time step. That is, we must limit the Courant number such that

$$c = \frac{u\Delta t}{\Delta x} \leq 1 \tag{12.19}$$

for stability. Another way to look at this is to say that we should not be able to advect more stuff out of a neighboring cell than is in there. However, we'll find out that the analysis done above is only part of the story. There's a more fundamental restriction, and one you wouldn't expect ...

12.4.2 The von Neumann stability analysis with diffusion only

Look again at the curve that blew up. It looks a lot like an oscilloscope trace. Not surprisingly, because the results have clearly oscillated out of control. This suggests that doing a Fourier analysis of the problem might be illuminating. It would be wise for you to review that chapter again to refresh your memory. Because it turns out to be simpler to understand, we'll start by doing the analysis for the FTCS with diffusion only. The FTCS for just diffusion is:

$$C_i^{n+1} = C_i^n + d\left(C_{i+1}^n + C_{i-1}^n - 2C_i^n\right) \tag{12.20}$$

Because we have been using i as an index of the location in space, we will have to use I to denote the imaginary number $\sqrt{-1}$ in the Fourier components. Note also that the Fourier analysis will be done in space rather than time, so the Fourier domain is scaled with *wavenumber* rather than frequency. We could write all the nasty summation signs, and make it look really impressive, but we want to keep it simple, and you'll see in a minute that it won't matter. When we Fourier transform the concentration distribution C we select only one Fourier component (i.e. only one term from the Fourier series) with wavenumber k_x on a discretized grid. A Fourier component for the concentration at time n and location i looks like:

$$C_i^n = V_n e^{Iik_x \Delta x} \tag{12.21}$$

where V_n is the amplitude function at time level n of this particular component and we have defined the wavenumber k_x as the inverse of the Fourier wavelength Λ:

$$k_x = \frac{2\pi}{\Lambda} \tag{12.22}$$

We've also broken the convention of using capital and lower case to represent Fourier pairs. And what we'll further do is to replace the wavenumber and grid spacing ($k_x \Delta x$) combination with an arbitrary angle $\theta \equiv k_x \Delta x$, so we can rewrite the Fourier components as:

$$\begin{aligned} C_i^n &\Leftrightarrow V_n e^{Iik_x \Delta x} = V_n e^{Ii\theta} \\ C_{i+1}^n &\Leftrightarrow V_n e^{I(i+1)k_x \Delta x} = V_n e^{I(i+1)\theta} \\ C_{i-1}^n &\Leftrightarrow V_n e^{I(i-1)k_x \Delta x} = V_n e^{I(i-1)\theta} \\ C_i^{n+1} &\Leftrightarrow V_{n+1} e^{Iik_x \Delta x} = V_{n+1} e^{Ii\theta} \end{aligned} \tag{12.23}$$

Now to get comfortable with the definitions above, we note that stepping along from node to node (e.g. from point i to point $i + 1$) means increasing the phase in the exponent. This is a characteristic of Fourier transforms, and reveals an interesting way of looking at the nature of finite difference grids that we'll discuss later. We can rewrite the purely diffusive FTCS algorithm of Equation (12.20) in terms of these Fourier components:

$$V_{n+1} e^{Ii\theta} = V_n e^{Ii\theta} + d \left(V_n e^{I(i+1)\theta} + V_n e^{I(i-1)\theta} - 2V_n e^{Ii\theta} \right) \tag{12.24}$$

and then group the Vs and divide through by $e^{Ii\theta}$ to get:

$$V_{n+1} = V_n \left[1 + d \left(e^{I\theta} + e^{-I\theta} - 2 \right) \right] \tag{12.25}$$

Remembering the well-known trigonometric identity

$$\cos\theta = \frac{e^{I\theta} + e^{-I\theta}}{2} \tag{12.26}$$

with a little manipulation Equation (12.25) reduces to:

$$V_{n+1} = V_n \left[1 - 2d \left(1 - \cos\theta \right) \right] \tag{12.27}$$

We've played a little with signs to make things a little clearer. Now you see why we didn't have to worry about *which* Fourier component: everybody knows that the cosine function can only vary between plus and minus 1, regardless of the size of its argument θ. You can think of the above equation like an amplification relationship. That is, we can express the amplitude of a Fourier component at time $n + 1$ as its amplitude at the previous time step n multiplied by a gain G:

$$V_{n+1} = G \cdot V_n \qquad (12.28)$$

where

$$G = [1 - 2d\,(1 - \cos\theta)] \qquad (12.29)$$

If the absolute value of the gain $|G|$ is greater than 1, the Fourier components will grow without bound and the solution will "explode". If $|G|$ is less than 1, then the solution will undergo *numerical damping*, which has less spectacular but certainly important consequences (we'll discuss these later). If $|G|$ is exactly 1, the Fourier components neither grow nor shrink with each time step. From the perspective of numerical stability, as long as the absolute value of the gain $|G|$ is less than or equal to 1, the Fourier components will not grow; i.e. V_{n+1} will be less than or equal to V_n, and the system will be stable. This is equivalent to saying that:

$$|G| \le 1 \qquad (12.30)$$

or

$$-1 \le [1 - 2d\,(1 - \cos\theta)] \le +1 \qquad (12.31)$$

Now careful inspection of the relation above shows that the right-hand inequality is always satisfied. You can see this by substituting -1 and $+1$ for the cosine. For -1, the center becomes $1 - 4d$, and since d is always positive, that's OK. For $+1$, the center expression becomes 1. Now, that leaves the left-hand relation. Substituting $+1$ of course gives 1 for the center expression, which is clearly greater than -1. And when you substitute -1, you have:

$$-1 \le 1 - 4d \qquad (12.32)$$

or

$$d \le \frac{1}{2} \qquad (12.33)$$

This is not particularly surprising; it's what we concluded from a far more intuitive and easier argument before. While it seems we've killed the proverbial fly with a sledgehammer here, it's the next steps that really should make you glad you kept with us on this.

12.4.3 The von Neumann stability analysis with diffusion and advection

When we build the full FTCS relation for diffusion and advection (given by Equation (12.9)) from the Fourier components, it becomes:

$$V_{n+1}e^{Ii\theta} = V_n e^{Ii\theta} - \frac{c}{2}\left(V_n e^{I(i+1)\theta} - V_n e^{I(i-1)\theta}\right)$$
$$+ d\left(V_n e^{I(i-1)\theta} + V_n e^{I(i+1)\theta} - 2V_n e^{Ii\theta}\right) \qquad (12.34)$$

Dividing through by $e^{Ii\theta}$ and grouping terms gives:

$$V_{n+1} = V_n \left[1 - \frac{c}{2} \left(e^{I\theta} - e^{-I\theta} \right) - 2d \left(1 - \cos \theta \right) \right] \tag{12.35}$$

where in addition to the cosine identity used for the diffusion term, we use a trigonometric identity for the sine for the advection term, giving:

$$V_{n+1} = V_n \left[1 - cI \sin \theta - 2d \left(1 - \cos \theta \right) \right] \tag{12.36}$$

Yes, you read it right, the term in square brackets contains an imaginary number, which in Fourier space really means a 90° phase shift in the Fourier component.

Similar to the diffusion-only case, we can define the gain G:

$$V_{n+1} = G \cdot V_n \tag{12.37}$$

where

$$G = [1 - 2d \left(1 - \cos \theta \right)] - Ic \sin \theta \tag{12.38}$$

In the above we've purposely separated the imaginary from the real parts. Now just because G is complex is no reason to panic. We simply require for stability that:

$$|G| \leq 1 \tag{12.39}$$

which is the modulus of complex G, or

$$G \cdot G^* \leq 1 \tag{12.40}$$

where G^* is the complex conjugate of G. Remember that you get the length (actually the square of the length) of a number on the complex plane by multiplying it by its complex conjugate. This becomes:

$$G \cdot G^* = [1 - 2d \left(1 - \cos \theta \right)]^2 + c^2 \sin^2 \theta \leq 1 \tag{12.41}$$

A rudimentary examination of this equation leads us to the same basic conclusions that we derived from the control volume analysis of stability discussed earlier: namely, for stability we at least need to have

$$d \leq \frac{1}{2} \tag{12.42}$$

$$c \leq 1 \tag{12.43}$$

but you'll find that this is not the complete story. To demonstrate this we created a MATLAB script vnsa.m to plot the value of G on the complex plane in Fig. 12.7. We compute G for a representative set of diffusion and Courant number pairs and with a range of $-\pi < \theta < \pi$.

In general, the gain curves will form ellipses that "kiss" the inside of the right-hand extreme of the unit circle. You can actually see this in Equation (12.41). Any ellipse that extends beyond the unit circle constitutes an unstable combination of diffusion and Courant numbers. Not surprisingly, $(d, c) = (0.35, 0.75)$ is stable (thin solid line in Fig. 12.7).

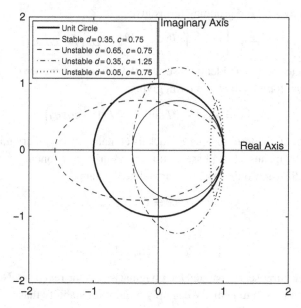

Figure 12.7 A plot of gain G for the FTCS algorithm on the complex plane for some example pairs of diffusion and Courant numbers. Note that if the trajectory falls outside the unit circle, the combination is unstable for FTCS. See the text for discussion.

$(d, c) = (0.65, 0.75)$ is diffusively unstable because the diffusion number exceeds 0.5, and $(d, c) = (0.35, 1.25)$ is advectively unstable because the Courant number exceeds 1. The surprise, however, is that the combination $(d, c) = (0.05, 0.75)$ is also unstable (note the ellipse extends outside the unit circle). Neither the diffusion nor the Courant number individually violate the criteria defined above, so what has happened here?

A more thorough analysis of Equation (12.41) reveals another constraint related to the fact that if the diffusion number is too small, then the gain ellipse can project outside the unit circle (above and below the **R**-axis; see the dotted line in Fig. 12.7) even though the Courant number is less than 1. Yet we didn't encounter any stability issue for small diffusion number in the purely diffusive case. Thus there is a sensitivity to the *ratio* of the two numbers as well, which can be derived by some algebraic and trigonometric manipulation of Equation (12.41) (we won't subject you to that here). This new, more restrictive constraint is related to the *cell Reynolds number* defined by $R_c = c/d$ and can be expressed as:

$$c \, R_c = \frac{c^2}{d} = \frac{u^2 \Delta t}{K} \leq 2 \tag{12.44}$$

This is not something that you'd readily anticipate from the control volume stability analysis we did earlier. The *cell Reynolds number* R_c is a dimensionless number. It represents the relationship of the strengths of the advection and diffusion terms (a large cell Reynolds number means you are advection dominated, a small cell Reynolds number

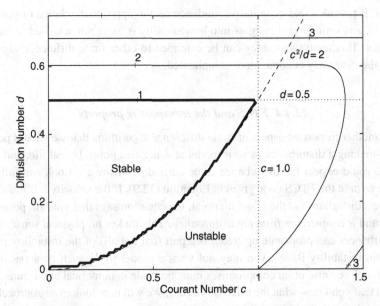

Figure 12.8 A plot of the squared gain $G \cdot G^*$ as function of diffusion and Courant number for the FTCS. The contour intervals are unitary, so the area of stability is inside the first contour only. The horizontal and vertical dotted lines indicate the individual constraints on those numbers, and the dashed curve line stipulates the cell Reynolds number constraint, cutting the region defined by the Courant and diffusion numbers approximately in half.

means you are diffusion dominated). You can convince yourself (by substitution) that the stipulation that c is less than or equal to 1 is satisfied if the diffusion number criterion and the cell Reynolds number criterion are simultaneously satisfied.

To get a better idea of the effect of this interaction, we devised the MATLAB script vnsa2.m to compute and contour the maximum value of $G \cdot G^*$ (for all θ) as function of diffusion and Courant number. The results are presented in Fig. 12.8. This is a useful complementary strategy, as you can explore "stability space" effectively this way. Note that the stability region defined by Equations (12.42 and 12.43) and outlined by the dotted vertical and horizontal lines is cut in half by the cell Reynolds number restriction (the dashed line).

What is quite remarkable is that the cell Reynolds number really tells you that for a purely advective system (that is, $K = 0$), the FTCS algorithm is unconditionally unstable. You can see this if you set d to zero in Equation (12.41), which gives:

$$G \cdot G^* = 1 + c^2 \sin^2 \theta \qquad (12.45)$$

Except when $\sin \theta$ is identically zero, the value of $G \cdot G^*$ is always greater than 1. Qualitatively, this says that the diffusion term tends to stabilize the second-order accurate advection term, which is unstable if insufficient diffusion is present.

We hope you can see that the type of stability analysis derived here is not simply an academic exercise, but can explain a number of things about the apparent stability of the FTCS algorithm. It provides real quantitative guidance on the appropriate choice of operational parameters, and some useful insights into how stability is, or is not achieved in numerical simulations. This analysis strategy can be extended to other finite difference algorithms, although the math may become more complicated.

12.4.4 FTCS and the transportive property

There is another important aspect of finite difference algorithms that we should point out. Consider making a disturbance in your model at some grid point. In real life, and without diffusion, you'd expect the disturbance to be carried downstream. Look carefully at the advective term in the FTCS code given in Equation (12.9). If the velocity u is positive, and you create a disturbance at the $(i + 1)$th point, the equation says that your ith point will be affected, and *it is upstream from the disturbance*. This makes no physical sense. The fact that disturbances can propagate upstream is a part (but not all) of the instability problem. Aside from instability issues, you may not want a model to do such unphysical things. We say "may" because often compromises must be made in numerical algorithms and it is important that you know what the trade-offs entail. We will now look to an approach which was designed to satisfy the transportive requirement.

12.5 Upwind differencing schemes

As the name suggests, these algorithms first became widely used in atmospheric science, but they may also be applied in oceanography.

12.5.1 The first upwind differencing method

We will try an alternate approach to formulating the advection term in our algorithm, while leaving the diffusion term alone. If we *insist* that the transportive property is satisfied, then the advection term changes to:

$$-c \left(C_i^n - C_{i-1}^n \right) \quad \text{for} \quad u > 0 \tag{12.46}$$

$$-c \left(C_{i+1}^n - C_i^n \right) \quad \text{for} \quad u < 0 \tag{12.47}$$

Actually this isn't difficult. You just need to make sure you know which way the wind is blowing when you set up your calculation. Now the complete equation (for positive u; you can figure out what it'll be for negative u) is:

$$C_i^{n+1} = C_i^n - c \left(C_i^n - C_{i-1}^n \right) + d \left(C_{i+1}^n + C_{i-1}^n - 2C_i^n \right) \tag{12.48}$$

The resulting algorithm is referred to as the *first upwind differencing method* or FUDM. It is designed to be *transportive*, since it only moves perturbations in the direction of flow. Moreover, it is *conservative*. You can test this by setting up the weights and seeing if they sum up to 1. You can even do the von Neumann stability analysis and find the stability criteria for this algorithm (sounds like an exercise for the student to us ...). So the FUDM seems like a simple, intuitive, and useful algorithm. However, you'll find that the FUDM doesn't appear in the literature very much. In fact Roache (1998) points out that a major computational fluid dynamics journal has an editorial policy that stipulates automatic rejection for any submitted manuscript that uses FUDM. You'll see one of the reasons for this in the next subsection.

12.5.2 First upwind differencing and numerical diffusion

By way of example, we'll do a simple numerical experiment with the FUDM. Actually, this is the kind of experiment you ought to consider doing for any finite difference scheme you plan on using. This is a general validation and testing strategy: feed the model something simple for which you know the ideal behavior, and see how the model responds. In this instance we start with 0 concentration everywhere in the model domain, and then step-increase the concentration instantly to a value of "1" at the model's upstream end and let the step propagate into the model domain. We'll start with no explicit diffusion ($d = 0$) in the model. Actually, that is a crucial part of our experiment. So the MATLAB code dostep.m is as follows:

```
% This is dostep.m
%
% does a first upwind differencing method
% calculation of pure advection of a step
% function in a linear domain
%
  InitializeFUDM            % assign constants, etc
  ComputeFUDMweights        % compute weight factors
  c=zeros(nx,1); cold=c;    % initialize arrays
  InitializePlot            % create initial plots
%
for i=1:nt                  % the time stepping
  PlotStepData              % periodically plot data
  c(1)=1;                   % force upstream cell to 1
  c(2:nx)=cold(2:nx)*w0 +  cold(1:nx-1)*wm;
  cold=c;                   % update holding vector
end
```

And what you see is in Fig. 12.9. The right-hand plot is a blow-up of the last step. What's happened? If the numeric code were working perfectly, we should see a vertical line at some fixed distance. It looks like a very diffusive process despite the fact that we had no explicit diffusion in the calculation!

We'll give you two explanations for this effect. The first is qualitative and intuitive, and the second is more rigorous and mathematical. First, consider a fluid particle being advected along your finite difference grid. Think of the finite difference algorithm as being a kind of "musical chairs" process, whereby every time there is a time step Δt and the

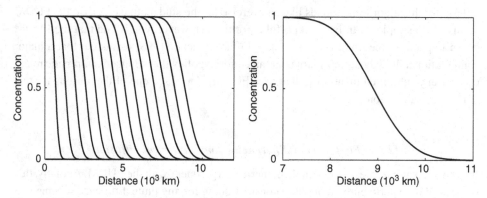

Figure 12.9 An example of the effect of numerical diffusion on a pure step function using the first upwind differencing method (FUDM). The model is run initially with zero concentration, and with a step of 1 in the concentration at the upstream end of the domain. The flow is from left to right, and each curve corresponds to the time-evolving step. The right-hand panel is a blow-up of the final curve.

music stops, the fluid particle has to drop onto one of the closest grid nodes (or in the control volume approach, into one of the cells). This is kind of a random process in that if the particle is more than half-way toward a grid point, it moves forward, and if it is less than half-way, it drops back. Essentially a kind of diffusion is associated with the discretization of the continuum domain. You can see this better if you think of a cluster of particles, and now you start putting some in one box and some in another. The net result is similar to a random-walk separation of particles, just like diffusion.

Now we'll try a more mathematical approach. The purely advective version of the FUDM for positive velocity can be written as:

$$\frac{C_i^{n+1} - C_i^n}{\Delta t} = -u \frac{\left(C_i^n - C_{i-1}^n \right)}{\Delta x} \tag{12.49}$$

We said before that this is only "first-order accurate in x". We start by examining the continuum equation and comparing the "truncated" version in Equation (12.49) with the Taylor series expansion in space (after Equation (12.14)). We express the concentration at the $(i - 1)$th grid point, since that is the point that affects the ith point:

$$C(x_i - \Delta x) = C(x_i) - \Delta x \frac{\partial C}{\partial x}\bigg|_{x_i} + \frac{(\Delta x)^2}{2} \frac{\partial^2 C}{\partial x^2}\bigg|_{x_i} - \frac{(\Delta x)^3}{6} \frac{\partial^3 C}{\partial x^3}\bigg|_{x_i} + \cdots \tag{12.50}$$

and we'll ignore any terms higher than quadratic, so we can express the right-hand side of the FUDM equation as:

$$-u \frac{C(x_i) - C(x_i - \Delta x)}{\Delta x} = -u \frac{\partial C}{\partial x} + u \frac{\Delta x}{2} \frac{\partial^2 C}{\partial x^2} \tag{12.51}$$

What we have done by using the FUDM is to add a hidden term associated with the truncation error that looks like a diffusion term, since it is multiplied by the second derivative of the concentration distribution.

But wait, there's more. By taking a finite time step, and approximating the time derivative in the equation above, we introduce an additional error, now in the time derivative. Doing a Taylor series expansion in time for the nth time step:

$$C(t_n + \Delta t) = C(t_n) + \Delta t \frac{\partial C}{\partial t}\Big|_{t_n} + \frac{(\Delta t)^2}{2} \frac{\partial^2 C}{\partial t^2}\Big|_{t_n} + \frac{(\Delta t)^3}{6} \frac{\partial^3 C}{\partial t^3}\Big|_{t_n} + \cdots \quad (12.52)$$

the left-hand side of the FUDM equation (again, only accurate to the quadratic terms) becomes:

$$\frac{C(t_n + \Delta t) - C(t_n)}{\Delta t} = \frac{\partial C}{\partial t} + \frac{\Delta t}{2} \frac{\partial^2 C}{\partial t^2} \quad (12.53)$$

Now what we really wanted, and what must be the case in the continuum, is:

$$\frac{\partial C}{\partial t} = -u \frac{\partial C}{\partial x} \quad (12.54)$$

and so we can rewrite the second derivative in the time relation for constant u as:

$$\frac{\partial^2 C}{\partial t^2} = \frac{\partial}{\partial t}\left(\frac{\partial C}{\partial t}\right) = -u \frac{\partial}{\partial t}\left(\frac{\partial C}{\partial x}\right) = -u \frac{\partial}{\partial x}\left(\frac{\partial C}{\partial t}\right) = u^2 \frac{\partial}{\partial x}\left(\frac{\partial C}{\partial x}\right) = u^2 \frac{\partial^2 C}{\partial x^2}$$
$$(12.55)$$

Notice that we've repeatedly inserted the relationship in Equation (12.54) in the preceding equation. So by substituting Equations (12.51) and (12.53) into the equation for FUDM we can show that the FUDM is actually equivalent in the second order to:

$$\frac{\partial C}{\partial t} + \frac{u^2 \Delta t}{2} \frac{\partial^2 C}{\partial x^2} = -u \frac{\partial C}{\partial x} + \frac{u \Delta x}{2} \frac{\partial^2 C}{\partial x^2} \quad (12.56)$$

or by grouping the second space derivitives as

$$\frac{\partial C}{\partial t} = -u \frac{\partial C}{\partial x} + K_n \frac{\partial^2 C}{\partial x^2} \quad (12.57)$$

where

$$K_n = \frac{u \Delta x}{2} (1 - c) \quad (12.58)$$

and where we'll remind you that the Courant number is defined as $c = u \Delta t / \Delta x$.

This K_n is an effective numerical diffusivity asssociated with the algorithmic discretization of space and time. That is, there is this additional term due to the truncation errors (to second order) that looks like a diffusivity. Even in a simulation with no *explicit* diffusion, sharp features like our step function test will be smeared out by an apparent diffusion. We deliberately formulated K_n in Equation (12.58) in terms of the Courant number. Note that in the special case where $c = 1$ the numeric diffusivity is exactly zero. This makes sense, because as the fluid particles are displaced exactly a distance Δx in the time Δt (if $c = 1$)

there is no diffusion. Thus another way of looking at the numerical diffusion is that during each time step, as material is advected from one "cell" to the next, there is an artificial division of material between the cells (if the movement isn't *exactly* one cell length), and that this "division" is a kind of diffusion.

Well, how much of a problem is it? For the example code in dostep.m, we have that:

$$\Delta x = 50\,000\,\text{m} \quad \Delta t = 200\,000\,\text{s} \quad u = 0.02\,\text{m s}^{-1}$$
$$c = 0.08 \qquad \text{and} \qquad K_n = 460\,\text{m}^2\,\text{s}^{-1} \tag{12.59}$$

If we had used the FUDM in the calculation we did for the FTCS pipe model (which had $u = 0.01$) the numerical diffusivity would have been $240\,\text{m}^2\,\text{s}^{-1}$. Considering that we were using explicit diffusivities ranging from 1000 to $8000\,\text{m}^2\,\text{s}^{-1}$, this would not be a huge error; but if we were interested in examining flows at lower diffusivities, it would be an issue.

If we apply the same type of analysis to the FTCS code, the numerical diffusivity is:

$$K_n = -\frac{u^2 \Delta t}{2} \tag{12.60}$$

The somewhat disturbing result is that this is a *negative* numerical diffusivity. This is a consequence of the character of the algorithm, and explains why the FTCS code is uncon-ditionally unstable for zero explicit diffusion. A negative diffusivity amplifies small errors rapidly, and causes an explosion. This also explains the size of the cell Reynolds number limitation (think about it!). In our pipe model simulation, the size of the numerical diffu-sion error using FTCS code was only $-10\,\text{m}^2\,\text{s}^{-1}$ (plug in the values to find this out). So it wasn't a big problem for us there. The reason why it is so small compared with FUDM is because the code is second-order accurate in space.

Surely, given this exact formulation of the scale of the problem, you would be inclined to actually put the calculation of numerical diffusivity in your code as a "diffusivity cor-rection" – that is, subtract it off your algorithm. Avoid this approach if you can. The reason we advise this is that by adding the extra term, your algorithm ceases to be *conservative*. However, there are problems that cannot be solved without it, but these are beyond the scope of this book.

There's one final complication to numerical diffusion you need to be aware of. The numerical diffusion will depend on whether the solution is for a transient or steady-state case. For steady-state solutions the time-dependent terms (Equations (12.52) through (12.55)) go to zero, and the resultant numerical diffusion will be correspondingly different. The FTCS numerical diffusion for steady state will be zero, while for the FUDM it will be:

$$K_n = \frac{u \Delta x}{2} \tag{12.61}$$

Compare this with the expression given by Equation (12.58) and you can see that the difference lies in the absence of the Courant number dependence. Roache (1998) provides an extended discussion of numerical diffusion (also referred to as "artificial viscosity") for a variety of algorithms.

12.5.3 The second upwind differencing method

The FUDM is both simple and easy to implement. It also appears attractive because it produces *smooth* results, largely due to its numerical diffusivity. It is often embedded in packaged software for general analysis for this reason. However, it is only first-order accurate in space, and it is possible to do better. One significant improvement can be made in the FUDM that allows us to recover the second-order accuracy in the advection term. This simply involves the more accurate assessment of the cell wall velocities by taking the average between the node points:

$$C_i^{n+1} = C_i^n - \frac{\Delta t}{\Delta x} \left(u_R C_R^n - u_L C_L^n \right) + d \left(C_{i+1}^n + C_{i-1}^n - 2C_i^n \right) \tag{12.62}$$

where

$$u_R = \frac{1}{2} \left(u_{i+1} + u_i \right) \quad \text{and} \quad u_L = \frac{1}{2} \left(u_{i-1} + u_i \right) \tag{12.63}$$

and we select the correct choice of concentrations depending on the *direction* of the cell wall face velocity

$$\begin{aligned}
C_R^n = C_i^n \quad \text{for} \quad u_R > 0 \quad \text{or} \quad C_R^n = C_{i+1}^n \quad \text{for} \quad u_R < 0 \\
C_L^n = C_{i-1}^n \quad \text{for} \quad u_L > 0 \quad \text{or} \quad C_L^n = C_i^n \quad \text{for} \quad u_L < 0
\end{aligned} \tag{12.64}$$

This algorithm is both *conservative* and *transportive*. The former you can prove by summing your weights; the latter you guarantee by the correct choice of concentrations above. Be careful, however: if you take the added "refinement" of defining the cell wall face concentrations by averaging (interpolating) the concentrations with adjacent node points, you actually end up with the FTCS technique, which is not transportive. The *second upwind differencing method* (SUDM) is already second-order accurate in space for the advection term.

There are many other algorithms out there. We hope that if you consider using them, you ask the hard questions about them that you should.

12.6 Additional concerns, and generalities
12.6.1 Phase errors and aliasing

There are a couple of other problems that we should mention. We won't go into great detail, except to alert you to the potential issues. The first relates back to our analysis of stability. Recognize that the Fourier components that we analyzed can propagate through the finite difference mesh at different rates. If you think of a concentration distribution (for example the leading edge of a concentration change with time) as a composite of different Fourier components, then the components will disperse at different rates. The net result will be that the edge will decompose as it propagates, with some faster-moving components leading, and slower ones lagging the front. This is a process analogous to dispersion in optics (or acoustics), where a pulse will distort as it moves because of a dependence of propagation velocity on frequency.

The second problem is one of aliasing. Yes, this is just the same problem as we talked about in Chapter 6. The finite difference grid will have a *Nyquist limit* governed by the separation of the grid points. The space-wavenumber equivalent of this limitation is just:

$$k_x^c = \frac{2\pi}{2\Delta x} = \frac{\pi}{\Delta x} \tag{12.65}$$

which simply says that you cannot resolve any wavelength shorter than two grid point spacings. Now the unfortunate thing about this is that you will get *aliasing* of these short wavelength features onto very long wavelengths in your mesh. If you are dealing with a relatively diffusive system (e.g. for large-scale tracer simulations) then the diffusion tends to strongly damp the high-frequency components, and you won't observe this problem. If you are looking at high-resolution systems (e.g. sub-mesoscale or frontal phenomena) or complex geometries (e.g. coastal systems or flow around a copepod), this can be a concern. Just be aware of the risks if you start dealing with highly advective, low-diffusivity systems with sharp transients. Your grid may "ring".

12.6.2 Summary of criteria and concerns about finite difference equations

We sum up the kind of considerations that you need to be concerned with when you deal with finite difference schemes. Choosing the right algorithm minimizes some of the problems at the expense of others. Compromise is the way of life here. It depends on the problem you are dealing with. You decide.

- **Truncation precision** – we use the term precision advisedly here. Some schemes are more precise than others in the sense of how many terms in the Taylor series they incorporate. A first-order accurate scheme will require a certain number of points to achieve the desired accuracy. Higher-order schemes will do better for the same number of grid points, or achieve the same precision with fewer points. FTCS and SUDM are second-order in space while FUDM is only first-order.
- **Conservative property** – does the algorithm conserve mass? This is *extremely* important for most problems. Some schemes don't conserve mass. Carefully deriving the algorithm from the control volume model more-or-less guarantees mass conservation. Check the sum of your weights. FTCS, FUDM, and SUDM do conserve mass.
- **Transportive property** – does a disturbance propagate downstream only? If it propagates upstream it is unphysical. This is important if you are dealing with transient problems. FTCS is *not transportive*. FUDM and SUDM are transportive.
- **Stability** – there are limits to your choice of space and time resolutions beyond which the algorithm blows up. The von Neumann stability analysis can be used to deduce what those limits are. In general, you require at a minimum that the diffusion number is less than or equal to $\frac{1}{2}$ and that the Courant number is less than or equal to 1. The cell Reynolds number constraint is more restrictive for the FTCS. You will find it unwise to skirt too close to those numbers in practice. That is, don't go above $c = 0.8$ or 0.9, and $d = 0.3$ or 0.4, and similarly for the cell Reynolds number. It is prudent to compute

or test the choice of time (or space) step in your code so that it automatically assures stability in case you unthinkingly change some parameter like diffusivity or velocity.

- **Phase errors and aliasing** – your grid has a Nyquist limit. Respect it.
- **Numerical diffusion** – as a result of truncation error, your code will have an implicit numerical diffusion which will tend to smear out your results. Taylor series analysis reveals the exact value of this numerical diffusivity for given algorithms. Be aware of the size of the effect and its relationship to your explicit diffusivity.

12.6.3 Some recommendations

We will leave you with some specific recommendations. The first is to keep it simple: choosing exotic and complicated schemes may seem clever at the time, but they can lead to exotic and complicated problems. The worst case scenario is a problem that you are unaware you have. Second, don't reinvent the wheel; there are lots of algorithms out there which do just fine. Check Roache (1998) to see what the algorithm does. There are some useful discussions in the literature (e.g. Rood, 1987, and Hecht *et al.*, 1995) that you can consult. Third, know your tools: be aware of your algorithm's properties (as outlined in Section 12.6.2). Fourth, carry along checksums/tests of model performance. For example, always keep track of the total amount of tracer in your model domain. Sum it up for each time step or at appropriate intervals to see if you are inadvertently creating or destroying mass. Fifth, test your model with idealized functions to see if it is behaving correctly (or at least as badly as you expect). This means, for example, forcing the model with a spike (delta function) or step function like we did in the last section to see what it does with it. This is a check not only on the fidelity of your numerics, but also on your coding (have you made a coding error?). Finally, compute and recognize the algorithmic limitations: stability, numerical diffusion, etc. Always keep them in mind when you look at your results. Are your results an artifact of your algorithm rather than physical reality?

12.7 Extension to more than one dimension

What happens when there is more than one dimension? That is, suppose you are interested in diffusion and advection in a two-dimensional ocean gyre? Another example might be modeling the chemical alteration or zonation of individual sediment grains. The extension of the techniques we have been using to two or more dimensions is relatively straightforward, subject to stability and feasibility considerations. You should also be aware that it may be possible to reduce the dimensionality of the problem by using symmetry.

12.7.1 Two-dimensional systems

Techniques such as the first and second upwind differencing methods can be extended to two or more dimensions in a simple fashion. The equation for the FUDM in two dimensions with positive u (x-direction velocity) and v (y-direction velocity) is as follows:

$$C_{i,j}^{n+1} = C_{i,j}^n - \frac{u\Delta t}{\Delta x}\left(C_{i,j}^n - C_{i-1,j}^n\right) - \frac{v\Delta t}{\Delta y}\left(C_{i,j}^n - C_{i,j-1}^n\right)$$
$$+ \frac{K_x\Delta t}{\Delta x^2}\left(C_{i+1,j}^n + C_{i-1,j}^n - 2C_{i,j}^n\right) + \frac{K_y\Delta t}{\Delta y^2}\left(C_{i,j+1}^n + C_{i,j-1}^n - 2C_{i,j}^n\right)$$

$$(12.66)$$

Note that here we now have *two* indices for the spatial dimensions, i for the x direction, and j for the y direction. We also allow diffusivity to differ in either direction, although this may not necessarily be the case. Note that this equation is only first-order accurate in space, and we have not included the cross-derivatives ($\partial^2/\partial x\partial y$) in the scheme. We then would rewrite this equation as:

$$C_{i,j}^{n+1} = C_{i,j}^n - c_x\left(C_{i,j}^n - C_{i-1,j}^n\right) - c_y\left(C_{i,j}^n - C_{i,j-1}^n\right) + d_x\left(C_{i+1,j}^n + C_{i-1,j}^n - 2C_{i,j}^n\right)$$
$$+ d_y\left(C_{i,j+1}^n + C_{i,j-1}^n - 2C_{i,j}^n\right)$$

$$(12.67)$$

where the Courant numbers c_x and c_y and the diffusion numbers d_x and d_y determine the stability of the algorithm. Using the von Neumann stability analysis in two dimensions, now, is cumbersome, but not very different from the one-dimensional analysis. What you arrive at for stability is:

$$c_x + c_y \leq 1 \quad \text{and} \quad d_x + d_y \leq \frac{1}{2} \tag{12.68}$$

That is, the sum of the Courant or diffusion numbers must be less than or equal to the restriction imposed on the equivalent one-dimensional problem. Actually, from the point of view of control volume analysis, this makes sense. If you are removing material from your control volume in two dimensions, you still don't want the total removed to exceed the contents of your cell. If u and v were the same, K_x and K_y the same, and Δx and Δy the same, we would have:

$$c = c_x = c_y \leq \frac{1}{2} \quad \text{and} \quad d = d_x = d_y \leq \frac{1}{4} \tag{12.69}$$

Thus you have to be *twice* as conservative in two dimensions as you do in one.

Now how do we set up the algorithm in MATLAB? Well, we group the terms to calculate weights, so that we have:

$$C_{i,j}^{n+1} = w_0 C_{i,j}^n + w_{x-}C_{i-1,j}^n + w_{x+}C_{i+1,j}^n + w_{y-}C_{i,j-1}^n + w_{y+}C_{i,j+1}^n \tag{12.70}$$

where we have:

$$w_0 = 1 - c_x - c_y - 2d_x - 2d_y$$
$$w_{x-} = c_x + d_x$$
$$w_{x+} = d_x \tag{12.71}$$
$$w_{y-} = c_y + d_y$$
$$w_{y+} = d_y$$

Notice a couple of things about the above weights. First, the stability limitations become apparent when you look at the structure of w_0 and recognize that it must be greater than or equal to zero. This also points out that it is not enough to satisfy the diffusion and Courant stability criteria individually, but rather you must satisfy their sum. This comes out of the von Neumann analysis. The second is that if you add the weights, they add up to 1. Finally, you now have five weights for two dimensions; and you would have seven weights in three dimensions.

Performing the calculation in MATLAB is fairly straightforward, except for one seemingly minor but important quirk. Remember that MATLAB refers to things in "row–column" space. If you picture a matrix with rows and columns, this means that the row position corresponds to the y direction, and the column direction corresponds to the x direction. Thus you refer to a matrix as $C(y, x)$, not $C(x, y)$. You could actually do it the other way, but you'd have to lay the matrix on its side (transpose it) to plot it. So in the rather trivial situation where you have constant velocity and diffusivity, you might have the following kind of code:

```
C=zeros(ny,nx); Cold=C;

w0  = 1 - u*dt/dx - v*dt/dy -2*kx*dt/dx/dx -2*ky*dt/dy/dy;
wxm = u*dt/dx + kx*dt/dx/dx;
wxp = kx*dt/dx/dx;
wym = v*dt/dy + ky*dt/dy/dy;
wyp = ky*dt/dy/dy;

for t=1:nt   % your main time step loop here

... some kind of code to set your boundary conditions here,
which involves the edges of your domain ...

C(2:ny-1,2:nx-1)= w0*Cold(2:ny-1,2:nx-1) + wxm*Cold(2:ny-1,1:nx-2) + ...
                 wxp*Cold(2:ny-1,3:nx) + ...
                 wym*Cold(1:ny-2,2:nx-1) + wyp*Cold(3:ny,2:nx-1);
C = Cold;
end
```

Note that the code above looks like the one we did in one dimension, but now we need a two-dimensional array to hold the concentration values. As before, we need to carry a second copy of the concentration array to do the bookkeeping. The actual iteration in the space domain is done with the same tricky indexing as before, but now with two indices. Read the line carefully and it should make sense. The trick is a little like making sub-arrays offset from the central index points by one element in each of the x and y directions. Also, remember that the above FUDM code is for positive and constant u and v.

12.7.2 Variable velocities

Another aspect of higher-dimensional systems is that velocities generally will vary spatially or temporally. It is hard to imagine in a one-dimensional system changing the velocity as a function of x, since that would not conserve mass. However, if you add another dimension, you can now change the velocity in the x direction and still conserve mass by changing the velocity in the y direction as well. Another example would be a quasi-one-dimensional system such as a plume where entrainment of water from outside the plume leads to a change in mass transport. Thus the code we described in the previous section is not very useful. You would, in general, have a u and v that are functions of x and y. Although we can vary the u and v velocities, we do not have absolute freedom to pick arbitrary values. In order to conserve mass we have to specify nondivergent velocity fields if our fluid is incompressible. That is, our choice of u and v must satisfy:

$$\frac{\partial u}{\partial x} + \frac{\partial v}{\partial y} = 0 \tag{12.72}$$

One way to guarantee that this is satisfied is to specify the velocity fields in terms of a *streamfunction* ψ such that

$$u = -\frac{\partial \psi}{\partial y} \quad \text{and} \quad v = \frac{\partial \psi}{\partial x} \tag{12.73}$$

Substituting the definitions for u and v into Equation (12.72) should convince you that deriving the velocities from a streamfunction guarantees non-divergence of the flow field.

For example, consider a simple symmetric gyre in two dimensions within a square domain of length L_x by L_y by defining the streamfunction with

$$\psi = \sin \frac{\pi y}{L_y} \sin \frac{\pi x}{L_x} \tag{12.74}$$

which gives a streamfunction shown in Fig. 12.10. The contours of the streamfunction are really *streamlines*, which may be viewed as paths followed by the fluid in the

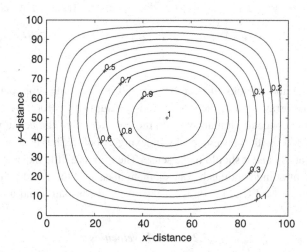

Figure 12.10 An example of a streamfunction for a circular gyre in a square array, computed from Equation (12.74).

flow field. Note also that there is one other level of consideration here. Because the velocities generally have defined values at the grid points themselves (at least as far as the algorithmic definitions are concerned), then because the velocities are computed from the spatial derivatives of the streamfunction, the streamfunction needs to be computed or defined midway between the grid points. Just another bookkeeping chore to be aware of.

With velocities varying in space, the corresponding weights would also. Thus you would calculate *weight matrices*. The problem becomes rather complicated for a number of reasons. The first is that the direction of the velocity components can change throughout the space domain, so if you are using the FUDM or SUDM schemes, you need to adjust the weight factors so that they reflect the *upwind* choice of the cells. Second, you have to be careful about a "registration" problem between the streamfunction and the velocity grids. Third, you need to compute the minimum stable time step based on the absolute *maximum* velocity component in your model domain.

Let's assume you have constructed your weight matrices for the problem, and for simplicity assume that $u(x, y)$ and $v(x, y)$ are positive everywhere (extremely unlikely, of course), and that if you were doing the *real* thing, you'd have to do it on a cell by cell basis, possibly with some for loops and if tests. Also, we'll assume that you've calculated the diffusivity matrices (possibly variable) as function of x and y. You'll end up with five matrices all of the same size, namely n_y by n_x.

```
c=zeros(ny,nx); cold=c;

w0  = ones(ny,nx) - u*dt/dx - v*dt/dy -2*kx*dt/dx/dx -2*ky*dt/dy/dy;
wxm = u*dt/dx + kx*dt/dx/dx;
wxp = kx*dt/dx/dx;
wym = v*dt/dy + ky*dt/dy/dy;
wyp = ky*dt/dy/dy;

for t=1:nt  % your main time step loop here

 ... some kind of code to set your boundary conditions here ...

c(2:ny-1,2:nx-1) = w0(2:ny-1,2:nx-1).*cold(2:ny-1,2:nx-1) +...
                 wxm(2:ny-1,2:nx-1).*cold(2:ny-1,1:nx-2) +...
                 wxp(2:ny-1,2:nx-1).*cold(2:ny-1,3:nx) +...
                 wym(2:ny-1,2:nx-1).*cold(1:ny-2,2:nx-1) +...
                 wyp(2:ny-1,2:nx-1).*cold(3:ny,2:nx-1);

c = cold;
end
```

Now notice in the above, rather complicated segment of MATLAB code that we've subindexed the weight arrays to match with the corresponding concentration nodes, and that the multiplications are the "dot" variety.

12.7.3 What to do with the boundaries

In the above code, we've glossed over the boundary portions of the calculation. You will need to take into account that the boundary "cells" will be adjacent to a reduced number of neighboring cells. For example, the cells with $i = 1$ will obviously not have neighbors at $i - 1$. Your weights will have to take this into account, and your computations ought not to seek non-existent neighbors. The same especially holds true for the corner cells. So in the case of a two-dimensional rectangular domain, you would need at least four lines of code computing the four boundaries, plus four lines of code computing the four corners. In all cases, the weights must sum to 1 for mass conservation.

This also relates to some fundamental decisions you must make about how you pose your model. Boundary conditions are an important and challenging aspect of model design. For example, do you consider the boundaries of your domain to be impermeable walls? If so, then you need to construct your weights and calculations accordingly. However, you might also construct a model that is infinitely periodic, where the $i = 1$ cells would be considered adjacent to the $i = n$ cells, and so on (i.e. the domain repeats itself, indefinitely). You might also imagine "permeable" walls subject to some kind of mass transport, for example if one wall were considered the ocean surface and gas were allowed to transfer to or from the atmosphere, or if one edge represented the outcrop of an isopycnal. If across-boundary flux were to be incorporated in your model, then construction of property inventories within your code (for checking numerics or other diagnostics) should track and account for the fluxes.

Often you are trying to model a region, such as a subtropical gyre, that is also connected to a larger system. How do you make this connection in a way that doesn't unduly influence or control your simulation? These are very knotty issues that ocean modelers have to contend with. One approach is to embed the model in a larger, coarser-resolution model that hopefully emulates the boundary values well enough that we can get on with our regional simulation. Another technique is either to set the boundary values to some "climatological" values, or to *relax* them toward the values using a restoring strategy. Bear in mind that this essentially creates artificial *in situ* sources or sinks along your boundaries, which means that you are no longer conserving mass.

Another strategy is to create a *sponge layer* within your model that effectively buffers the region of interest from the "*terra incognita*" that you wish to avoid. This could be done by specifying regions with zero velocity (i.e. constant streamfunction) but with non-zero diffusivity. Such regions tend to introduce long-term drifts in model behavior which may or may not be important, depending on your objectives. The key point is not to lose sight of the impact of your decisions!

12.7.4 Non-uniform grid spacing

We consider non-uniform grids to lie beyond the scope of the text, but you should be aware of their use. The gyre shown in Fig. 12.10 may be modified to crudely emulate the westward intensification of boundary currents, for example like the Kuroshio or Gulf Stream. To resolve these narrow currents, the grid spacing would need to be much finer and the additional computational expense may become substantial if you are attempting to model an entire subtropical gyre with a uniformly fine grid. One solution is to use a non-uniform grid spacing, where the spacing is finer in the regions where it is needed, and coarser in the regions where it is not.

Using a non-uniform grid introduces an additional level of complexity, however, in that from the viewpoint of the control volume analysis the cell volumes and interfacial areas become spatially variable. Choice of a simple functional form to the grid spacing sometimes leads to an easier analysis. The longest stable time step will generally be limited by smallest grid point spacing, which will likely be near the highest velocities in the domain, or where property gradients are sharpest. Additional care must be taken to ensure that the algorithm conserves mass and that it is stable against transients. Consideration of numerical diffusion and Fourier component dispersion also becomes more demanding.

12.7.5 Dimensional reduction using symmetry

Sometimes problems are of an inherently lower dimensionality because of symmetry – for example if you were modeling the evolution of a spherical sediment grain subject to chemical diffusive exchange with the surrounding pore water medium. You might have reason to expect that the evolution of concentrations in the grain would be spherically symmetric (although mineral grains in general tend to be non-isotropic, you might ignore this concern initially). In this instance it is not mandatory to run a full three-dimensional calculation to model the distribution, but it is necessary to use the appropriate coordinate transformation. You can treat the problem quite sensibly using the control volume approach in spherical coordinates, provided that you are careful and meticulous about your bookkeeping. In the case of the spherical volume, you would have volume elements whose volumes vary as $r^2 \Delta r$, and cell interfaces whose areas vary as $r \Delta r$.

12.8 Implicit algorithms

The codes presented in this chapter have all been *explicit* codes. That is, we have only used concentrations from the previous time step to calculate the next time step. Mathematically, we have been computing:

$$C^{n+1} = f\left(C^n\right) \tag{12.75}$$

where we have purposely left off the space subscripts. Now think about the control volume approach. During the time step Δt, we have "held C constant" over that interval while we diffuse and advect material between the cells. Wouldn't it be more "accurate" to allow C to change over that time? The simplest approach might be to take the average concentration over that time (i.e. the average between time n and time $n + 1$). Well, that seems a little tricky, because now we have:

$$C^{n+1} = f\left(C^n, C^{n+1}\right) \tag{12.76}$$

This is not as intractable as it may seem. What we have now is a set of simultaneous equations that must be solved for each time step. This isn't too bad in one dimension; the solution to the equations involves the inversion of a tridiagonal matrix because the concentrations only depend on neighboring cells, which is relatively fast. The result is that you can take larger time steps because the methods are accurate to a higher order in time. In more than one dimension, however, things become more difficult since the equation solutions become eigenvalue problems associated with the solution of block tridiagonal matrices. Thus the computational expense per step goes up, but for the larger 3D models the benefits pay off. Our advice is to avoid the implicit codes for most simple problems that you may encounter.

There is one other reason to stick to explicit codes. Implicit codes involve the *simultaneous solution* of equations. This is equivalent to an infinite propagation speed for information through your domain. How this manifests itself mathematically is that many Fourier modes are propagated at anomalously high velocities. You get Fourier decomposition, phase aliasing, and dispersion. This problem goes away if you are looking at steady state problems, since all of this stuff washes away. If you are dealing with a transient (time evolving) system, or you are interested in the actual evolution toward steady state, you are advised to use explicit methods. When there's a mixture of steady-state and transient phenomena, or in other words problems that have both fast and slow timescales (e.g. the movement of tracer by internal waves), *blended* algorithms may be used. Most state-of-the-art numerical models *do* use implicit codes, because the efficiency gains are mandatory when you are working on a computationally limited problem. If you get to this stage, you will have moved far beyond the simple approaches you are learning here.

12.9 Problems

12.1. Using the von Neumann stability analysis technique, derive the appropriate stability criteria for the advective–diffusive FUDM and the SUDM. In addition, use a MATLAB script similar to `vnsa.m` and `vnsa2.m` to visualize the stability fields as a function of the diffusion and Courant number.

12.2. Derive for yourself the expression giving the numerical diffusivity for the FTCS algorithm.

12.3. Use the following numerical experiment to demonstrate that Equation (12.58) accurately predicts the numerical diffusivity for the FUDM. Modify the purely advective FUDM code `dostep.m` so that you inject a single spike in the middle of the domain (i.e. by setting one grid point's value to 1 for the first time step only) and devise a plotting routine to monitor the shape of this spike as a function of time for 15 years with a velocity of $1\,\mathrm{cm\,s^{-1}}$. You will have to choose appropriate Δx and domain length so that your spike doesn't escape off the end of your model domain (think carefully on how to do this!). Have the code compute an appropriate Δt based on your choice. The spike should evolve as a Gaussian as a function of time such that

$$C(x) = C_{\max} e^{-\left(\frac{x-x_0}{\sqrt{K_n t}}\right)^2} \tag{12.77}$$

where x_0 is the location of the center of the peak. You could then take the final curve and use a nonlinear least squares fitting routine to calculate the peak width and thereby the value of K_n. Do this for two values of the Courant number, and for the case where $c = 1$.

12.4. You will have noted that by setting a single cell to unity for one time step is equivalent to injecting an amount of tracer equal to Δx (1 times the area of step). For the example above, we note that the area under a Gaussian is given by

$$A = \int_{-\infty}^{\infty} C_{\max} e^{-\left(\frac{x-x_0}{\sqrt{K_n t}}\right)^2} dx = \pi C_{\max} \sqrt{K_n t} \tag{12.78}$$

so if your model conserves mass (it does, doesn't it?), you have

$$A = \pi C_{\max} \sqrt{K_n t} = \Delta x \tag{12.79}$$

or by squaring this, you have

$$\left(\frac{1}{C_{\max}^2}\right) = \left(\frac{\pi^2 K_n}{\Delta x^2}\right) t \tag{12.80}$$

So by plotting $\left(\frac{1}{C_{\max}^2}\right)$ as a function of time, the slope would give K_n with

$$K_n = \frac{\Delta x^2}{\pi^2}\text{Slope} \tag{12.81}$$

Perform the tests you did in the previous problem, but this time plot and compute K_n from a linear fit of the relationship. How do the results compare?

12.5. For the previous problem, install code in the model that keeps track of the total amount of tracer in the domain as a function of time and monitor this result. Does your model conserve mass? Look at the residual by subtracting the model's initial tracer inventory (at the first time step) and explain what you see. Now modify the algorithm by putting in a "flux corrector" that subtracts off the numerical diffusion term and run the model. Assuming you had a stable run, what happens to your total tracer inventory?

13

Open ocean 1D advection–diffusion models

Faith and doubt both are needed – not as antagonists, but working side
by side to take us around the unknown curve.

Lillian Smith

13.1 Rationale

Our main objective in studying one-dimensional, open-ocean advection–diffusion models
is pedagogical. The fact that they have relatively simple analytical solutions makes them a
useful starting point for studying ocean models. In fact, you may find yourself turning to
these more idealized representations as a tool for building intuition about the behavior of
more complex models. That is, you might build a "model of your model" to explore what is
happening within it. Perhaps more important to the student of modeling, they represent an
elegant example of how we can use spatial distributions to illuminate underlying physical
and biogeochemical dynamics.

They're not really considered "state-of-the-art", having been extensively exploited by
geochemists starting in the 1950s and used by others many decades before that. In truth,
there are few parts of the ocean that can be regarded as truly satisfying the assumptions
and requirements of this class of model. Even then it is highly debatable how generalizable
the parameters derived from such models really are to the rest of the world. However, it
is instructive to think of the abyssal ocean in terms of simple one-dimensional balances
because it helps build intuition about open-ocean processes. Certainly it is an interesting
historical stage in the evolution of geochemical ocean modeling, and has much to offer as
a learning tool for understanding the process of ocean modeling.

We will focus our discussion around a type of analysis that was featured in a paper
entitled *Abyssal carbon and radiocarbon in the Pacific*, by Harmon Craig (1969). The
paper is not particularly easy to read, but it is a rather complete and in-depth discussion
of one-dimensional advection–diffusion equations for a variety of different tracers in the
deep Pacific. It is a refinement and correction of an earlier paper called *Abyssal recipes*
by Walter Munk (1966), which appears to be the first attempt to use ^{14}C in a 1D model
to estimate upwelling rates and vertical turbulent diffusivity in the deep ocean waters.

Although Munk's attempt must be regarded as ground-breaking, there were two errors in his analysis. The first was that he ignored the fact that ^{14}C was, after all, *carbon*, and hence was not conserved in deep waters because of *in situ* oxidation of organic material. His second error was that he modeled $\Delta^{14}C$, which is a kind of *isotope ratio anomaly* and not a genuine molar quantity. A geochemically correct treatment of the problem requires the treatment presented a few years later by Craig, which we develop in our own inimitable way here.

13.2 The general setting and equations

The basic situation is thus: there is a depth range in the deep Pacific between the very deep core of incoming *Common Water* – a mixture of North Atlantic Deep Water (NADW), Circumpolar Waters, and Antarctic Bottom Water (AABW) that enters the South Pacific at around 3500 to 4000 m depth – and the low salinity core associated with Antarctic Intermediate Water (AAIW) at around 1000 m depth. The premise is that water properties are maintained at the endpoints of this range by horizontally flowing water, which fixes the concentrations of the various properties "at the ends". In between, it is assumed that the influence of horizontal flow is weak and that the property concentrations are determined by a combination of vertical advection, vertical (turbulent) diffusion, *in situ* processes of biogeochemical production or consumption, and radioactive decay. That is, deep water formation in regions distant from this location results in a generalized upwelling of deep water in the abyss, and the vertical distribution of properties during this upwelling is sustained by a combination of vertical turbulent mixing, possibly *in situ* consumption or production, and in some cases radioactive decay. This depth range was defined as an *advective–diffusive subrange*. Throughout this range we therefore argue that the processes affecting tracer distributions are purely vertical, and hence subject to one-dimensional modeling.

Not all tracers exhibit the same behavior in the ocean, as they are controlled by different physical and biogeochemical processes. We can characterize tracers in the following ways:

Passive/active – depending on whether they play a role in determining the density of the water. Temperature and salinity are active tracers, while nitrate and radiocarbon are passive tracers. For this study, however, because the equation of state is nearly linear over the subrange we are dealing with, we can get away with treating temperature and salinity as passive.

Steady state/transient – if the tracer distribution is changing in time, it is considered transient. In the deep Pacific, oxygen, temperature, and salinity may be regarded as steady state. Chlorofluorocarbons (CFCs) and bomb-produced tritium are transient.

Conservative/non-conservative – if a tracer is consumed or produced by biological or chemical processes, it must be regarded as non-conservative. Dissolved oxygen, nutrients, and inorganic carbon are non-conservative. Salinity and temperature are conservative.

Radioactive/stable – whether a tracer radioactively decays or not. Tritium, radiocarbon, and radium are radioactive. Oxygen, iron, and barium are not.

Clearly the classifications here are operational, depending on the application. For example, temperature and salinity are not truly passive tracers, and on long enough timescales (e.g. glacial–interglacial), all tracers must change with time.

The biogeochemical behavior of a tracer could be described by a combination of the above four characteristics, but it will not be, in general, possible to separate these processes by measuring the profile of a single tracer. The general differential equation for a passive, steady state, non-conservative, radioactive tracer in one dimension is:

$$\frac{\partial C}{\partial t} = 0 = K\frac{\partial^2 C}{\partial z^2} - w\frac{\partial C}{\partial z} + J - \lambda C \tag{13.1}$$

where z is the depth (assumed positive upward, $z = 0$ at the bottom and $z = z_m$ at the top), K is the vertical turbulent diffusivity (assumed to be constant with depth), w is the vertical velocity (also positive upward, and assumed to be constant), J is the *in situ* production rate (negative for consumption), and λ is the radioactive decay constant. Although we've called it a radioactive term it could represent any first-order (concentration-dependent) consumption/production process. This is the general form of the equation. For example, a conservative tracer would have $J = 0$, and a stable, non-conservative tracer would have $\lambda = 0$.

As a general note, because we are dealing here with steady state tracers ($\partial C/\partial t = 0$), the above partial differential equation becomes an ordinary differential equation, and hence can be solved with the standard ODE methods we explored earlier (see Chapter 8). However, for pedagogical and historical purposes, our approach will be to seek analytical solutions.

The objective is to determine the rates of upwelling and diffusion in the deep Pacific within this advective–diffusive subrange, and to determine the rate of oxygen consumption/nutrient remineralization. If you look at the equation, you will see that there are three basic unknowns, K, w, and J. The only constant really "known" in the equation is the λ, so that we will have to use some radioactive tracer (radiocarbon is the one) to somehow determine the other terms. The problem is that radiocarbon is non-conservative (it has a non-zero J), so we really need to develop three equations for the three unknowns (K, w, and J). Thus we need to use three different tracers.

Our general approach, as taken in the Craig paper, will be to use T and S to solve for the ratio of K/w, use oxygen to solve for the ratio of J/w (and then to infer the carbon J), and then use the radiocarbon distribution to solve for w. We can then work backwards to get J and K.

13.3 Stable conservative tracers: solving for K/w

The first step is to constrain K/w. First, let's think about what is happening in the deep Pacific. On the global scale, we have waters that are "formed" by density modification in the polar/subpolar oceans. This water is made very dense by extraction of heat (colder water is denser than warmer) and removal of fresh water by evaporation and formation of sea ice (salty water is denser), and sinks into the abyss. North Atlantic Deep Water (NADW) is formed predominantly in the Norwegian and Greenland Seas, and overflows the sills between Iceland and Greenland and between Iceland and Scotland. This water flows southward, being modified by mixing with low-salinity Labrador Sea Water and salty Mediterranean Outflow Water. NADW flows all the way to the Antarctic Circumpolar Current, where it is further modified and combined with Weddell Sea Bottom Water and other bottom waters from other Antarctic regions. It enters the deep Pacific along the western boundary, and ultimately upwells through the water column to return at shallower depths. About 20–40 $\times \cdot 10^6 \, \text{m}^3 \, \text{s}^{-1}$ of water enters the Pacific, and if the area of the Pacific is about 2–3 $\times 10^{14} \, \text{m}^2$, then the upwelling rate must be of order $10^{-7} \, \text{m s}^{-1}$ (order $3 \, \text{m y}^{-1}$). Not very fast, but because the deep Pacific is so old, even this very slow rate is significant.

We begin by writing the one-dimensional, steady state advection–diffusion equation for a stable, conservative tracer:

$$0 = K \frac{\partial^2 C}{\partial z^2} - w \frac{\partial C}{\partial z} \tag{13.2}$$

This one you can solve in your head! In fact the solution should be just a constant plus an exponential term; it is just

$$C = a_1 + a_2 e^{z/z^*} \tag{13.3}$$

where a_1 and a_2 are constants of integration, and z^* is the scale height defined by:

$$z^* = \frac{K}{w} \tag{13.4}$$

You can convince yourself that C in Equation (13.3) is a solution by substituting its expanded form into Equation (13.2). You can also demonstrate that z^* is indeed a length by dimensional analysis: in mutually consistent units you might have K expressed in $\text{m}^2 \, \text{s}^{-1}$ and w in m s^{-1} so the dimensions reduce to m. As defined in Equation (13.4), z^* is a measure of the relative strengths of mixing and advection. The bigger z^* is, the more important diffusion is, and the flatter the exponential curve is – that is, the more it looks like a straight line. The smaller z^* is, the more "bowed" the curve becomes.

Now what are a_1 and a_2? Mathematically, they arise from the solution, and must be set so that the function C matches the observed concentrations at either end of the subrange. That is, they arise from the requirement of satisfying the *boundary conditions* of the problem. Put another way, the functional form of C (i.e. the fact that it is an exponential) comes from the structure of the differential equation. The constants that go into the function come from the boundary condition. This makes sense when you think about it; "the equation represents the theory and the constants represent the experiment".

Now we have the fact that $C = a_1 + a_2$ when $z = 0$ (the bottom of the subrange), and we have that $C = a_1 + a_2 e^{z_m/z^*}$ when $z = z_m$ (remember that z_m is the height of the top of the subrange). Thus we can solve for a_1 and a_2 in terms of C_0 (the concentration at the bottom) and C_m (the concentration at the top). This gives the form that Craig reports:

$$C = C_0 + (C_m - C_0) f(z) \tag{13.5}$$

where

$$f(z) \equiv \frac{e^{z/z^*} - 1}{e^{z_m/z^*} - 1} \tag{13.6}$$

which is not too bad. We leave it as an exercise for the reader to see how that form arose from the previous constraint. (Give it a try, it isn't hard at all!)

Let's get a feeling for this business. If the subrange is purely diffusive (i.e. w is 0), then the z^* becomes infinite, and the exponentials tend to 1. It is a little hard to see with Craig's function $f(z)$ since you have a funny ratio: as z^* tends to infinity, the exponents become small, so we can expand the exponentials as infinite Maclaurin series, so that we have for large z^*:

$$f(z) = \frac{1 + \dfrac{z}{z^*} + \dfrac{1}{2}\left(\dfrac{z}{z^*}\right)^2 + \cdots - 1}{1 + \dfrac{z_m}{z^*} + \dfrac{1}{2}\left(\dfrac{z_m}{z^*}\right)^2 + \cdots - 1} \Rightarrow \frac{z}{z_m} \tag{13.7}$$

which degenerates the solution to a linear equation. Another way of seeing this is to look at the original equation, and you'll see that setting w to zero gives you:

$$K\frac{\partial^2 C}{\partial z^2} = 0 \tag{13.8}$$

whose solution is similar to Equation (13.3):

$$C = a_1 + a_2 z \tag{13.9}$$

a straight line. Oh yes, and two constants again: no surprise. You can then solve for the constants in the same way you did for the advective–diffusive case.

Let's look at actual temperature and salinity profiles in Fig. 13.1, taken from the World Ocean Circulation Experiment (WOCE) line P6 at $32°$ S, $160°$ W in the South Pacific. Notice the overall upward bowing of the deep temperature profile below the warm surface layers, suggestive of abyssal upwelling. This would be expected from the fact that the deep Pacific is a more-or-less enclosed basin to the north, and that inflowing bottom water from the Antarctic must somehow upwell in the interior in order ultimately to leave the basin. There is also evidence of this upwelling in the salinity profile, but the low salinity core of Antarctic Intermediate Water at about 1000 m depth causes a positive downward salinity gradient and a more subtle curvature.

Something can be learned from relating two stable conservative tracers. You can demonstrate using Equation (13.5) that two stable conservative tracers must be linear functions of one another over an advection diffusion subrange (an exercise for the reader – see Problem 13.1). So looking at the T–S relationship, we should look for a linear region

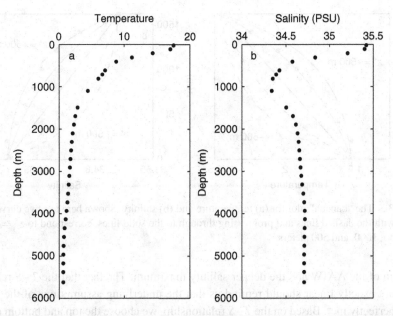

Figure 13.1 (a) Temperature and (b) salinity profiles from WOCE line P6 at 32°S and 160°W. The assumption is that the water mass lying between the salinity minimum of Antarctic Intermediate Water at 1000 m depth and the bottom water is maintained by a vertical advective–diffusive balance.

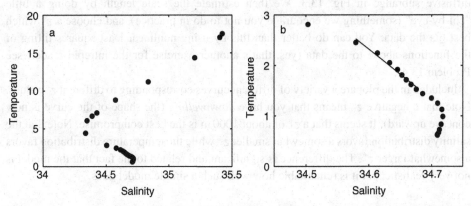

Figure 13.2 The $T-S$ diagram of the profiles shown in Fig. 13.1, (a) for the entire depth range and (b) a close-up showing the so-called "linear subrange" (the thin line).

between the salinity minimum of the AAIW at about 1000 m and the incoming AABW at about 3500 m.

The $T-S$ relationship on the right-hand side of Fig. 13.2 is a blow-up of the full $T-S$ diagram on the left, and shows an approximately "linear" region between the salinity

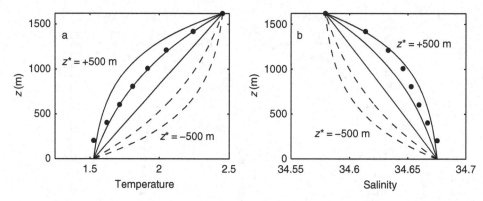

Figure 13.3 The "banana" plot for (a) temperature and (b) salinity. Shown here are five curves that, starting with the dashed lines and progressing through to the solid lines, correspond to $z^* = -500$, -1000, ∞, 1000, and 500 meters.

minimum of the AAIW and the deeper salinity maximum. The fact that the T–S relationship is not exactly linear should remind us that the underlying assumptions of the model are not perfectly met. Based on the T–S relationship, we choose the top and bottom depths of the subrange to be 1690 and 3350 m respectively. Remember that by our convention, $z = 0$ m at 3350 m depth (the bottom of the advective–diffusive subrange), and $z = 1660$ m at 1690 m depth (the top of the subrange).

We now plot T and S as a function of depth referenced to the bottom of the advective–diffusive subrange in Fig. 13.3. We then estimate the scale length by doing a little "chi-by-eye" (something we encourage you not to do in practice), and choose a z^* which best fits the data. You can do better than this by using nonlinear least squares fitting of the functions above to the data (yes, that's another exercise for the intrepid reader; see Problem 13.4).

Included in the plot are a variety of different curves corresponding to different z^* values. Note that a negative z^* means that you have *downwelling* (the shape of the curve is now concave upward). It seems that a z^* of about 1000 m is the best compromise. Note that the salinity distribution favors a somewhat smaller z^* while the temperature distribution favors a somewhat larger z^*. The difference is significant and related to the fact that the model is not very realistic, but it is remarkable how well such a simple model does.

13.4 Stable non-conservative tracers: solving for J/w

The next step in our quest is to deal with non-conservative tracers. We have to do this to avoid Munk's first mistake (ignoring the non-conservative behavior of ^{14}C). In his 1969 paper, Craig used dissolved oxygen because at that time total carbon was not measured well enough to be useful. Having obtained J/w for oxygen, Craig used "Redfield stoichiometry" (the overall correlation between oxygen consumption and inorganic carbon production in the deep ocean) to compute the equivalent J/w for carbon. Here, however, we will directly model the total inorganic carbon because it is now measured well enough to do this. The general equation for a stable, non-conservative tracer is:

$$0 = K\frac{\partial^2 C}{\partial z^2} - w\frac{\partial C}{\partial z} + J \qquad (13.10)$$

whose general solution is:

$$C = a_1 + a_2 e^{z/z^*} - \frac{J}{w}z \qquad (13.11)$$

which exploits the relationship between homogeneous and inhomogeneous ODEs (see Section 8.3).

You can prove this by substituting the function back into the equation. The form of this equation is not difficult to understand. You have the first part being the conservative advective–diffusive functionality, and you've added to it a linear non-conservative portion. There are still only two constants to be determined by requiring that the function satisfies the two boundary conditions. Now we go through the process of matching up the boundary conditions, i.e. make two equations for $C(0) = C_0$ and $C(z_m) = C_m$. Notice that when we do this, the J/w part creeps into the constants, since they need to be adjusted to make ends meet in the presence of consumption/production. Then we can express the solution in terms of C_0 and C_m as:

$$C = C_0 + (C_m - C_0)f(z) + \frac{J}{w}\left[z - z_m f(z)\right] \qquad (13.12)$$

Comparing this to the stable-conservative equation, you see the added term on the RHS associated with consumption/production. It doesn't look exactly like the form we showed in the previous equation, but if you expand it out, it *does* reduce to the simpler form.

Now let's look at the example station. Figure 13.4 is a plot of the total dissolved inorganic carbon profile for the whole water column (left panel) as well as in this subrange (right panel). Along with the data (the dots) are a number of model curves, corresponding to different values of the *scaled production term* $J^* = J/w$. Bear in mind that the units of J^* will be concentration divided by length, and this will only work if you use *consistent* time units for both J and w. Notice that the effect of positive J^* is to enhance the curvature, while a negative J^* reduces it. We included negative values only for instruction purposes because *in situ* consumption of dissolved inorganic carbon does not occur in the abyss outside of some unusual chemosynthetic environments. Although the fit is less than perfect (remember our initial assumptions and reservations), it's clear that the data are consistent with some positive value of *in situ* production less than $0.1\ \mu\text{mol}\,\text{kg}^{-1}\,\text{m}^{-1}$. Applying our "chi-by-eye", we would argue for an optimal value of J^* to be approximately $+0.06\ \mu\text{mol}\,\text{kg}^{-1}\,\text{m}^{-1}$ (again, you could do much better with nonlinear least squares).

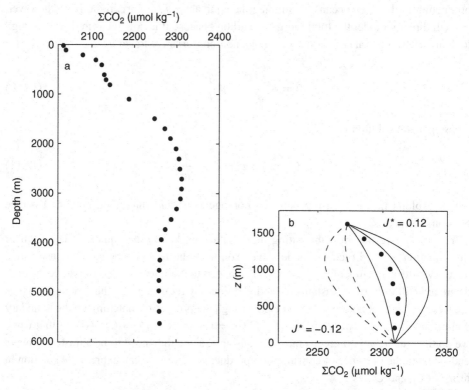

Figure 13.4 (a) The total dissolved inorganic carbon profile for the study station, and (b) the same plot over the advective–diffusive subrange. The model curves drawn correspond to $J^* = -0.12$ through $+0.12 \, \mu\text{mol} \, \text{kg}^{-1} \, \text{m}^{-1}$ (contour interval of $0.06 \, \mu\text{mol} \, \text{kg}^{-1}$) with the dashed line representing consumption and the solid line representing positive (or no) production.

13.5 Radioactive non-conservative tracers: solving for w

Now comes the challenging part. We need to solve for the more complicated case for radiocarbon, a non-conservative radioactive tracer:

$$0 = K \frac{\partial^2 C}{\partial z^2} - w \frac{\partial C}{\partial z} + J - \lambda C \tag{13.13}$$

Now you may be surprised to find that the solution to this equation is deceptively simple looking:

$$C = a_1 + a_2 e^{\alpha z} \tag{13.14}$$

which should look vaguely familiar: it resembles the stable, conservative tracer solution. However, the complexity of the solution is hidden in the α. If you substitute this solution back into Equation (13.13) you get the following:

$$Ka_2\alpha^2 e^{\alpha z} - wa_2\alpha e^{\alpha z} + J - \lambda\left(a_1 + a_2 e^{\alpha z}\right) = 0 \qquad (13.15)$$

For the above relation to be true over all z, we must separate the terms by how they depend on z, which means we must have both $J - \lambda a_1 = 0$ and

$$K\alpha^2 - w\alpha - \lambda = 0 \qquad (13.16)$$

which we arrived at by dividing the other $e^{\alpha z}$ terms by $a_2 e^{\alpha z}$. This is just a quadratic equation, so for $C = a_1 + a_2 e^{\alpha z}$ to be a solution of Equation (13.13) we must have

$$\alpha = \frac{w \pm \sqrt{w^2 + 4K\lambda}}{2K} \quad \text{and} \quad a_1 = J/\lambda \qquad (13.17)$$

where both roots of α are demonstrably real. Having two roots just means that the general solution includes terms containing both roots of α.

One benefit of finding an analytic solution is it gives the ability to explore solution behavior when you vary the intrinsic parameters. Remember that Equation (13.14) is the general solution for a tracer that is influenced by advection, diffusion, consumption, and decay. For example, if one were to set $J = 0$ and $\lambda = 0$ in Equation (13.17) then α reduces to a single root, i.e. $\alpha = w/K$, which is simply the solution to Equation (13.2) (try it).

The real complexity comes when we have to match the boundary conditions. You do this by solving for a_1 and a_2 in two equations containing the endpoint concentrations. This process becomes algebraically rather more complicated because you have competing effects of decay and consumption to deal with when matching things up. We won't go through the algebraic contortions to do this here, but rather present you with the result. When we express the solution in the boundary matched form, we have:

$$C = \frac{J}{\lambda} + \frac{\left(C_m - \dfrac{J}{\lambda}\right)e^{-(z_m - z)/2z^*}\sinh\left(\dfrac{Az}{2z^*}\right) + \left(C_0 - \dfrac{J}{\lambda}\right)e^{z/2z^*}\sinh\left[\dfrac{A(z_m - z)}{2z^*}\right]}{\sinh\left(\dfrac{Az_m}{2z^*}\right)} \qquad (13.18)$$

where the definition of the hyperbolic sine is given by:

$$\sinh(x) \equiv \frac{e^x - e^{-x}}{2} \qquad (13.19)$$

and where

$$A = \sqrt{1 + \frac{4\lambda z^*}{w}} = \sqrt{1 + 4\lambda^* z^*} \qquad (13.20)$$

because we've defined $\lambda^* = \lambda/w$. Note that unit-wise, λ is in inverse time units so that λ^* is an inverse length scale. Qualitatively, $1/\lambda^*$ can be thought of the distance over which radiocarbon can move before it decays significantly. We can also replace the J/λ

in Equation (13.18) with J^*/λ^* because the w in the numerator cancels out the w in the denominator.

Although a little intimidating, the above solution is actually functionally very simple (recall the form of Equation (13.14)). Notice also that the definition of A in Equation (13.20) is related to the square root term in α/w.

The next step has to do with avoiding the second mistake that Munk made in dealing with radiocarbon profiles in the Pacific. Munk did not model the actual ^{14}C concentration, but rather used $\Delta^{14}C$, which is close to being an isotope ratio anomaly, expressed in "permil" (parts per thousand) deviation from the standard atmospheric reference value. This is a common error in geochemical modeling that must be avoided at all costs. The temptation is to model the anomalies and not the actual concentrations, but nature advects, diffuses, consumes, and decays moles (or atoms, or concentration), not anomaly. Sometimes it's OK to deal in anomalies, for the equations sometimes work out, but in this and many other cases it doesn't. Although the radiocarbon isotope ratio anomaly apparently decreases with depth, the absolute activity may not, since the total CO_2 tends to increase with depth.

We can calculate the ^{14}C concentration from the $\Delta^{14}C$ and the total dissolved inorganic carbon (ΣCO_2) using

$$\left[^{14}C\right] = \left(1 + \frac{\Delta^{14}C}{1000}\right) R(\Sigma CO_2) \qquad (13.21)$$

where R is the reference atmospheric isotope ratio $^{14}C/^{12}C$ (a very small number, around 10^{-12}). We can safely work in terms of $\left[^{14}C\right]/R$ since the R would cancel out in the equations if we carried it through.

There is one final geochemical nuance; the carbon that is growing into the deep waters arises from the *in situ* degradation of organic material that is falling from the near-surface, sunlit zone of the ocean. The alert reader may note that in Fig. 13.5 the surface $\Delta^{14}C$ exceeds +100 permil. However, surface radiocarbon levels have only recently reached these elevated values due to the production and release of radiocarbon from atmospheric nuclear weapons testing in the 1950s and 1960s. Since the abyssal carbon signal has built up over a number of centuries, most of the carbon is "pre-bomb". Prior to the weapons tests, surface water $\Delta^{14}C$ was generally about 5% depleted relative to the atmospheric isotopic ratio (and what we define as the reference level for the isotope ratio anomaly). The reason it is depleted has to do with the fact that radiocarbon exchanges very slowly between the atmosphere and the ocean. The net result is that we have $J^*\left(^{14}CO_2\right) = 0.95J^*(CO_2) = 0.057\,\mu\text{mol}\,\text{kg}^{-1}\,\text{m}^{-1}$, not a large change considering the other uncertainties in our calculations, but we may as well be as precise as we can.

The lower right-hand plot in Fig. 13.5 compares our $\left[^{14}C\right]/R$ data (dots) to model results (the curves) computed from the boundary matched solution given in Equation (13.18) using the value of z^* obtained from the T and S profiles, J^* obtained from our fit to the ΣCO_2 data and corrected above, and a range of λ^* values from 3×10^{-5} to $12 \times 10^{-5}\,\text{m}^{-1}$. Again

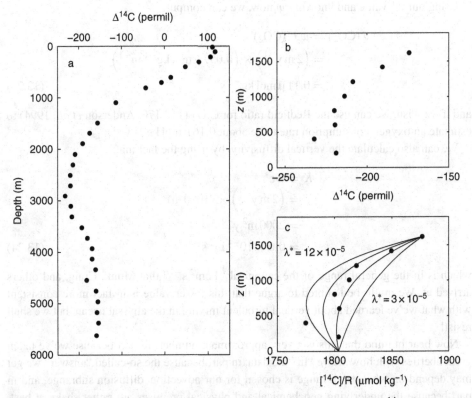

Figure 13.5 (a) The $\Delta^{14}C$ profile at the WOCE station, as well as (b) $\Delta^{14}C$ plotted over the advective–diffusive subrange (upper-right plot). (c) The concentration of ^{14}C divided by R, the reference isotope ratio of $^{14}C/^{12}C$ (dots) and model fits (curves) for various values of λ^* (see text).

using our reprehensible chi-by-eye strategy, it appears that $\lambda^* = 6 \times 10^{-5} \, \text{m}^{-1}$ is close to optimal.

13.6 Denouement: computing the other numbers

Now we can begin to calculate the numbers that we were actually after. Knowing the half-life of radiocarbon is 5730 years, we have $\lambda = 1.21 \times 10^{-4} \, \text{y}^{-1}$ so that we can compute w from:

$$w = \lambda/\lambda^* = \frac{1.21 \times 10^{-4} \, \text{y}^{-1}}{6 \times 10^{-5} \, \text{m}^{-1}} = 2 \, \text{m} \, \text{y}^{-1} \tag{13.22}$$

which is roughly similar to what we estimated from basic principles. Notice that we are being careful to include units in our calculations. This is an important habit to get into, as it prevents silly mistakes like using time units of years for one parameter and seconds for another.

Using our J^* value and knowing w now, we can compute:

$$J(CO_2) = wJ^*(CO_2)$$
$$= \left(2\,\mathrm{m\,y^{-1}}\right) \times \left(0.057\,\mu\mathrm{mol\,kg^{-1}\,m^{-1}}\right)$$
$$= 0.11\,\mu\mathrm{mol\,kg^{-1}\,y^{-1}} \tag{13.23}$$

and if we wish, we can use the Redfield ratio for $C:O_2$ (117:170, Anderson *et al.*, 1994) to estimate an oxygen consumption rate to be about $0.16\,\mu\mathrm{mol\,kg^{-1}\,y^{-1}}$.

We can also calculate the vertical diffusivity, by using the fact that:

$$K = wz^*$$
$$= \left(2\,\mathrm{m\,y^{-1}}\right) \times (1000\,\mathrm{m})$$
$$= 2000\,\mathrm{m^2\,y^{-1}}$$
$$= 0.6 \times 10^{-4}\,\mathrm{m^2\,s^{-1}} \tag{13.24}$$

which is in the general range of the canonical "$1\,\mathrm{cm^2\,s^{-1}}$" that Munk, Craig, and others arrived at. We might be tempted to argue that this lower value is in fact more consistent with what we've learned about vertical turbulent mixing in the abyssal ocean, but we shall resist!

Now bear in mind that this is a very approximate number; in part because we've taken great liberties with how we've "fit" the data, in part because the so-called "answer" we get may depend on what depth range is chosen for our advection–diffusion subrange, and in part because the underlying geochemical and physical premises are rather shaky at best. Even with this all taken into account, you will likely get different results from another profile taken elsewhere in the abyssal Pacific. Moreover, as we discussed in Chapter 11, one would expect K to vary spatially as a function of vertical stratification, velocity shear, and proximity to topographic roughness. Similarly, J will vary as a function of depth owing to vertical changes in substrate abundance, and horizontally owing to regional changes in biological productivity. Assuming some functional form for K and J would lead to a more complicated solution to the advection–diffusion equations, which could be solved either analytically or numerically, but added *degrees of freedom* would require additional tracer-based constraints to arrive at unique solutions. Accommodating this spatial variability would lead us to both higher-dimension models and eventually to inverse techniques, which we discuss in Chapter 18.

13.7 Problems

13.1. Using the solutions for two separate stable conservative (passive) tracers, demonstrate algebraically that one will be a linear function of the other. For one stable conservative tracer and one stable non-conservative tracer, algebraically show how the curvature relates to J when you plot the non-conservative tracer as a function of the conservative one. (Hint: use the boundary-matched solutions to the equations, e.g. Equation (13.5).)

13.2. Using the methodology in this chapter, derive the general and boundary matched solutions for a radioactive, conservative tracer.

13.3. Demonstrate algebraically that Equation (13.18) reduces to C_0 for $z = 0$ and C_m for $z = z_m$. Show that when you set $J = 0$, the equation reduces to that for a radioactive conservative tracer.

13.4. Using the station data contained in the file C14Station.mat, redo the analysis presented in this chapter using nonlinear regression techniques developed in previous chapters. The variables stored in this file have self-explanatory names. When you construct an appropriate cost function, weight the model–data differences by the data uncertainties (the sigma array), and include both temperature and salinity for estimating z^*. How do your final results differ if you exclude one or the other? There appears to be one errant point in the $\Delta^{14}C$ profile (the third from the bottom of the subrange). How do your estimates change when you exclude this point?

13.5. What happens to your results if you set $J = 0$ for ^{14}C?

13.6. Repeat the analysis in Problem 13.4 changing the advection–diffusion subrange from what was used in this chapter to the values used by Craig (1969). Bear in mind that you will need to interpolate your profiles to those depths. How do your results change?

14

One-dimensional models in sedimentary systems

> My own brain is to me the most unaccountable of machinery – always
> buzzing, humming, soaring roaring diving, and then buried in mud.
>
> *Virginia Woolf*

We treat one-dimensional models of sedimentary systems separately in this book because of the added complication that they contain two phases – solid material and pore waters – that not only can interact biogeochemically, exchanging chemicals, but also can move in relation to one another. In fact, with a reference system fixed at the sediment–water interface, the solid phase is actually moving owing to a combination of sedimentation (addition of material at the interface) and compaction. If you're thinking that this makes the construction of models a little more complicated, those are exactly our sediments!

We will be talking about the process of *diagenesis*, i.e. the sum total of all processes that bring about changes to sediments after they have been deposited on the seafloor. This includes everything from bioturbation through chemical transformation to compaction and pore water extrusion. The general topic of diagenesis extends to even longer timescale processes that include metamorphism and weathering of sedimentary rocks after uplift, but we will focus on *Early Diagenesis* (Berner, 1980), which encompasses changes that occur at or near the sedimentary surface or in the upper portion of the sedimentary column.

14.1 General theory

To begin with, we summarize terms that we will be using to distinguish among solids, pore fluids, and whole sediment:

ω = burial rate of solids

v = burial rate of pore fluids

C = mass of dissolved constituent per unit volume pore fluid

C_s = mass of solid constituent per unit mass of total solids

$\widehat{C}_d = \phi C$ mass of dissolved constituent per unit volume total sediment

$\widehat{C}_s = (1 - \phi)\bar{\rho}_s C_s$ mass of solid constituent per unit volume of total sediment

Here ϕ is porosity and $\bar{\rho}_s$ is the average sediment density.

14.1.1 Diagenesis in a sinking coordinate frame

It is important to keep in mind that in sedimentary diagenetic models your coordinate frame is apparently moving with respect to sedimentary particles. Consider the simplest case of depth profile concentration of some property C associated with sediment grains:

$$\widehat{C}_s = C(z, t) \tag{14.1}$$

where z represents depth (positive downward) and t represents time.

If we were to inject a layer of dyed particles in the sedimentary column and wait long enough, we would see the layer apparently move downward as more sediment is deposited on top. This raises the question as to which level we define $z = 0$, a fixed layer within the sediment column or the seawater–sediment interface? Choosing some sediment layer is analogous to deciding on a *Lagrangian* frame of reference (see Chapter 11) where the conservation equations are intrinsically simpler. But when faced with actual field situations your primary frame of reference is the sediment–water interface,[1] which is an Eulerian frame. To translate between these two choices we take advantage of the definition of the full (Lagrangian) derivative:

$$\left.\frac{dC}{dt}\right|_{\text{fixed layer}} = \left.\frac{\partial C}{\partial t}\right|_{\text{fixed depth}} + \left.\frac{\partial (\omega C)}{\partial z}\right|_{\text{fixed time}} \tag{14.2}$$

where for emphasis we have subscripted the derivatives (using the vertical bars) to remind you that the partial derivative with respect to time is evaluated at fixed depth and the partial derivative with respect to depth is done at fixed time. Also, ω is the rate of sediment burial (related to deposition rate) and could in general vary with depth.

We generally work in a fixed-depth coordinate system with the seawater–sediment interface as $z = 0$, positive downward. In this situation Equation (14.2) is best arranged as:

$$\frac{\partial C}{\partial t} = \frac{dC}{dt} - \frac{\partial (\omega C)}{\partial z} \tag{14.3}$$

where we have dropped the subscripting for simplicity. Now if we assume (for the moment) that ω is constant with time and depth, which is equivalent to ignoring effects like compaction, Equation (14.3) simplifies to

$$\frac{\partial C}{\partial t} = \frac{dC}{dt} - \omega \frac{\partial C}{\partial z} \tag{14.4}$$

Within this framework, we can examine two scenarios that cover most of the processes we are likely to be interested in: burial without diagenesis and burial with diagenesis. In the first case, with no diagenesis, we have $\frac{dC}{dt} = 0$, meaning that the concentration of a given layer does not change after it is buried; so we can rewrite Equation (14.4) in a simpler form:

$$\frac{\partial C}{\partial t} = -\omega \frac{\partial C}{\partial z} \tag{14.5}$$

---

[1] On sufficiently small scales, even this reference point is "fuzzy".

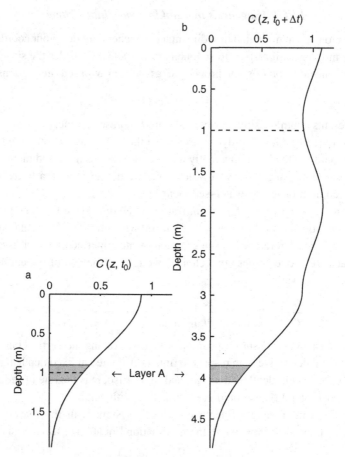

Figure 14.1 (a) At time t_0, for a fixed depth (say 1 meter), the property C has a different value than it does at (b) time $t_0 + \Delta t$, but the value for C at the descending layer A stays the same.

where we are describing the apparent downward motion of the layers of sediment (and their properties) due to deposition, and possible changes in the profile due to changes of sediment properties at the time of deposition. That is, sedimentary burial produces an apparent advection term when we fix our coordinate system to the sediment–water interface. An example of what you might see is displayed in Fig. 14.1 and shows a situation that typically applies to the solid phase of the sedimentary column.

Now suppose the concentration of C in the material delivered to the top of the sediment column remains historically constant but as it is buried, i.e. with the passage of time, the property C in the sediment grains changes. This gives rise to the concept of steady-state diagenesis:

$$\frac{\partial C}{\partial t} = 0 \tag{14.6}$$

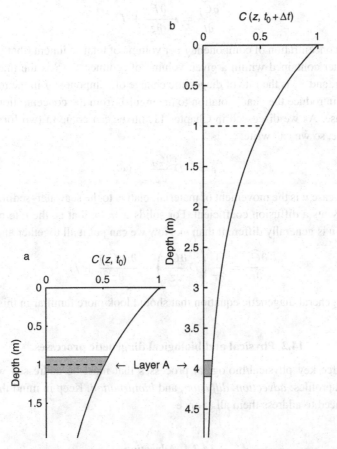

Figure 14.2 In this situation the value at a fixed depth (say 1 meter) remains constant and it is the value of layer A that changes from (a) time t_0 to (b) $t_0 + \Delta t$ owing to diagenetic changes in C.

which means that for any depth z in the sedimentary column, C remains constant and the profile looks the same no matter when you look at it. Thus Equation (14.4) becomes:

$$\frac{dC}{dt} = \omega \frac{\partial C}{\partial z} \tag{14.7}$$

Figure 14.2 shows an example of a case where diagenetic processes serve to remove material (C) with time, that is where dC/dt is negative. In this situation the value at a fixed depth remains constant and it is the value associated with a (sinking) layer that changes.

14.1.2 The general diagenetic equation

When we consider the mass balance of material in a box of sediment as simply the difference between what goes in and what comes out, possibly augmented/depleted by what is produced/consumed on the premises, we can write:

$$\frac{\partial \widehat{C}_j}{\partial t} = -\frac{\partial F_j}{\partial z} + J_j \tag{14.8}$$

where $\widehat{C}_j$ is concentration of component j per volume of total sediment (that is, solids plus any pore water contained within a given volume of sediment), F_j is the material flux of component j, and J_j is the rate of diagenetic change of component j in the total sediment volume. We introduce the "hat" notation to distinguish from the concentration in the solid or water phase. As we discussed in Chapter 11, fluxes can come in two forms, diffusive and advective, so we can write:

$$F_j = -D\frac{\partial \widehat{C}_j}{\partial z} + v\widehat{C}_j \tag{14.9}$$

where in this case v is the movement of material relative to the seawater–sediment interface and D represents a diffusion coefficient. For solids $v = \omega$, that is, the rate of burial. For pore waters, v is generally different than ω. Now we can put it all together and obtain:

$$\frac{\partial \widehat{C}_j}{\partial t} = \frac{\partial}{\partial z}\left(D\frac{\partial \widehat{C}_j}{\partial z}\right) - \frac{\partial(v\widehat{C}_j)}{\partial z} + J_j \tag{14.10}$$

which is the general diagenetic equation that should look more familiar in this form.

14.2 Physical and biological diagenetic processes

There are three key physical/biological processes that must be addressed when modeling sediment profiles: *advection*, *diffusion*, and *bioturbation*. Keep in mind that you don't necessarily need to address them all at once.

14.2.1 Advection

Advection is the movement of solids or pore water with respect to the frame of reference, i.e. the seawater–sediment interface. *Compaction* is the loss of water from a sediment layer due to compression caused by the weight of the overlying sediments. These two concepts are linked to the concept of *porosity*, ϕ, which is defined as the ratio of the volume of interconnected water to the volume of total sediment. This is not quite the same thing as *total porosity*, but for recently buried sediments it is sufficiently close. Total porosity includes those isolated pockets of water in the sediment through which water cannot flow. In the types of sediments that we will be talking about here these extra voids are very small and can be ignored.

Another concept that needs to be defined is s the *deposition rate*, the thickness of a sediment layer divided by the deposition time interval. Burial rate equals the surface deposition rate only if the sedimentation rate is constant with time and layer compaction is minimal. If there has been compaction, we can write:

$$\omega(z, t) = s(0, t) + \int_0^z \frac{1}{s(z, t)}\frac{ds(z, t)}{dt}dz \tag{14.11}$$

where ds/dt is the change in layer thickness after deposition. At the surface the two can be set equal to each other, $s(0, t) = \omega(0, t)$, but it is still difficult to evaluate (14.11) because we would need to know ds/dt over time. So we generally assume both constant sedimentation rate and steady state compaction, so that finding the rate of deposition becomes a matter of correcting ω for the effects of steady state compaction. This can be written as:

$$S = \bar{\rho}_s \omega (1 - \phi) \tag{14.12}$$

where S is the sedimentation rate in mass per unit area per unit time and $\bar{\rho}_s$ is the average density of the solids in the sediment.

14.2.2 Diffusion

In pore water, the process of molecular diffusion operates, but with some additional complications due to the presence of the solids. Let's consider how "having to go around" the sediment solids can affect Fick's laws. In an aqueous environment we write:

$$F_j = -D_j \frac{\partial C_j}{\partial z} \tag{14.13}$$

$$\frac{\partial C_j}{\partial t} = \frac{\partial}{\partial z} \left(D_j \frac{\partial C_j}{\partial z} \right) \tag{14.14}$$

for Fick's first and second laws. Here F_j is the diffusive flux of component j in mass per unit area per unit time and D_j is the aqueous molecular diffusivity.

In sediments, corrections must be made not only for geometry but also for electrical effects, so we have:

$$F_j = -D_j \left(\frac{\partial C_j}{\partial z} + \frac{Z_j C_j \widehat{F} \frac{\partial E}{\partial z}}{RT} \right) \tag{14.15}$$

where D_j is the aqueous molecular diffusion coefficient of component j, Z_j is the valence of component j (can be positive or negative), T is the absolute temperature, R is the gas constant, $\partial E/\partial z$ is the charge gradient set up through the movement of ions, and $\widehat{F}$ is Faraday's constant (we put a "hat" over it to distinguish it from the material flux F_j). To correct for electrical effects we employ one more constraint, *electroneutrality*:

$$\sum_j Z_j F_j = 0 \tag{14.16}$$

Electrical effects may be appreciable, but more important are the effects of *tortuosity*. This is a reflection of the tortuous path ions must follow when diffusing through the pore fluids, owing to the presence of the solids. The ions are not free to diffuse in any direction; they collide with the particles and we must compensate for this. Tortuosity is defined as:

$$\theta = \frac{d\ell}{dz} \tag{14.17}$$

where $d\ell$ represents the actual path length the ion must travel over a sediment depth of dz. Since $d\ell \geq dz$ we must have $\theta \geq 1$. Because diffusion is modeled as a *random walk* process, the diffusion coefficient in sediments, corrected for tortuosity, can be expressed as:

$$D_s = \frac{D}{\theta^2} \tag{14.18}$$

Here D_s refers to the *whole sediment* diffusion coefficient. Unfortunately θ is difficult to evaluate or measure, so we often use the following relationships:

$$\theta^2 = \phi \mathcal{F} \tag{14.19}$$

$$\mathcal{F} = \phi^{-n} \tag{14.20}$$

where $\mathcal{F}$ is the *formation factor*, given by the ratio of the whole sediment electrical resistivity to the pore water electrical resistivity. The underlying rationale is that electrical transport is limited by molecular diffusion in the sediment, and thus resistivity must be inversely proportional to it. If n equals 2 in the above formulation it is called *Archie's law*, approximating sands and sandstones well. Using D_s in Fick's laws they become:

$$F_s = -\phi D_s \frac{\partial C}{\partial z} \tag{14.21}$$

$$\frac{\partial C}{\partial t} = \frac{1}{\phi} \frac{\partial}{\partial z} \left(\phi D_s \frac{\partial C}{\partial z} \right) \tag{14.22}$$

It should be noted here that so far we have been talking about the *pore water* concentration C; as we shall see, these equations can be made to look more like the ones we're used to seeing by recasting them in terms of the whole sediment.

14.2.3 Bioturbation

Bioturbation refers to mixing effects associated with the activities of benthic organisms in oxic sediments. Although largely concentrated at the sediment–water interface, bioturbation can penetrate tens of centimeters into the sediment column. As the creatures burrow, sediment grains are displaced, effectively mixing the solid phase. Sometimes their influence includes changes in pore-fluid movement resulting from structural changes in the sedimentary "matrix", and from *bioirrigation*, where these creatures actively flush their burrows with seawater.

Because each organism is unique in its impact on the sediment environment, modeling of bioturbation can be difficult. Often biomixing processes are lumped together and described as a random mixing process, much like diffusion. For solids, the bioflux is:

$$F_B = -D_b \frac{\partial \widehat{C}_s}{\partial z} \tag{14.23}$$

where $\widehat{C}_s$ equals the mass of solid per volume of total sediment and D_b is the biodiffusional coefficient. For pore waters, the resultant bioflux is:

$$F_{BI} = -D_b \frac{\partial(\phi C)}{\partial z} - \phi D_I \frac{\partial C}{\partial z} \tag{14.24}$$

where in this case the subscript I stands for irrigation. It should be mentioned that bioturbation may lead to fundamentally *non-Fickian* transport of materials, so some caution must be exercised in interpreting and modeling these processes.

14.2.4 Non-Cartesian coordinate systems

Although you are used to dealing with *Cartesian* (i.e. x,y,z) coordinate systems, there are may be reasons to consider these equations in other forms. For example, concretion growth is better handled in spherical coordinates, since precipitation occurs around nucleation sites in a spherically isotropic fashion. In *spherically symmetric geometries* Equation (14.14) (Fick's second law) becomes:

$$\frac{\partial C_j}{\partial t} = \frac{1}{r^2}\frac{\partial}{\partial r}\left(r^2 D_j \frac{\partial C_j}{\partial r}\right) \tag{14.25}$$

where r is the distance from the center of the concretion (or sediment grain). This is a simplified version of the full three-dimensional spherical coordinate expansion that we'll be discussing in Chapter 17.

Another situation that warrants a non-Cartesian coordinate system is where you are modeling the detailed effects of burrowing animals in the sediment. In this instance, you might expect the disturbance to radiate outward from the burrow in a radially symmetric fashion. Thus Equation (14.14) for *cylindrically symmetric geometries* becomes:

$$\frac{\partial C_j}{\partial t} = \frac{1}{r}\frac{\partial}{\partial r}\left(r D_j \frac{\partial C_j}{\partial r}\right) \tag{14.26}$$

where r is the distance from the axis of the cylinder.

14.3 Chemical diagenetic processes

In this section we will discuss some of what goes into the J_j term in Equation (14.10). You will recall that this term represented a number of source and sink processes that may be occurring during diagenesis. These can be broken down into five broad categories of biogeochemical reactions: equilibrium, radioactive decay, microbial, precipitation, and authigenic processes. We will consider each in its turn.

14.3.1 Chemical equilibrium

If we assume that we have a simple chemical reaction of the form:

$$AB(\text{solid}) \rightleftharpoons A + B \tag{14.27}$$

then we can write the diagenetic equation for A as:

$$\frac{\partial(\phi C_A)}{\partial t} = \frac{\partial}{\partial z}\left[D_b\frac{\partial(\phi C_A)}{\partial z} + \phi(D_s + D_I)\frac{\partial C_A}{\partial z}\right] - \frac{\partial(\phi v C_A)}{\partial z} + \phi J_{A_{sol}} + \phi J'_A \quad (14.28)$$

and similarly for B, keeping in mind that when we write C_B we are referring to the concentration of B but when we write D_b we are referring to "biodiffusion". In these equations $J_{A_{sol}}$ refers to the rate of dissolution/precipitation of AB and J'_A refers to all other, slow, non-equilibrium reactions affecting A and B. Also:

$$J_{A_{sol}} = J_{B_{sol}} \quad (14.29)$$

and we can write the thermodynamic equilibrium constant as:

$$K = C_A C_B \quad (14.30)$$

ignoring, for the time being, the activity coefficients. We can now difference the diagenetic equations for A and B, arriving at:

$$\frac{\partial\left[\phi\left(C_A - \frac{K}{C_A}\right)\right]}{\partial t} = \frac{\partial}{\partial z}\left[D_b\frac{\partial\left[\phi\left(C_A - \frac{K}{C_A}\right)\right]}{\partial z} + \phi(D_s + D_I)\frac{\partial\left(C_A - \frac{K}{C_A}\right)}{\partial z}\right]$$
$$- \frac{\partial}{\partial z}\left[\phi v\left(C_A - \frac{K}{C_A}\right)\right] + \phi\left(J'_A - J'_B\right) \quad (14.31)$$

Although this equation looks complicated, it is structurally similar to the kinds of equations we have dealt with before; there are just a few more factors.

14.3.2 Radioactive decay

Like chemical reactions in the sediment, radioactive decay can also change the concentrations of components within the sediment column. Considering the simple first-order decay equation, the conservation equation for radioactivity on sediment particles now becomes:

$$\frac{\partial\left[(1-\phi)\bar{\rho}_s C_s\right]}{\partial t} = \frac{\partial}{\partial z}\left[D_b\frac{\partial\left[(1-\phi)\bar{\rho}_s C_s\right]}{\partial z}\right] - \frac{\partial\left[(1-\phi)\bar{\rho}_s\omega C_s\right]}{\partial z}$$
$$- (1-\phi)\bar{\rho}_s\lambda C_s + (1-\phi)\bar{\rho}_s J'_s$$
$$(14.32)$$

where λ is the radioactive decay constant and ω represents the burial rate of the solids.

14.3.3 Microbial processes

Typically, in sediments, we will be looking at *catabolic* processes, i.e. processes that are involved in the decomposition of organic matter, as opposed to *anabolic* (assimilation) processes. Because of the incredibly complex biological activities taking place, we generally simplify our modeling efforts by making an analogy to familiar, well characterized laboratory enzyme reaction studies. One of the most basic of such reaction models is the *Michaelis–Menton* kinetic equation:

$$\frac{dC}{dt} = -\frac{R_{\max}C}{K_M + C} \tag{14.33}$$

where C is the concentration of the metabolite, $R_{\max}$ is the maximum rate of reaction, and K_M is known as the Michaelis constant. Then by analogy we can write the fundamental equation that describes Robert Berner's (1980) well-known "multi-G" model:

$$\frac{dG_T}{dt} = -\sum_j \left(\frac{R_{\max}G_j}{K_{M_j} + G_j} \right) \tag{14.34}$$

where G_j is the concentration of complex organic matter of type j and $G_T = \sum_j G_j$ represents the total decomposable organic matter in the sediment. Of course, this can be substituted into one of those J' terms. In Section 14.4 we'll make a very simple demonstration of this and apply it to the anoxic sediments from the Long Island Sound FOAM (Friends Of Anoxic Mud) site.

14.3.4 Precipitation

The mathematics that lie behind precipitation diagenetics are based largely on what we've learned about the energetics of crystal formation from aqueous solutions. Take, for example, the free energy of formation of a crystal, ΔG, which can be broken into two components:

$$\Delta G = \Delta G_{\text{bulk}} + \Delta G_{\text{interface}} \tag{14.35}$$

where ΔG is not to be confused with Berner's multi-G in the last subsection. The terms of Equation (14.35) can be written as:

$$\Delta G_{\text{bulk}} = -nk_B T \ln(\Omega) \tag{14.36}$$

where n refers to the number of atoms or ions precipitated, Ω is the ratio of initial ion activity product of supersaturation to equilibrium, k_B is the Boltzmann constant, and T is the absolute temperature. The second term of Equation (14.35) can be written as:

$$\Delta G_{\text{interface}} = \sigma A \tag{14.37}$$

where σ is the specific interfacial free energy between crystal and solution and A is the surface area of the crystal.

There are several processes to consider when considering the kinetics of precipitation, but there are two (at least) that are fundamentally important: nucleation and growth rate. Nucleation requires an increase in free energy before the nucleation process can happen. When considering growth rate it is important to identify whether the growth is controlled by transport or surface reaction rates. In transport-controlled growth, the rate is determined by how quickly ions can get to the active surface. In an idealized model one might approximate the surface by assuming spherical grains and then use the spherical coordinate frame form of Fick's second law (Equation (14.25), expanded by differentiation by parts):

$$\frac{\partial C}{\partial t} = D_s \left[\frac{\partial^2 C}{\partial r^2} + \frac{2}{r}\frac{\partial C}{\partial r} \right] \tag{14.38}$$

Surface-reaction-controlled growth occurs when the rate limiting step is the formation of a flat 2D crystalline layer one ion thick on a smooth crystal surface. Dissolution can, in many ways, be considered the reverse of precipitation.

14.3.5 Authigenesis

An authigenic mineral is one that formed in the sediment after burial, i.e. **via** diagenesis. There are a number of authigenic processes: for example *cementation* is a process where the mineral forms or crystallizes in the pore space of the sediments. Another process is *replacement* where the mineral replaces another that has dissolved. Finally, if the mineral dissolves in one part of the sediment and migrates elsewhere to reprecipitate, this is known as *diagenetic redistribution*.

14.4 A modeling example: CH$_4$ at the FOAM site

Some fraction of organic material produced near the sea surface by photosynthesis ultimately reaches the sediment surface. As this carbon is buried it is oxidized by bacteria. Depending on circumstances, all the available oxygen dissolved in the pore waters may be used up, so that at some depth (sometimes even at the sediment–water interface) the sediments become anoxic. Being rather clever beasts, bacteria then turn to other sources of oxidative potential, using a progressive series of organic carbon oxidation pathways with decreasing thermodynamic energy yields ranging from nitrate reduction, through sulfate reduction, and ending up with fermentation (the last of which produces methane). Figuring out what is happening and where is a big challenge, in part because oxidants and oxidative products will subsequently intermingle via pore-fluid advection and diffusion. An application of sedimentary modeling is to disentangle where these processes occur in the sediment column, at what rates, and in what combination. In this final section we will develop an example of 1D modeling in anoxic sediments found in the Long Island Sound. This example is taken from a paper by Martens and Berner (1977). Of interest to these investigators, among many things, was the presence or absence of SO_4^{-2} where CH$_4$ could be found in the sediment column.

14.4.1 The setting

The FOAM (Friends Of Anoxic Mud) site is located near-shore of Connecticut where the interstitial waters exhibit incomplete to complete sulfate reduction in the upper 10 or so centimeters. The sediment cores were collected by gravity coring and the pore fluids were squeezed out by an inert gas-filtration system. The pore waters were analyzed for gases and sulfate.

Of particular interest is the small overlap (in the sediment column) of CH$_4$ and SO$_4^{-2}$ in the muds they collected. At that time, evidence had accumulated suggesting that methane did not form in sediments until all sulfate had been removed. In their sediment cores they found a small region of overlap with non-zero concentrations of both constituents (see Fig. 14.3). In the absence of evidence for actual bubble migration of methane into the overlying sediments Martens and Berner formulated the following hypotheses (and we quote):

1. Methane is produced at roughly the same rate throughout the sediment column but is consumed by sulfate-reducing bacteria and associated microorganisms where sulfate is present.
2. Methane is produced only in the absence of bacterial sulfate reduction and the limited coexistence of the two is due to interdiffusion.
3. Methane is produced only in the absence of bacterial sulfate reduction but on diffusing upward it is consumed by sulfate-reducing bacteria and associated microorganisms.
4. Methane is produced to a limited extent in the presence of sulfate-reducing bacteria but is not utilized by them. The production is limited to sulfate-free microenvironments.

14.4.2 The equations

Given the measured profiles, hypothesis 3 seems plausible, and in order to test this hypothesis in a 1D model we need a set of equations that describe the diffusion and microbial consumption of CH$_4$ in the sediment column. For our purposes here we'll assume that CH$_4$ production occurs deeper in the column and treat it as a bottom boundary condition. As CH$_4$ diffuses upward, it will be consumed. As methane and sulfate diffuse into different oxidative regimes, they will be subject to consumption. The following reaction was suggested to account for this:

$$CH_4 + SO_4^{-2} \rightarrow HS^- + HCO_3^- + H_2O \tag{14.39}$$

which is thermodynamically favorable under the conditions at this location.

Given the lack of bioturbation in the sediment core and the a priori knowledge of the sedimentation rate ($0.3 \, \text{cm} \, \text{y}^{-1}$) and whole sediment methane diffusion coefficient ($2 \times 10^{-6} \, \text{cm}^2 \, \text{s}^{-1}$) in this area from other work, they constructed a steady state diagenetic equation to describe methane in the sediment column:

$$\frac{\partial C}{\partial t} = D_s \frac{\partial^2 C}{\partial z^2} - \omega \frac{\partial C}{\partial z} - k_1 C = 0 \tag{14.40}$$

where k_1 is the first-order methane consumption rate ($0.24 \, \text{y}^{-1}$) estimated from "jar" experiments performed by other investigators and consistent with their own incubations. The idea here was if k_1 could "bend" the methane curve downward enough to agree with the data, then the sulfate-reducing bacteria were consuming the methane that diffused upward into the sulfate reduction zone. This equation must be solved using an upper boundary condition of zero methane concentration ($C_{upper} = 0$) and a lower boundary condition of methane at saturation concentration ($C_{lower} = 2.6\,\text{mM}$). While this equation has an analytical solution, it is also an excellent place to flex our numerical muscles and apply a new implicit algorithm.

14.4.3 A finite difference formulation

We want to fit Equation (14.40) to the methane profile shown in Fig. 14.3. We will apply the first upwind differencing (FUDM – see Chapter 12) scheme. The implementation is especially simple in this example because D_s and ω are assumed constant (as is ϕ), so the discretized form of Equation (14.40) becomes:

$$\frac{C_i^{n+1} - C_i^n}{\Delta t} = -\frac{\omega}{\Delta z}(C_i^n - C_{i-1}^n) + \frac{D_s}{\Delta z^2}(C_{i+1}^n - 2C_i^n + C_{i-1}^n) - k_1 C_i^n = 0 \quad (14.41)$$

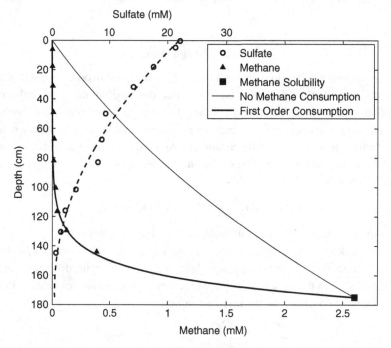

Figure 14.3 Methane and sulfate concentrations vs. depth for a core collected at the FOAM site reported by Martens and Berner (1977). Two solid lines are shown: no consumption of methane (the light solid line) and consumption via first-order kinetics (heavier solid line).

where we remind you that n refers to the time step and i is the spatial index. Notice that unlike the FUDM formulation in Chapter 12 we retain the Δt in the denominator of the left-hand side, because we will use the fact that the solution is steady state, so the left-hand side (i.e. the time derivative) will be zero. For steady state, we can rewrite Equation (14.41) in a simpler form, in what is referred to as an *interior point equation*:

$$w_- C_{i-1} + w_0 C_i + w_+ C_{i+1} = 0 \qquad (14.42)$$

where we have dropped the time index (n) as no longer necessary, and where the weights are defined respectively as:

$$w_- = \frac{D_s}{\Delta z^2} + \frac{\omega}{\Delta z}$$

$$w_0 = -\left(2\frac{D_s}{\Delta z^2} + \frac{\omega}{\Delta z} + k_1\right) \qquad (14.43)$$

$$w_+ = \frac{D_s}{\Delta z^2}$$

Note that the weights no longer sum to one because we have folded in a chemical loss term k_1.

By calling Equation (14.42) an interior point equation, we are emphasizing that the concentration C at any interior point in the model domain depends in a systematic way on only the concentrations at neighboring points. Imposing restrictions or boundary conditions on the grid points at the ends of the model domain combines with this relation to allow us to compute all the interior values throughout the domain. The precise nature of the imposed boundary conditions (i.e. whether they are fixed concentrations, defined fluxes, or some combination) depends on the nature of the model equations and ultimately on the physical configuration of the system being studied. Not all choices may be valid.

We can think of the steady-state finite difference solution as being a set of N discrete layers in the sediment column between the two boundaries (the surface and the bottom of the domain), which in turn can be described as a set of N simultaneous equations shown below:

$$\begin{pmatrix} w_0^{UBC} & w_+ & & & & \\ w_- & w_0 & w_+ & & & \\ & w_- & w_0 & w_+ & & \\ & & & \ddots & & \\ & & & w_- & w_0 & w_+ \\ & & & & w_- & w_0^{LBC} \end{pmatrix} \begin{pmatrix} C_1 \\ C_2 \\ C_3 \\ \vdots \\ C_{N-1} \\ C_N \end{pmatrix} = \begin{pmatrix} 0 \\ 0 \\ 0 \\ \vdots \\ 0 \\ 0 \end{pmatrix} \qquad (14.44)$$

where all of the elements in the design matrix that are not on the primary or the adjacent diagonals are zero. The matrix can thus be described as a *tridiagonal matrix*.

So if in this example we specify the top-most and bottom-most concentrations (our *Dirichlet* or fixed concentration boundary conditions) $C_1 = C_{\text{upper}}$ and $C_N = C_{\text{lower}}$, and know the parameters ω, D, and k_1, we can solve this set of equations to compute the entire curve. However, we may not know one or more of the parameters a priori, as we will likely want to determine them by matching model curves to data points. Since this will involve a nonlinear regression, we would need to embed the model calculation within an iterative solver (see Chapter 3). This raises the issue that even using sparse system solving methods, this can be a time-consuming computation. For example, inverting an $N \times N$ matrix requires approximately N^3 divide/multiply operations. Fortunately, the structure of the problem shown in Equation (14.44) lends itself to a very efficient algorithmic solution (one that scales as N) appropriately called the *tridiagonal algorithm*.

14.4.4 Tridiagonal algorithms

The tridiagonal algorithm was developed for a more general problem and is more commonly referred to as the *Thomas algorithm*. Basically, it is an optimized *Gaussian elimination* procedure.[2] Let's consider the more general problem where, for each of the N grid points, we have:

$$\alpha_i C_{i-1} + \beta_i C_i + \gamma_i C_{i+1} = \delta_i \qquad (14.45)$$

where α, β, γ, and δ could have either positive or negative values. This gives a system of equations in matrix form:

$$\begin{pmatrix} \beta_1 & \gamma_1 & & & & \\ \alpha_2 & \beta_2 & \gamma_2 & & & \\ & \alpha_3 & \beta_3 & \gamma_3 & & \\ & & & \ddots & & \\ & & & \alpha_{N-1} & \beta_{N-1} & \gamma_{N-1} \\ & & & & \alpha_N & \beta_N \end{pmatrix} \begin{pmatrix} C_1 \\ C_2 \\ C_3 \\ \vdots \\ C_{N-1} \\ C_N \end{pmatrix} = \begin{pmatrix} \delta_1 \\ \delta_2 \\ \delta_3 \\ \vdots \\ \delta_{N-1} \\ \delta_N \end{pmatrix} \qquad (14.46)$$

where it should be noted that there is no α_1 or γ_N in the left-most matrix.

The method described by Press *et al.* (2007) for solving a tridiagonal system is a valid implementation of the Thomas algorithm for solving Equation (14.46), but unfortunately it is not suitable for our application. The reason is that we want to solve the system subject to specific boundary conditions. In the case we are considering here, these are fixed concentration, or Dirichlet, boundary conditions. Thus we are really solving the following problem:

$$\begin{pmatrix} \alpha_2 & \beta_2 & \gamma_2 & & & \\ & \alpha_3 & \beta_3 & \gamma_3 & & \\ & & & \ddots & & \\ & & & \alpha_{N-1} & \beta_{N-1} & \gamma_{N-1} \end{pmatrix} \begin{pmatrix} C_1 \\ C_2 \\ C_3 \\ \vdots \\ C_{N-1} \\ C_N \end{pmatrix} = \begin{pmatrix} \delta_2 \\ \delta_3 \\ \vdots \\ \delta_{N-1} \end{pmatrix} \qquad (14.47)$$

---

[2] We won't go into the details of this procedure, but it is basically how you used to solve simultaneous equations "manually" in algebra.

which must be solved subject to $C_1 = C_{\text{upper}}$ and $C_N = C_{\text{lower}}$. Since the design (left-most) matrix in Equation (14.47) will have a rank of $N - 2$, adding the two boundary constraints should yield a unique solution. The method we follow is taken from Appendix A in Roache (1998), which appears superficially similar to the method described in Press *et al.* (2007), but incorporates the boundary condition application described above. We also deviate slightly from Roache's approach *vis-à-vis* sign conventions in the interior point equation and algorithm.

We begin by recognizing that the tridiagonal matrix can be represented as three vectors (α, β, and γ) of length N corresponding to the three diagonals (including α_1 and γ_N). Now we can postulate the existence of two more vectors ε and ζ also of length N such that for the ith interior point:

$$C_i = \varepsilon_i C_{i+1} + \zeta_i \tag{14.48}$$

and if we move to the previous (i.e. the $(i - 1)$th) grid point

$$C_{i-1} = \varepsilon_{i-1} C_i + \zeta_{i-1} \tag{14.49}$$

Substituting Equation (14.49) into Equation (14.45) and rearranging allows you to create an equivalent to Equation (14.48):

$$C_i = \frac{-\gamma_i}{\beta_i + \alpha_i \varepsilon_{i-1}} C_{i+1} + \frac{\delta_i - \alpha_i \zeta_{i-1}}{\beta_i + \alpha_i \varepsilon_{i-1}} \tag{14.50}$$

implying by comparison to Equation (14.48) that:

$$\varepsilon_i = \frac{-\gamma_i}{\beta_i + \alpha_i \varepsilon_{i-1}} \tag{14.51}$$

and

$$\zeta_i = \frac{\delta_i - \alpha_i \zeta_{i-1}}{\beta_i + \alpha_i \varepsilon_{i-1}} \tag{14.52}$$

which means that all of ε_i and ζ_i can be expressed as a function of the interior point coefficients (which we already know) and the preceding values of ε and ζ (that is to say, their $(i - 1)$th points).

Using Equation (14.48) we can use the top boundary condition to set the values for ε_1 and ζ_1 because the relation $C_1 = \varepsilon_1 C_2 + \zeta_1$ must hold for all C_2, which requires that $\varepsilon_1 = 0$ and $\zeta_1 = C_{\text{upper}}$. Following this, we can step through from $i = 2$ to $i = N - 1$ to compute the interior values of ε and ζ. Next, we apply the bottom boundary condition $C_N = C_{\text{lower}}$, and then march backward through the C vector (from $i = N - 1$ to $i = 2$) to compute the interior point concentrations (C_i) using Equation (14.48) once again.

An implementation

We provide a MATLAB implementation of this code in the form of `tridiagde.m`, which we list below:

```
function C = tridiagde(alph,bet,gam,del,C1,CN)
%    A function that solves a tridiagonal system for an unknown
%        concentration vector C subject to Dirichlet boundary
%        constraints, i.e., setting top (C1) and bottom (CN)
%        concentrations
    N=length(alph);                    % length of input vectors
    E=zeros(N,1); F=zeros(N,1);    % internal coefficients
    C=zeros(N,1); C(1)=C1; C(N)=CN;
    F(1)=C1;                           % upper boundary condition
    for i=2:N-1
        E(i)=-gam(i)/(bet(i)+alph(i)*E(i-1));
        F(i)=(del(i)-alph(i)*F(i-1))/(bet(i)+alph(i)*E(i-1));
    end
    for i=(N-1):-1:2
        C(i)=E(i)*C(i+1)+F(i);
    end
```

You might invoke this as follows (assuming you have defined the appropriate constants, etc.):

```
alph = (D/deltaZ^2 + w/deltaZ) * ones(N,1);
bet = -(2*D/deltaZ^2 + w/deltaZ +k1)*ones(N,1);
gam = (D/deltaZ^2) * ones(N,1);
del = zeros(N,1);
Cupper=0;
Clower=2.6;
C=tridiagde(alph,bet,gam,del,Cupper,Clower);
```

Stability and accuracy issues

In general, the process of "solving" these N equations involves division of terms that can lead to near-singular behavior. It turns out that this is generally not an issue in the kind of problems we describe here, and a sufficient (but not necessary) condition for stability of the tridiagonal algorithm is *diagonal dominance*, i.e. that $|\beta| > |\alpha| + |\gamma|$ (see Press *et al.*, 2007). Examination of Equations (14.42) reveals that for $k_1 > 0$ this is satisfied.

As with any algorithm, accumulation of roundoff errors due to subtraction of large but comparable numbers that result in small numbers can amplify the total error in the tridiagonal method. As a general rule, roundoff error is kept under control if $\alpha_i > 0$, $\beta_i < 0$, $\gamma_i > 0$. Again, examination of Equations (14.42) reassures you that this is the case.

14.4.5 Discussion

Based on diagenetic modeling of their sediment core, Martens and Berner (1977) concluded that hypothesis 3 was the most likely explanation for the small, but persistent, region of overlap between CH_4 and SO_4^{-2}. While methane is only produced where the sulfate reduction has gone to completion, the small amounts of methane that diffuse upward are consumed by sulfate-reducing bacteria. It would be instructive to convince yourself that you can replicate the FOAM profile by building on the above code. We suggest you create a function file `tridiagde.m` and build on the code fragment listed above. You will need the data file `ch4.dat` available on our website. We encourage you to download it and try changing the k_1 value (e.g. set it to zero) in your program and see how this changes the model results.

While Martens and Berner (1977) reached their conclusion based on analytical solutions to Equation (14.40), we come to essentially the same conclusion based on our numerical solution. We note that the one-dimensional modeling problem in Chapter 13 (Craig, 1969) was also a two-point boundary value problem and could have been solved numerically. The modeling techniques you adopt will depend largely on the problem being studied and the assumptions one must make in order to make either the analytical or numerical solution tractable.

14.5 Problems

14.1. Using the nonlinear optimization methods in Chapter 10, construct a suitable cost function (χ_r^2 minimization of model–data fit) for the FOAM problem, and embed the model in an optimization program to find the optimal value of k_1.

14.2. Apply this methodology to solve for K, w, and J in Chapter 13.

14.3. Adapt the `tridiagde.m` algorithm code to incorporate Neumann (gradient) boundary conditions (you might call it `tridiagnm.m`) on each end. Your starting point would be to recognize that requiring $(C_2 - C_1)/\Delta z = s_{upper}$ allows you to set the values of ε_1 and ζ_1.

14.4. Use the tridiagonal algorithm to solve a boundary value problem involving the distribution of ^{222}Rn in the sediments of the Bering Sea. The equation used for this problem has similarities to the Martens and Berner CH$_4$ problem, namely it contains a first-order consumption term. When the activity of ^{222}Rn is equal to the activity of ^{226}Ra (its parent), it is said to be in secular equilibrium. The sediment profiles of the southeastern Bering Sea are mostly sandy sediment with little or no compaction and a porosity (ϕ) of approximately 0.7 over the upper 30 cm.

(a) If the ^{226}Ra concentration can be considered constant over this depth at 9.614×10^7 atoms cm^{-3} whole sediment, the sediment diffusion coefficient is also constant at 2.66×10^{-6} cm^2 s^{-1}, and ^{222}Rn concentration at the seawater–sediment interface is 283.1 atoms cm^{-3} whole sediment, use the diagenetic equation and the tridiagonal algorithm (`tridiag.m`) to produce a profile of ^{222}Rn and ^{226}Ra down to 30 cm. The diagenetic equation in this case would look like:

$$\frac{\partial \hat{C}}{\partial t} = D_s \frac{\partial^2 \hat{C}}{\partial z^2} - \omega \frac{\partial \hat{C}}{\partial z} - \lambda_{Rn}\hat{C} + \lambda_{Ra}[Ra] \qquad (14.53)$$

where $\hat{C}$ and [Ra] represent the ^{222}Rn and ^{226}Ra concentrations respectively in atoms cm^{-3} whole sediment, $\lambda_{Rn} = 2.098 \times 10^{-6}$ s^{-1}, and $\lambda_{Ra} = 1.373 \times 10^{-11}$ s^{-1}. Use the above value for the sediment diffusivity, already corrected for temperature and tortuosity. From ^{210}Pb profiles the burial rate has been estimated to be $\omega = 9.6 \times 10^{-10}$ cm s^{-1}. Produce a figure with the radon and radium plotted together in terms of their activity versus depth.

(b) Correctly modeled, the above diagenetic equation fails in two, interrelated, ways: the depth of the deficit is not deep enough and the size of the implied radon flux is too low. The deficit in actual sediment cores is observed to extend 25 to 30 cm below the seawater–sediment interface. In steady state, the size of the radon deficit is presumed to be maintained by a balance between the flux out of the sediment and radioactive decay. Integrating the ^{222}Rn deficit with a quadrature such as "trapz" will allow you to compare the size of the radon flux obtained in part (a) with independent estimates from actual sediment cores taken from the southeastern Bering Sea shelf, which yielded 3.5×10^{-3} atoms cm^{-2} s^{-1}. What was the size of the deficit in part (a)? If the ^{210}Pb profiles tell you that there could not be enough bioturbation to homogenize the upper centimeters of the core and the biologists tell you that there are large colonies of tube-dwelling worms living on the shelf, can the process of bioirrigation account for the extra ^{222}Rn deficit? Model a radon/radium profile as above but with an additional term representing bioirrigation in the form of:

$$- V_{BI} \frac{\partial \hat{C}}{\partial z} \tag{14.54}$$

How much bioirrigation (V_{BI}) is necessary to account for the above radon flux? How well does it compare to estimates from the Washington continental shelf of approximately 2 cm d^{-1}? Why is V_{BI} divided by porosity?

15

Upper ocean 1D seasonal models

Autumn to winter, winter into spring,
Spring into summer, summer into fall, –
So rolls the changing year, and so we change;
Motion so swift, we know not that we move.
Dinah Maria Mulock

15.1 Scope, background, and purpose

So far in our models the physical characteristics of velocity and diffusivity have been specified or "hard-wired" into the calculations. The next step to consider is allowing them to respond to changing conditions. In this chapter, we will be developing and exploring a class of models aimed at simulating the seasonal behavior of the upper ocean in response to changing atmospheric forcing. We subsequently will extend this model to simulate the response of dissolved gases in the upper ocean. This approach can be more generally applied to other shallow water column properties (including bio-optical modeling, particle dynamics, etc.) with very minor modifications. What we're trying to show you here is not just how to design, build, and extend the model, but more importantly how to figure out what the model is actually doing, and how to compare its performance quantitatively with actual observations.

15.1.1 Generalizations

There are two general types of upper ocean models (although there are hybrids of these two as well). There are the *bulk mixed layer models* which, as the name suggests, treat the mixed layer as a homogeneous, well-mixed box, within which properties including chemical species, temperature, salinity, and physical momentum are uniformly distributed. The other class is referred to as the *turbulence closure models* which attempt to explicitly model the detailed turbulent processes occurring within the mixed layer. Because the timescales for turbulent motions are so short (often seconds to minutes) and small-scale, these models are computationally very "expensive" and technically harder to formulate and run. We will be dealing with the simpler, bulk models here, although it must be recognized that if our focus were on more detailed behavior in the mixed layer, particularly on short timescales, we would need to turn to the more demanding models. As you will sense as we proceed, however, we have our hands full with even these "simple" bulk mixed layer models.

You may ask, "Why bother with all this physics stuff?" We're glad you asked. The main goal here is to understand the character and rates of biogeochemical processes. Nature provides us with a dynamic laboratory in the upper ocean where we observe the biogeochemical response to changing forcing, and this response is a fundamental signal that we can use to figure out what mechanisms and rates are operating behind the scenes. The challenge, however, is to separate the physical phenomena from the biogeochemical effects, and that is where the model comes in. In some sense, we are making the supposition that if the model moves some things around in a believable (and testable) fashion, then it must move other things, in particular biogeochemically affected substances, in a similarly realistic fashion.

Bear in mind that models are fundamentally abstractions of reality. We do not have the data, knowledge, or computational power to construct a computer calculation that incorporates all physical, chemical, and biological processes at all space and timescales. What we have to do is deliberately simplify the problem to the point where we can construct a model that contains just enough detail, and just the important processes to capture the bulk behavior of the system. This is obviously a fuzzy, philosophical issue, and you have to be very careful that you don't throw away important mechanisms in the simplification process, or that you don't parameterize unresolved processes incorrectly. Further, you may get yourself in a situation of having the model successfully mimic the observations for all the wrong reasons! In other words, when a model seems to be working, it may be two or more wrongs making a right.

What's more, model output generally tends to be more structured, more smoothly varying, and more appealing than ugly, gritty data. Thus you should be your own harshest critic when you get into this area, for models often produce attractive and believable results that look close enough to reality to convince you of the wrong things. In fact, there is a school of thought which claims that you only learn something from a model when it fails. In that case, you have successfully ruled out candidate processes as being responsible for the observations – a kind of null hypothesis test!

15.1.2 Biogeochemical context

During the spring and summer in the subtropical oceans, the warmer days and gentler winds cause surface waters to heat up and form a relatively thin surface mixed layer. Below this layer, the temperature quickly decreases downward through what is known as the *seasonal thermocline*. By the summer months, an oxygen supersaturation (i.e. concentrations exceeding solubility equilibrium with atmosphere at *in situ* temperature and salinity) builds up. A classic paper by Shulenberger and Reid (1981) noted this buildup and attributed it to primary production; photosynthetic oxygen released by phytoplankton is trapped in the stratified waters just below the summer mixed layer. Serious questions were raised about that interpretation, however; could the oxygen maximum also be produced by physical processes? One could picture bubble trapping by breaking waves at the sea surface creating a supersaturation, for example. Another possibility is that temperature changes due to the seasonal heating cycle, while not changing the absolute oxygen concentration below the mixed layer, may produce an apparent supersaturation because gas solubilities generally decrease with increasing temperature.

Well, these effects can be tested. Nature has been kind enough to provide us with an *abiogenic analog* of oxygen. Argon (Ar) is a noble gas, and is thus chemically and biologically inert. It conveniently has physical characteristics – molecular diffusivity and solubility – very close to that of molecular oxygen. Thus if physical processes were greatly affecting the oxygen distribution, it would show up in the argon concentrations as well. Figure 15.1 shows contour plots of the *saturation anomaly* of oxygen and argon as a function of depth and time at a location near Bermuda in the Sargasso Sea. The saturation anomaly is the percent deviation of the concentration of a gas from the concentration expected for equilibrium with the atmosphere at the pressure and temperature of the water.

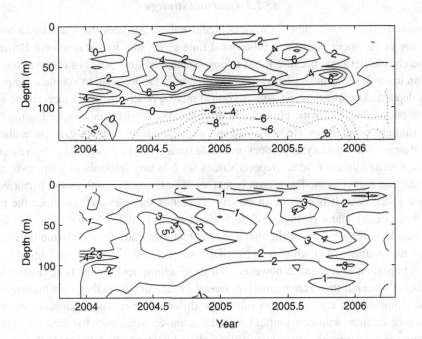

Figure 15.1 A contour plot of the saturation anomalies as a function of depth and time for dissolved oxygen (upper panel) and dissolved argon (lower panel) for samples taken at the Bermuda Atlantic Time Series site (BATS). The data were kindly provided by Dr. Rachel Stanley (see Stanley, 2007).

The two gases show some definite similarities; note for example the summertime subsurface maxima for both gases at about 70 meter depth, and the double maxima – one shallow and one deeper – in late summer of 2005. However, there are some differences. The deeper oxygen is under-saturated owing to biological consumption. Also note that the contour interval for oxygen in the upper panel is 2%, double that of argon in the lower panel. The oxygen maximum is of order 8–9%, while argon peaks at 4–5%. We can conclude from this that the biological effects are comparable to or greater than the physical influences for oxygen.

This is all well and good, but how do you go from qualitative observations of the summertime development of oxygen and argon maxima to quantitative estimates of oxygen production rates? The problem is complicated, because not only are things changing on a seasonal basis, but the upper part of the water column is not a closed system in that gases are lost to, or gained from the atmosphere by gas exchange processes that are time-, wind-, and temperature-dependent. Further, possible vertical mixing processes may play a role in moving gases around in a way which we cannot quantify. Some "back of the envelope" types of calculations (e.g. Jenkins and Goldman, 1985) indicate that all of these processes contribute at some level, and that we need a more quantitative model to do a better job.

15.1.3 Goal and strategy

We want to construct a simple physical model which emulates the seasonal to annual structure of the upper ocean at a subtropical time series site, in this case near Bermuda. We've chosen Bermuda because it has a long-established time series of observations that we can use to test our model's performance. In particular, we want to simulate the mixed layer depth and temperature, as well as vertical mixing processes. We'll incorporate some simple but relatively realistic physics and drive the model with atmospheric forcing (heat flux, sunlight, wind stress, etc.). We will use a realistic form of forcing, in particular the NCEP atmospheric reanalysis product (available from http://www.cdc.noaa.gov/ncep_reanalysis), which reports values for 6-hourly intervals globally over more than the past half century. The reason is that we'll be trying to match model performance against actual data. It is a bit of an exercise to extract the relevant data, since the pertinent files occupy many gigabytes of storage and are in NetCDF format. We've included an example m-file on our website that does such an extraction for the Bermuda area, but realize that getting MATLAB to read NetCDF files efficiently is not for the faint of heart.

It is important to recognize, however, that these atmospheric products (and there are a number of them out there) are themselves *model outputs* subject to their own uncertainties. Some climatologies are better than others for different aspects of forcing, so you need to become familiar with how particular products are derived, and what their strengths or weaknesses appear to be. There is a well-developed literature that discusses the pros and cons of the various products, and we urge you to look carefully into whatever products you choose to use.

You should also be aware that purely vertical processes (in particular vertical heat fluxes) may not be in balance, whereas we know that averaged over several years the upper ocean is more or less in some kind of balance. Examination of the distribution of temperature in the Sargasso Sea, for example, indicates that there must be a significant horizontal transport of heat by ocean currents. This is the primary weakness of one-dimensional models. However, for the purposes of instruction, we'll continue onward.

Once we have a physical model that behaves well enough to simulate observed trends in temperature and mixed layer depth, we will put in inert gases. This involves parameterizing the gas exchange processes at the sea surface, and ultimately matching some inert gas observations, especially Ar, which is a close analog of oxygen, against the model. This gives us a model which moves gases around in the right way. After that we put in oxygen, with some specified oxygen production profiles, and see what kind of productivity is required to explain the observations.

15.2 The physical model framework

15.2.1 What we need to simulate

Remember that we're not trying to do any ground-breaking physics here, just do the physics well enough to create a comfortable home for our gases. There are really three different things we need to do well in the model.

The temperature history will be critical for determining the dissolved gas solubilities, so we must be able to simulate the surface water temperature as a function of time. It is the surface water temperature that controls the solubility-equilibrium gas concentrations, and hence the flux of gases (which are modeled as a function of concentration difference from equilibrium) between the ocean and atmosphere. The subsurface temperature distribution plays no direct role in controlling dissolved gas concentrations, but does greatly influence the temporal evolution of surface temperatures.

The mixed layer depth must be simulated because the timescale of gas exchange equilibrium is related to the mixed layer depth; the deeper the mixed layer, the longer it takes to reach equilibrium for a given gas exchange rate. If the model mixed layer is too shallow, then the mixed layer will too quickly reach equilibrium, and the gas exchange flux will shut off. Also, too small a part of the water column would be involved with the exchange process.

Vertical mixing – and in general vertical transport – is likely to be important. Although vertical mixing tends to be very small in the main thermocline (e.g. at 300 m depth), there is a lot of wind and thermal energy being put in at the ocean surface which will drive turbulent motions and hence mixing. Further, there is a net convergence of surface water in the area due to wind stress patterns. This is, in fact, characteristic of all subtropical gyres. The net result is a downwelling of surface waters which will affect property distributions as well. Getting vertical transport right will not only be important for getting the temperature distributions and evolution right but also play a role in the dissolved gas transport.

15.2.2 How we set up the model

We'll use a one-dimensional model similar to Price *et al.* (1986) which we'll call a PWP model after the authors' initials. We'll set it up with a depth range of 500 meters and a vertical resolution of 2 meters. The idea is to make the domain relatively large, say a factor of two, compared with the region that is of most interest. Since winter convection occurs to a depth of 150–200 m, then a boundary at 500 m is sufficiently "far away" from the region of most action that arbitrary or unrealistic boundary conditions (e.g. fixing the bottom temperature and salinity or assuming a no-flux condition) won't strongly influence the simulation. Certainly the proximity to the "bottom" of our model suggests that we don't want to run the model for too long, to avoid having our choice of boundary condition coming back to haunt us.

The choice of the vertical resolution is a compromise between truly resolving thin mixed layers (which can be as shallow as 10 m) and computational expediency; numerical stability will be limited most by vertical mixing, so computation time increases roughly quadratically as the inverse of Δz.

Actually, running such a model is relatively simple, providing you do your homework right and pay attention to the details. The model involves a vertical, one-dimensional domain which we must force at the surface with heat exchange and wind stress, adjust the vertical structure (mixed layer depth) with simple physical criteria, and vertically advect/diffuse material (temperature, salinity, and momentum). We need to keep track of a few things: temperature and salinity – they control the density of the water, which is important for regulating the mixed layer depth – and the horizontal velocities, which are also important for controlling mixed layer depth. For a more complete discussion, see Price, Weller, and Pinkel (1986) (thus PWP model).

We start the model with an observed initial profile of temperature and salinity using Bermuda Atlantic Time Series (BATS) bottle data obtained in March 2003 interpolated onto the model grid. We also start with zero horizontal velocities; this is a safe starting point because a very thick mixed layer characteristic of winter months is not strongly accelerated by wind stress. Horizontal velocities don't play a significant role in controlling mixed layer depth this time of year – we'll demonstrate this later. Here's how a model time step would look:

1. Add/remove surface heat, and deposit solar heat in upper water column;
2. Vertically mix the mixed layer according to density stratification;
3. Apply wind stress to mixed layer;
4. Adjust mixed layer depth for dynamic stability;
5. Vertically mix and advect whole water column.

That's all there is to it. Now for those picky details.

15.3 Atmospheric forcing

We could apply long-term averaged forcing, such as the Oberhuber Climatology, to the model if we wanted to look at the average behavior of the upper ocean. However, because we want to compare the model simulation with real data, we will use NCEP Reanalysis heat fluxes and wind products for a specific period (March 2003 through March 2006). We have extracted and interpolated this forcing to the Bermuda location and made it available on our website as `NCEPforcing.mat`, which contains the relevant heat fluxes, precipitation rate, wind stresses, and wind speed at 6-hour intervals for the relevant time period.

15.3.1 Heat fluxes

The heat flux between the ocean and the atmosphere arises from a combination of four types of heat flux (see Fig. 15.2) :

1. Latent heat flux (evaporation taking away heat);
2. Sensible heat flux (thermal conduction driven by a temperature difference between sea and air);
3. Net outgoing longwave radiation (black-body radiative cooling); and
4. Incoming shortwave radiation (solar heating).

The first three are predominantly from the ocean to the atmosphere, and occur at the very surface of the ocean. Latent heat transfer is by far the largest of the three. As expected, the greatest heat loss from these three occurs in the winter months. The sensible heat flux is unique among the three in that it can be negative (i.e. heat entering the ocean) as well as positive. The fourth heat flux, incoming shortwave radiation, largely balances the other three. Notice that the curve is very dark because it falls to zero each night. Unlike the others, solar heating is stronger in the summer months.

While the latent, sensible, and outgoing longwave radiation heat fluxes apply to the very top of the ocean, the radiant heating is distributed over a vertical distance. As light penetrates the ocean, it is attenuated as a function of depth. This attenuation is the conversion from light to heat. The spectrum of the downwelling irradiance also changes with depth. The character of this attenuation is governed by a variety of factors, primarily biological in origin. There is a classification of water types by optical properties compiled by Jerlov. The region where we are modeling is a type Ia in the Jerlov classification (Jerlov, 1968). Paulson and Simpson (1977) have modeled the net downwelling solar irradiance as a double exponential function of depth, which for Jerlov type Ia waters is given by:

$$I(z) = I(0) \left(0.62 e^{-1.67z} + 0.38 e^{-0.05z} \right) \tag{15.1}$$

The first term represents the incoming longer-wave component, which attenuates very rapidly and is deposited in the uppermost meter or two. The second term represents the shorter-wave component, which is distributed over the top 25 to 50 m. Computationally,

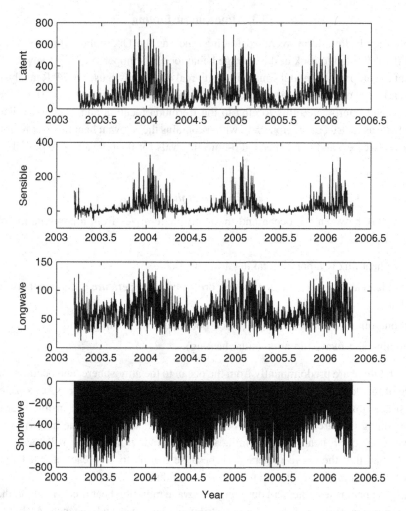

Figure 15.2 The NCEP Reanalysis heat flux time series for the Bermuda region. From top to bottom panels, they are Latent, Sensible, Net Longwave, and Net Shortwave. The fluxes are in $W\,m^{-2}$ where positive is from the ocean to the atmosphere.

what we do is to use Equation (15.1) at the beginning of the program to calculate the fraction of surface irradiance arriving at the top and the bottom of each cell in our model. Then we compute the difference between the top and the bottom as the fraction of surface irradiance deposited in each cell. This is then used during the simulation as a multiplicative factor, along with the heat capacity and mass of water in the model cell, to calculate the resultant temperature change during a time step with a given surface irradiance.

Finally, it's fair to ask the question, "Can (or should) the model work here?" The answer is a very definite "Maybe!" What really happens when we look at the time evolution of a profile at a fixed location is that we are seeing water advecting in from some place

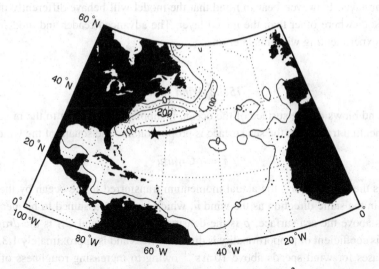

Figure 15.3 A map of the net ocean–atmosphere heat flux (in $W\,m^{-2}$) in the North Atlantic for the 10-year period 1997–2006 based on the NCEP Reanalysis products. The contour interval is $50\,W\,m^{-2}$, and positive values mean net heat transfer from the ocean to the atmosphere. Note the maximum situated over the Gulf Stream off the east coast of North America. Bermuda is indicated by the black star, and the heavy line extending northeastward from it approximates a 1-year upstream trajectory of a fluid parcel.

upstream of our site. You can think of the water column moving by, so setting up forcing and expecting it to work is somewhat optimistic since we should be looking upstream of the site for the forcing. Bermuda may be one of the few places where things might work out. The reason is that the flow lines of the circulation actually run *parallel* to the contours of total heat flux. In Fig. 15.3 we present the total heat flux averaged over the decade from January 1, 1997 through December 31, 2006 from the NCEP Reanalysis for the North Atlantic. The BATS site is designated as a star, and we've drawn a "tail" pointing upstream that corresponds to about 1 year of travel. Remarkably, the trajectory parallels the nearest heat flux contour. Thus a hypothetical water column tends to see the same seasonal heat flux cycle over a 1-year trajectory. In some crude sense, we're banking on the water column "forgetting" prior years so that a local approximation can work for us.

Note that the upstream trajectory shown in Fig. 15.3 is outside the $50\,W\,m^{-2}$ contour. This means that the water column experiences an approximately constant cooling, but at a rate much lower than within the Gulf Stream region to the north. This suggests that its heat budget must be closed by lateral convergence of heat. However, some fraction of this offset may be related to unknown biases in the NCEP climatologies. This is a continuing point of debate. We will correct for this offset by simply adding the "missing heat" back in at the surface. Computationally, this is easily accomplished; simply average the heat budgets over several complete annual cycles and subtract the mean. Adding the missing heat at the surface is reasonable, since the bulk of the lateral heat convergence likely occurs in the Ekman layer. However, bear in mind that the model will behave differently if we add the heat somewhere other than the mixed layer. The advanced reader and modeler might consider experimenting with this later on.

15.3.2 Wind stress

As the wind blows on the surface of the ocean, it supplies momentum to the mixed layer. The momentum transfer to the sea surface is proportional to the square of the wind speed.

$$\vec{\tau} = \rho C_D |\vec{u}| \vec{u} \tag{15.2}$$

Here, $\vec{\tau}$ is the *wind stress* – the actual momentum transferred to the ocean by the wind – which is in the same direction as the wind $\vec{u}$, which in turn is assumed to be measured at 10 meters above the sea surface, ρ is the density of seawater, and C_D is the *drag coefficient*. This coefficient of proportionality is dimensionless and is approximately 1.3×10^{-3}, but increases for wind speeds above $10 \, \text{m s}^{-1}$ owing to increasing roughness of the sea surface. Rather than calculating the wind stress "from scratch", we use the meridional and zonal components of the wind stress in the NCEP Reanalysis product, which is included in NCEPforcing.mat. In Fig. 15.4 we plot the 6-hourly wind speed (upper panel) and the corresponding magnitude of the wind stress (lower panel). As expected, wind speed and wind stress are greatest in the winter months and least in the summer.

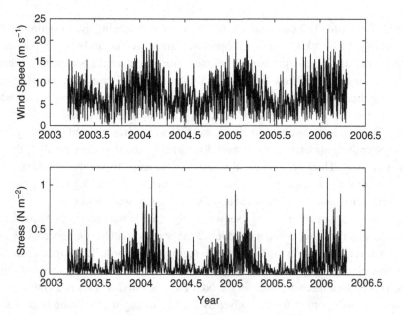

Figure 15.4 The NCEP Reanalysis wind speed (upper panel, in m s^{-1}) and the magnitude of the wind stress (lower panel, in N m^{-2}). Note that the winter intensity is more pronounced with the wind stress owing to its quadratic dependence on wind speed (see Equation (15.2)).

Wind stress will be important for the model since it serves to accelerate the mixed layer. This acceleration can cause instabilities that serve to deepen the mixed layer, or at least maintain a minimum thickness during spring and summer months. As we perform the simulation, we need to keep track of the water column flow direction in relation to the wind stress vector. We do this by allowing the model's horizontal velocity vectors (for each depth cell) to "rotate" with inertial motion (remember that we are on a rotating planet).

15.3.3 Ekman pumping

Next, we need to apply Ekman pumping to the model. There is a convergence of surface waters driven by large-scale patterns of wind stress in the region, and this forces water downward at the surface. This flow does not simply extend all the way to the bottom, but because of the dynamical balance of the subtropical gyre, the vertical velocity attenuates downward. We do this in our simulation by simply tapering the vertical velocity from its maximum value at the sea surface to zero at the bottom of the model domain.

We could try to calculate the wind stress curl directly from the NCEP Reanalysis output using the wind stress field near Bermuda, but that kind of calculation is particularly sensitive to noise. Instead, we use a smoothed monthly product, the Trenberth climatology from `http://ingrid.ldeo.columbia.edu/SOURCES/.TRENBERTH/.Monthly/` `.curl/`, interpolated to the location of Bermuda.

Now we can get from the wind stress curl to the rate of Ekman pumping w_E by noting that:

$$w_E = \frac{\nabla \times \vec{\tau}}{\rho 2\Omega \cos(\text{lat})} \tag{15.3}$$

and in Fig. 15.5 we've converted w_E to $\text{m}\,\text{y}^{-1}$ (not exactly mks units, we're afraid, but commonly used) to give you a feeling for its significance. Thus over the course of the year, there is a net downward pumping of about 20–30 meters. Most of this occurs in the winter, when the mixed layer is deep or getting deeper, but there is a net downward flow of about 10–20 m y^{-1} during the rest of the year. The climatology has no simple functional form, so

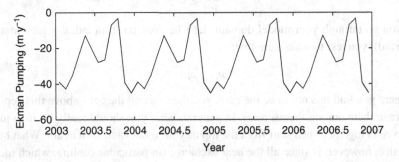

Figure 15.5 The Ekman pumping velocity (negative means downward) in meters per year computed from the Trenberth monthly wind stress climatology interpolated to Bermuda.

we have repeated the annual cycle and linearly interpolated values to the model time grid. To give you a perspective on its importance, Ekman pumping may move a fluid parcel over the approximate scale of the average mixed layer depth in a year, and over about a fifth of the model domain during a 3-year simulation. Thus it may be significant, but likely not a dominant factor.

Now there's one difficulty introduced by the Ekman pumping that we need to consider. From the perspective of conservation of mass, the warm water introduced in the model at the top of the water column by Ekman pumping replaces colder water further down in the water column that must be removed by horizontal flow divergence. Thus this process acts as an extra heat flux that must be compensated for if the model is to maintain an approximate steady state. Given the complicated seasonal variation of the Ekman pumping rate and the changing vertical temperature profile, it's virtually impossible to calculate the required heat offset accurately without actually running the model. So we must experimentally adjust this term to match observed trends in heat content. We have now inherited one adjustable parameter in the model.

15.4 The physical model's internal workings

Here we talk about those things going on *inside* the model that adjust the distributions of properties and that set the mixed layer depth. The internal workings are actually rather simple. After adding the forcing, we apply three criteria for vertical stability (i.e. whether water should mix vertically, and whether the mixed layer should deepen). After that, we apply vertical advection and vertical diffusion to the properties of the water column.

15.4.1 Vertical stability criteria

The first stability criterion, and the one that will prove the most important in our model, is *static stability*. In fact, it accounts for about 80% of the action. Quite simply put, you cannot have denser water overlying lighter water. This means that you must have:

$$\frac{\partial \rho}{\partial z} \geq 0 \tag{15.4}$$

Thus you go through your model domain (let i be your position index, with i increasing downward), you test to make sure that:

$$\rho_i \leq \rho_{i+1} \tag{15.5}$$

and where you find this not to be the case, you then mix all the cells above this depth (that is, average them among themselves). In general, what you should really do is to just mix the two cells together, then start from the top of your model and do it again. What happens in practice, however, is since all the heat exchange (in particular cooling, which increases the density) takes place at the top of the model, you will always find that the effect of this instability is to mix all the way back to the top. So you may as well do it the first time. Also, we tend to deal in σ_t, which is the density anomaly (the density minus $1000\,\mathrm{kg}\,\mathrm{m}^{-3}$), which works just the same. Note also that we are using a linearized equation of state; that is, we compute the density as a linear function of temperature and salinity about a mean value. Since we have to mix temperature and salinity and then recalculate density for every grid cell at every time step, this saves an enormous amount of time over doing the full-blown calculation involving the equation of state.

The second stability criterion is the *bulk Richardson number stability*. This arises because if the mixed layer gets going too fast (i.e. the wind stress is allowed to accelerate it to too great a speed), it tends to "stumble" over itself. What actually happens is that if there is too much velocity shear at the base of the mixed layer, it will become unstable and mix downward. This effect, determined by field and laboratory experiments, is such that the mixed layer deepens if the bulk Richardson number goes below a critical value:

$$Ri_B \equiv \frac{gh\Delta\rho}{\rho(\Delta v)^2} \geq 0.65 \tag{15.6}$$

where g is the gravitational acceleration ($9.8\,\mathrm{m}\,\mathrm{s}^{-2}$), h is the height (thickness) of the mixed layer, $\Delta\rho$ is the density contrast between the mixed layer and the water below, and Δv is the difference in horizontal velocity between the mixed layer and the underlying water. This effect tends to be important when the mixed layer becomes very thin, because a thin mixed layer becomes easily accelerated by wind stress, and the inverse quadratic nature of the dependence makes for a strong damping. The relative activity of this process is about 20% of the static instability.

The third stability criterion is based on the *gradient Richardson number* and has the effect of stirring together layers where the velocity gradient becomes too great. You can think of this as the different water layers "rubbing" against the waters underneath. This largely has the effect of blurring the transition between the mixed layer and the seasonal thermocline below, which would otherwise become rather sharp. Laboratory experiments indicate that there is a critical gradient Richardson number, below which stirring occurs:

$$Ri_G \equiv \frac{g\dfrac{\partial\rho}{\partial z}}{\rho_0\left(\dfrac{\partial v}{\partial z}\right)^2} \geq 0.25 \tag{15.7}$$

This turns out to be a not very vigorous process, but becomes a little more important in the absence of any explicit turbulent vertical diffusion below the mixed layer.

Now before we get down to actually running the model, we'll show the relative behavior of these profile adjustment processes during different parts of the annual cycle (Fig. 15.6). Keep in mind that there are no "free parameters" in this formulation: you cannot "dial" the behavior of the model by tuning these processes. They are fundamental physical processes constrained by laboratory observations.

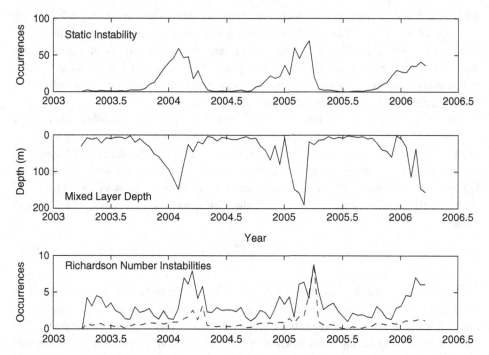

Figure 15.6 A comparison of the three types of stability criteria in terms of their relative action frequencies in relation to the model mixed layer depth (middle panel). Note the difference in scales between the static instability rates (top panel) and the two Richardson number criteria in the bottom panel. The bulk Richardson term is the solid line in the bottom panel, while the gradient Richardson frequency is the dashed line.

How did we get these numbers? It's really quite simple. The model is run with counters inside the three stability adjustment subroutines that are incremented each time one of the subroutines is called and the results tallied at regular intervals. This is an important technique that you can use to see what is happening inside the model. Otherwise, you wouldn't have much of a clue as to what is going on.

The middle panel shows the depth of the mixed layer during the model run. The static instability becomes extremely important during the cooling part of the seasonal cycle, and serves to steadily deepen the mixed layer from about October onward. At the end of the winter, when warming starts up, the static instability shuts off since the water column is being heated from above. As it shuts down, you see the bulk Richardson instability Ri_B take over, spiking in spring, but maintaining a plateau during the summer months when the mixed layer is thin. The plot of the mixed layer depth here is actually a little deceptive, because what really happens (and you can see the Ri_B adjustment responding) is that the early spring mixed layer starts off as instantaneously very shallow, but is rapidly destroyed by Ri_B processes, so that average mixed layer depth only gradually shoals. The gradient Richardson number Ri_G instability is at its most active during the rapid spring shoaling of the mixed layer, for reasons related to the Ri_B effect. Here again is an important lesson. Because we've sampled the model output every 2 weeks (to save storage space), high-frequency responses won't be picked up so easily. As with everything, there's a trade-off. You might consider sampling at a much higher rate to observe such features, or carrying along some other statistics (such as RMS depth variation) as a probe of model behavior.

15.4.2 Turbulent mixing below the mixed layer

Another adjustable and unknown parameter in this model is the vertical mixing coefficient in the stratified region below the mixed layer. From the viewpoint of the physics, it turns out that one is necessary, because if the model is run without vertical mixing, there is strong trapping of heat in the mixed layer, and it becomes extremely hot and thin in the summer time. Moreover, vertical mixing is needed to carry heat down into the seasonal thermocline and for ultimately eroding it during the autumn. We have no a priori way of determining the "right" vertical diffusivity for this model, except by numerical experiment, because the model subsumes a multitude of sins (unresolved processes). We will need to adjust it until we see the "right" behavior of the temperature field. Certainly, one might want to refine the formulation of the vertical diffusivity, but we choose in this case the very simple approach of a constant value. We have experimented with ones that are inversely proportional to some power of N^2 (Brunt–Väisälä frequency) or Ri_G but this actually yields worse results. We'll leave it for a more adventurous reader to experiment with that one.

15.4.3 How long can we run it?

We have already mentioned that we're running the model for a 3-year period, and that the vertical model domain we've designed is 500 meters. The two are to some extent linked. That is, there may be a long-term drift driven by the much slower change of the deeper part of the water column (the part not directly exposed to the winter convection, say below 200 m). How long might changes in the lower, isolated part of the model take? If it were left to vertical diffusion alone, then the characteristic timescale would be:

$$t_{\text{diff}} \approx \frac{L^2}{K_z} = \frac{(300\,\text{m})^2}{10^{-4}\,\text{m}^2\,\text{s}^{-1}} = 10^9\,\text{s} \approx 30\,\text{y} \qquad (15.8)$$

Here we've used a "canonical" vertical mixing rate of $10^{-4}\,\text{m}^2\,\text{s}^{-1}$, which turns out to be close to what we find works well in this case. That is, any sins that we have committed by fixing the temperature and salinity of the lower boundary of our model will come back to haunt us in a decade or two, so we had better not run our model longer than a decade or so.

You might think that we also have the help of vertical advection, which in our scheme is of the order of tens of meters per year near the surface, pushing the bottom influence back. But by the 200 m mark, it is down to about $10\,\mathrm{m\,y^{-1}}$ or so, so the time constant for advection is:

$$t_{\mathrm{adv}} \approx \frac{L}{w} = \frac{300\,\mathrm{m}}{10\,\mathrm{m\,y^{-1}}} \approx 30\,\mathrm{y} \tag{15.9}$$

Thus they are comparable in magnitude, but don't count on them cancelling out. What we have, then, is a deep water relaxation timescale that is longer than the time length of our model runs. Thus it turns out not to affect our conclusions greatly here, but obviously it bears thinking about.

This, in fact, is a classical problem in numerical models, where you try to compartmentalize a part of the ocean because it's too difficult to model the entire thing. You avoid undue influence associated with poorly known or ill-constrained boundary conditions by relying on a "buffer zone", but you run the risk of having the interesting region "dragged along" by a much slower responding part of the system in some subtle way.

15.5 Implementing the physical model

15.5.1 The code and how it works

Well, enough fooling around, let's do it! First, let's look at the code. It consists of the main program m-file, pwp.m, plus a series of m-files which do specific tasks each time step. They are:

- LoadForcing.m – loads in the NCEP forcing and BATS data
- inifctr.m – initializes common factors and forcing vectors, etc.
- inihydro.m – initializes the model data vectors (temperature, etc.)
- dostins.m – does static instability adjustment, adjusts mixed layer depth
- addmom.m – adds momentum imparted by wind stress to mixed layer
- dobrino.m – does bulk Richardson number adjustment
- dogrino.m – does gradient Richardson number stirring
- advdif.m – advects and diffuses properties vertically
- dooutput.m – prints/accumulates output data at intervals
- pwcompare.m – compares the model to BATS data and computes/plots diagnostics
- pwsave.m – saves output data to a file in a subdirectory

These m-files, in turn, use a couple of utility functions to get some of the basic work done:

- mlmix.m – mixes up the mixed layer
- stirit.m – stirs two density layers when Ri_G is unstable

and the initialization m-files (inifctr.m and inihydro.m) use two forcing or data files in MATLAB binary format:

- NCEPforcing.mat – forcing data
- BATSdata.mat – the BATS bottle data for comparison

Now before you rush into this, we advise you to work with a specific directory structure on your computer so the routines will work properly. You can build this directory by expanding the zip (or tar) file from our website. If you choose to do it manually, first create a directory (call it PWPmodel or something like that). In this directory you can download the two data files listed above. Next create a directory within this directory called "Physical", and download the m-files into it. Within the Physical directory, create a directory called "Runs". This will be where your model output is stored.

Have a look through the individual m-files to get an idea of how they work. We've tried to put enough comments in to be instructive. This is generally a useful strategy even if the only person you're instructing is yourself years later. We've tried to compartmentalize the model functions so that it's reasonably obvious what the code is trying to accomplish. This is what the main program pwp.m looks like:

```
LoadForcing          % load the NCEP forcing, etc
EkmHeatConv = 12;    % Ekman Heat Convergence Correction (w/m2)
Kz = 1.25e-4;        % vertical diffusivity (m2/s)
   inifctr;          % initialize useful factors & forcing vectors
   inihydro;         % initialize water column profile
%
% Main time step loop
%
  for it=1:nt
       T(1)=T(1)+thf(it);     % add sensible + latent heat flux
       S(1)=S(1)*FWFlux(it);  % flux fresh water (latent/precip)
       T=T+rhf(it)*dRdz;      % add radiant heat to profile
       dostins;               % do static instability adjustment
       addmom;                % add wind stress induced momentum
       dobrino;               % do bulk Ri No Adjustment
       dogrino;               % do gradient  Ri No Adjustment
       advdif;                % advect and diffuse properties
       dooutput;              % periodically save data
  end
pwcompare;           % compare model with data
pwsave;              % save results to a file
```

The first thing that pwp does is to select the two adjustable parameters that we mentioned would have a bearing on model behavior: the vertical diffusivity coefficient and the Ekman heat convergence required to balance the local heat budget. We've chosen ones here that work reasonably well, but we will vary them later to see what happens.

The model takes less than a minute to run the 3-year simulation on a 2 GHz processor, so you won't grow old waiting. When the code runs, there is a significant period of time while it is initializing (inside the routines inifctr and inihydro). There it is building the forcing vectors, etc. Note that both the heat flux (thf) and the salinity adjustments (FWFlux) terms are pre-calculated inside inifctr with all the appropriate factors. It is computationally advantageous in models to do as much as possible "up front" so that the calculations are easily vectorizable. Also, inifctr computes the stable time-step size, which is a function of vertical mixing. The initial vertical profiles of temperature and salinity are specified within routine `inihydro`.

After the initialization, the main time-stepping loop takes place. The cell nearest the surface is first cooled by the heat flux (sensible + latent + outgoing longwave radiation), then fresh water flux is applied to adjust the salinity. Next, radiant heating is applied to the whole water column. Once this forcing is imposed, thereby altering the density, then the water column is adjusted for static stability by (`dostins`). The reason for doing that here is that the cooling process tends to produce density instabilities. Next, wind stress is applied to the mixed layer and the water column allowed to inertially rotate (`addmom`). After that, the bulk Richardson number stability is applied, and then the gradient Richardson number mixing. Finally, vertical advection–diffusion is applied. The routine `dooutput.m` contains code which prints to the screen as the model progresses (it's nice to see something happening), and saves the model output at fixed intervals from the run in a form which is usable for analysis of the run in the end. Go ahead! Why not try it now?

After the simulation is complete, `pwcompare` compares the model output to observations. This is where the science hits the keyboard. We'll dwell on this a bit later, but `pwcompare` gives you three example plots: a contour plot of the model temperature, a contour plot of model–data differences, and a graph of mixed-layer temperatures and depths. The m-file `pwsave` is a single line that simply saves the model output in a specially constructed file name in a subdirectory called "Runs". What is saved is the accumulated temperature and salinity profiles (`Ta` and `Sa`), the mixed layer depth history (`zmld`), and counters indicating the number of times adjustments are made to the water column (see previous section). You can build in other things that you think might be informative or interesting, and save them as well. The idea is that you can return to any particular simulation to re-analyze it in new ways without having to re-run the calculation. It doesn't matter much with this model, since it takes only a minute to run, but in more complex situations, it makes a lot of sense. The downside, however, is that storing such "history" may involve much larger amounts of storage with more complex models.

15.5.2 *Some physical model runs and experimentation*

In Fig. 15.7, we show the output of a "typical" run, which uses the vertical mixing coefficient and Ekman heat convergence correction numbers included in the listing above. The behavior seems quite reasonable; a seasonal summer mixed layer and deep winter mixing seem to occur in about the right place and time. There are some interesting details that appear consistent with observations, for example the outcropping of the 19 °C isotherm in March 2005. There are, however, other worrisome signatures such as the steady deepening of the 18 °C isotherm that may be a signature of model drift caused by unrealistic boundary conditions at the bottom of the model domain.

Let's face it, it's tough enough to get a model running and to get some kind of credible output. The next step might be to peer at a comparable plot of actual data from the BATS time series and declare a victory, but we're going to do better than that. Comparison of the

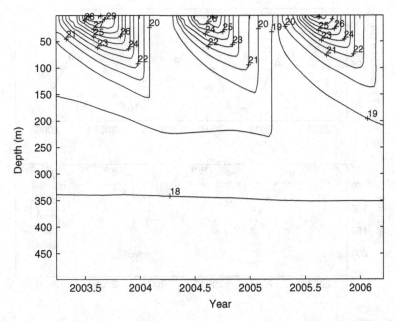

Figure 15.7 Contour plots of temperature and salinity on a time–depth grid from our PWP model.

model results with observations is a critical phase in the process of modeling. The tricky part is to find the appropriate data to make the comparison, and then to choose the right aspects of the model output to compare with the data. What are the critical modes of model behavior? What is important to match with reality and what can we safely ignore? The answers are complex, and rooted in your goals. The only tunable parameters in the model formulated here are the vertical mixing rate K_z and the Ekman heat divergence, so we would like to see what combination of the two yields the best match to data.

One measure of success would be how close the model gets the mixed layer temperature and depth compared to the data. Figure 15.8 is a comparison of mixed layer temperature (upper panel) and mixed layer depths (lower panel) for the model and observations. The plots indicate that although the match is not perfect, it is nonetheless fairly respectable. The model overestimates temperature in the first year, underestimates it in the second year, and is very close in the third. The mixed layer emulation is also reasonable, particularly in the second winter. Mixed layer depth comparison is particularly challenging in the early spring (note the early spring of 2004) because rapid, intermittent stratification can occur and is difficult to match between model and data. This again may be a problem of undersampling model output that we discussed earlier.

Much like in the case of looking at residual deviations of data from regression fits, it is often useful to do a similar thing for the comparison of model simulations and data. We won't show it here, but the m-file `pwcompare` generates a time-series contour plot of the model–observation temperature differences throughout the water column, and it's

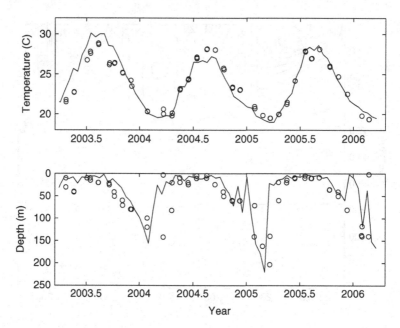

Figure 15.8 Time series comparison of model *versus* observed mixed layer temperature (upper plot) and depth (lower panel) for the simulation described above.

useful to examine the space-time structure of these offsets. Are there any regions where the model does poorly, or is the map of residuals a patchwork of quasi-random deviations? Examination of such plots is useful, because it can often point to some structural deficiency in the model's ability to match reality.

If we are seriously to attempt to find the combination of adjustable parameters that optimally simulate reality, we need to construct some sort of numeric *cost function* that is a measure of model fidelity. The pwcompare function computes four such measures. These are

1. The average model–data temperature offset in the mixed layer;
2. The root mean square (RMS) model–data offset in mixed layer temperature;
3. The average model–data temperature offset in the water column; and
4. The root mean square (RMS) model–data temperature offset in the water column.

These choices are a reflection of what we think are the important things to simulate correctly. There is some redundancy (the mixed layer is after all part of the water column), but that is a measure of the importance of the mixed layer in determining gas concentrations. We'd better get that right if we're going to model gases whose solubilities are temperature-dependent. Moreover, the mixed layer temperature is sensitive to vertical mixing rate, so it remains an important diagnostic. The offset for the whole water column is a valid test of whether we've got the overall heat budget right for our simulation (and hence the Ekman heat divergence parameter). The balance is fairly complicated, since we're playing off heat extraction at the surface against a combination of radiative heating, Ekman heat divergence, and downward mixing. The latter two must to some extent play off against one another, so we need to explore the optimum combination of those two parameters.

Notice that we've studiously avoided looking at the salinity fields as diagnostic of model performance. One reason is that salinity variations apparently don't play a big role in controlling the mixed layer and upper ocean stratification near Bermuda (see Doney, 1996). Another is that we wanted to keep our example relatively simple. It could be an exercise for the reader to consider implementing a salinity-based diagnostic as well.

We're going to optimize the model fidelity by a "brute force" technique. Since the model runs fairly quickly, we'll embed the model in nested for loops and run a two-dimensional grid in parameter space. But where do we start? Experience dictates that the original "Munk number" (Chapter 13), i.e. $10^{-4}\,\text{m}^2\,\text{s}^{-1}$, is a reasonable starting point for K_z, and we can scan a factor of two either side of that, at least to begin with. The approximate value of the Ekman heat convergence can be calculated in the following manner: assuming an average temperature difference between surface and subsurface as about 5 °C and an average downwelling rate of $30\,\text{m}\,\text{y}^{-1}$, this gives a heat excess of about $20\,\text{W}\,\text{m}^{-2}$. The actual value may be less than this because vertical mixing also fluxes heat (in fact the two reinforce one another), so we choose a range of 0 to $30\,\text{W}\,\text{m}^{-2}$ as safe. So here's the modified code (now called mpwp.m to distinguish from the first version):

```
% Note: we have left out some code where the comment line begins with %////
       %//// Load physical forcing ////

for Ehc=1:11                       % goes from 0 to 30 watts/m2
  EkmHeatConv = 3*(Ehc-1);         % Ekman Heat Convergence Correction
  for ndiff = 1:7                  % goes from 2 x 10^-4 down to 0.5 x 10^-4
    Kz = 2.0e-4-(ndiff-1)*2.5e-5;  % set vertical diffusivity

       %//// code to initialize and run the model here ////

      mdooutput      % print out results
      mpwcompare     % compare to data
      pwsave                          % save the run
   end
end
pwplot                               % plot the diagnostics
```

Note that we've created a Results array to hold the diagnostic data, and we had to change dooutput to mdooutput so that it doesn't put a lot of stuff on the screen (fine for a single run, but tedious for multiple ones). We've also modified the pwcompare to become mpwcompare so that it saves the diagnostics to the Results array. Finally, we've added a routine pwplot that plots the data up nicely (see Fig. 15.9).

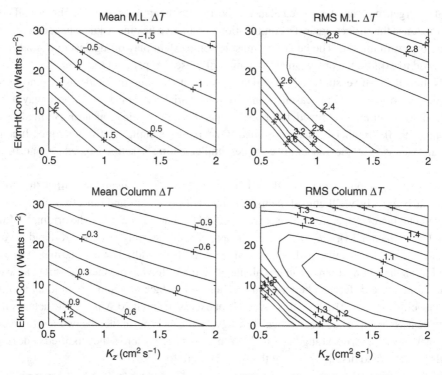

Figure 15.9 Contour plots of model–data temperature difference. Upper left, mean difference for the mixed layer; upper right, RMS difference for mixed layer; lower left, mean difference for the water column; lower right, RMS difference for water column. In each case, the temperature difference is shown as a function of vertical diffusivity (horizontal axis) and Ekman heat convergence correction for the model simulation.

When we said "brute force", we really meant it. The model is run 77 times to fill in parameter space with some reasonable degree of resolution. In our case, the total runs took about a half-hour on a 2 GHz computer. If we were trying to optimize with more parameters, such a strategy would quickly get time-intensive. Note also that each model run saves about a megabyte of data, so if storage is a problem you might either want to reduce the number of time slices saved (less desirable), or just not save the data to disk.

Examination of the plots reveals a few interesting trends. First, there is a clear trade-off between the Ekman heat divergence and vertical mixing; increasing one is compensated for by decreasing the other. We expected this when we talked about the mechanics of these processes. The tendency for the trends to mirror our expectations is certainly comforting and helps build confidence in the integrity of our model simulations. One should in general be able to explain the basic trends based on intuitive arguments, but not always; that's why we run the models. However, if the behavior runs counter to expectation, closer scrutiny is always advised. Is there a coding error somewhere, or do we not really understand what's going on?

There appears to be a "valley of the optimum" (a minimum) in the RMS mixed layer temperature offset extending diagonally from low K_z, high $EkmHtConv$ through to high K_z, low $EkmHtConv$. A similar feature is observed in the water column RMS offset, but the latter has a somewhat better defined shallow minimum centered around $K_z = 1.25 \times 10^{-4}\,\mathrm{m^2\,s^{-1}}$, $EkmHtConv = 12\,\mathrm{W\,m^{-2}}$. Also, and not particularly surprisingly, the "zero lines" for the mean temperature offsets appear to run close to the RMS valley minima. Thus our cost function points to an optimal set of adjustable parameters that best suit the model–data fidelity.

A final question is whether the sizes of the RMS temperature offsets are acceptable. The RMS scatter is about 2 °C in the mixed layer and about 1 °C throughout the water column. Such deviations are about 10–20% of the total range of the observed temperature changes, so in a qualitative sense, the model explains roughly 80–90% of the story. Not exact, but considering its simplicity, that's really not too bad. In summary, it looks like we have a physical model which does a more than half-way decent job of pretending it's an ocean. Now we'll go ahead and try to put gases into it.

15.6 Adding gases to the model

15.6.1 Noble gases as probes

You may wonder, why bother with all the noble gases? Why not just use argon, which we've argued to be very close in physical behavior to oxygen? Bear in mind that the match isn't exactly perfect. Moreover, we would like to use whatever information we have to model and understand as completely as possible, and as sensitively as possible, the nature of the physical processes affecting oxygen, and for that matter all gases in the surface ocean. In a sense, we're looking for "leverage" from the relatively wide range of physical characteristics of the noble gases. This comes from the fact that there is a wide range in both the solubility of the noble gases and their molecular diffusivity (see Fig. 15.10). It turns out that both properties will play an important role in changing upper-ocean gas concentrations.

The noble gases span approximately an order of magnitude in solubilities, with xenon (Xe) being about 10 times as soluble as helium (He) in seawater. This will be important in bubble-trapping (air injection) processes; for every mole of air forced into solution, He will be affected approximately 10 times more than the much more soluble Xe. There are also significant differences in temperature dependence; Xe solubility changes by nearly a factor of 2 over the temperature range depicted, yet He changes by only 4%. This will prove important in the surface layer of the Sargasso Sea, which changes by approximately 10 °C on a seasonal basis. This swing drives large solubility changes, and hence potential inventory shifts, in Xe but not in He. Note also the close similarity between O_2 and Ar. The noble gases also span a large range in molecular diffusivities, with He diffusing roughly 7-fold faster than Xe. This will play a role in air–sea gas exchange processes, which are ultimately limited by molecular diffusivity. Note again the close correspondence between O_2 and Ar.

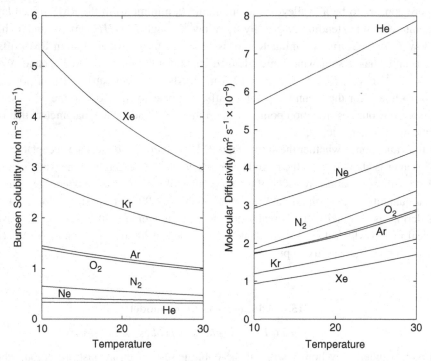

Figure 15.10 The Bunsen solubilities of the noble gases, nitrogen, and oxygen as a function of temperature (left panel), and their molecular diffusivities as a function of temperature (right panel).

Thus noble gases provide us with a large range in physical properties, and respond to seasonal timescale physical forcing very differently. They have one other, very clear advantage; they participate only in physics, and are unaffected by biological or chemical reactions. Thus we should be able to fingerprint the underlying physical processes both sensitively and unambiguously. At least that's what we hope!

15.6.2 *All you need to know about gas exchange*

Gas exchange between the ocean and the atmosphere is a complex and contentious topic. It is still a very active area of research, and we will make no attempt to advance the state of the art here. For the purpose of instruction, we'll work with a very simple model of gas exchange suitable for our calculations. If you really want to work on the leading edge, we encourage you to start with a recent publication (e.g. Stanley *et al.*, 2006, 2009) and explore the literature from there.

We will choose an existing gas exchange formulation, treat it as a given (subject to a multiplicative scale factor of order 1), and then evaluate model–data agreement as a function of that parameter. In this way, we hope to demonstrate that our observations can be used to constrain the gas exchange rate in some quantitative way. Although our objective is to learn something about biological production of oxygen in the upper ocean, being able to make a seasonal timescale estimate of gas exchange rates is actually a useful scientific goal in its own right.

To accomplish this, we'll break gas exchange between the ocean and the atmosphere into two separate (but in many ways linked) components:

- A molecular diffusively limited gas exchange that serves to restore dissolved gases to their solubility equilibrium concentrations with the atmosphere; and
- A bubble-mediated air injection component that forces air into the water by hydrostatic compression, thereby creating *supersaturations*.

We'll discuss the first component in this subsection, and consider the second component in the next. Although both are technically *gas exchange* processes, we'll use a shorthand terminology whereby we refer to the first component as "gas exchange" and the second as "air injection". In truth, the two are entangled at the basic physico-chemical process level, but for the sake of simplicity and our model, we'll artificially separate the two.

For reasonably insoluble and relatively weakly reactive gases – and this includes all of the gases we'll be considering here – diffusively mediated gas exchange is generally modeled as a first-order process where the flux of gas between the ocean and atmosphere F is assumed to be proportional to the degree of under- or supersaturation of the gas with respect to equilibrium conditions:

$$F = k(C - C_{sat}) \tag{15.10}$$

where C is the measured gas concentration in suitable units, and C_{sat} is the saturation (expectation) value in equilibrium with the atmosphere. If we were working in mks units, C would be expressed in $mol\,m^{-3}$, and for a flux expressed in $mol\,m^{-2}\,s^{-1}$, the coefficient of proportionality k will have units of $m\,s^{-1}$, like a velocity. In fact, k is generally referred to as a *piston velocity* or a *gas transfer velocity* for that reason. Conceptually, it may be visualized as the speed by which gas is "pushed out" (if supersaturated) or "pulled into" the water column (if supersaturated), as if by a giant piston.

The size of the gas exchange velocity depends on a variety of factors, many of which are not completely known. These include the wind speed, the intensity, duration, and shape of waves, the presence or absence of surfactants, and the state of turbulence in the water below the surface. In the absence of this detailed information, investigators resort to a compromise model that characterizes the coefficient as a function of the molecular diffusivity for the gas, the kinematic viscosity of the water, and the wind speed.

Various determinations of gas exchange rates have been made under a large range of circumstances (in the laboratory, and in rivers, lakes, and oceans) using a variety of techniques, including tracer release experiments, radon deficit measurements, and global-scale radiocarbon balance calculations. The results are subject to a significant scatter, but in general suggest that gas exchange velocities increase strongly with wind speed. That is, the wind speed dependence appears steeper at higher wind speed. Thus quadratic (Nightingale,

2000) or even cubic (Wanninkhof and McGillis, 1999) dependencies have been proposed. Here we use the quadratic relationship proposed by Wanninkhof (1992):

$$k = 0.31 u_{10}^2 (Sc/660)^{-1/2} \tag{15.11}$$

where k is the piston velocity, which in Wanninkhof's equation above is expressed in units of $cm\,h^{-1}$. You will also see piston velocities expressed in $m\,d^{-1}$, in which case the numeric constant in Equation (15.11) would be 0.0744. In our model, we'll stick to mks units, and simplify the equation to

$$k = 2.212 \times 10^{-5} \frac{u_{10}^2}{Sc^{1/2}} \tag{15.12}$$

by rolling the $660^{1/2}$ into the constant. In Equation (15.12) the gas exchange velocity k would be in $m\,s^{-1}$. This is a very important point; use consistent units and your life will be simpler and less error-prone! Also note that in Equations (15.11) and (15.12), u_{10} is the "short term" wind speed measured at 10 meters above the ocean surface. Wanninkhof also presents a separate equation for use with *climatological* or long-term averaged winds. The reason this is necessary is that since gas exchange depends on the square of the wind speed in this formulation, calculating the gas exchange rate from average winds using Equations (15.11) or (15.12) will give an underestimate – the square of the mean of any data set will always be less than the mean of the squares.

The Sc in Equation (15.12) embodies the dependency of gas exchange on the characteristics of the gas and the thickness of the seawater boundary layer. It is a unitless quantity called the *Schmidt number*, defined by:

$$Sc = \frac{\nu}{D} \tag{15.13}$$

that is, the ratio of kinematic viscosity of the water (ν) to the molecular diffusivity of the gas (D). If you picture a thin boundary layer in the water in contact with the atmosphere, transfer of gas will be limited by molecular diffusion. If this boundary layer is permanent, which it might be at very low wind speeds, then the flux will be proportional to the diffusivity, and hence would vary inversely with Sc. Where ν comes in is that the thickness of this layer is related to the water's kinematic viscosity. The gas exchange flux will be inversely related to the layer's thickness. At more than very low wind speeds, however, the layer is constantly being sheared away and replaced, so on average, there is a weaker dependence. We suggest you consult an introductory textbook (e.g. Emerson and Hedges, 2008) for a more complete discussion. You may have wondered about the strange factor of 660 in Equation (15.11). It is just a normalization factor and has no deep significance; it is the Schmidt number of CO_2 gas at $20\,°C$. As you may have guessed, this gas is of considerable interest regarding climate change nowadays.

Table 15.1 contains Sc numbers for different gases at different temperatures, which serves also to underline part of our strategy here. Note the similarity in Sc numbers between oxygen and argon. Since gas exchange depends on the square root of Sc, then the difference

Table 15.1 *Schmidt numbers vs temperature for some gases*

Temp.	Helium	Neon	Nitrogen	Oxygen	Argon	Krypton	Xenon
18	173	324	466	538	547	741	932
22	148	271	377	442	450	603	753
26	125	226	304	358	365	488	607
30	105	186	244	287	292	392	486

in behavior physical behavior between the two gases ought to be less than 1%. Thus Ar looks like a good proxy for oxygen, i.e. an abiogenic analog of oxygen. This means that observation and successful modeling of Ar will do a good job of tracking abiotic effects in oxygen. But note the large range in Sc between He on the one extreme to Xe: almost a factor of 5. By emulating the entire range of gases we will be able to test and constrain our model more stringently and account better for the relatively small physical differences between Ar and O_2.

So all we need to do for gas exchange in our model is, for each time step, to construct a flux of gas to or from the ocean surface box driven by the gas transfer velocity multiplied by the difference between the actual gas concentration and the equilibrium gas concentration. Thus we have for the ith gas:

$$M^i = F^i A = k^i \left[C^i - C_{sat}^i(T, S) \right] A \qquad (15.14)$$

where A is the area through which the gas must pass (typically assumed to be one square meter) and where we calculate the gas exchange rate k from Equation (15.12). Note that k depends on the specific gas, and also depends on the temperature and salinity through the Schmidt number dependence in Equation (15.12). Also, C_{sat} is the equilibrium concentration calculated as a function of temperature and salinity. If we examine the resultant concentration change due to this flux over the time step (Δt) and spread over the mixed layer depth (h), we have for the ith gas:

$$\Delta C^i = k^i \left[C^i - C_{sat}^i(T, S) \right] \frac{\Delta t}{h} \qquad (15.15)$$

The area A cancels out because the volume of the water column is the height times the area. Thus our budgeting amounts to accounting for a gas flux over each time step due to disequilibrium with the atmosphere. We could have simply done the flux adjustment for the top cell in the model grid, but that introduces another restriction on the computation; the time step must be small enough in the case of outgassing that the amount of gas exchanged out of the box does not exceed its contents. In essence, this is a kind of Courant number for the gas exchange velocity. By increasing the number of boxes involved to the depth of the mixed layer, this relaxes the requirement by increasing the effective Δz. The convenient aspect of this is that the mixed layer is generally at its deepest when the gas exchange is at its highest.

15.6.3 Air injection

An additional aspect of gas exchange between the ocean and the atmosphere is the formation and dissolution of bubbles by wave action. Because bubbles formed by waves (as typified by white-caps) are forced downward into the water column by wave motion and turbulence, they are subjected to increasing hydrostatic pressure, and will partially or completely dissolve. The net effect is a transfer of gases from the atmosphere to the ocean, and a concomitant generation of supersaturation. Actually, the process results in a two-way exchange of gases because bubbles rising back toward the surface, if not completely dissolved, will exchange gases with surrounding water and act as vehicles of gas transport. To some extent, the nonlinear nature of the gas transfer velocity's dependence on wind speed may be a reflection of increased bubble activity at high wind speeds.

The impact of bubbles on upper ocean gas fluxes is extremely complicated, because the surface ocean is generally characterized by a broad spectrum of bubble sizes, and the effects of turbulence, buoyancy, and surface tension vary with bubble size. Moreover, the internal gas pressure and the gas composition of individual bubbles evolve over the course of its trajectory, changing its interaction with surrounding water. The interested reader should consult the literature for further information. In the spirit of teaching you modeling rather than geochemistry per se, we will choose a highly idealized model of air injection, namely a one-way flux of unfractionated air into the water column. That is, we are effectively assuming that all bubbles, once entrained by wave action, are subducted and forced completely into solution. For a more sophisticated and realistic treatment we suggest you consult recent work by Rachel Stanley (Stanley, 2007; Stanley *et al.*, 2009).

We will use a simple formulation which treats the volume flux of air F_{Air} per unit area into the surface ocean as a function of the fractional white-cap coverage proposed by Keeling (1993):

$$F_{Air} = A_C V_A W_A \qquad (15.16)$$

where A_C is a kind of "efficiency coefficient" that is roughly 1.4×10^{-3}, V_A is the air entrainment velocity, estimated to be approximately $0.01 \, \mathrm{m \, s}^{-1}$, and W_A is the fractional surface area covered by white-caps, estimated by Monahan and Torgersen (1991) to be:

$$W_A = 1.9 \times 10^{-6}(u_{10} - 2.27)^3 \qquad (15.17)$$

where u_{10} is the velocity in $\mathrm{m \, s}^{-1}$ at 10 m height (the standard meteorological height for wind measurements and reporting) and $W_A = 0$ for $u_{10} < 2.27 \, \mathrm{m \, s}^{-1}$. Combining the two equations and converting from cubic meters of air to moles of air, we get the flux F_{Air}:

$$F_{Air} = 1.16 \times 10^{-9}(u_{10} - 2.27)^3 \, \mathrm{mol \, m}^{-2}\mathrm{s}^{-1} \qquad (15.18)$$

In addition to the crudeness of the models (no reflection on the proponents, it's just that the actual physical processes are extraordinarily complex) the "constants" used in these relations are subject to considerable uncertainty. Our strategy will be to use the existing model as a point of departure, and to scale the net injection rate by a constant (initially 1), which we'll adjust to match the observations. Since the effect of the injection on gas saturation is inversely proportional to solubility of the gases involved, and since He and Ne are many times less soluble than Ar and O_2, we have a "very long lever" on the magnitude of the process.

15.7 Implementing the gas model

We are going to build on the physical model to make it gas-enabled. We can inherit most of the physical code without much modification, if any. This convenience arises from our strategy of compartmentalizing different functions. There are a few gas-specific routines, of course.

15.7.1 Gas model m-files

Below are the m-files constituting the gas-enabled PWP model, which we've called pwpg.m. It contains most of the same m-files as the original, non-gas version of pwp with a few additional files. Since the model must carry along, operate on, and store more variables (all the gases), it will run noticeably move slowly than the purely physical model. Still, it takes less than a minute for a full run on 2 GHz laptop. The extra m-files and data sources are:

- inigas.m – initializes gas profiles and factors
- gasexch.m – does gas exchange and air injection at surface

There is an addition utility function, listed below

- gassol.m – computes gas solubilities

Also, we need some additional "forcing" files:

- GasInfo.mat – used for solubility and Schmidt number calculations
- GasData.mat – gas observation data (courtesy Dr. Rachel Stanley) for comparison

The basic code is augmented by the addition of these routines in the following way:

```
            //// load the physical forcing and initialize ////
AirInjAmp=1;        % set Air Injection Amplitude
GasExAmp=1;         % set Gas Exchange Amplitude
inigas;             % initialize water column gases
%
%                         Main time step loop
%
for it=1:nt              % begin time loop
           //// heat & FW forcing, physical stability adjustments ////
   gasexch;                   % exchange gases
```

```
advdif;                             % advect and diffuse
dooutg;                             % periodically save data
end                     % end time loop
pwgcompare                          % compare model and data
pwgsave                             % save results
```

Like with the physical model, we have added two routines (`pwgcompare.m` and `pwgsave.m`) that show some plots, calculate diagnostics, and save the model run. You'll notice, not surprisingly, that the data files stored are bigger because of the added gas data. The comments inside the routines should tell you what is going on.

One note. You'll find that we've done some sleight of hand in computing the Schmidt numbers, solubilities, and various other gas-related factors. The actual, full precision calculations are multiparameter semi-thermodynamic formulae which involve logs and exponents. These are computationally very expensive, so we have done the calculations up front, and stored them in `GasInfo.mat` as grids or "look-up tables" in temperature and salinity space. We then use linear interpolation to compute the relevant properties, which is intrinsically much faster. An enterprising reader might experiment with evaluating first how large an error is introduced by this approach (it's small, we've checked!) and second how much the model would be slowed down with the full-blown solubility calculations.

15.7.2 Comparing the gas model to observations

Figure 15.11 shows the gas concentrations (He, Ar, and Xe) for a typical, but not necessarily optimal run. Note that the calculations are always done in terms of mass, and therefore with gas concentrations in units of $mol\,m^{-3}$ and with fluxes – whether advection, diffusion, gas exchange, or air injection – in terms of $mol\,m^{-2}\,s^{-1}$. We cannot overemphasize the importance of this. *Never* run your model in terms of saturation anomalies or with inconsistent units. The former leads to subtle nonlinearities, the latter can lead to silly errors.

All three gases, which span the "spectrum" of noble gas characteristics (see Fig. 15.10), show qualitatively similar patterns: higher near-surface values in the winter relative to the summer. This happens for two very different reasons. For He, the winter increase is due in large part to increased air injection – remember that we characterized air injection as a cubic function of wind speed – whereas the differences for Xe are largely driven by thermal changes: decreased wintertime temperatures lead to increased solubility. The span of the changes differ as well: He has a range of variation approaching 4 or 5%, whereas Xe changes by about 20%. Argon appears to be in between, but perhaps a bit more like Xe than He.

It is instructive to look at gas saturation anomalies, that is, the percentage deviation from solubility equilibrium with the atmosphere (Fig. 15.12), which is defined for Xe (for example) by

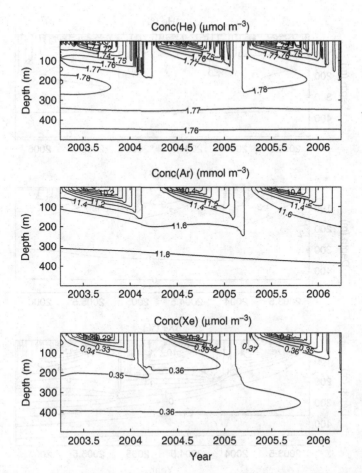

Figure 15.11 A three-year depth vs. time contour plot of He (upper), Ar (middle), and Xe (lower) concentrations. Note that Ar, because of its relatively high atmospheric abundance, is in mmol m^{-3} instead of the usual µmol m^{-3} for the other gases. Their seasonal evolutions show a tendency for lower concentrations in the summer than in the winter, but for different reasons.

$$\Delta \text{Xe} = 100 \times \left(\frac{C(\text{Xe})}{C_{\text{sat}}(\text{Xe})} - 1 \right) \tag{15.19}$$

so that a gas at 110% of saturation would have $\Delta = 10\%$ and a gas at 95% of saturation would have $\Delta = -5\%$. Figure 15.12 shows the saturation anomaly plots for the same three gases and the same model run as Fig. 15.11.

Now specific distributional differences appear between the gases that more closely reflect some of their characteristic responses to the underlying processes. For example, Ar and Xe show distinct summertime saturation anomaly maxima at approximately 50 meter depth, which clearly appear in mid-summer owing to solar warming, and He shows a maximum in winter owing to air injection processes.

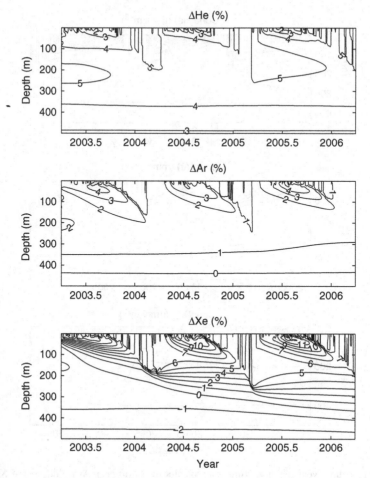

Figure 15.12 A three-year depth versus time contour plot of the saturation anomalies of He (upper),
Ar (middle), and Xe (lower) from the same model run.

These model trends are also apparent in surface water saturation anomalies (the solid
line in Fig. 15.13). We show the surface water saturation anomalies separately because they
are important – remember Equation (15.10)? The gas exchange flux of gases is controlled
by the surface saturation anomalies, and we must get this right if we are to get the water
column budgets right. Put another way, these saturation anomalies are particularly sensitive
to gas exchange and air injection processes, so getting them right means we get the gas
exchange and air injection processes right. Note that for the first time, we are daring to show
real data – courtesy of Rachel Stanley (Stanley, 2007). Now we see persistent summertime
saturation anomaly maxima in Ar and Xe, and minima in He. Thinking about the processes,
this suggests that the ocean is breathing out He throughout the year (balancing the air
injection), but also breathing more Ar and Xe out in the summer months, owing to the

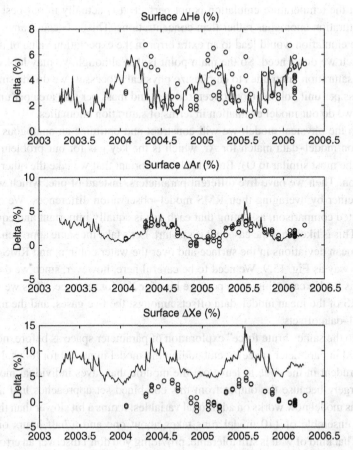

Figure 15.13 Model (solid line) and observed (open circles) surface water saturation anomalies of He (upper), Ar (middle), and Xe (lower). The observations are courtesy of Dr. Rachel Stanley (Stanley, 2007).

warming effect. But something has become very clear here; our baseline model simulation does a reasonable job for He and Ar, but not nearly good enough for Xe. We need to adjust our model to get a better "fit" to observation. We need to optimize our choice of gas exchange and air injection rates.

15.7.3 Constructing a cost function

What are the criteria? Other than getting the model to run "right", construction of a cost function is the most important part of the process. Here is where important scientific as opposed to technical decisions must be made. We have at our disposal some 570 measurements of dissolved gas concentration for each noble gas. That is, we have over 2800 observational constraints. Thinking about the nature of the simulation, and in particular the fact that the temperature emulation is not perfect, it is actually in our best interest to compare saturation anomalies rather than concentrations. This is because any error in the temperature emulation would lead to an extra error in the expectation value of the concentration, which we don't need. So the main point is that although we may look at the data in terms of saturation anomalies to emphasize physical processes, we do the simulation in units of mass per unit volume (for concentrations) and mass per unit area per unit time for fluxes, and we do our model evaluation in terms of saturation anomalies.

Like with the physical model, we will construct and optimize an analogous cost function based on model–data match for Ar, which is the key gas for our problem since it is physically the most similar to O_2. But it is also important that we take the other gases into consideration. Then we have five different parameters, instead of one, which we can just "lump" together by averaging their RMS model–observation differences. We can choose an unweighted comparison, assuming that each gas is equally important and equally well-measured. This is likely not the case, but it's a start. Let's take the same approach of looking at lumped mean deviations in the surface and over the water column, and RMS deviations in the same way as Fig. 15.9. We need to be careful here, however, since we don't want a negative bias for He canceling out a positive bias for Xe, for instance. Thus we will calculate the RMS of the mean model–data offsets amongst the five gases, and the mean of the RMS model–data offsets.

We can do the same "brute force" exploration of parameter space as before, now over an 11 by 10 grid in parameter space by embedding the model in nested for loops. Note that it would be prudent in the code to leave out the module that saves individual model output datasets, largely because the output from this combined set approaches half a gigabyte. Since the gas model now works on additional variables, it runs a bit slower than the physical model; the ensemble of 110 model runs takes about one and a half hours on a 2 GHz computer. That kind of wait is still tolerable, providing you don't discover an error half-way through the run – or worse yet 90% of the way through.

```
% Note: we have left out some code where the comment line begins %////
        %//// load the physical forcing ////

Ainj=[0.5:0.5:5]; GasEx=[0.5:0.1:1.5];   % parameter vectors
Results=zeros(length(Ainj),length(GasEx),4,ngas);   % storage for results
for iAI=1:length(Ainj)          % loop over air injection parameters
    AirInjAmp=Ainj(iAI);        % set Air Injection Amplitude
    for iGE=1:length(GasEx)     % loop over gas exchange parameters
        GasExAmp=GasEx(iGE);    % set Gas Exchange Amplitude

        %//// gas model code run here ////

    pwgcompare              % compare model output to data
    end                     % end air injection loop
end                         % end gas exchange loop
save Results Results Ainj GasEx  % save diagnostics file 'Results.mat' for later
```

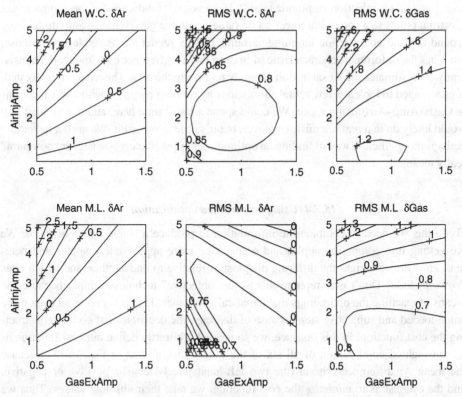

Figure 15.14 Contour plots of six diagnostics comparing model–data gas saturation anomalies as a function of amplitudes of air injection (y axis) and gas exchange velocity (x axis). The six diagnostics are the mean water column Ar difference (model minus data, top left), the RMS water column Ar difference (top middle), mean RMS water column difference for all the noble gases (top right), the mean mixed layer Ar difference (model minus data, bottom left), the RMS mixed layer Ar difference (bottom middle), and the mean RMS mixed layer difference for all the noble gases (bottom right). The observations are courtesy of Dr. Rachel Stanley (Stanley, 2007).

The RMS and average model–data differences are computed and stored in a multi-dimensional array called "Results" in `pwgcompare` – we encourage you to look at the model code to see how that is done – and the array is stored at the end of the run. We then contour the diagnostics using the routine `pwgplot`, which are presented in Fig. 15.14.

Be warned that the following discussion ventures into the realm of observational geochemistry, but we suggest you follow along with a view that this is an example of how the interpretation of model diagnostics might progress. Examination of the contour plots in Fig. 15.14 hints at some of the trade-offs associated with model fidelity. If we were to aim blindly at simulating Ar alone, which after all is part of our biogeochemical motivation, the "topography" of the Ar cost functions (the left-hand and middle panels of Fig. 15.14) suggests an optimum near the middle of the range, say with a gas exchange velocity amplitude near 1 and an air injection amplitude near 2. However, the landscape for the complete gas spectrum points to a somewhat lower set of values, with the gas exchange amplitude being around 0.9 and air injection amplitude around 1. This seems a reasonable compromise, but it has the unfortunate characteristic of biasing the Ar differences: i.e. the model consistently overestimates the Ar saturation anomaly relative to the data. The reader, if interested, is encouraged to look at a few actual simulations by running pwpg.m with a few locations in GasExAmp–AirInjAmp space. We could spend a lot of time here, and that is what you would likely do in a real scientific endeavor, but it's time to move on. We need a better and perhaps more efficient way of finding an optimum set of parameters – whatever "optimum" really means.

15.7.4 *Using constrained optimization*

To do this we need to combine the information to produce a single figure of merit. We are getting into comparing apples and oranges, or more appropriately apples and petunias. How do we weight the differing diagnostics relative to one another, and which ones do we include? Don't ever refer to this as an "objective" technique simply because the means of reaching the optimum is mathematically automated. This becomes an even more value-loaded and subjective step because of the scientific decisions that go into constructing the cost function. In this instance, we somewhat arbitrarily define our cost function as an unweighted combination of all six of the diagnostics plotted in Fig. 15.14. Because the mean Ar difference criteria (the two left-hand panels) can be positive or negative, and the exercise is to *minimize* the cost function, we take their absolute values. Thus we have:

$$J = \frac{1}{6}\left\{ \sqrt{\frac{1}{N}\sum \delta C_3^2} + \frac{1}{N}\left|\sum \delta C_3\right| + \sqrt{\frac{1}{N_S}\sum_{\text{Surf}} \delta C_3^2} + \frac{1}{N_S}\left|\sum_{\text{Surf}} \delta C_3\right| \right.$$
$$\left. + \frac{1}{5}\sum_{i=1}^{5}\left[\sqrt{\frac{1}{N}\sum \delta C_i^2}\right] + \frac{1}{5}\sum_{i=1}^{5}\left[\sqrt{\frac{1}{N_S}\sum_{\text{Surf}} \delta C_i^2}\right] \right\}$$

(15.20)

where the summations are over the N_S surface or N model–data pairs, and δC_i is the percent model–observation concentration difference defined by

$$\delta C_i = \left(\frac{C_i}{\widehat{C}_i} - 1\right) \times 100$$

(15.21)

where C_i and $\widehat{C}_i$ correspond to observation and model concentrations for the ith gas and where $i = 1 - 5$ corresponding to He, Ne, Ar, Kr, and Xe respectively. In a crude sense, the expectation value of the cost function is related to the RMS scatter of the model–data concentration differences. If we think that the combined measurement and sampling error is of order 0.5%, then we would expect a cost function of comparable magnitude if the model is a good fit to the data. Bear in mind that one should look for space and time structure in the residuals (model–data differences) because we can delude ourselves that the model does approximate reality when significant excursions may occur in important places.

We modify mpwpg to become costpwpg, a self-contained MATLAB function that now returns the cost function value TheCost within CostCalc:

```
%         the cost function version of the pwpg program
%         takes as input a parameter vector
%         Parameters = [AirInjAmp GasExAmp]
%         and returns a cost function value based on
%         model-data differences
%
function TheCost = costpwpg(Parameters)
GlobalBlock         % declare  the global storage
AirInjAmp=Parameters(1);
GasExAmp=Parameters(2);

        //// initialize and run the model ////

CostCalc            % compare model to data
                    % and assign output to TheCost
end
```

There is one additional complication: to avoid reloading the forcing for each model run, we put the forcing data into global storage, which occurs in GlobalBlock. This way it is accessible from cost-calculating routine costpwpg, which needs to know about forcing. The driver routine OptimizePWPG performs the optimization and uses the MATLAB Optimization Toolbox routine fmincon, which you learned about in Chapter 10:

```
%         the driver program that runs the cost     %
%         function to optimum                        %
%
GlobalBlock       % declare the global storage

    //// load forcing initialize physics ////

Pmin=[0.1 0.1];     % we choose these limits based on our earlier
Pmax=[4 2];         % explorations
Pstart=[1 1];       % our best guess at a starting point
options = optimset('LargeScale','off','Display','iter', ...
        'MaxFunEvals',30,'MaxIter',10,'TolCon',1e-4);
warning off all
[Poptim,fval,exitflag,output,Lambda,Grad,Hessian]= ...
        fmincon(@costpwpg,Pstart,[],[],[],[],Pmin,Pmax,[],options);
```

Here we have to declare the global data block, load the forcing, set the physical parameters, set the appropriate minimum and maximum acceptable values for the gas parameters, and give the optimizer a sensible starting point. We also set a limit on the maximum number of model runs (30) and iterations (10), and some kind of convergence limit for the cost function improvement (`1e-4`).

The optimization completes after 7 iterations (32 model runs), since it doesn't stop until it exceeds one of the above limits or meets the convergence criterion above. Interestingly, it comes close to the optimum point in 22 runs, and then spends 10 runs improving the cost function by less than 1% of its value. The optimization runs in about 35 minutes. This clearly is faster than the brute force approach, where we scoped out parameter space with 110 model runs. Also, it produces well-defined optimum values for the parameters rather than making us pick them from a graph. The minimum in the cost function corresponds to a gas exchange velocity amplitude of 0.67 and an air injection amplitude of 2.4. We won't dwell on the geochemical implications, but the fact that they are not hugely different from the original estimates lends some confidence that nothing bizarre is going on in modelworld. (A factor of 2.4 in the air injection parameter is small compared with the large uncertainties in our starting point.) The optimum cost function of 0.62 suggests that the average disagreement between model and data is of that order (i.e. about 0.6%), which is crudely comparable to our measurement uncertainties. Considering the scale of variation in the gases (Fig. 15.12) this suggests a "model confidence interval" of the order of 10% for the rates and fluxes obtained from it.

The final step in the gas saga is to re-run the model (`pwpg` with the optimal parameters) and have a close look at the model plots it produces. Better than any "objective" measure like our little cost function, this should give a real feeling about how well the model is doing, and where it might be improved. Predictably, the model does best for Ar, but has to compromise between the He and Kr ends of the spectrum. Clearly an improved model could be considered, likely with a more sophisticated form of air injection, but for our purposes, we'll declare this a victory and move on.

15.8 Biological oxygen production in the model

The next step is to take our optimal model system and apply a simple oxygen production function that has one or two adjustable parameters. In the spirit of full disclosure, the choice of the character of the oxygen production function – in particular its spatial and temporal structure – is critical to the results. For example, if we choose an oxygen production function that is relatively large very near the surface, then much more of its production will be lost through gas exchange, than for a function that is larger below the mixed layer. Further, an oxygen production function that is large during wintertime deep convection, when gas exchange rates are large because of high wind speeds, will also have to be very vigorous to leave the observed oxygen "signal". This could prove to be a long and contentious discussion venturing far into the details of biological production, so we'll cut things short and simply specify an idealized form that may not be considered too ludicrous. After all, we're trying to teach you modeling skills here, not biogeochemistry.

15.8.1 An oxygen production/consumption function

We will specify a purely schematic oxygen production function that is zero right at the ocean surface (since production would be virtually invisible there anyway) and shaped like the first half of a sine wave whose maximum is at 50 meters and zero below 100 meters. We then finish it off with a linearly decreasing oxygen consumption curve below this depth that tapers to zero at the bottom of the model domain. For the sake of mass conservation, we stipulate that the area under the composite curve will be zero. Rather than trying to picture this, look at Fig. 15.15. The production curve, although a pure schematic, makes some sense because photo-inhibition likely suppresses production very near the surface, and production is known to decrease toward the base of the euphotic zone. The oxygen consumption part of the curve also roughly fits with our understanding of what is going on: oxygen demand is greatest just below the euphotic zone and decreases downward.

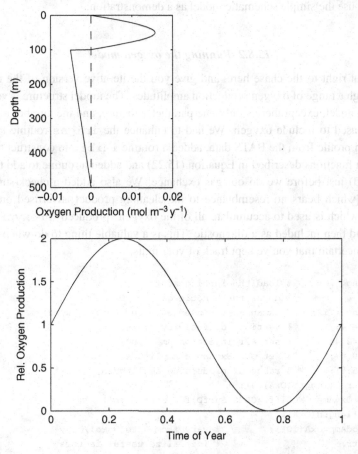

Figure 15.15 The model oxygen curve as a function of depth (upper panel) and time (lower panel). Note that the area under the depth curve is defined to be zero, and the mean time curve fixed to 1.

In Fig. 15.15 we also show a hypothetical timing curve, which is a sinusoid whose average is unity. The shape also makes sense in a schematic way, since we expect a maximum in early spring (in essence a spring bloom), and decreasing rates over the summer months owing to stratification and low nutrient fluxes. This is consistent with sediment trap measurements, for example those by Deuser (1986). Total production is given by the product of the depth and time curves:

$$P(t, z) = f(t)g(z) \tag{15.22}$$

i.e. assume the variables are separable. Now there is no fundamental basis on which we expect this to be true, because the depth distribution of production will be driven by nutrient availability, which will change in depth structure throughout the year. Clearly this is an interim choice, and one which would need to be reconsidered if one were to do fully prognostic modeling of the system (i.e. make a predictive model of biological production driven by nutrient transport, light profiles, species succession and grazing). However, for now, we'll use the simple schematic model as a demonstration.

15.8.2 Running the oxygen model

We will cut right to the chase here, and give you the iterative version of the model that runs through a range of oxygen production amplitudes. The model structure is very similar to the gas model, except there's only one parameter to vary, and the number of gases has been increased to include oxygen. We had to enhance the `inigas` routine to initialize the oxygen profile from the BATS data, added a routine `inibio` to construct the oxygen production functions described in Equation (15.22) and added a routine to add the oxygen (`oxyprod`) just before we do our gas exchange. We also added a check-sum variable `OxySum` (which bears no resemblance to the cleaning product advertised on the cable channels) which is used to accumulate all of the oxygen "produced" in `oxyprod` during the run, and then included as a diagnostic. This is a valuable thing to do when you are in the least uncertain that you've kept track of your units.

```
LoadForcing            % load the forcing stuff
ngas=6;                % doing the 5 noble gases + Oxygen
EkmHeatConv = 12;      % Ekman Heat Convergence Correction (w/m2)
Kz = 1.25e-4;          % vertical diffusivity (m2/s)
AirInjAmp=2.44;        % set Air Injection Amplitude
GasExAmp=0.67;         % set Gas Exchange Amplitude
OxA=[0:0.5:5];         % range of production amplitudes
Results=zeros(length(OxA),5);
fprintf(' OxyAmp  WCDiff WCRMS SurfDiff  SurfRMS\n')
for iOP=1:length(OxA)
    OxyProdAmp=OxA(iOP);        % vary from 0 to 5 mol/m2/y
    inifctr;                    % initialize useful factors
    inihydro;                   % initialize water column profile
```

```
inigas;              % initialize water column gases
inibio;              % initialize the oxygen production curves
%
% Main time step loop
%
for it=1:nt
    T(1)=T(1)+thf(it);      % add sensible + latent heat flux
    S(1)=S(1)*FWFlux(it);   % flux fresh water (latent/precip)
    T=T+rhf(it)*dRdz;       % add radiant heat to profile
    dostins;                % do static instability adjustment
    addmom;                 % add wind stress induced momentum
    dobrino;                % do bulk Ri No Adjustment
    oxyprod;                % produce the oxygen (before gas exchanging it)
    gasexch;                % exchange gases at the surface
    dogrino;                % do gradient  Ri No Adjustment
    advdif;                 % advect and diffuse
    dooutb;                 % periodically save data
end
    pwbcompare              % compare model and data -> diagnostics
end
save Results Results OxA    % save diagnostics for later
```

The alert reader of the complete code will notice that we've introduced yet another pair of cost functions, this time the mean and RMS model–data differences in the top 100 m. The argument is that we need to evaluate model fidelity in the top 100 m and wish to avoid "contamination" of the result from a poorly constructed oxygen *consumption* curve below the euphotic zone. You might argue that the surface water diagnostic should give this, but recall that the oxygen concentrations in the surface water will be affected strongly by air injection and gas exchange, and hence may not respond as clearly to oxygen production – particularly since our production profile is deliberately zero at the surface!

So in Fig. 15.16 we plot the RMS model–data difference as a function of the oxygen production amplitude. The result is a broad minimum in the vicinity of $2.7 \, \text{mol m}^{-2} \, \text{y}^{-1}$. We would be hesitant to draw any biogeochemical conclusions from this number, given the arbitrary nature of the oxygen production curves that we used. The cost function minimum varies only weakly as function of the oxygen production amplitude. This is both an indication of the relatively broad confidence intervals on the estimate, and an indication that the quality of the fit is likely not very good: the RMS misfit is in excess of 2%, whereas we know that the oxygen measurement precision is significantly better than 0.5%. The fact that we have significantly better fit qualities for the noble gases suggests that we can do more to improve things with the oxygen production curves, but we have reached the end of what we wanted to accomplish here.

Perhaps the most important thing to do is to *look at the model output itself.* We have the "optimum" solution, and we can then run the model at the optimum oxygen production

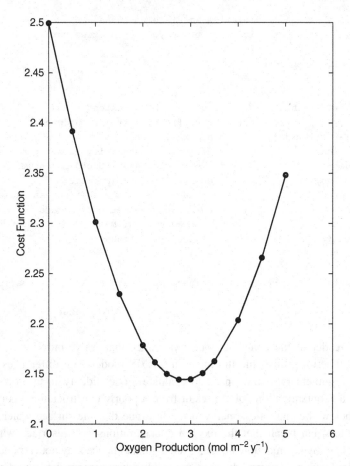

Figure 15.16 A plot of the RMS model–data saturation anomaly difference for upper 100 m.

amplitude, plot the results (Fig. 15.17), and compare its oxygen and argon output to the observations (remember Fig. 15.1). Yes, this is the Rorschach test that we have repeatedly cautioned you not to rely on, but the visual comparison is often a powerful tool for seeing just why something is not working right, and it may give you useful clues as to how to fix things.

We've demonstrated an approach here that we hope you'll find generally applicable to a variety of modeling problems. We refer not just to the numerical tools associated with building models and running them efficiently and correctly, but also with *diagnosing* what they're doing, and how, where, and perhaps why the models do or don't do well. It is very important to build probes to understand what's going on inside the models, as more often than not, you're the only one who may know or have access to the detailed code. The one thing that is worse than a model that won't work is one that isn't working right – but you don't know it!

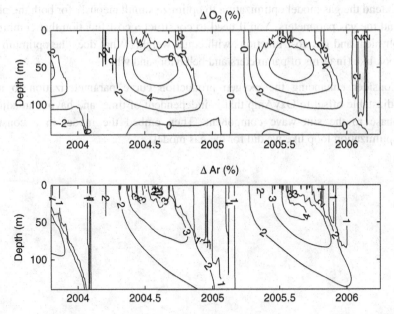

Figure 15.17 Contour plots for the model oxygen and argon output for upper 140 m. Compare this with Figure 15.1.

15.9 Problems

15.1. Apply the optimization technique used on the gas model to the physical model in order to optimize Kz and EkmHeatConv. You will need to make the same kind of modifications as we made for the gas model optimization. In the same direc-tory as the Physical directory, create a separate directory (call it something like PhysOptim) and copy over the physical m-files. Use the gas optimization discussion and copies of the m-files to do the job.

15.2. Run the gas model (pwpg) with the optimized parameters and describe the character of the model–data mismatch. Can you think of any additional processes that you could invoke that might mitigate the problem? Please, we will not consider cold fusion or Xe-eating bacteria!

15.3. Convince yourself that the gas cost function is at a global optimum by running the optmization from radically different starting locations in AirInjAm versus GasEx-Amp space. Try the following starting points: (0.1, 5), (2, 0.1), (2, 5), and (0.1, 0.1). Do you get to the same optimum? Does the optimizer take much longer? You will need to increase the maximum number of model runs from 30 to 50, and try setting it up to run overnight as a batch. You'll have to get the m-file that runs the four opti-mizations to save the resultant Poptim to a disk file. You can use the MATLAB diary function to track convergence.

15.4. Extend the gas model optimization to optimize simultaneously for both the physical and the gas parameters. You'll need to construct a cost function that combines the physical and gas cost functions with equal weight. What does the optimum model look like (in terms of parameters and behavior) and why?

15.5. Consider enhancing the oxygen production curve parameterization to add an adjustable offset to OxyAmp that is independent of time, and having an adjustable phase to the sine-wave component. Then embed the model in a constrained optimization loop like we did for the gas model.

16

Two-dimensional gyre models

What is art but life upon the larger scale, the higher. When, graduating
up in a spiral line of still expanding and ascending gyres, it pushes
toward the intense significance of all things, hungry for the infinite?

Elizabeth Barrett Browning

16.1 Onward to the next dimension

Although one-dimensional models provide useful insight into basic biogeochemical processes, we are forced to admit that the world is made of more than one spatial dimension. The addition of an extra dimension to a model often does more than "fill space", but rather imbues the model with behavior that is qualitatively different from its lower-dimensional analogue. The opportunity presented by the extra dimension is that more interesting, and perhaps more "realistic" phenomena may be modeled. This opportunity brings with it challenges, however, that are not just computational in nature. The choices of model geometry, circulation scheme, and boundary conditions become more complicated. Seemingly innocuous choices can have subtle or profound effects on how your model behaves. Moreover, matching model results to observations often requires decisions about whether features result from intrinsic processes of interest, or are mere artifacts of the choices made in model configuration.

For instructional purposes, we'll stick to a genre called *gyre models* which, as you might guess, are characterized by a quasi-circular flow on a plane. Such models have utility in the subtropics – at least that's where we'll be dwelling here – but can be used in many other parts of the ocean. They are useful for two reasons. First, the shallow circulation of the ocean tends to organize itself into regional gyres. Second, because the ocean is density stratified, horizontal motion is strongly favored over vertical. Density tends to be conserved away from the sea surface so we can model a "horizon" of constant density (an *isopycnal*) along which water moves. Given the evidence of weak turbulent diapycnal mixing within the thermocline alluded to in Chapter 11, this is a reasonable first step as we continue "graduating up in a spiral line" of increasing model complexity.

16.1.1 An ocean application: gyre ventilation

Within the subtropics, the warm surface waters are separated from the colder intermediate and deep waters by a permanent thermal transition layer – called the *main thermocline* – that spans a range of depths from a few hundreds of meters to about a kilometer. This "barrier layer" separates those waters that are affected by the changing seasons from those waters that participate on decade to century timescale global circulation. If you picture one of the isopycnal surfaces that form the main thermocline, you would see that it does not have a uniform depth but is "bowl shaped", in that it is deepest in the center of the subtropical gyre and shallower along the edges. This shape results from the balance between Coriolis force (associated with a planet's rotation) and horizontal pressure gradients. Figure 16.1 is a contour plot of the depths for just such an isopycnal surface in the subtropical North Atlantic in the month of March. Note that the isopycnal *outcrops*; that is, it reaches the sea surface north of about 40° N (the gray area in Fig. 16.1). It reaches its deepest point (~700 m) just east of the Gulf Stream, which creates a steep "cliff" in response to high flow rates. The isopycnal begins to shoal in the tropics south of the subtropical gyre, but doesn't outcrop again until thousands of kilometers away, poleward (south) of the South Atlantic subtropical gyre.

Below the sunlit zone of the ocean, and in particular within the main thermocline, dissolved oxygen is constantly being consumed by bacteria that are oxidizing the downward rain of particulate organic material produced at the sea surface by photosynthesis. Given

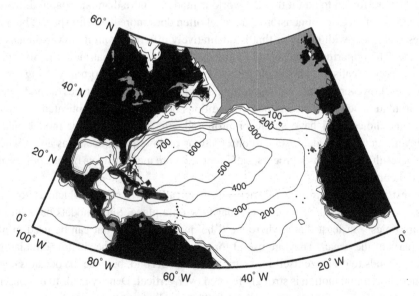

Figure 16.1 A contour plot of the average depth (in meters) of the isopycnal horizon whose density is $1026.8 \, \mathrm{kg \, m^{-3}}$ during the month of March. Data were taken from the World Ocean Atlas (Garcia *et al.*, 2006). The gray area to the north is where the water column is denser than this isopycnal, so its southern edge represents the isopycnal's outcrop.

that this oxygen demand has been occurring for a very long time, one may wonder why the dissolved oxygen content within the thermocline has not been depleted to zero. This does happen in a few special, isolated places, but not in most of the open ocean. That is, in ocean waters away from the sea surface, oxygen is generally undersaturated, but certainly not zero. So we must conclude that the oxygen levels we do see in the main thermocline, since they appear to be relatively constant, must result from a balance between the oxygen demand and the rate of supply ultimately from the atmosphere. The latter process we call *ventilation*.

It has been understood for a long time that ventilation of the main thermocline occurs largely along isopycnal horizons from where they outcrop at the sea surface. You can see this in Fig. 16.2, where we've contoured the percent oxygen saturation on the isopycnal horizon featured in Fig. 16.1.

Note that where the isopycnal outcrops (north of 40° N), the saturation value is close to 100% because of exchange with the atmosphere. Further into the gyre, the saturation value decreases southward owing to *in situ* demand for oxygen. The shape of the 70% contour in the western part of the subtropical North Atlantic (in the vicinity of 25–40° N by 50–70° W) is suggestive of the clockwise gyre circulation "winding up" the saturation contours. A southwest-trending transition zone (from ~20° N off the coast of Africa to ~15° N in the west) characterized by a strong oxygen saturation gradient demarcates the southern boundary of the subtropical gyre. The saturation value reaches a minimum off

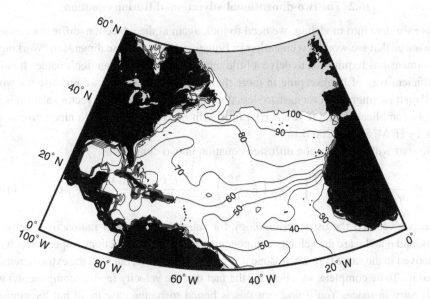

Figure 16.2 A contour plot of the percent oxygen saturation on the isopycnal horizon whose density is 1026.8 kg m^{-3} during the month of March. Data were taken from the World Ocean Atlas 2005 (Garcia *et al.*, 2006). Note that where the isopycnal reaches the ocean surface (see Fig. 16.1) saturation reaches 100%.

the west coast of North Africa, in large part due to high biological productivity spurred by upwelling, but possibly also due to a stagnation point in the gyre circulation.

At any arbitrarily chosen location on this density horizon, the rate at which oxygen is being removed by bacterial oxidation must be exactly countered by the advective and diffusive transport of oxygen to that point from elsewhere. Our goal is to use our modeling techniques to make a determination of this ventilation rate, which in turn would allow us to estimate the oxygen consumption rate. The approach we will use is to simulate the invasion of an anthropogenic transient tracer – bomb-produced tritium – onto this horizon, and by comparing the model results with observations, we hope to constrain ventilation rates. Instinct tells you, however, that given we have at least two unknowns (velocity and turbulent diffusion rate), we have an *under-determined* situation; that is, we might be able to emulate observations with more than one combination of flow and mixing. Moreover, we will use a highly simplified model geometry (a rectangular domain), so an exact comparison between the simulation and observation is problematic. However, we'll also model the daughter product of this isotope, namely [3]He, and compare not only the "geometries" of the tracer distributions between the simulations and observations, but also the tritium–[3]He relationships. Much as this relationship gave us insight into the relative importance of advection versus diffusion in one dimension in Chapter 11, we expect that this combination will be of value in two dimensions.

16.2 The two-dimensional advection–diffusion equation

Before we dive into modeling, we need to look again at the advection–diffusion equation for a tracer that we worked with earlier in Equation (13.1) for one dimension. Working in two dimensions requires us to delve a little into *vector calculus*, but don't panic! It is just an efficient way of bookkeeping in more than one dimension, and we promise we won't dwell on it any more than we have to (less than three pages). Actually, vector calculus is not terribly complicated, and if you need a little help here, you might try a nice introductory book by H. M. Schey (2005).

We start with the advection diffusion equation in two dimensions:

$$\frac{\partial C}{\partial t} = \frac{\partial}{\partial x}\left(K_x \frac{\partial C}{\partial x}\right) + \frac{\partial}{\partial y}\left(K_y \frac{\partial C}{\partial y}\right) - \frac{\partial(uC)}{\partial x} - \frac{\partial(vC)}{\partial y} + J - \lambda C \qquad (16.1)$$

where we've used the usual terminology for source/sink (J) and radioactive decay (λ) terms, and u and v are the velocity components in the x and y directions respectively. It can be derived in the same way we've done it in previous chapters, but with the extra dimension added in. To be complete, we allow for the fact that the velocity (and its components) will likely vary in space. You'll find that this is bound to be the case in all but the simplest models. Second, it is possible that K_x and K_y may differ from each other (i.e. that turbulent diffusion may be *anisotropic*), and could vary spatially. For simplicity, we will assume that K is *isotropic* and spatially constant.

16.2.1 A brief foray into vector calculus

Assuming a constant, isotropic diffusivity, along with some compact vector notation, we can arrive at a simpler looking equation,

$$\frac{\partial C}{\partial t} = K\nabla^2 C - \nabla \cdot (\mathbf{u}C) + J - \lambda C \tag{16.2}$$

Here we've used the standard definition of the *gradient operator*, traditionally represented by the *nabla symbol* ∇, which is named after a Hebrew stringed instrument of similar shape. It represents a vector operator defined by

$$\nabla = \frac{\partial}{\partial x}\hat{\mathbf{x}} + \frac{\partial}{\partial y}\hat{\mathbf{y}} \tag{16.3}$$

where $\hat{\mathbf{x}}$ and $\hat{\mathbf{y}}$ are unit vectors in the x and y directions respectively. It follows that for some scalar property C you can define its *gradient*

$$\nabla C = \frac{\partial C}{\partial x}\hat{\mathbf{x}} + \frac{\partial C}{\partial y}\hat{\mathbf{y}} \tag{16.4}$$

which is analogous to a one-dimensional *slope*, but is a vector quantity having both direction and magnitude.

You can multiply a vector by a constant, and this is done by multiplying both components by the same number. For example, multiplying a vector $\mathbf{a}$ by a constant c is done with

$$\mathbf{a} = a_x\hat{\mathbf{x}} + a_y\hat{\mathbf{y}}$$
$$c\mathbf{a} = ca_x\hat{\mathbf{x}} + ca_y\hat{\mathbf{y}} \tag{16.5}$$

where, for example, a_x is the component of $\mathbf{a}$ in the x direction. Geometrically, this has the effect of "stretching" the vector, i.e. changing its length but not its direction. So if $c = 2$, then the vector's length is doubled. You can convince yourself that this is true by noting that the vector's length is equal to the square root of the sum of the squares of its components (really just the Pythagorean Theorem). Also, since the angle θ between the vector $\mathbf{a}$ and $\hat{\mathbf{x}}$ (the x axis) is

$$\theta = \arctan(a_y/a_x) \tag{16.6}$$

you can demonstrate that multiplication of a vector by a scalar (as defined in Equation (16.5)) doesn't change its direction.

Multiplication of two vectors is a little more complicated. There are a couple of ways of defining vector multiplication, and each has a very distinct meaning and application. These are *vector* and *scalar* multiplication. We'll deal only with the latter here because that's all we'll need. The scalar product of two vectors is denoted with a "·" between them (hence it is often referred to as the "dot product") ...

$$\mathbf{a} \cdot \mathbf{b} = (a_x\hat{\mathbf{x}} + a_y\hat{\mathbf{y}}) \cdot (b_x\hat{\mathbf{x}} + b_y\hat{\mathbf{y}})$$
$$= a_xb_x\hat{\mathbf{x}} \cdot \hat{\mathbf{x}} + (a_xb_y + a_yb_x)\hat{\mathbf{x}} \cdot \hat{\mathbf{y}} + a_yb_y\hat{\mathbf{y}} \cdot \hat{\mathbf{y}}$$
$$= a_xb_x + a_yb_y \tag{16.7}$$

The last step reduces nicely because $\hat{\mathbf{x}}$ and $\hat{\mathbf{y}}$ are unit length vectors that are orthogonal (they're *orthonormal*), so their mutual product is zero and their self-products are both one. True to its name, the scalar vector product is a scalar.

A geometric interpretation of the scalar product of vectors is a little more difficult, but we'll try anyway. Scalar multiplication of two vectors produces a scalar whose value is the product of the length of one vector with the *projection* of the second vector on the first. That is, if the two vectors **a** and **b** are separated by an angle θ, then the projection of **b** on **a** is given by $b \cos \theta$, and their scalar product is given by

$$\mathbf{a} \cdot \mathbf{b} = ab \cos \theta \tag{16.8}$$

which you can convince yourself by doing a little drawing. What is immediately evident from this relation is that the scalar vector product generally has a magnitude less than the product of the two vector lengths because you always have $|\cos \theta| \leq 1$. The scalar product of two vectors is a maximum if they are parallel ($\theta = 0$, so $\cos \theta = 1$). If the vectors are at right angles (*orthogonal*), we have $\cos \theta = 0$, which is simply saying that the projection of **b** on **a** (or **a** on **b**) must be zero, so the scalar product is zero. We'll leave it to the industrious student to use a little algebra and trigonometry to convince him- or herself that Equations (16.8) and (16.7) are the same.

The "dot product" of ∇ with a vector works like a scalar vector multiplication. For example, we have

$$\nabla \cdot \mathbf{a} = \frac{\partial a_x}{\partial x} + \frac{\partial a_y}{\partial y} \tag{16.9}$$

and this is known as the *divergence* of **a**. It is a scalar property, and it denotes how the vector **a** is changing in space. It has a special meaning when associated with a vector velocity field of a fluid, in that it is a measure of the degree of fluid accumulation (if negative) or removal (if positive). If you are dealing with an *incompressible* fluid, you must have a zero divergence everywhere. We'll discuss this important point in the next section.

Finally, we explore the two-dimensional second derivative which is really the *divergence of the gradient* or the *Laplacian* of a scalar quantity. Remembering from Equation (16.4) that the gradient is a vector quantity, as is ∇, the Laplacian is defined by a scalar vector product:

$$\begin{aligned}
\nabla^2 C &= \nabla \cdot \nabla C \\
&= \nabla \cdot \left(\frac{\partial C}{\partial x} \hat{\mathbf{x}} + \frac{\partial C}{\partial y} \hat{\mathbf{y}} \right) \\
&= \frac{\partial^2 C}{\partial x^2} \hat{\mathbf{x}} \cdot \hat{\mathbf{x}} + 2 \frac{\partial^2 C}{\partial x \partial y} \hat{\mathbf{x}} \cdot \hat{\mathbf{y}} + \frac{\partial^2 C}{\partial y^2} \hat{\mathbf{y}} \cdot \hat{\mathbf{y}} \\
&= \frac{\partial^2 C}{\partial x^2} + \frac{\partial^2 C}{\partial y^2}
\end{aligned} \tag{16.10}$$

where we've exploited the orthonormal qualities of $\hat{\mathbf{x}}$ and $\hat{\mathbf{y}}$ yet again. We could have achieved the same result by first multiplying $\nabla \cdot \nabla$ but we'll let you convince yourself that it works. Like its one-dimensional analog, the Laplacian of a scalar is a measure of how sharply its gradient is changing. This applies to our Fick's law of diffusive flux divergence embodied in Equation (16.2). Note also that it is a scalar quantity.

16.2.2 Streamfunctions and non-divergent flow

Flow in two dimensions is more complicated than in one. You cannot vary flow arbitrarily in one dimension without considering flow in the other. This is because seawater must be regarded as basically incompressible when it comes to our models and mass balance calculations, especially on an isopycnal surface. This quite simply means that if you picture a two-dimensional box in your model domain, water flow into it must equal water flow out of it. If it didn't, the box would overflow or empty. Put mathematically, the two dimensional flow **u** must be *non-divergent*, that is $\nabla \cdot \mathbf{u} = 0$ everywhere in the model.

So the challenge is to specify a flow field that satisfies the non-divergence requirement. This turns out to be simpler than you might think – at least in two dimensions – as we will always define our velocity field using the *streamfunction*. Given a streamfunction ψ, you can derive the velocity components u and v for the x (longitude) and y (latitude) directions respectively using

$$u = -\frac{\partial \psi}{\partial y} \tag{16.11}$$

and

$$v = \frac{\partial \psi}{\partial x} \tag{16.12}$$

Please remember the minus sign in Equation (16.11). You may infer from these relationships that the streamfunction ψ must be in units of $m^2 s^{-1}$ if **u** is in units of $m s^{-1}$. Although this seems an added layer of complexity, it has an important benefit. If we define and compute our velocity fields from the streamfunction, we can guarantee that the velocity fields are non-divergent. Here's the proof:

$$\nabla \cdot \mathbf{u} = \frac{\partial u}{\partial x} + \frac{\partial v}{\partial y} = \frac{\partial}{\partial x} \left(-\frac{\partial \psi}{\partial y} \right) + \frac{\partial}{\partial y} \left(\frac{\partial \psi}{\partial x} \right) = -\frac{\partial^2 \psi}{\partial x \partial y} + \frac{\partial^2 \psi}{\partial x \partial y} = 0 \tag{16.13}$$

where we have swapped the order of differentiation in the last term (yes, it's legal!).

Note that it is possible to define the streamfunction with the opposite sign convention (i.e. with $u = \partial \psi / \partial y$ and $v = -\partial \psi / \partial x$). Substitution of this alternate definition in Equation (16.13) shows that this also satisfies the non-divergence condition. We will stick with the definition embodied in Equations (16.11) and (16.12) because it is commonly used in oceanography and in meteorology.

The somewhat odd-looking relationship between ψ and **u** defined by Equations (16.11) and (16.12) may seem a little strange. However, the streamfunction is not as unfamiliar as you might think. You've likely seen it before; in a contour plot of the streamfunction, the contours are *streamlines*. Such a map shows you the pathways that water takes: that is, along (parallel to) streamlines. There is another thing you can see. Given the relationship between ψ and $\mathbf{u}$, wherever the streamlines are "bunched up", the velocity will be high, and where the streamlines are far apart, the flow will be weak.

Before we show you some example streamfunctions, there are a couple of considerations we'd like to share. First, to keep things simple, we are going to stipulate that we won't have water entering or leaving our model domain. This is not mandatory, but we'd like to keep things simple. This means that all flow normal to the boundaries of our model must be exactly zero. When you then look at the definition of ψ, this requires that its value be the same at all boundaries. Second, remembering that the components of $\mathbf{u}$ are defined as *derivatives* of ψ, then the streamfunction can have an arbitrary constant added to it, and the velocity field will be unaffected. What's important is how ψ changes spatially. Thus we can define ψ to vary between, say, 0 and $10\,\text{m}^2\,\text{s}^{-1}$ or equivalently between 100 and $110\,\text{m}^2\,\text{s}^{-1}$. Finally, integrating a two-dimensional streamfunction vertically yields the *transport*, which is the cumulative flux of water (in $\text{m}^3\,\text{s}^{-1}$).

We will define $\psi = 0$ along the boundary of our model domain. This is convenient because we can easily use sinusoids to define a streamfunction that is zero along all the edges. For example, defining our spatial coordinates so that $0 \geq x \geq 1$ and $0 \geq y \geq 1$, we can construct a streamfunction as

$$\psi = 10\sin{(\pi x)}\sin{(\pi y)} \tag{16.14}$$

where we've arbitrarily chosen a maximum amplitude of ψ to be 10. Examining this relationship reveals that at least one of the multiplicative terms is zero when either x or y equals 0 or 1; that is, on the boundaries. This streamfunction is contoured in Fig. 16.3a. Note that as you might expect from the equation, it is a maximum in the center of the gyre. Topographically, the streamfunction looks like a hill centered in the model domain.

We know that this represents a symmetric gyre, but what is the direction of rotation? Looking at Equations (16.11) and (16.12), you can deduce that u is negative in the lower half of the model domain because $\partial\psi/\partial y > 0$ there, and it is positive on the top half. Because $\partial\psi/\partial x > 0$ in the left half, v is positive (remember Equation (16.12)), and conversely v is negative in the right half of the domain. The net result is that water is flowing in a *clockwise* fashion in this gyre. This sense of rotation is the same as the subtropical gyre in the North Atlantic, which is bounded on the west (the left) by the northward flowing Gulf Stream, on the north by the eastward flowing Azores Current, and on the south by the westward flowing North Equatorial Counter Current. Setting $\psi = -10\sin{(\pi x)}\sin{(\pi y)}$ gives a "pit" rather than a "hill", as shown in Fig. 16.3b. Similar reasoning would convince you that it describes a *counter-clockwise* gyre, such as you would find in the subtropical South Pacific.

16.2.3 A westward intensified gyre

Beyond the sense of rotation, the resemblance of the schematic circulation depicted in Fig. 16.3a to the actual North Atlantic subtropical gyre is not very good. The most obvious

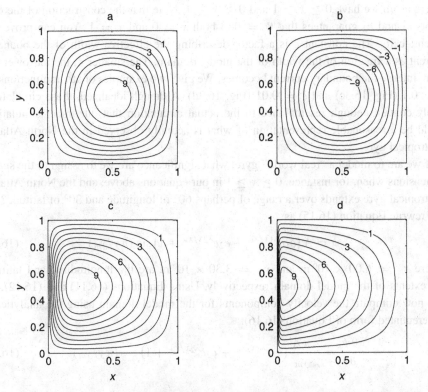

Figure 16.3 Some example streamfunctions described in the text. These include (a) a clockwise symmetric gyre ($\psi \geq 0$), (b) a counter-clockwise symmetric gyre ($\psi \leq 0$), (c) a clockwise Stommel gyre ($\epsilon = 0.1$; see text), and (d) a clockwise Stommel gyre with a "tighter" boundary current ($\epsilon = 0.01$).

lack is that there's no equivalent to the Gulf Stream in the model circulation. In the symmetric gyre, the width of the northward flow is half of the domain (many thousands of km), while the actual Gulf Stream is around 200 km in width (perhaps 5% of the width of the domain). We can improve things by introducing a westward intensified boundary current by using a schematic circulation scheme called a *Stommel gyre*. This is an analytic solution for idealized wind forcing introduced by Henry Stommel (1948). The form of the streamfunction is given by

$$\psi = A \left(c_1 e^{\lambda_1 x} + c_2 e^{\lambda_2 x} + 1 \right) \sin (\pi y) \qquad (16.15)$$

where we have

$$c_1 = \frac{1 - e^{\lambda_2}}{e^{\lambda_2} - e^{\lambda_1}} \qquad \lambda_1 = \frac{-1 + \sqrt{1 + 4\pi^2 \epsilon^2}}{2\epsilon}$$

$$c_2 = -(1 + c_1) \qquad \lambda_2 = \frac{-1 - \sqrt{1 + 4\pi^2 \epsilon^2}}{2\epsilon}$$

where again we have $0 \geq x \geq 1$ and $0 \geq y \geq 1$. Note that the construction of the constants c_1 and c_2 guarantees that $\psi = 0$ at both $x = 0$ and $x = 1$. You can prove this algebraically. The constant ϵ is a factor describing the effective width of the boundary current as a fraction of the width of the model domain. The smaller ϵ, the narrower and more intense the boundary current becomes. We give two examples of streamfunctions for $\epsilon = 0.1$ (Fig. 16.3c) and $\epsilon = 0.01$ (Fig. 16.3d). Although idealized, these circulations likely contain enough resemblance to the actual circulation that our tracer simulations could be a reasonable representation of what is actually observed in the North Atlantic subtropical gyre.

If we are to model a "real world" gyre, what significance are we to assign to the spatial dimensions when, for instance, $0 \geq x \geq 1$ in our equations above, and the North Atlantic subtropical gyre extends over a range of perhaps 60° of longitude and 30° of latitude? We can rewrite Equation (16.15) as

$$\psi = A \left(c_1 e^{\lambda_1 x/x_m} + c_2 e^{\lambda_2 x/x_m} + 1 \right) \sin \left(\pi y/y_m \right) \qquad (16.16)$$

where $x_m = 6.00 \times 10^6$ m and $y_m = 3.30 \times 10^6$ m are the longitudinal and latitudinal extents of the model domain respectively. Using Equations (16.11) and (16.12), we can now compute the velocity components for the model domain using the analytically differentiated form of Equation (16.16):

$$u = -\frac{\partial \psi}{\partial y} = -\frac{\pi A}{y_m} \left(c_1 e^{\lambda_1 x/x_m} + c_2 e^{\lambda_2 x/x_m} + 1 \right) \cos \left(\pi y/y_m \right) \qquad (16.17)$$

and

$$v = \frac{\partial \psi}{\partial x} = \frac{A}{x_m} \left(c_1 \lambda_1 e^{\lambda_1 x/x_m} + c_2 \lambda_2 e^{\lambda_2 x/x_m} \right) \sin \left(\pi y/y_m \right) \qquad (16.18)$$

What value are we to use for the streamfunction amplitude A in Equation (16.16)? Although we may ultimately adjust the A to match tracer observations, what would be a sensible starting point? Observations suggest flows of the order of 10^{-2} m s^{-1} are broadly typical of the south/southwestward circulation in the eastern half of the gyre. Assigning a characteristic velocity scale $V = 10^{-2}$ m s^{-1} and using half the basin width, we can estimate $|\psi_{max}| = 0.5 V x_m = 3 \times 10^4$ m^2 s^{-1}. Translating this to a value for A in Equation (16.16) can be done with some algebra and depends on the choice of ϵ. This can be rather tedious, so we instead choose $\epsilon = 0.05$ and use the following MATLAB code (called PsiDef.m) to determine the amplitude:

```
xm=6.0e6; ym=3.3e6;                        % domain dimensions
eps=0.05;                                  % boundary current scale
nx=ceil(10/eps); ny=ceil(nx*ym/xm); dx=xm/nx; dy=ym/ny;
x=[0:dx:xm]; y=[0:dy:ym];                  % x & y coordinates
[X,Y]=meshgrid(x,y);                       % create X & Y  grids
lam1=(-1+(1+4*pi^2*eps^2)^0.5)/2/eps;      % Lambda-1
lam2=(-1-(1+4*pi^2*eps^2)^0.5)/2/eps;      % Lambda-2
c1=(1-exp(lam2))/(exp(lam2)-exp(lam1));    % C-1
c2=-(1+c1);                                % C-2
```

```
A = 3.0e4;                                % first guess amplitude
Psi=A*(c1*exp(lam1*X/xm)+c2*exp(lam2*X/xm)+1).*sin(pi*Y/ym);
Actual = max(Psi(:));                     % find actual extremum
A = A*A/Actual;                           % estimate new amplitude
Psi=A*(c1*exp(lam1*X/xm)+c2*exp(lam2*X/xm)+1).*sin(pi*Y/ym);
```

Note that since we cannot guarantee that nx*ym/xm is an integer, we employ the ceil function to round up to the nearest whole number. We've used the function meshgrid to convert the two domain-defining vectors x and y to their corresponding grids. That is, meshgrid creates two $n_y \times n_x$ arrays of x and y values from the n_x- and n_y-length vectors. Thus we now have an array of x and y coordinates for each of the $n_y \times n_x$ grid cell nodes. Note also that we must use *array multiplication* (the " .* ") in our calculations. The last line recomputes the new value of $A = 9.42 \times 10^4 \, m^2 \, s^{-1}$ required to give $\psi_{max} = 3 \times 10^4 \, m^2 \, s^{-1}$. The fact that the new amplitude is approximately π times the maximum value is only a happy coincidence (honest!).

16.3 Gridding and numerical considerations

16.3.1 Grid setup and alignment

The first order of business in setting up the model grid requires that we think carefully about the relationship between the grid cells and the boundaries. It seems logical to us that the "walls" of cells along each edge of the model ought to coincide with the model domain boundaries. This is a natural arrangement from the perspective of the control volume approach discussed in Chapter 12, especially for "hard wall" boundaries where zero flux of material (and fluid) is expected. In this scheme, the *node* (the center of the cell) in the lower left-hand corner of the domain should be situated at $(\Delta x/2, \Delta y/2)$ and the one in the upper right-hand corner at $(x_m - \Delta x/2, y_m - \Delta y/2)$. We can create our velocity fields for the Stommel gyre by putting the following in a new script (VelCalc.m):

```
x=[0.5*dx:dx:xm-0.5*dx];                  % nodes offset by 1/2 step
y=[0.5*dy:dy:ym-0.5*dy];                  % nodes offset by 1/2 step
[X,Y]=meshgrid(x,y);                      % create X & Y node grids
u = -(pi*A/ym)*(c1*exp(lam1*X/xm)+c2*exp(lam2*X/xm)+1).*cos(pi*Y/ym);
v = (A/xm)*(c1*lam1*exp(lam1*X/xm)+c2*lam2*exp(lam2*X/xm))...
                                          .*sin(pi*Y/ym);
```

where we have now redefined our velocity grid node positions (the X and Y arrays) to align the cell edges with the domain boundaries. This grid offset (relative to cell center) is necessary from the definition of second upwind differencing, the numerical technique used for our gyre model. The computed streamfunction and velocities are contoured in Fig. 16.4.

Inspection of Fig. 16.4b shows that we have made a reasonable choice for the streamfunction amplitude; the magnitudes of u and v are in the vicinity of $0.01 \, m \, s^{-1}$ in the "eastern" (right-hand) portion of the model domain. The western edge of the domain is

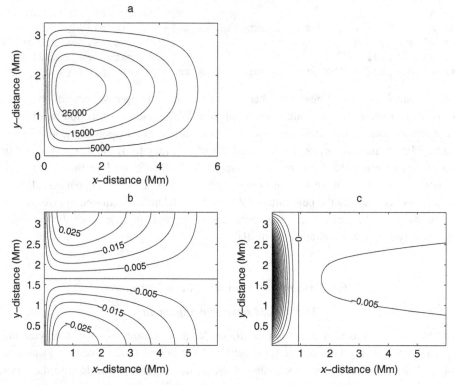

Figure 16.4 Contours of a Stommel gyre (a) streamfunction computed with $\epsilon = 0.05$ and $A = 9.42 \times 10^4 \, \mathrm{m^2 \, s^{-1}}$, (b) u, the zonal (x-direction) velocity and (c) v, the meridional (y-direction) velocities. Note that the contour interval for both u and v is $0.005 \, \mathrm{m \, s^{-1}}$, and that the positive contours for v are unlabeled to avoid clutter.

dominated by the westward intensified boundary current (note we have not labelled the positive velocity contours in the figure, to avoid crowding), where v exceeds $0.1 \, \mathrm{m \, s^{-1}}$.

For this simple model we can achieve satisfactory results by using a uniform grid and the *second upwind differencing scheme* as our finite difference algorithm. It is reasonably efficient and robust, and as importantly both *conservative* and *transportive*. More sophisticated schemes exist, but they come with their own challenges. The reader will recall that upwind differencing involves choosing which cell "donates" material based on the flow direction (hence the name). With first upwind differencing, the mass flux is calculated as the product of the concentration and the velocity at the node point of the donor cell. With second upwind differencing, the velocity used is the velocity at the interface between the donor and receiver cells.

Resolution becomes a more important consideration because computational demands tend to rise more sharply with resolution than in the one-dimensional case. For example, doubling the resolution in one dimension leads to an approximately four-fold increase in computing because a higher spatial resolution requires a shorter time step for stability.

Doubling resolution in a two-dimensional model would result in an eight-fold increase in computation. To resolve the western boundary current, we need maybe 10 grid points (in the x direction) across it. This is arbitrary, but we can evaluate the impact later on. Given $\epsilon = 0.05$, we choose a domain resolution of 110×200 cells, giving the same spatial increment in both directions. With a domain of $y_m = 3.3 \times 10^6$ by $x_m = 6.0 \times 10^6$ m, this corresponds to $\Delta y = \Delta x = 3.0 \times 10^4$.

Now we implement the basic equations for the second upwind differencing method in two dimensions with

$$C_{i,j}^{n+1} = C_{i,j}^n - \frac{\Delta t}{\Delta x} \left(u_R C_R^n - u_L C_L^n \right) - \frac{\Delta t}{\Delta y} \left(v_A C_A^n - v_B C_B^n \right)$$
$$+ \frac{K_x \Delta t}{\Delta x^2} \left(C_{i+1,j}^n + C_{i-1,j}^n - 2C_{i,j}^n \right) + \frac{K_y \Delta t}{\Delta y^2} \left(C_{i,j+1}^n + C_{i,j-1}^n - 2C_{i,j}^n \right)$$

$$(16.19)$$

where now we have two pairs of donor cells to deal with, and the subscripts R, L, B, and A corresponding to right, left, below, and above respectively. For second upwind differencing, the definitions of the interfacial velocities are usually given by

$$u_R = \frac{1}{2} \left(u_{i+1,j} + u_{i,j} \right) \qquad u_L = \frac{1}{2} \left(u_{i-1,j} + u_{i,j} \right) \qquad (16.20)$$

$$v_A = \frac{1}{2} \left(v_{i,j+1} + v_{i,j} \right) \qquad v_B = \frac{1}{2} \left(v_{i,j-1} + v_{i,j} \right) \qquad (16.21)$$

However, because we have an analytical form to calculate the velocity fields, we have the luxury of knowing the interfacial velocities more precisely. Thus we can compute them directly by inserting the following at the end of `VelCalc.m`:

```
xr=[dx:dx:xm];  xl=[0:dx:xm-dx];          % left & right x-coord
ya=[dy:dy:ym];  yb=[0:dy:ym-dy];          % below & above y-coord
[Xo,Yo]=meshgrid(xr,y);                   % offset grid coords
ur=-(pi*A/ym)*(c1*exp(lam1*Xo/xm)+c2*exp(lam2*Xo/xm)+1).*cos(pi*Yo/ym);
[Xo,Yo]=meshgrid(xl,y);                   % offset grid coords
ul=-(pi*A/ym)*(c1*exp(lam1*Xo/xm)+c2*exp(lam2*Xo/xm)+1).*cos(pi*Yo/ym);
[Xo,Yo]=meshgrid(x,ya);                   % offset grid coords
va= (A/xm)*(c1*lam1*exp(lam1*Xo/xm)+c2*lam2*exp(lam2*Xo/xm))...
                                   .*sin(pi*Yo/ym);
[Xo,Yo]=meshgrid(x,yb);                   % offset grid coords
vb= (A/xm)*(c1*lam1*exp(lam1*Xo/xm)+c2*lam2*exp(lam2*Xo/xm))...
                                   .*sin(pi*Yo/ym);
```

where, to save storage space, we've reused the offset coordinate arrays Xo and Yo for each calculation. Note that the interfacial velocities are defined at the center of each cell wall, which means that the v-component velocities are defined along the node points (non-offset coordinates) in the x-direction, and the u-component velocities are defined along the node points (non-offset coordinates) in the y-direction. You may want to look back at Fig. 12.1,

which shows this in one dimension. In upwind differencing the donor concentrations are
defined by the sense of the velocity directions:

$$
\begin{array}{llll}
C_R^n = C_{i,j}^n & \text{for} \quad u_R > 0 & \text{and} \quad C_R^n = C_{i+1,j}^n & \text{for} \quad u_R < 0 \\
C_L^n = C_{i-1,j}^n & \text{for} \quad u_L > 0 & \text{and} \quad C_L^n = C_{i,j}^n & \text{for} \quad u_L < 0 \\
C_A^n = C_{i,j}^n & \text{for} \quad v_A > 0 & \text{and} \quad C_A^n = C_{i,j+1}^n & \text{for} \quad v_A < 0 \\
C_B^n = C_{i,j-1}^n & \text{for} \quad v_B > 0 & \text{and} \quad C_B^n = C_{i,j}^n & \text{for} \quad v_B < 0
\end{array}
\tag{16.22}
$$

16.3.2 Choosing a stable time step

We discussed the extension of numerical stability criteria to two dimensions in
Section 12.7. This seems relatively straightforward, in that *Courant* and *diffusion number* stability requirements must still apply. Consideration of the control volume approach
in two dimensions points to more stringent restrictions because there are now more boundary fluxes to consider when ensuring that no more than the total cell inventory of a property
is allowed to be transported during a time step. We define the dimensionless Courant and
diffusion numbers where, for example, $c_x = u\Delta t/\Delta x$ is the zonal (x-direction) Courant
number, and $d_y = K_y\Delta t/\Delta y^2$ is the meridional (y-direction) diffusion number. We can
then apply the *von Neumann analysis* (see Section 12.7.1) to examine the stability of the
method. Stability of any finite difference scheme requires that the amplitude of the effective
"noise gain", G, is a complex number associated with moving from one time step to the
next (i.e. for a given Fourier component V of the concentration function, $V^{n+1} = GV^n$)
and must be less than or equal to unity for all frequency components. For second upwind
differencing, we thus must have:

$$
|G| = |1 - (c_x + 2d_x)(1 - \cos\theta) - (c_y + 2d_y)(1 - \cos\phi) - I(c_x\sin\theta + c_y\sin\phi)| \leq 1
\tag{16.23}
$$

where we remind you that we're breaking with convention by using $I = \sqrt{-1}$, as we
did in Chapter 12. The vertical bars indicate the enclosed function's amplitude on the
complex plane, θ and ϕ are the von Neumann phase space angles for the x- and the
y-directions respectively, and for convenience we've assumed positive u_L, u_R, v_B, and
v_A. This relationship imposes the requirement that

$$
c_x + c_y + 2d_x + 2d_y \leq 1
\tag{16.24}
$$

For spatially constant diffusivity, the minimum time step is dictated by the maximum
velocity in the model domain. Operationally, we typically choose a spatial resolution (Δx
and Δy) that would adequately resolve the main features of the model circulation (in this
case, the western boundary current), and then compute the maximum time increment that is
stable for this circulation and diffusion. A useful strategy is to include the calculation of the
acceptable time step in the model code, by finding the most restrictive case on a cell-by-cell
basis. Thus we construct a MATLAB script `WeightCalc.m` that begins with:

```
Kx=1000*ones(size(X)); Ky=Kx;                    % isotropic, constant
dt=min(min(1./(abs(u/dx)+abs(v/dy)+2*Kx/dx.^2+2*Ky/dy.^2))));
dt=0.9*dt;                        % conservative reduction
nyrs=50;tyear=365.25*24*60*60;          % run time, seconds in year
nt=ceil(nyrs*tyear/dt);                 % total # of time steps
ntyr=ceil(tyear/dt);                    % # steps in a year
```

which gives us $\Delta t \approx 1.1 \times 10^5$ s, i.e. a little more than a day. We want you to notice the peculiar way that we've structured this calculation. We have used array operations where appropriate; that is, ./ and .^ construction. We've chosen to scale the time step down by a modest amount (10%) so that we don't work too close to the stability limit. Finally, we've derived some numbers dependent on Δt that will be used later.

16.3.3 Computing weight matrices

In setting up the model calculations, we can also take advantage of the fact that the circulation (velocity) and diffusion fields are steady, so we can recast Equation (16.19) in terms of weights:

$$C_{i,j}^{n+1} = w_0 C_{i,j}^n + w_{x-} C_{i-1,j}^n + w_{x+} C_{i+1,j}^n + w_{y-} C_{i,j-1}^n + w_{y+} C_{i,j+1}^n \qquad (16.25)$$

and save computing time by calculating the weighting factors in advance. There are now five weight arrays. We first compute the diffusion components of the weights:

```
Dx=Kx*dt/(dx^2);                  % diffusion number x-dir
Dy=Ky*dt/(dy^2);                  % diffusion number y-dir
wxm=Dx;                           % from x-minus
   wxm(:,1)=0;                    % not through west wall
wxp=Dx;                           % from x-plus
   wxp(:,nx)=0;                   % not through east wall
wym=Dy;                           % from y-minus
   wym(1,:)=0;                    % not through south wall
wyp=Dy;                           % from y-plus
   wyp(ny,:)=0;                   % not through north wall
```

where we avoid problems on the very edges of the model by accommodating the fact that there is no diffusive loss through the model boundaries, so we set the appropriate diffusive weights to zero. You may think that this is unnecessary as, when we implement the advection–diffusion calculation in the next section, we will not use, for example, wxm for those grid cells on the western boundary because they would refer to cells outside the model domain. However, we must ensure that the wxm(:,1) are set to zero because of the way we compute w0 below; otherwise the model will not conserve mass.

Then we apply the selection criteria embodied in Equation (16.22) for the advection parts in Equation (16.19) by adding to `WeightCalc.m`:

```
wxp(ur<0)=wxp(ur<0)-dt*ur(ur<0)/dx;    % if upstream
wxm(ul>0)=wxm(ul>0)+dt*ul(ul>0)/dx;    % if upstream
wyp(va<0)=wyp(va<0)-dt*va(va<0)/dy;    % if upstream
wym(vb>0)=wym(vb>0)+dt*vb(vb>0)/dy;    % if upstream
w0=1-wxp-wxm-wyp-wym;                  % SUDM conservative!
```

where, for example, wxp(ur<0) selects only those weights that have $u_R < 0$. The reader may be confused by the fact that we seemingly have only implemented half of the decisions enumerated in Equation (16.22); actually only those assignments associated with adjacent cell concentrations. If you look at how we have computed w0 you will see that the *complementary* cases, which involve $C_{i,j}$, are accounted for in a single step. That is, we have taken advantage of the fact that the second upwind difference scheme is conservative; i.e. the sum of the weights within each cell must add exactly to unity.

16.3.4 Implementing boundary conditions

The first boundary condition we deal with is the initial concentrations. For the moment, we'll define the initial concentrations to be zero with a script InitConc.m:

```
C=zeros(ny,nx); Cnew=C;         % initialize conc & storage
```

which serves the dual purpose of initializing the concentration array and also reserving storage for both the new and old concentration arrays. For more interesting situations, this will be the place where we choose more complicated initial concentration distributions.

We'll deal first with the interior using offset indexing like we did in one dimension with the following code, which is included in an initial version of AdvDiff.m:

```
ry=[2:ny-1]; rx=[2:nx-1];       % ranges define interior cells
for it=1:nt                     % main time loop
   Cnew(ry,rx)=w0(ry,rx).*C(ry,rx) ...
                  + wxm(ry,rx).*C(ry,rx-1) ...
                  + wxp(ry,rx).*C(ry,rx+1) ...
                  + wym(ry,rx).*C(ry-1,rx) ...
                  + wyp(ry,rx).*C(ry+1,rx);
      C=Cnew;                   % update separate storage
end                             % end time loop
```

where we've saved some typing by pre-defining the range of indices rx and ry in the *x* and *y* directions, respectively, that specify the interior cells of the model. Note also the concentration arrays' layout matches the model domain layout. That is, the *y*-coordinate corresponds to the row number. Irrespective of what we do with our boundary cells, the model run described above does absolutely nothing: the concentration starts at zero and if our code is working correctly, it will stay zero. The boundary conditions will be key to making things more interesting.

Boundary conditions vary depending on the property and problem. You may wish to specify fluxes on one boundary and concentrations on another. Because we have specified a constant (zero) streamfunction on the domain boundary, we are intrinsically stipulating zero advective flux through the edges of the model. This makes sense, at least for our "western" and "eastern" boundaries, since they correspond to continental "walls". For simplicity, we will also assume that the southern and northern boundaries are also walls. Thus we can deal with the edges by inserting the following code into the for-loop in AddvDiff.m:

```
Cnew(ry,1)=w0(ry,1).*C(ry,1) ...                    % west wall
             + wxp(ry,1).*C(ry,2) ...
             + wym(ry,1).*C(ry-1,1) ...
             + wyp(ry,1).*C(ry+1,1);
Cnew(ry,nx)=w0(ry,nx).*C(ry,nx) ...                 % east wall
             + wxm(ry,nx).*C(ry,nx-1) ...
             + wym(ry,nx).*C(ry-1,nx) ...
             + wyp(ry,nx).*C(ry+1,nx);
Cnew(1,rx)=w0(1,rx).*C(1,rx) ...                    % south wall
             + wxm(1,rx).*C(1,rx-1) ...
             + wxp(1,rx).*C(1,rx+1) ...
             + wyp(1,rx).*C(2,rx);
Cnew(ny,rx)=w0(ny,rx).*C(ny,rx) ...                 % north wall
             + wxm(ny,rx).*C(ny,rx-1) ...
             + wxp(ny,rx).*C(ny,rx+1) ...
             + wym(ny,rx).*C(ny-1,rx);
```

which handles the model edges, but excludes the corners of the model because these need to be handled separately. We deal with corners accordingly by adding:

```
Cnew(1,1)=w0(1,1).*C(1,1) ...                       % southwest corner
             + wxp(1,1).*C(1,2) ...
             + wyp(1,1).*C(2,1);
Cnew(1,nx)=w0(1,nx).*C(1,nx) ...                    % southeast corner
             + wxm(1,nx).*C(1,nx-1) ...
             + wyp(1,nx).*C(2,nx);
Cnew(ny,1)=w0(ny,1).*C(ny,1) ...                    % northwest corner
             + wxp(ny,1).*C(ny,2) ...
             + wym(ny,1).*C(ny-1,1);
Cnew(ny,nx)=w0(ny,nx).*C(ny,nx) ...                 % northeast corner
             + wxm(ny,nx).*C(ny,nx-1) ...
             + wym(ny,nx).*C(ny-1,nx);
```

Because we are doing a ventilation model, we imitate an outcrop boundary by setting the northernmost n_w cell rows to some value by inserting the following code into the same for loop in AddvDiff.m:

```
for iw=1:nw                                 % do this over nw rows
    Cnew(ny-iw+1,1:nx)= Cnorth(it);         % northern B.C.
end
```

where you'll note that we've admitted to the possibility that the surface boundary condition may be time-varying. We will return to the northern boundary condition as we develop our simulations. Note that you will need to set up the northern boundary condition function with a script called `SetNorthBndy.m`, which could look like:

```
Cnorth=ones(1,nt);              % constant 1
nw=0;                           % turns off n-bndy fixing
```

where setting $n_w = 0$ turns off the northern boundary concentration setting by insuring that the i_w `for` loop never gets executed. Here again, more interesting simulations would result from more code in this script.

So in summary, we can now construct a model simulation using the code segments we've presented here. The resultant program would look like `StommelGyre.m`, which we've listed below:

```
%       Runs a Stommel Gyre tracer simulation in a
%       rectangular domain using second upwind differencing.
%
    PsiDef        % set up grid & compute streamfunction
    VelCalc       % compute velocity fields
    WeightCalc    % compute weight matrices
    SetNorthBndy  % set up northern boundary condition history
    InitConc      % initialize concentration fields
    AdvDiff       % do advection-diffusion calculation
```

but don't do it just yet!

16.4 Numerical diagnostics

Before running your first serious simulation with the model code that you have built, it is important that you fully understand the limitations and behavior of your model. This can be done both by theoretical analysis (see Subsection 16.4.4) and by running diagnostics on your model to make sure that it is performing as expected. There are three reasons that you need to do this. The first is that the more complicated the model becomes, the more lines of code you have to write, and the greater the risk that there is an error in that code. No matter how careful and fastidious you are, or how elegantly you structure your programs, even the best programmers make mistakes. MATLAB is very good at catching the obvious semantic errors like unbalanced parentheses and using uninitialized variables (for example because of spelling errors), but subtler problems caused by incorrect logic or inadvertently transposed indices may slip through. You also may have committed some structural error in formulating your boundary conditions along the way.

Second, there are intrinsic limitations to whatever finite difference algorithm you have used, such as truncation errors and numerical diffusion. There are theoretical approaches to analyse these effects, but more complicated algorithms and model geometries make them more challenging to perform and interpret. Third, there is an "organic" interplay between the model structure (geometry, circulation, and mixing), the imposed boundary conditions, and possibly even the numerics. It is important to identify (and if possible eliminate) those features that emerge as a result of your model construction and implementation rather than from nature. That is, the diagnostics are aimed at assessing model fidelity. In this section we will give examples of some diagnostics that you can perform.

16.4.1 Monitoring your model output

Perhaps the most effective tool in diagnosing your model is to apply some simple initial condition and watch your model output as it evolves. In the two-dimensional case, the best way is to contour the property values as the simulation progresses. You don't want to do this every time step, unless something really dramatic is happening. We'll insert some code at the beginning of the for-loop (just after the for statement) in AdvDiff.m to do a plot once a year.

```
if floor(it/ntyr) > nyr              % is it a new year?
    nyr=floor(it/ntyr);              % increment year count
    theSums(nyr+1)=sum(C(:));        % track C inventory
    figure(1)                        % select a figure
    clabel(contour(x,y,C));          % look at distribution
    title(['Year ' sprintf('%.2f',it/ntyr)]); % date your work
    pause(0.1)                       % pause to look
end
```

To get this started, we've also inserted "nyr = -1;" just before the for statement in AdvDiff.m. How this works is that ntyr is the number of time steps in a year, so that once it/ntyr exceeds the current year number (nyr), the code snippet executes, and nyr is updated so that it only happens once a year. Since nyr starts at −1, we make sure that we get a plot of the initial concentration distribution. Note that we've also titled the plot with the year number so you can see what time it represents, and we've inserted a very short pause statement to force MATLAB to update the figure on each loop. Finally, we slipped a statement in the "annual checkup" that computes the inventory of C as a diagnostic. Notice that we've used nyr+1 as the index because nyr will be zero the first time, and MATLAB starts all its indices with 1. Because C is now two-dimensional, simply using sum(C) would give us a vector containing the column-wise sums, so we use sum(C(:)). Alternately, you could use sum(sum(C)).

16.4.2 Testing the code

One of the most basic requirements of the models we are dealing with is that they conserve mass. One simple approach to testing the model code (and the correctness of the boundary conditions) is to configure the model so there is no material transfer across the boundaries, start with some initial "mass" of material, and track the total in the model as it progresses. Remember that this only works because we have defined the structure of our model such that no material moves through the boundaries. We can test that this actually occurs by initializing the concentration everywhere to unity in the script InitConc.m with:

```
C=ones(ny,nx); Cnew=C;              % initialize C-matrix to ones
theSums=zeros(nyrs+1);              % initialize storage for sums
```

and where we've also created a storage vector theSums for keeping track of the model inventory over time. We can then plot the fractional deviation of the inventory from its starting value as a function of time with the following:

```
plot([0:nyrs],theSums/theSums(1) - 1);    % plot deviation
```

Examining the results shows that there is no discernable trend in the total mass, even after 50 years of running (not shown here). However, this doesn't rule out the possibility of material accumulating in some areas at the expense of other areas in the model. So the next level of diagnosis might be to look at the spatial distribution of deviations. You can do this by contouring the deviation of C from unity, and perhaps make a histogram of the deviations with:

```
subplot(2,2,1)
   [Ch,h]=contour(x/1e6,y/1e6,(C-1)*1e15);
subplot(2,2,2)
   hist(1e15*(C(:)-1))
```

which we'll describe, but not present here. The sizes of the deviations from unity concentration are extremely small ($\sim 10^{-15}$) and are due to accumulated roundoff errors, which for most computers are of the order 10^{-16}. This tells us two things: first that we've not made any egregious errors in our coding, and second that our the algorithm is intrinsically mass-conserving. This follows from our insistence on defining w0 the way we did. It also shows that there are no apparent "leaks" at the model domain boundaries and corners. Another characteristic of the error field is that the deviations are greatest near the western boundary, where the velocities are highest. Finally, we should realize that this test will catch only a subset of possible coding errors and algorithmic limitations, so further diagnostics are warranted.

16.4.3 Examining algorithmic artifacts

To probe limitations in the accuracy of our algorithm's ability to transport material, we can introduce spatial patterns in the initial conditions with one of the following lines:

```
C=zeros(ny,nx); C(ny/2+1:ny,:)=1;    % northern half set to 1
C=zeros(ny,nx); C(1:ny/2,:)=1;       % southern half set to 1
C=zeros(ny,nx); C(:,1,nx/2)=1;       % western half set to 1
C=zeros(ny,nx); C(:,nx/2+1:nx)=1;    % eastern half set to 1
```

which first sets the entire model domain's concentrations to 0, then selectively sets one half to unity. By monitoring the total mass as a function of time, we can observe and diagnose deviations from ideal behavior. This kind of test is especially stringent because it presents the model with an unnaturally sharp concentration front, so that inaccuracies in the algorithm will be vividly highlighted.

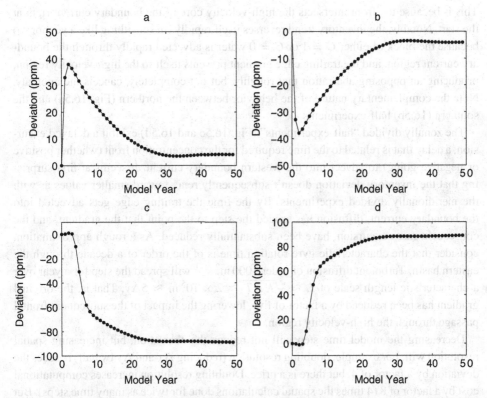

Figure 16.5 Time history of the deviation (in parts per million) of the model inventory for four initial states. (a) Northern half set to 1, (b) southern half set to 1, (c) eastern half set to 1, and (d) western half set to 1.

Figure 16.5 shows the deviation of model inventory due to algorithmic truncation errors plotted as a function of model year. We define the *model inventory deviation* as

$$\Delta \mathrm{Inv}(t) = \frac{\Sigma C_{ij}(t)}{\Sigma C_{ij}(0)} - 1 \qquad (16.26)$$

which we compute once a year in the model run and conveniently express in parts per million. The four step-function simulations displayed in Fig. 16.5 correspond to setting (a) the northern half, (b) the southern half, (c) the eastern half, and (d) the western half concentrations respectively to unity while the remainder of the model domain starts at zero concentration.

The non-conservative behavior arises because the upwind differencing code is only accurate to first order in Δx, and discrepancies occur in proportion to the concentration gradient ∇C, the velocity, and the square of the spatial resolution Δx^2. The maximum deviation, and hence the sharpest change in inventory, in the "meridional-half" experiments (i.e. when you set either the northern or southern half to 1, as in Fig. 16.5a and b) occurs immediately.

This is because the front intersects the high-velocity core of the boundary current right at the start. Notably, the inventory trend reverses itself rapidly (in something like a year or so) because the block of either $C = 1$ or $C = 0$ water is advected rapidly through the boundary current region, and the trailing edge gradient presents itself to the high-velocity region, producing an opposing aberration that roughly, but not completely, cancels the anomaly. Note the complementary nature of the behavior between the northern (Fig. 16.5a) and the southern (16.5b) half-experiments.

The zonally divided "half-experiments" (Fig. 16.5c and 16.5d) exhibit a delayed excursion, a delay that is related to the time required for the concentration front (whether positive or negative going) to advect into the western boundary current. It seems at first surprising that the inventory deviation doesn't subsequently relax toward smaller values as with the meridionally divided experiments. By the time the trailing edge gets advected into the boundary current, diffusion has eroded the step to the point that the gradient, and the consequent mass excursion, have been substantially reduced. As a rough approximation, consider that the characteristic gyre rotation time is of the order of a decade through the eastern basin. Turbulent diffusion of order $1000 \, \text{m}^2 \, \text{s}^{-1}$ will spread the step for a year over a characteristic length scale of $L \approx \sqrt{K \Delta T} \approx 2 \times 10^5 \, \text{m} \approx 5 \Delta y$. That is, the original gradient has been reduced by a factor of five, lowering the impact of the subsequent frontal passage through the high-velocity region.

Decreasing the model time step will not reduce the aberration, but increasing spatial resolution will. For example, doubling resolution (reducing Δx and Δy by half) reduces the deviation by a factor of 4, but there is a price. Doubling resolution increases computational cost by a factor of 8 (4 times the spatial calculations done for twice as many time steps). But how serious is this problem? First, we have performed an unusually harsh step-experiment. So-called "real world" changes will not be so severe. Moreover, an error in bookkeeping of less than 100 ppm (that's less than 0.01%) is small compared with the other sins that we've been committing.

16.4.4 Numerical diffusion

Before we actually do a simulation, we need to consider an important numerical artifact. You can demonstrate, by comparing a Taylor series expansion of the continuum equation with the finite difference form, that upwind differencing, whether first or second, has an apparent *implicit algorithmic* or *numerical* diffusivity that is given by (Equation (12.58)):

$$K_{nx} = \frac{u \Delta x}{2} (1 - c_x) \quad \text{and} \quad K_{ny} = \frac{v \Delta y}{2} \left(1 - c_y\right) \qquad (16.27)$$

If characteristic zonal velocities are of order $0.01 \, \text{m} \, \text{s}^{-1}$ with $\Delta x = 3 \times 10^4 \, \text{m}$, we obtain an approximate numerical diffusivity of about $150 \, \text{m}^2 \, \text{s}^{-1}$. The maximum meridional velocity encountered in our gyre is about $0.11 \, \text{m} \, \text{s}^{-1}$, and with a $\Delta y = 3 \times 10^4 \, \text{m}$, this gives a maximum meridional along-stream numerical diffusivity of about $1000 \, \text{m}^2 \, \text{s}^{-1}$. Note

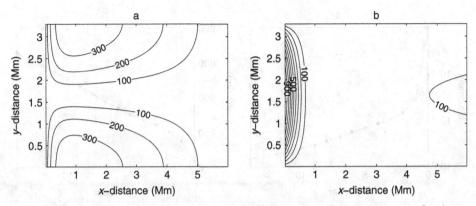

Figure 16.6 Contours of the calculated numerical diffusivity (a) K_{nx} and (b) K_{ny} in $m^2\,s^{-1}$ for a second upwind differencing simulation of a Stommel gyre.

that the K_n has been reduced from $v\Delta y/2$ because c_y approaches 1 in the high-velocity core. You can map out the numerical diffusivity components with the following MATLAB code:

```
cx=abs(u)*dt./dx;
cy=abs(v)*dt./dx;
Knx=(abs(u).*dx/2).*(1-cx);
Kny=(abs(v).*dy/2).*(1-cy);
```

which we've plotted in Fig. 16.6.

Throughout most of the eastern half of the gyre, the numerical diffusivity is modest, being about 10–15% of the explicit mixing coefficient. Things change as we approach the western boundary current, where velocities become large. Although disturbingly large (the maximum is comparable to our explicit diffusion coefficient) and anisotropic, it may not be such a severe limitation in our model simulations in at least the eastern part of the gyre. This numerical diffusion may unintentionally mimic enhanced lateral mixing in the vicinity of the Gulf Stream that has been suggested in the literature. In any case, we need to be aware that the model exhibits an intrinsic diffusiveness beyond what we explicitly stipulate.

We can experimentally evaluate the effects of numerical diffusion. Basically, you set the entire model domain to zero concentration, then set a single cell to unity concentration. A reasonable starting location would be in the relatively quiet part of the gyre, say at $y = 0.75y_m$ and $x = 0.75x_m$. You then run the simulation of this evolving "spot" with zero explicit diffusivity and monitor its spread and attenuation over time. How would you do this? An example code snippet might be something like:

```
TheMax(nyr+1)=max(C(:));
TheWidth(nyr+1)=dx*(length(C(C>0.5*TheMax(nyr+1))))^0.5;
```

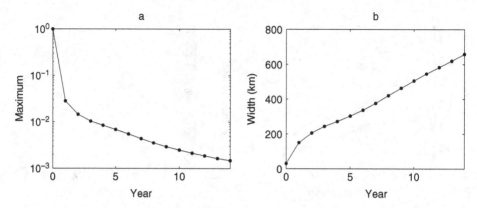

Figure 16.7 The evolution of a "spot" placed in the northeastern quadrant of the model with zero explicit diffusion. Plots show (a) the maximum value as a function of time and (b) the width at half maximum.

where we've assumed that the peak width is defined by those values greater 50% of the maximum peak concentration for that particular time step. The evolution of the spot is plotted in Fig. 16.7.

We should point out that the simulation is only run for a little more than a decade because after this the spot becomes a "streak" as it enters the western boundary current and becomes teased out by velocity shear. The subsequent distortion would make the simple geometric interpretation in the previous code snippet difficult to support. It is also instructive to visualize the spot as it evolves in the model, so you might consider replacing the annual contour plot in AdvDiff.m with contours of the concentrations normalized to the spot maximum:

```
clabel(contour(x/1e6,y/1e6,C/max(C(:)),CVs));
```

where you pre-define the contour values in CVs to some arbitrary set of values like [0.1:0.1:0.9]. Fixing the relative contour intervals throughout a visualization rather than relying on MATLAB to arbitrarily assign them helps regularize the output for comparative purposes.

Observing the spot evolution over time this way, especially its changing shape, is instructive about the nature of numerical diffusion. After the first year, the peak height decreases in a roughly logarithmic fashion, and the peak width increases approximately linearly. The latter trend at first seems odd, since our expectation would be for the length scale to increase more slowly, i.e. $L \sim \sqrt{Kt}$. However, as the spot moves into higher flow regions, the numerical diffusivity increases.

To obtain a rough estimate of the apparent numerical diffusivity, we note that after 10 years the spot width has increased to about 500 km (corresponding to a half-width of 250 km). Using the approximation that $\kappa \approx L^2/T$ we obtain an effective numeric diffusion of roughly $200\,\text{m}^2\,\text{s}^{-1}$, which is entirely consistent with the analysis done earlier in this subsection (see the numerical diffusivity contours in the eastern half of Fig. 16.6).

Basically, then, the model appears to be performing much as we expect, given the limitations of the finite difference algorithms we are using. The aforementioned diagnostics exercises give us some confidence that we haven't made any serious coding errors (but be ever vigilant!). Performance-wise, our model is stable, conserves mass (there are no "boundary leaks"), has reasonable transport accuracy even with strong gradients, and has a significant but acceptable numerical diffusivity. It looks like we're ready to progress to an actual tracer simulation.

16.5 Transient tracer invasion into a gyre

Our objective is to run our gyre simulation with a transient tracer, tritium (^{3}H), and its daughter (^{3}He), "tuning" the physical parameters of the model so that the simulated tracer fields best match what observations we have. Given the best combination of circulation and mixing (which in combination constitute ventilation), we then can run a simulation for oxygen with a view to estimating the *in situ* oxygen consumption rate that best matches the observed oxygen fields. We begin by describing the transient tracers and constructing their advection–diffusion equations.

16.5.1 Background and equations

We have already discussed the "geochemistry" of tritium and its daughter ^{3}He when developing one-dimensional pipe model simulations in Chapter 12. Tritium is an ideal hydrologic tag; as the heaviest isotope of hydrogen it occurs virtually solely in nature as part of the water molecule. Its natural background (produced by cosmic ray interactions in the stratosphere) was dwarfed by massive production from atmospheric nuclear weapons testing in the late 1950s and early 1960s. Tritium is radioactive, with a half-life of 12.36 years (MacMahon, 2006), and decays to ^{3}He, a stable, inert isotope of helium. Ignoring the background ^{3}He dissolved from the atmosphere,[1] we define *tritiugenic* ^{3}He as only the ^{3}He produced by *in situ* decay of tritium. It is lost to the atmosphere by gas exchange at the sea surface.

Before wading into the detailed mechanics of the simulations, we ought to look at the data to understand the basic framework that we're working within. We consider observations that were taken on a number of cruises carried out within a year or two of 1981. The choice of timeframe is driven by the occurrence of a major tracer survey in the summer of 1981 called *Transient Tracers in the Ocean* (TTO). The coverage of this survey was complemented by several smaller cruises that also took place around this time. Because tritium will have decayed significantly over the time span of the observations, we take the

---

[1] Atmospheric ^{3}He can be corrected for by measuring the sister isotope ^{4}He.

additional step of decay-correcting the tritium data, measured in tritium units (TU), to a common date (January 1, 1981) according to

$$C_{1981} = C_{\text{obs}} e^{\lambda(t_{\text{obs}} - 1981)} \tag{16.28}$$

where C_{obs} is the tritium concentration observed at the time t_{obs} and λ is the decay constant for tritium expressed in y^{-1}. Because ^3He doesn't decay, it is not decay-corrected, and is reported in TU. We used the tritium and ^3He data from samples that had a density close to the target density horizon in Fig. 16.1, that is $1026.75 \leq \rho \leq 1026.85 \, \text{kg m}^{-3}$. We thereby obtained values for approximately 140 locations in the subtropical North Atlantic, which we contour in Figs. 16.8 and 16.9.

In Fig. 16.8 we see evidence of the penetration of bomb tritium from the northeast corner into the gyre interior, a relatively homogeneous tritium distribution throughout the middle of the gyre, the bending of tritium isopleths northeastward near the top of the gyre by the westward intensified boundary current, and a steady southward decline in tritium concentrations (by as much as a factor of four) on the southern boundary of the gyre. Tritium concentrations are much lower in the south, and in particular the Southern Hemisphere, because the atmospheric nuclear weapons testing occurred mostly in the Northern Hemisphere. The southern hydrographic boundary of the gyre is characterized by a similar transition in a variety of properties, including salinity and dissolved oxygen

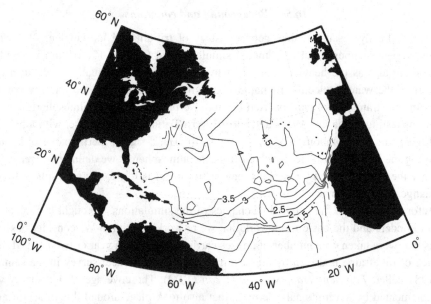

Figure 16.8 North Atlantic tritium (in TU_{1981}) observed in the early 1980s on the density horizon where $1026.75 \leq \rho \leq 1026.85 \, \text{kg m}^{-3}$. Because tritium decays at a rate of $\sim 6\% \, \text{y}^{-1}$, we have decay-corrected the tritium data to the same point in time (January 1, 1981) using Equation (16.28).

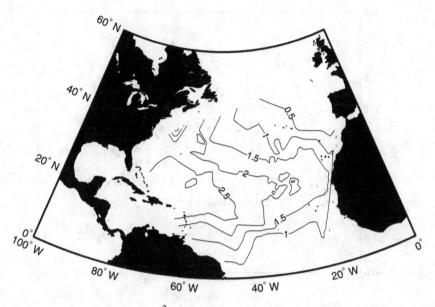

Figure 16.9 North Atlantic tritiugenic ^{3}He (in TU) observed in the early 1980s on the density horizon where $1026.75 \leq \rho \leq 1026.85\,\mathrm{kg\,m^{-3}}$.

(see for example Fig. 16.2), and extends from approximately 15° N in the west to roughly 20° N in the east.

The ^{3}He distribution has its lowest values in the northeast, where tritium values are highest, indicating the "ventilation window" for this surface, presumably where the isopycnal horizon intersects the ocean surface. There is a steady, positive, southwestward gradient in ^{3}He into the gyre with the greatest ^{3}He values in the southwestern corner. Like tritium, there is a steady ^{3}He decrease (but only a factor of two) along the southern flank of the gyre. We hope to emulate these gross features (in both ^{3}He and tritium) with our gyre model.

In addition to the spatial patterns of these isotopes, the relationship between them can be considerd diagnostic of the nature of the ventilation process. We show a scatter plot of tritium and tritiugenic ^{3}He in Fig. 16.10. Note the characteristic hook-shaped pattern that we saw in the one-dimensional simulations in Chapter 12. The structure can be rationalized as follows. Freshly ventilated waters have low ^{3}He and high tritium, forming the lower right-hand part of the hook. As water enters the gyre, ^{3}He increases downstream with only a modest decrease in tritium, reaching a ^{3}He maximum of ~3 TU with a tritium concentration around 4 TU. Transitioning to less ventilated waters, both isotopes decrease along an approximately straight line. Comparison to the maps in Figs. 16.8 and 16.9 shows that the latter linear trend is across the property gradients on the southern gyre boundary. Thus our objective is to match not only the approximate spatial patterns of these two tracers, but also the broader characteristics of their inter-relationship.

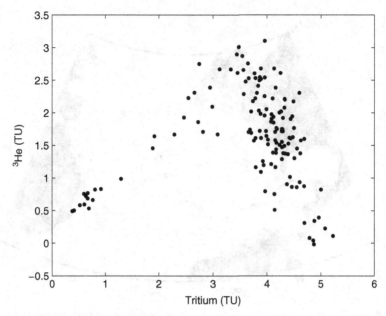

Figure 16.10 North Atlantic tritiugenic ^{3}He (in TU) vs. tritium (in TU$_{1981}$) observed in the early 1980s on the density horizon where $1026.75 \leq \rho \leq 1026.85 \, \mathrm{kg\,m^{-3}}$.

The two-dimensional advection–diffusion equations for these tracers are given by:

$$\frac{\partial C}{\partial t} = K \nabla^2 C - \nabla \cdot (\mathbf{u}C) - \lambda C \qquad (16.29)$$

$$\frac{\partial H}{\partial t} = K \nabla^2 H - \nabla \cdot (\mathbf{u}H) + \lambda C \qquad (16.30)$$

where C is the tritium concentration and H is the ^{3}He concentration. Note that the second equation is linked to the first by the last term; that is, ^{3}He is just "dead tritium". Finally, in our model we will report both tritium and ^{3}He in *TU*, short for *tritium units*. Surface ocean water tritium concentrations were typically about 0.5 TU before the bomb tests, and peaked at around 20 TU in the mid-1960s before decreasing back toward pre-bomb levels many decades later.

Implementing these equations requires extending our advection–diffusion code to incorporate two sets of calculations (one for C and one for H) into a new version, which we now call `THAdvDiff.m`. This is straightforward, and we won't include that here, but there is one additional step that is required to account for the decay of tritium and ingrowth of ^{3}He. We define two variables in the new version of our weight calculation routine `THWeightCalc.m` that pre-calculate the amount of decay and growth during a time step with

```
Lambda=1.784e-9;            % tritium decay constant in sec^-1
TDecay=exp(-dt*Lambda);     % tritium decay in one time step
HGrowth=1-TDecay;           % 3He growth in one time step
```

where `Lambda` is the tritium decay constant. The first constant, `TDecay`, is just a scaling factor by which the tritium concentration is reduced by radioactive decay over one time step. The second, `HGrowth`, gives the amount of tritium that has decayed, which of course is how much ³He grows in. Given these numbers, we can modify the following statements at the end of the `for` loop (before the `end` statement):

```
C=CNew*TDecay;              % decay the tritium
H=Hnew+CNew*HGrowth;        % in-grow the 3He
```

Other than making some decisions about how to initialize tracer concentrations in our gyre and to implement boundary conditions along the northern edge, we are almost ready to go.

16.5.2 Revisiting boundary and initial conditions

We now reconsider our northern boundary conditions, i.e. how we represent the connection between the gyre's isopycnal horizon and the sea surface. We refer to this as the *gyre outcrop*, which is analogous to where a geological stratum reaches the land surface. We will simulate the outcrop by setting the "nw" northernmost rows of cells to surface conditions. That is, the tritium concentrations will be set to the appropriate surface water concentration (for that model year) and the ³He concentrations will be set to zero. Increasing nw serves to increase the outcropping area for the gyre, which can have a significant influence on the ventilation of our gyre (see Fig. 16.11).

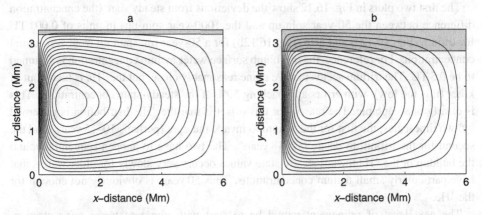

Figure 16.11 Emulating the gyre outcrop for two example scenarios, (a) where only three of the northernmost cell rows are set to surface conditions and (b) where 17 of the northernmost rows of cells are reset to surface conditions.

The two examples shown probably bracket the range of conditions we might consider for density horizons within the main thermocline. The first example, Fig. 16.11a, is for out-cropping three rows. In this instance, only one of the contoured streamlines flows into and out of the outcrop region. Compare this with the second example (Fig. 16.11b) where seven streamlines enter and leave the outcrop region characterized by 17 rows. A much larger volume of gyre water is "reset" in this scenario than in the first case. We would argue that the streamlines entering the outcrop region in the west represent the *obducted* Gulf Stream waters that are brought to the surface by heat extraction and vertical mixing in the northwest Sargasso Sea, especially during the winter. The streamlines leaving the outcrop region in the central and eastern part of the gyre are those waters *subducted* by a combi-nation of wind stress convergence and an oceanic process called *lateral induction* (e.g. see Qiu and Huang, 1995). As you might expect, the choice of outcrop fraction impacts our tracer simulation and thus is another parameterization — in addition to gyre strength and mixing coefficients — that might be adjusted in our simulation experiments.

We also need to think about how to initialize our model to "pre-bomb", steady-state concentration levels for both tracers. In the one-dimensional model case, we could use a simple steady-state analytic solution, but that would be a little more difficult here. We'll take another approach. We start by simply setting the interior concentrations to zero, then running the model forward (that is, *spinning up*) for 50 years with the northern boundary condition set to pre-bomb tritium concentration and zero ^{3}He concentration. Since the effective recirculation timescale is of order 15–20 years, this means that the gyre would recirculate about three times during our spin-up. The question is whether this is enough time to get the distributions close enough (where "close" is clearly an operational concept) for our purposes. In this simple model we have the luxury of being able to run the model for some ridiculously long time (say 1000 years model time) to see what the distributions should look like in steady state.

The first two plots in Fig. 16.12 show the deviations from steady state (the concentration difference between the 50-year spin-up and the 1000-year spin-up) in units of 0.001 TU for tritium (Fig. 16.12a) and ^{3}He (Fig. 16.12b) for a 50-year run spun up from zero initial concentrations. For reference, the pre-bomb surface water tritium concentration is assumed to be 0.5 TU. The tritium seems to have come reasonably close, but the ^{3}He is badly out of kilter (it is too low), with errors approaching 50% of the steady-state concentrations. This is clearly unacceptable. The reason for this offset is that the ^{3}He distribution is governed by two timescales, the first for the tritium to invade the gyre (perhaps 10–15 years) and the second for the tritium to decay (12–18 years). The two have to happen sequentially, and the buildup of the ^{3}He toward steady-state values occurs very slowly for older waters that have particularly small tritium concentrations. Thus 50 years is obviously not enough for the ^{3}He.

The next level of refinement would be to start with non-zero tracer concentration. We start by imagining some characteristic ("average") north–south advection velocity v_0 of order 0.01 m s^{-1}, and use this to impose a simple north–south gradient in tracer distributions according to:

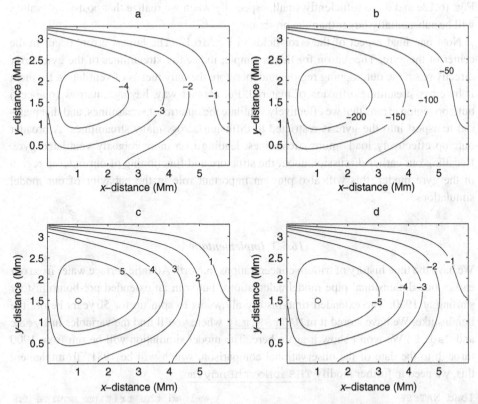

Figure 16.12 Two examples of the "error fields" (in 0.001 TU) between a 50-year spin-up and the steady-state distribution estimated from a 1000-year run: (a) tritium and (b) ^{3}He spun up from zero concentrations, and (c) tritium and (d) ^{3}He spun up from a "quasi-analytical" state.

$$C(y_o) = C_0 e^{-\lambda y_o / v_0} \tag{16.31}$$

$$H(y_o) = C_0 - C(y_o) \tag{16.32}$$

where y_o is the distance from the outcrop, and λ is the tritium decay constant. The second expression simply assumes that any water parcel that has entered the gyre from the outcrop retains its ingrown ^{3}He *in situ*.

This is neither a very clever nor a particularly realistic representation of the impact of gyre circulation on the tracer fields, but it effectively emulates the gross, first-order characteristic of the regional distributions. Starting our spin-up from this "quasi-analytical" state makes the ^{3}He field for a 50-year spin-up much closer to the steady state (see Fig. 16.12d), reducing the error by a factor of 40. Interestingly, the error in the tritium field changes sign (from negative to positive in the core of the gyre). This is a result of the fact that the model starts out with too much tritium in these regions, and is approaching steady state from the opposite direction. Regardless of the sign, however, the errors observed in

Fig. 16.12c and d are sufficiently small, especially when we realize that post-bomb values will be substantially larger than the pre-bomb.

Note one final aspect of the error fields in Fig. 16.12. The largest errors occur in the center of the gyre. The reason for this is simple; the center streamlines of the gyre never directly reach the outcropping region, and therefore the only means of ventilation for them is by cross-streamline diffusion of material. Put another way, having a narrow (one-cell) outcrop region means that we effectively ventilate the outermost streamlines, and that material transport into the gyre is restricted to diffusion across many streamlines. A broader outcrop effectively loads more streamlines, leading to a more robustly ventilated gyre. Recalling our earlier discussion about the structure and functioning of our outcrop region in the gyre model, this will also play an important role in the behavior of our model simulations.

16.5.3 Implementation

We have the time history of tritium concentrations in North Atlantic surface water from our earlier one-dimensional pipe model calculations, but with an extended pre-bomb history starting at 1900. The extended time-history allows us to spin up for 50 years before the bomb spike. We have stored it in NATrSF.mat where you'll find the variables in tyear and Csurf . We won't show it again here. The model simulation will be run from 1900 through to the date of the observational comparison, which will be 1981. To implement this, we need to further modify THSetNorthBndy.m:

```
load NATrSF                          % load the tritium source fn
ModelTime=1900+[1:nt]/ntyr;          % make model time vector
Cn=interp1(tyear,Csurf,ModelTime);   % interp TSF onto model time
for iw=1:nw                          % map tritium onto outcrop
    Cnorth(iw,:)=Cn;
end
Hnorth=zeros(nw,nt);                 % set 3He to zero
```

where nw determines the outcrop width and is now set in the main body of code. This allows us to modify this parameter systematically to observe how the tracer simulations are affected by it. Next, we further modify THWeightCalc.m to allow for systematic alteration of the diffusivity with a parameter Diff that allows us to adjust the diffusivity in steps of $500\,\mathrm{m}^2\,\mathrm{s}^{-1}$:

```
Kx=Diff*500*ones(size(X)); Ky=Kx;    % compute K
```

and we further modify THPsiDef.m to allow for scaling the velocity field by introducing a variable Amp such that:

```
A = Amp*1e4;                         % first guess amplitude
```

We can now systematically march through a range of operational parameters. For example, the following code in THStommelGyre.m runs the simulations for a range of diffusivities (500–2000 m^2 s^{-1}), streamfunction amplitudes (30 000 to 50 000 m^2 s^{-1}), and outcrop widths (1–5 cells):

```
for Diff=1:4
    for Amp=3:5
        for nw=1:2:5
            THPsiDef            % set up grid, compute streamfunction
            VelCalc             % compute velocity fields
            THWeightCalc        % compute weight matrices
            THSetNorthBndy      % set up northern boundary condition
            THInitConc          % initialize concentration
            TheDiff=max(Kx(:));ThePsi=max(Psi(:));
            THAdvDiff           % do advection-diffusion calculation
            save(sprintf('Runs/THnw%.0fd%.0fa%.0f.mat', ...
                nw,Diff,Amp),'x','y','C','H','TheDiff','ThePsi','nw');
        end
    end
end
```

where the last line within the embedded for-loops saves the simulated 1981 fields to code-named files in a subdirectory called "Runs". For example, THnw3d2a4.mat corresponds to a simulation where the outcrop is three rows wide, the diffusivity is 1000 m^2 s^{-1} and the streamfunction maximum amplitude is 40 000 m^2 s^{-1}. Each simulation takes a few minutes on a reasonably fast desktop computer, so the above would take at least a few hours to run.

16.5.4 A Rorschach test

Before doing an exhaustive exploration of parameter space, however, we ought to compare a representative model simulation to the observations. The reason we should do this will become apparent shortly, but the main motivation is to convince ourselves that we are at least working with a viable model geometry. We call this kind of approach a *Rorschach test*, which is a reference to the "ink blot" pattern interpretation test used for psychological evaluation. We call it this to remind you that there is a great deal of interpretation involved in looking at model simulation patterns, and although it has some value, it is hardly quantitative in nature. Nevertheless, it is a good common-sense test that is perhaps most useful in (a) gaining intuition about how a model is behaving, and (b) seeing whether its behavior is so outlandish that it cannot possibly be viable.

While the model simulation (Fig. 16.13) bears a vague resemblance to the observations (Figs. 16.8 and 16.9), the shapes of the tracer isopleths appear all wrong in the southern portion of the model gyre. Rather than decreasing southward and southeastward like the observations, they actually do the opposite. Although varying model parameters over what might be considered a reasonable range does indeed change the details of the patterns, this basic behavior is common to all the simulations, and we must conclude that in some fundamental way our model has failed the Rorschach test.

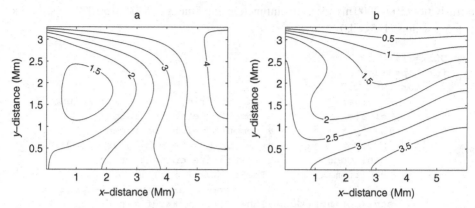

Figure 16.13 A representative simulation of the invasion of (a) tritium and (b) ^{3}He onto our Stommel gyre surface. Here we have used an outcrop width of one row of cells, a lateral diffusivity of $2000\, m^2\, s^{-1}$, and our standard circulation streamline amplitude of $3 \times 10^4\, m^2\, s^{-1}$.

When we think about our model's southern boundary condition, this begins to make sense. We have treated the southern edge of our subtropical gyre as a solid wall (no flux condition). This allows tracer to "pile up" along the southern edge of model. While it makes sense to do this for the western and eastern boundaries, which actually correspond to continents, this doesn't match what we are trying to model in the real ocean. The southern edge of the subtropical gyre is bounded not by a "solid wall", but rather by a counter-clockwise circulating tropical gyre. Rather than piling up, our tracer should be mixing southward into more tracer-impoverished tropical waters, producing the negative southward tracer gradients observed in Figs. 16.8 and 16.9. This immediately suggests to us how we might improve our model.

16.6 Doubling up for a better gyre model

We will construct a double gyre model with the northern clockwise gyre representing our subtropical gyre and the southern gyre corresponding to a tropical counter-clockwise circulation. Mathematically, it is easy to make a streamfunction that represents such a scheme. We simply use the Stommel gyre streamfunction described by Equation (16.15) and expand the latitudinal range to $-y_m \geq y \geq y_m$. This is done by modifying the appropriate line in `PsiDef.m` to define y over the larger range, i.e. `y=[-ym:dy:ym]`, creating a new routine `DblPsiDef.m` (we won't list that here, because it's basically the same except for that one change). This produces a pair of antisymmetric gyres with a reflection point at $y = 0$ shown in Fig. 16.14.

The price paid for this extension is that it doubles the size of the model domain, and consequently doubles the time it takes to run. We are not trying to simulate the tracer evolution in the southern gyre in this model, but rather using it as a kind of "buffer zone" that may be crudely representative of the tropical gyre that bounds the subtropical gyre's

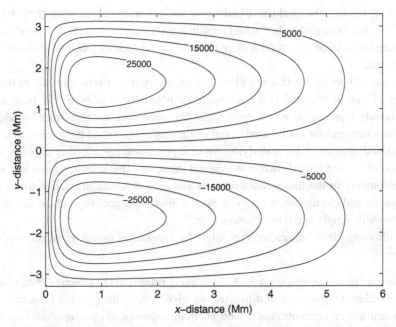

Figure 16.14 An idealized double gyre streamfunction. Note that the positive streamfunction in the northern half describes a clockwise circulation and the negative streamfunction in the southern half describes a counter-clockwise gyre.

southern edge. Thus we are pushing the problem associated with embedding a finite sized, isolated model domain inside a much larger ocean farther away.

16.6.1 Rorschach test or cost function?

We have already used a pattern recognition strategy to reject our single gyre model based on irreconcilable differences with the observed property distributions. Certainly, it is worthwhile to continue comparing the basic structures of the tracer distributions between simulation and observation. But attempting to assess whether one simulation is actually better than another, particularly when the differences are subtle, becomes increasingly difficult. Moreover, "manually" searching for an optimal simulation in a multi-dimensional parameter space becomes rather time-consuming (see Chapter 10). While we won't abandon these visual comparisons (you ought to keep coming back to pattern visualization), we need a more quantitative measure of model performance. For this we turn to a cost function.

The fact that our model has an idealized shape (it is rectangular) that does not directly match the shape of the real subtropical gyre in the North Atlantic makes the comparison between model simulation and observations problematic. Choosing a blanket measure of model–data mismatch, e.g. root mean square difference between model and observations, is not simple because we would have to develop some operational *mapping* of model to real space. This may be possible, but would require a good deal of thought and justification. It would perhaps be better to opt for a more realistic model geometry before going down that particular path.

Optimistically assuming that our choice of geometry hasn't seriously biased the gross behavior of our model, and that our choice of boundary conditions has been at least approximately representative, we must settle for some simple "first-order" model–data comparison metrics that can be used to refine our choices of model parameters. Such metrics must be diagnostic of the underlying physical processes, e.g. allowing us to distinguish the effects of mixing from diffusion in material transport. This selection must be based on an understanding of the tracer characteristics and their behavior in relation to the nature of ventilation and circulation in the gyre. Such a discussion goes beyond the scope of this text, so we will simply assert our choices here.

The following "bulk" characteristics might be considered useful diagnostics of model behavior:

- Maximum ^{3}He concentration = 3.11 TU is likely related to the effective *Peclet number* (i.e. the relative importance of diffusion and advection) of the gyre circulation.
- Minimum tritium concentration = 2.66 TU in the subtropical gyre represents the degree to which some portion of the subtropical gyre remains isolated from the outcrop region. Inasmuch as those streamlines that do not outcrop must be ventilated by cross-streamline diffusion, this property responds to the strength of diffusion.
- Average ^{3}He concentration = 1.68 TU for the subtropical gyre is a measure of the average circulation timescale, which is related to the streamfunction amplitude, and also to the loss of ^{3}He by mixing and obduction.
- Average tritium concentration = 4.12 TU is another integrated measure of gyre ventilation, responding both to streamfunction amplitude and mixing.

We obtain the target (observed) values for these characteristics from the observational data set stored in <u>obs1981.mat</u> within the subtropical gyre (north of 20° N) using

```
load obs1981.mat              % load in observations
ObsHe3Max = max(h(y>20));     % maximum 3He in subtropical gyre
ObsHe3Avg = mean(h(y>20));    % average 3He in subtropical gyre
ObsTritMin = min(t(y>20));    % minimum tritium
ObsTritAvg = mean(t(y>20));   % average tritium
```

We can then construct a cost function that is the weighted RMS difference between the simulation and observed values for these four constraints:

$$J_{CF} = \frac{1}{N-3} \sum_{i=1}^{N} \left(\frac{C_i^{mod} - C_i^{obs}}{\sigma_i} \right)^2 \qquad (16.33)$$

which resembles a reduced chi-squared, in that we've corrected for the fact that we're using three parameters to "fit" our observations. Thus an optimal fit should yield a cost function of order unity. We assume an uncertainty σ_i of 0.2 TU for all of these properties. This latter choice is somewhat arbitrary, and need not be the same for each of the cost function components. Like the selection of the parameters themselves, the choice of their weights (uncertainties) could significantly influence the outcome of the optimization. If possible, the implications of these seemingly harmless decisions ought to be explored experimentally to gain an understanding of how robust the model results truly are.

In computing the northern gyre properties, we select the grid points only within the northern (subtropical) gyre, and avoid the southernmost 10 grid rows within the northern gyre to compute the cost function:

```
ng=[120:220];                      % northern gyre grid points
Hn=H(ng,:);                        % select only northern gyre 3He
Cn=C(ng,:);                        % select only northern gyre tritium
DelHe3Max = (max(Hn(:)) - ObsHe3Max)/0.2;    % maximum 3He
DelHe3Avg = (mean(Hn(:)) - ObsHe3Avg)/0.2;   % average 3He
DelTritMin = (min(Cn(:)) - ObsTritMin)/0.2;  % minimum tritium
DelTritAvg = (mean(Cn(:)) - ObsTritAvg)/0.2; % average tritium
JCF = (DelHe3Max^2 + DelHe3Avg^2 + DelTritMin^2 + DelTritAvg^2)/1;
```

The reason for this decision is that we wish to exclude as far as possible any artifacts associated with our less realistic southern boundary condition.

16.6.2 Optimization

To perform the optimization we embed the model simulation within a MATLAB function constructed to compute the cost function and return its value for a given set of parameters. In this case, the three parameters that we would seek to adjust are the streamfunction amplitude (which controls the gyre circulation strength), the horizontal diffusivity, and the outcrop area. The first two may be regarded as continuum variables, but the latter, at least for this model, is discretized by the number of grid cell rows. We could force this discretization by allowing the optimizer to select any value but then rounding this parameter in the code to the nearest integer. This could lead to awkward behavior with some optimizers, so we will do something different. We will find the model's optimal circulation amplitude and diffusivity by minimizing the cost function for fixed outcrop area. We then plot the cost function versus outcrop area and select the area with the minimum cost function. While any outcrop area, up to the size of the northern gyre itself, could be used, excessively wide outcrop regions would make the resultant circulation cease to resemble a gyre. Therefore we set a limiting width of 15 rows, or about 10–15% of the gyre's extent.

We begin by constructing the cost function calculator:

```
function CF = CalcCF(P)
    global nw
    Amp=P(1); TheAmp=Amp;
    Diff=P(2); TheDiff=Diff;
```

```
DblPsiDef          % set up grid & compute streamfunction
DblVelCalc         % compute velocity fields
THWeightCalc       % compute weight matrices
THSetNorthBndy     % set up northern boundary condition history
THInitConc         % initialize concentration
CFAdvDiff          % do advection-diffusion calculation
ng=[120:220];      % rows corresponding to subtropical gyre
H=H(ng,:); C=C(ng,:);   % select only subtropical gyre values
Hmax=max(H(:)); Cmin=min(C(:));
Havg=mean(H(:)); Cavg=mean(C(:));
CF=( ((Hmax-3.11)/0.2)^2 + ((Cmin-2.66)/0.2)^2 ...
     + ((Havg-1.68)/0.2)^2 + ((Cavg-4.12)/0.2)^2);
```

where the routine CFAdvDiff.m is just a modified version of THAdvDiff.m with the annual contour plots and inventory calculations removed. The streamfunction amplitude and diffusivity are passed to the cost function routine in the parameter array P, and we have introduced the outcrop width nw as a global variable.

We use constrained optimization to limit the range over which the circulation and diffusivity can vary. This makes sense because on the basis of oceanic observations, we have some idea of what "reasonable" values for those parameters ought to be. Recognizing that the idealized character of our model simulation may result in somewhat less realistic values, we set the range of acceptable choices relatively broadly. We then find the optimal parameters and cost function as a function of the outcrop width using GyreOptimize.m:

```
global nw                              % needed for cost function
options = optimset('LargeScale','off', 'Display', ...
                     'iter','MaxFunEvals',50,'MaxIter',50);
lb=[1 500]; ub=[10 5000]; X=[5 2500];  % set lower/upper/start vals
CFs = [];                              % storage for CF values
for iw=1:15                            % step outcrop widths
    nw=iw;                             % set outcrop width
    Parameters = X;                    % start with prev. optimum
    [X,fval,exitflag,output,Lambda,Grad,Hessian] ...
        = fmincon(@CalcCF,Parameters,[],[],[],[],lb,ub,[],options);
    CFs=[CFs fval];                    % accumulate cost functs
    save(sprintf('Optim/Optim%.0f',iw))  % save to file
end
```

which saves the resultant optimal parameters and cost functions in separate files in a subdirectory. Note that for efficiency, we start in the middle of the parameter range, and that in subsequent optimizations, we use the previous set of values under the assumption that they will change only gradually with nw.

Be warned, however, that the script above will take a very, very long time to run on all but the most powerful computers. Here's the reasoning: on a moderately fast desktop, a single simulation will take about 10 minutes, depending on the streamfunction amplitude

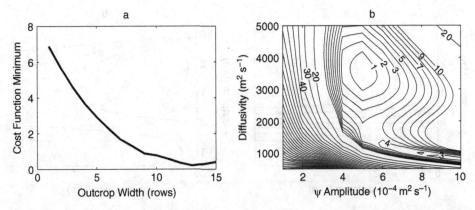

Figure 16.15 (a) The minimum cost function as defined by Equation (16.33) as a function of outcrop width (in number of rows). (b) The topography of the cost function as a function of streamfunction amplitude and diffusivity for an outcrop width of 13 rows. The contour interval is 1 for values less than 10, and the interval is 10 for values greater than 10.

(maximum velocity) and diffusivity. We've set the maximum number of function evaluations to 50 (more might be needed) so the program will execute at least 1000 times, which means that the program would run for about a week!

We've done the hard work so you won't have to. In Fig. 16.15a we see a plot of the cost function minimum against outcrop width. The cost function decreases with increasing outcrop width but approaches a broad minimum at an outcrop width of 13 rows. We interpret this to signify the importance of outcropping streamlines for "loading" the bomb-tritium into the gyre flow lines (see Fig. 16.11).

Figure 16.15b shows the general "topography" of the cost function as a function of streamfunction amplitude and diffusivity for an outcrop width of 13 rows. Note there is a well-defined minimum in the vicinity of $\psi_{max} = 5 \times 10^4\,\text{m}^2\,\text{s}^{-1}$ and $\kappa = 3500\,\text{m}^2\,\text{s}^{-1}$. You should note also that there is another, less intense *local minimum* around $\psi_{max} = 10 \times 10^4\,\text{m}^2\,\text{s}^{-1}$ and $\kappa = 700\,\text{m}^2\,\text{s}^{-1}$, but we will choose the *global minimum* as it is about an order of magnitude smaller. The value of the minimum cost function is approximately 0.2 (not shown, owing to contour interval), that is, somewhat better than the order unity we expected. This suggests that within the limits of its idealized form, the model matches the first-order characteristics of the observations reasonably well, or that we have been somewhat overly pessimistic about our choices of σ_i in Equation (16.33). Recall our discussion of cost functions in Chapter 3.

The steepness of the cost function in ψ_{max}–κ space near the minimum gives us a relatively "tight" constraint on the effective circulation strength and diffusivity required to simulate the transient tracer distributions. Indeed if we were able to assign a "normal (*Gaussian*)" expectation functionality to the cost function, we could assign a standard confidence interval of approximately 20% to these parameters based on the change over which the cost function increases significantly (doubles). That is, we would argue that we've

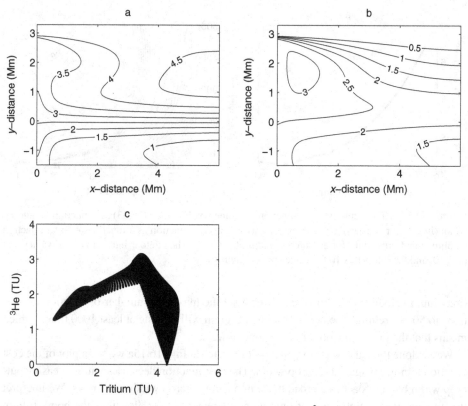

Figure 16.16 a) The contoured distributions of (a) tritium and (b) ^{3}He for the optimum simulation for the double Stommel gyre with an outcrop width of 13 rows (see text). Panel (c) is a plot of the model tritium versus ^{3}He distribution.

determined $\kappa = 3500 \pm 700\, \mathrm{m^2\, s^{-1}}$ and $\psi_{\mathrm{max}} = (5 \pm 1) \times 10^4\, \mathrm{m^2\, s^{-1}}$. However, these estimates only have meaning within the context of this particular simulation. Notably, the diffusivity is significantly higher than might be expected for the open ocean, but is likely a result of our highly idealized circulation scheme.

It is reassuring to return to our Rorschach roots and compare the simulated tracer distributions with the observations. In Fig. 16.16a and b we see the corresponding tritium and ^{3}He patterns for the optimal case, as well as the tritium–^{3}He relationship (Fig. 16.16c). Note that we've deliberately avoided plotting the lower portion of the southern gyre, as our main interest lies in the tracer distributions within the northern (subtropical) gyre. These need to be compared with maps of the tracer observations (Figs. 16.8 and 16.9) and the tritium–^{3}He scatter plot of Fig. 16.10. Certainly one has to squint to see the similarities, but the overall features are indeed there. Bearing in mind the liberties we've taken with the geometry and boundary conditions, the comparison is quite favorable (or are we being hopeless romantics?).

Certainly, we could amuse ourselves with endless further refinements in our model, but at this point, we will stop and declare a victory. That is, we have satisfactorily mimicked the invasion of a pair of transient tracers onto an isopycnal horizon within the main thermocline of the North Atlantic subtropical gyre. In this way we have been able to constrain the physical ventilation processes responsible for not only the invasion of these tracers, but also the supply of oxygen onto this horizon.

16.7 Estimating oxygen utilization rates

The final step in this exercise is to return to the reason why we're going to all this effort. Given the determination of physical parameters (circulation strength, diffusivity, and out-crop width) that are consistent with the observed tritium and ^{3}He distributions, we can now perform a model simulation of the oxygen saturation in the gyre. In reality, we should model the oxygen *concentration*, not its saturation value; the model moves mass (concentration). It could be argued that temperature and salinity variations on an isopycnal horizon will be small enough that the errors introduced by working with saturation values rather than actual concentrations are probably small compared with the other liberties we have taken with model construction.

We formulate our double gyre model with the northern boundary condition that sets the saturation value to precisely 100%, and the oxygen saturation S is subject to a constant consumption rate $-J$ (here J is a positive quantity) such that

$$\frac{\partial S}{\partial t} = 0 = K\nabla^2 S - \nabla \cdot (\mathbf{u}S) - J \qquad (16.34)$$

i.e. the two-dimensional equation for a steady-state, non-conservative tracer. In this simple example, we assume J to be a constant, but there is every reason to assume that there is significant spatial variation in this property. Oxygen consumption on a subsurface horizon arises from the zooplankton respiration and bacterial oxidation of falling organic material, which in turn is a function of the magnitude of *export primary production*. Export production is known to vary regionally from the relatively productive subpolar regions through the characteristically oligotrophic ("barren") central subtropics, to the dynamic upwelling regions off the coast of Africa. Attributing some spatial structure to J would certainly be an interesting next step, but we're going to settle here for a simple first-order estimate of the average consumption rate. Considering the simplicity of the geometry and circulation scheme of this model, that's about all we can reasonably expect.

As with the challenge associated with initializing the pre-bomb tritium distribution, we need to start with some sensible, quasi-analytic approximation to the oxygen saturation field prior to iterating the saturation field to steady state. The motivation for this is that the spin-up problem is made more acute by the timescale associated with the diffusive coupling of the otherwise isolated southern gyre. We will use a simple north–south saturation gradient imposed by a characteristic meridional velocity v_0 such that $\partial S_{\text{init}}/\partial y = -J/v_0$, whose solution for $S = 100$ along the northern boundary is $S_{\mathrm{init}} = 100 - J y_0/v_0$. Given this intial approximation we then run the model with the following code from OxyGyre.m

```
Amp=5.0; Diff=3500;nw=13; % optimal physical characteristics
J=1.0;                     % example consumption rate (1% per year)
DblPsiDef                  % set up grid & compute streamfunction
DblVelCalc                 % compute velocity fields
OxWeightCalc               % compute weight matrices
OxSetNorthBndy             % set up northern boundary condition history
OxInitConc                 % initialize concentration
OxAdvDiff                  % do advection-diffusion calculation
```

and then run the model long enough to reach a virtual steady-state distribution. By "virtual" we mean that the long-term drift in the distribution needs to be substantially smaller than the consumption term: that is, $|\partial S/\partial t| \ll J$. Thus we need to monitor the drift in our model. As an experiment, we include in OxAdvDiff.m a code snippet with the "annual checkup" to monitor not only the evolving distribution, but also the *difference field* (expressed as a fraction of the consumption term) between two consecutive years:

```
subplot(2,2,1)
clabel(contour(x/1e6,y/1e6,C,[0:10:100]));      % look at distribution
xlabel('X-Distance (Mm)');ylabel('Y-Distance (Mm)')
title(sprintf('J= %.2f y=%.0f',J,nyr))
if nyr>0
    subplot(2,2,2)
    clabel(contour(x/1e6,y/1e6,(C-Cold)/J));
    xlabel('X-Distance (Mm)');ylabel('Y-Distance (Mm)')
    title(sprintf('J= %.2f y=%.0f',J,nyr))
end
Cold=C;
```

where one could also accumulate mean statistics, for example including the average and maximum offsets over the entire domain.

Visually, the contour patterns appear to form and stabilize after a few decades, and certainly within the first century of the run. Figure 16.17a shows a plot of the percent saturation change per century as a function of time for a 500-year run. To all intents and purposes, a satisfactory convergence is achieved within a couple of centuries, with the average drift dropping below 0.01% per century. Thus we can be confident that running the model simulation for 300 years approaches steady state closely enough.

Simple experimentation with a choice of J reveals that a value of $1.1\%\,\mathrm{y}^{-1}$ (Fig. 16.17b) gives a reasonable representation of the observed distribution (Fig. 16.2), at least for the subtropical gyre. Note especially the similarity in the 60–90% contours. We have deliberately ignored the behavior in the southern (tropical) gyre, as this is regarded as a only buffer zone, but it is remarkable that the basic shape, if not the exact intensity, of the oxygen minimum zone is reasonably well represented. On the whole, given the simplicity of the model's structure, the overall correspondence to the data is encouraging.

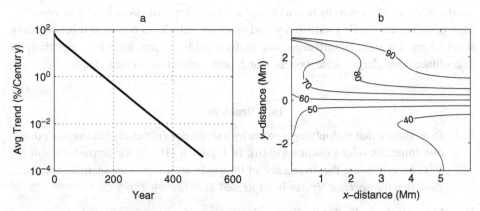

Figure 16.17 (a) The average model oxygen saturation drift in percent per century as a function of time. (b) The distribution of oxygen saturation for a consumption rate of $1.1\% \, y^{-1}$.

Given this "optimal" value, how might we set uncertainty limits on it? Experimentation with our choice of J and liberal application of the Rorschach criterion suggests that the character of the simulation's contours deviate significantly from observations for changes in J as little as 10–15%. However, this assumes no error in the "physics" of our model, and if we perform similar experiments over the approximately 20% range in our physical parameterizations, we obtain a net uncertainty of order 25% in our estimate of J. With a typical saturation value of $255 \, \mu\text{mol kg}^{-1}$ associated with the average temperature and salinity of this isopycnal horizon, this corresponds to an oxygen demand of $2.8 \pm 0.7 \, \mu\text{mol kg}^{-1} \, y^{-1}$. It turns out that this is roughly consistent with other estimates for the North Atlantic subtropical gyre (e.g. see Jenkins and Wallace, 1992).

16.8 Non-uniform grids

We leave you with a final note about grid construction. With the highly idealized coarse-resolution gyre model that we've been exploring, a uniform rectilinear grid suits our purpose, in that we can specify a fine enough grid that we can satisfy our resolution requirements over the whole model domain. However, extending this approach to larger-scale, more complicated, or higher-resolution simulations can become computationally costly. Recognizing that higher resolution may be needed in regions where streamlines are "bunched up", or conversely that we can get away with coarser resolution in regions where velocities are slower and less rapidly changing, we might be tempted to use a non-uniform grid. For example, one might specify a smaller Δx near the western boundary with some sort of quasi-exponential dependence on x. The trade-off is between an increase in algorithmic truncation precision (for a given total number of x-direction grid points) and additional computational complexity. The complexity arises from the fact that the finite difference algorithm must account for the spatially changing grid-cell volumes and interfacial areas. Moreover, there is an added level of complexity associated with consideration of numerical stability, diffusivity, and other artifacts. Thus one must think carefully about taking this step. Nevertheless, non-uniform grids become necessary, especially in three-dimensional, larger-scale models, which is the subject of our next chapter.

16.9 Problems

16.1. Demonstrate that multiplying a vector by a scalar does not change the vector's direction (hint: consider Equation (16.6)). In Equation (16.10) we derived a result for $\nabla^2 C$ by expanding the divergence of the gradient of C. As an alternate approach, show that the operator ∇^2 can be expressed as $\partial^2/\partial x^2 + \partial^2/\partial y^2$.

16.2. Show algebraically that the Stommel gyre streamfunction does indeed go to zero on the boundaries.

16.3. Construct a model based on the code and equations provided in this chapter for a symmetric counterclockwise gyre with a square domain 1000 km by 1000 km, a spatial resolution $\Delta x = \Delta y = 10$ km, and an average speed of $1\,\text{cm}\,\text{s}^{-1}$. Initialize the concentrations to zero, then set the southern boundary concentrations to 10 units at zero time. Run the model for 10 years with explicit mixing coefficients of 0, 500, 1000, 1500, and $2000\,\text{m}^2\,\text{s}^{-1}$. Throughout the simulations, monitor the effective tracer inventory by summing C over the model domain and plot it as an annual function. Plot also the time history for the concentration at the very center of the gyre, as well a center meridional section (N–S section through the center of the gyre). Explain as quantitatively as possible how the diffusivity affects the time evolution of the center concentration, and how the center meridional section behaves. Why and how does the material inventory depend on diffusivity?

16.4. Using the symmetric model simulation above, leave the southern boundary concentration at zero, set the explicit diffusivity to zero, and then perform four separate 10-year experiments by applying a point pulse concentration (i.e. setting $C = 1$ at $t = 0$ only) for the following (x, y) locations (in km): (100, 100) (300, 300) (500, 500), (750, 500). Plot the maximum concentration as a function of time, and then determine the geometric mean of the numerical diffusion from the slope of these curves using

$$\sqrt{K_x K_y} = \frac{S}{\pi^2} \Delta x \Delta y \qquad (16.35)$$

where S is the slope of $1/C_{max}$ versus time. Comment on the apparent spatial/temporal dependence of the diffusivities thus obtained, and explain what you see. Specifically, compare the observed evolution over the first 2–3 years with what is observed over the full 10-year simulation. Can you predict what you see? Explain the temporal evolution in terms of what we know about numerical diffusion.

17
Three-dimensional general circulation models (GCMs)

People don't understand the earth, but they want to, so they build a
model, and then they have two things they don't understand.

Gerard Roe

So far, we have introduced many of the elements of ocean modeling but in simplified situations with reduced dimensionality (i.e. box models, vertical 1D models, 2D gyre models). Here we pull all of the elements together, introducing the topic of 3D ocean *general circulation models* (GCMs). As you might expect, the topic is complex, and our GCM tour will be necessarily brief and focused. While you probably won't be able to construct your own GCM, you should at least be able to understand the conversation and perhaps even utilize 3D GCM output. Several good review articles and books have been written on ocean GCMs that the reader can refer to for more details (e.g. Haidvogel and Beckmann, 1999; Griffies *et al.*, 2000; Griffies, 2004). While our emphasis is on marine systems, most of the fundamental concepts about GCMs are applicable to a wide range of environmental fluid systems, from the atmosphere to mantle convection to groundwaters.

Several themes emerge when considering ocean GCMs. First, no matter how fast technology develops, the cutting edge of ocean modeling is always "compute bound", which is why you won't be able to build a decent GCM using MATLAB. Ocean GCM development is linked to the evolution of supercomputers, and in fact GCMs are commonly used to test new supercomputers. Second, the histories of ocean, atmosphere and climate modeling are intertwined (e.g. Smagorinsky, 1963; Arakawa, 1966; Bryan, 1969; Cox, 1984). The same general principles and basic numerical frameworks are used in both ocean and atmosphere GCM codes. Variants of some physical parameterizations, such as for mesoscale eddy and surface boundary layer mixing, find common application in both domains. GCM model codes are recycled as often as soda cans.

Third, the current generation of GCMs are complicated pieces of software, and the creation and long-term maintenance of state-of-the-art models often involve teams of scientists and software engineers, working for many years. In the past, the effort was often centered at national laboratories but it is increasingly distributed over a virtual community of researchers similar to the open-source code movement. Fourth and related, most people nowadays do not write their own GCM code from scratch, and you will note that we have not attempted to build one here in MATLAB. Given that GCMs are typically work-ing at the limits of computational capability, they are written often in computer language and compiler code that is heavily optimized for specific computer architecture (though this is changing with the growing availability of Linux clusters and more powerful desktop machines). While MATLAB is a useful computational tool for the purposes of our course, it is not well suited for constructing GCMs.

There are a number of good ocean 3D models (and model output) that are publicly available and can be used either straight "out of the box" or with some modifications. The models usually come with documentation – sometimes extensive, sometimes not – and they are often designed to allow for user modification by, for example, subdividing the code into distinct modules for different processes and dynamics (e.g. a module for mesoscale eddy mixing). Finally, models have different strengths and weaknesses, and the right model for one application may not be appropriate for another. It will sound familiar by now but we want to remind you to choose the right tool based on the science questions you are trying to answer.

17.1 Dynamics, governing equations, and approximations

In contrast to the 2D gyre models discussed earlier, the velocity field in a 3D ocean GCM is prognostic: that is, its behavior is predicted from some set of dynamical equations. One such set of relationships commonly used is called the *primitive equations*, a com-plete mathematical description of the local balances of momentum, energy, and mass for a fluid (e.g. Gill, 1982; Pedlosky, 1992; McWilliams, 2006). The primitive equations con-sist of linked conservation equations for momentum, mass, energy (heat), and salt together with an *equation of state* that relates density to temperature and salinity. Using these con-straints, we can calculate the wind driven and thermohaline (density) driven circulation as a response to the current state of the ocean and to forcing of heat, wind, and fresh water at the ocean surface. The underlying physics for many ocean (and atmosphere) GCMs is based on an approximated version of the primitive equations. Why would you want to make approximations when you can use the complete equations? Well, you may want to focus on a particular process or processes and other dynamics simply gets in the way; be careful, though, because sometimes approximations have unintended and unforeseen consequences. Remember Einstein's dictum "as simple as possible, but not simpler".

The choice of horizontal scale for an ocean GCM introduces different types of approxi-mations. The dynamics of ocean flow are fundamentally nonlinear, which manifests itself as flow instabilities on many scales from centimeters to tens of kilometers (Chapter 9). Turbulence associated with fronts and mesoscale eddies (roughly 10–300 km) strongly influences material transport over ocean gyres and basins. The character of ocean GCMs depends greatly on whether the effects of mesoscale eddies are resolved explicitly or rep-resented by sub-grid-scale parameterizations. Processes on even smaller scales, such as internal wave and planetary boundary layer mixing, will also need to be parameterized (Section 17.4). We start with some housekeeping on spherical coordinates and the gradient operator. We then present the frequently used, continuous form of the hydrostatic primitive equations, discussing some of the common approximations along the way.

17.1.1 Spherical coordinates and the material derivative

Many ocean GCMs span a large enough portion of the globe that we can no longer simply use the familiar flat x–y–z Cartesian coordinates for three dimensions. Instead, we have to admit that the world is indeed curved and use *spherical coordinates* instead. In spherical coordinates, the "origin" is the center of the Earth, and position is given in terms of an alternative three dimensions, longitude λ, latitude ϕ, and the radial distance from the center of the Earth r (Fig. 17.1). The corresponding components of the 3D velocity vector **u** in spherical coordinates are u, v, and w, respectively. Okay, they are the same variable names as Cartesian coordinates, but the meaning is somewhat altered; u, for example, is the velocity oriented now along lines of constant latitude, which are curved not straight.

The center of the Earth is an inconvenient origin for most ocean model grids, unless you are a great fan of Jules Verne, and you must forgive the use of depth z (usually defined as zero at the ocean surface) rather than r in most ocean models. The typical thickness of the ocean ($\sim 5 \times 10^3$ m) is small relative to the average radius of the Earth ($r_e \approx 6.38 \times 10^6$ m).

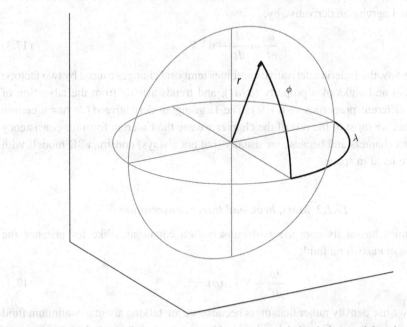

Figure 17.1 Spherical coordinate system, used in most general circulation models, where position is given in terms of longitude λ, latitude ϕ, and the radial distance from the center of the Earth r.

With very little error, we can use the *thin shell approximation* ($r \approx r_e$) to neglect variations in radius in the equations below, using a constant value r_e instead.

The gradient operator ∇ in spherical coordinates is also somewhat more cluttered than the one introduced for the 2D gyre:

$$\nabla = \frac{1}{r_e \cos \phi} \frac{\partial}{\partial \lambda} \hat{\lambda} + \frac{1}{r_e} \frac{\partial}{\partial \phi} \hat{\phi} + \frac{\partial}{\partial z} \hat{z} \qquad (17.1)$$

where $\hat{\lambda}$, $\hat{\phi}$, and $\hat{z}$ are the unit vectors along the three directions, with $\hat{z}$ the unit vector in the **r** direction. The terms $1/(r_e \cos \phi)$ and $1/r_e$ account for the planet's curvature, and ϕ and λ are in radians rather than degrees. The advection operator in spherical coordinates is:

$$\mathbf{u} \cdot \nabla = \frac{u}{r_e \cos \phi} \frac{\partial}{\partial \lambda} + \frac{v}{r_e} \frac{\partial}{\partial \phi} + w \frac{\partial}{\partial z} \qquad (17.2)$$

Using the shorthand of the gradient operator ∇ greatly simplifies the appearance of the equations below by hiding the curvature terms, but don't forget that they are there. While spherical coordinates are essential for ocean basins, the corresponding x–y–z Cartesian 3D advection operator is sufficient for small enough model domains and idealized conditions.

Earlier, in Chapter 11, we introduced the concept of the Lagrangian derivative d/dt, the rate of change calculated following a water parcel (Equation (11.5)). We should note that the Lagrangian derivative goes by many names including the "material derivative" and the "substantial derivative" as well as other notations, D/Dt being common. In three dimensions, the time rate of change at a fixed point in space $\partial/\partial t$ (Eulerian derivative) is related to the Lagrangian derivative by:

$$\frac{\partial}{\partial t} = \frac{d}{dt} - \mathbf{u} \cdot \nabla \qquad (17.3)$$

Put another way, the Eulerian derivative combines temporal changes caused by two factors: actual sources and sinks of a property (d/dt), and trends arising from the advection of water with different properties ($-\mathbf{u} \cdot \nabla$). The Lagrangian derivative d/dt has a certain simplicity, but for most of the rest of the chapter we use the Eulerian form for consistency with previous chapters and because we usually (but not always) construct 3D models with grids that are fixed in space.

17.1.2 Mass, heat, and tracer conservation

Fluid dynamics has at its core a set of conservation equations. Take for instance the *conservation of mass* for a fluid:

$$\frac{\partial \rho}{\partial t} + \nabla \cdot (\mathbf{u}\rho) = 0 \qquad (17.4)$$

Notice that we use density rather than mass because we are talking about a continuum fluid rather than a point object. The right-hand side is zero because away from the surface, water is not directly added or removed from the ocean. We neglect molecular diffusion of heat and salt which can alter water parcel density because in most applications it is orders of magnitude smaller than turbulent diffusion, which we'll take care of later. The second term on the left-hand side can be expanded using our recently acquired knowledge of vector calculus from Chapter 16:

$$\nabla \cdot (\mathbf{u}\rho) = \mathbf{u} \cdot \nabla \rho + \rho \nabla \cdot \mathbf{u} \tag{17.5}$$

This relationship applies regardless of the coordinate system. The first term on the right-hand side reflects the advection of density gradients, and the second term reflects net convergence/divergence of flow.

We can then recast the mass conservation or *continuity equation* in Lagrangian form:

$$\frac{d\rho}{dt} + \rho \nabla \cdot \mathbf{u} = 0 \tag{17.6}$$

That is, density (mass) is conserved following a parcel except for the effects of flow divergence. To good approximation, seawater is *incompressible*:

$$\nabla \cdot \mathbf{u} \approx 0 \tag{17.7}$$

such that the continuity equation reduces to:

$$\frac{d\rho}{dt} = \frac{\partial \rho}{\partial t} + \mathbf{u} \cdot \nabla \rho \approx 0 \tag{17.8}$$

The same is not true, of course, for a compressible gas like the atmosphere, but we won't cover that complication in this book (see for example Gill, 1982; McWilliams, 2006). From here on we assume an incompressible fluid (and neglect molecular diffusion).

The conservation equation for the concentration of a generic tracer C (in $mol\, m^{-3}$) is:

$$\frac{\partial C}{\partial t} + \mathbf{u} \cdot \nabla C = J_C \tag{17.9}$$

where the sources and sinks, accumulated into a single term J_C, include internal chemical and biological transformations. Boundary fluxes at the air–sea and sediment–water interfaces can be applied as additional source/sink terms in top, side, and bottom grid boxes. Similarly in the ocean interior, salinity is conserved:

$$\frac{\partial S}{\partial t} + \mathbf{u} \cdot \nabla S = 0 \tag{17.10}$$

and at the surface J_S terms would need to be added to the right-hand side reflecting the boundary conditions for fresh water; salinity is reduced by freshwater sources (precipitation, river runoff and sea-ice melt) and elevated by freshwater sinks (evaporation and sea-ice formation).

The next thing we need to worry about is *conservation of heat* (or potential temperature θ):

$$\frac{\partial \theta}{\partial t} + \mathbf{u} \cdot \nabla \theta = \frac{Q_{solar}}{\rho_0 c_p} \tag{17.11}$$

where c_p is specific heat ($\mathrm{J\,K^{-1}\,kg^{-1}}$) and Q_{solar} is the internal volumetric heating rate ($\mathrm{W\,m^{-3}}$ or $\mathrm{J\,s^{-1}\,m^{-3}}$). As mentioned in the 1D upper ocean modeling, Chapter 15, most heating terms can be applied as a boundary condition (e.g. sensible and evaporative cooling, geothermal heating) but solar radiation often penetrates tens of meters below the surface and would be represented by an internal heating term.

We use *potential temperature* θ rather than temperature in Equation (17.11). Potential temperature is defined as the temperature a water parcel would have if it were moved to the surface adiabatically (i.e. without exchange of heat with the surroundings). Seawater does in fact compress slightly under increasing pressure, and in the process thermodynamic work is done on a water parcel as it shrinks. This ends up heating water parcels slightly if they move downward in the water column (a few tenths of a degree going from the surface to typical deep water pressures). In a model it is more convenient to use a conservative temperature tracer that doesn't change as water parcels change depth.

The density of seawater changes with temperature and salinity, a fact that we used in the 1D upper ocean model (Chapter 15) to diagnose vertical convection and vertical stability. The *equation of state* for seawater relates density to temperature, salinity, and pressure:

$$\rho_\theta = f(\theta, S, p_0) \tag{17.12}$$

Actually the equation gives the *potential density*, which is used in most models following the same arguments as for potential temperature and where $p_0 = 0$ is the surface pressure. Like potential temperature, potential density is the density a water parcel would have if it were moved adiabatically to the sea surface. When analyzing intermediate and deep waters, we may be interested in referencing density to other pressures (e.g. we could define a ρ_2 referenced to 2000 db). In most, but not all, models potential temperature and salinity are the prognostic variables and potential density is computed diagnostically (the exception is isopycnal models, discussed in the next section).

The seawater equation of state can be computed in a simplified, linearized form using the *thermal expansion coefficient* α and *haline contraction coefficient* β:

$$\alpha = -\frac{1}{\rho}\frac{\partial \rho}{\partial \theta} \quad \beta = +\frac{1}{\rho}\frac{\partial \rho}{\partial S} \tag{17.13}$$

This linear form was computationally expedient in early GCMs and is a reasonable approximation for modest changes in temperature and salinity. Most modern GCMs use a more exact, full empirical equation of state expressed as a series of polynomial equations in θ, S, and p. There is a MATLAB seawater toolbox that contains many useful routines for analyzing ocean hydrographic data and ocean GCM output. It includes m-files for computing θ, ρ, α, β, and other ocean hydrographic properties. The toolbox is free and is available from CSIRO in Australia and the SEA-MAT project at the US Geological Survey in Woods Hole, a collaborative effort to organize and distribute MATLAB tools for the oceanographic community.

17.1.3 Momentum, Coriolis force, and velocities

The momentum of a water parcel, $\rho\mathbf{u}$, is a key property in fluid dynamics. *Conservation of momentum* allows us to say how water parcels accelerate in response to external forces (recall Newton's equation of motion, *force = mass × acceleration*). For reasons explained below, we'll often split the 3D velocity vector into horizontal and vertical components:

$$\mathbf{u} = \mathbf{u}_H + w\hat{\mathbf{z}} \tag{17.14}$$

where $\mathbf{u}_H$ is the horizontal velocity vector oriented normal to $\hat{\mathbf{z}}$. Recalling the definition of the 2D streamfunction ψ (Equations (16.11) and (16.12)), we can also write the horizontal advection operator in terms of ψ, so for a property C:

$$\mathbf{u}_H \cdot \nabla C = u\frac{\partial}{\partial x}C + v\frac{\partial}{\partial y}C = -\frac{\partial \psi}{\partial y}\frac{\partial}{\partial x}C + \frac{\partial \psi}{\partial x}\frac{\partial}{\partial y}C = \mathcal{J}[\psi, C] \tag{17.15}$$

The final term on the right is the *Jacobian operator*, a common notation for 2D horizontal flow.

The rate of change of the horizontal velocity:

$$\frac{\partial \mathbf{u}_H}{\partial t} + \mathbf{u} \cdot \nabla \mathbf{u}_H + f\hat{\mathbf{z}} \times \mathbf{u}_H = -\frac{\nabla p}{\rho_0} \tag{17.16}$$

contains acceleration terms reflecting the Earth's rotation due to the Coriolis force ($f\hat{\mathbf{z}} \times \mathbf{u}_H$) and pressure gradients ($\nabla p/\rho_0$) where f is the Coriolis parameter (17.19). The *vector cross-product* for two generic vectors is given by:

$$\mathbf{a} \times \mathbf{b} = ab\sin\theta\,\hat{\mathbf{n}} \tag{17.17}$$

where θ is the angle between the vectors, a and b are their magnitudes, and $\hat{\mathbf{n}}$ is the unit vector perpendicular to both $\mathbf{a}$ and $\mathbf{b}$ (the direction set by the right-hand rule). For Coriolis acceleration:

$$f\hat{\mathbf{z}} \times \mathbf{u}_H = -fu\hat{\mathbf{y}} + fv\hat{\mathbf{x}} \tag{17.18}$$

which gives the familiar Coriolis deflection of currents to the right in the Northern Hemisphere and to the left in the Southern Hemisphere.

A uniform reference density ρ_0 is used in the momentum equation following the *Boussinesq approximation*. This approximation basically states that seawater density perturbations ($\rho' = \rho - \rho_0$) are small in a fractional sense ($\rho' \ll \rho_0$). The density perturbations ρ' are retained only for vertical stability calculations. Notice the advection term $\mathbf{u} \cdot \nabla \mathbf{u}_H$ on the left-hand side of Equation (17.16). Yes, this means velocity can be transported by the fluid just as are chemical tracers. Remember, though, that velocity is a vector quantity; an east–west current transports both the east–west and north–south components of velocity.

The *Coriolis force* is an "apparent" force that arises because we are on a rotating planet and our "fixed" model grid (or latitude/longitude positions) is not actually fixed in space. We mentioned in Chapter 16 that it controls the shape of the subtropical gyre but buried the details for later (well, actually now). Newton's laws hold in a *fixed inertial frame*. One way

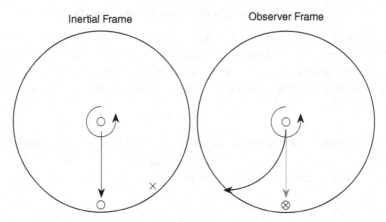

Figure 17.2 Schematic showing how an "apparent" Coriolis force arises for a simplified 2D rotating system (a merry-go-round). The person at the center of the merry-go-round throws a ball to a person on the edge. In an external inertial frame (left) the ball flies straight (trajectory marked by the black arrow) but the person moves from o to × as the merry-go-round rotates counter-clockwise. In a rotating observer frame (right) the observer is stationary (o and × stay in the same spot) and the ball's initial straight trajectory (gray arrow) appears to curve to the right (black curved arrow), which requires an apparent force acting at right angle to the ball's velocity.

to think about the effect of rotation is to imagine trying to throw a ball with a friend while standing on a moving merry-go-round (Fig. 17.2). Let's say you stand in the center and your friend stands on the edge. You throw the ball straight at your friend, but in the time it takes for the ball to arrive at the outer edge of the merry-go-round your friend has moved. From your friend's perspective you have terrible aim, and the trajectory of the ball actually curves to the left or the right (depending on which way the merry-go-round is spinning). From your frame of reference, the ball appeared to be pushed to the side by a mysterious force. In fact the ball really did fly straight, something you could verify by looking down from above, say from a tree (i.e. in a non-rotating inertial framework). It's the fact that the observer (your friend) is rotating that messes up everything. The same is true for the rotating Earth.

The strength of the apparent rotational force is encapsulated in the *Coriolis parameter* f, which varies with latitude (ϕ):

$$f = 2\Omega \sin \phi \tag{17.19}$$

where Ω is the planet's angular velocity. The Coriolis parameter is largest at the poles and goes to zero at the equator, where the Earth's rotation does not project as strongly onto horizontal motions. The planet's rotation bends the motion of a water parcel to the right in the Northern Hemisphere and to the left in the Southern Hemisphere. In many parts of the ocean, the horizontal momentum equation (17.16) can be reduced to a steady-state, *geostrophic balance* between two dominant terms, the Coriolis force and the horizontal pressure terms:

$$f\hat{\mathbf{z}} \times \mathbf{u}_H \approx -\frac{\nabla p}{\rho_0} \tag{17.20}$$

a topic that we will revisit in more detail in Section 18.3.2.

A similar dynamical balance occurs in the vertical but this time between gravity (accelerating water parcels downward) and the vertical gradient in water pressure (accelerating water parcels upward). Many ocean GCMs go so far as to neglect all other vertical acceleration terms, leading to the *hydrostatic approximation*:

$$\frac{\partial p}{\partial z} \approx -\rho g \tag{17.21}$$

where z is positive upward. Pressure at any depth can be found from density by simply vertically integrating (17.21) from the surface to that depth. Some of the neglected vertical acceleration terms can be important in certain circumstances. So for example if you are studying the details of deep convection, you may prefer to use a *non-hydrostatic* model. And that's basically the continuous primitive equations, or at least an approximated set, that can be used to calculate the general circulation of the ocean. The next section describes the sometimes challenging problems of how to discretize the equations and then solve them numerically.

Before we move on, we should mention some other dynamical ocean model variants. Not every problem requires the full primitive equations, and more simplified (idealized) equation sets are often sufficient, computationally tractable, and easier to understand (Haidvogel and Beckmann, 1999). The 2D *shallow water* equations arise by vertically integrating the primitive equations and are commonly used for tracking tsunamis and coastal storm surge. In the β-*plane* approximation, the Coriolis parameter is taken as linear with latitude proportional to the local gradient of planetary vorticity $f = f_0 + \beta\phi$. The β-plane equations, often in non-dimensional form, are a favorite idealized playground for geophysical fluid dynamists.

The *quasi-geostrophic* (QG) equations develop from further scaling arguments, perhaps the most important being that for large-scale motion, horizontal inertia typically is small relative to planetary rotation. This is encapsulated in the non-dimensional *Rossby number Ro*:

$$Ro = \frac{U}{2\Omega L} \tag{17.22}$$

where U and L are characteristic horizontal velocity and length scales. The Rossby number is small ($Ro \ll 1$) for ocean flows on spatial scales from mesoscale eddies up to basin-scale gyres. This means that to lowest order, horizontal velocities are given by a balance between the horizontal pressure gradients and Coriolis terms (17.20). The small departures from geostrophy are important, however, hence the designation quasi-geostrophic equations.

17.1.4 Vorticity

Fluid rotation is an important property of geophysical flows embodied in a property called *vorticity*. The *total vorticity* is a measure of flow rotation and is given by the curl of the velocity

$$\nabla \times \mathbf{u} = \left(\frac{\partial w}{\partial y} - \frac{\partial v}{\partial z}\right)\hat{\mathbf{x}} + \left(\frac{\partial u}{\partial z} - \frac{\partial w}{\partial x}\right)\hat{\mathbf{y}} + \left(\frac{\partial v}{\partial x} - \frac{\partial u}{\partial y}\right)\hat{\mathbf{z}} \qquad (17.23)$$

See Schey (2005) for the definition of the curl of a vector. The total vorticity is made of two parts. The *relative vorticity* ζ reflects the local rotation of the flow and is equal to the vertical component of the velocity curl:

$$\zeta = \frac{\partial v}{\partial x} - \frac{\partial u}{\partial y} \qquad (17.24)$$

The *planetary vorticity* f is the vorticity contribution due to the rotating reference frame. Their sum, the *absolute vorticity* $(\zeta + f)$, is dominated by the planetary component for most large-scale ocean flows (i.e. for all but the most vigorous flows), as we would expect from the comments above about typical Rossby numbers. A related quantity, the *potential vorticity*, is given by $(\zeta + f)/h$, where h is the thickness of a water parcel.

Potential vorticity is approximately conserved following a water parcel:

$$\frac{d}{dt}\frac{(\zeta + f)}{h} \approx 0 \qquad (17.25)$$

neglecting surface vorticity sources (e.g. wind stress curl) and frictional sinks, leading to intriguing dynamical consequences. For example, as a water parcel moves poleward in the Northern Hemisphere, f increases; to conserve potential vorticity the water parcel must either stretch in the vertical (increasing h) and/or begin to spin more clockwise (decreasing ζ). Currents that reach the seafloor are steered by topography; currents like the Antarctic Circumpolar Current are deflected equatorward over ridges (lower h and f) and poleward over troughs (higher h and f). Vorticity conservation also helps explain the westward intensification of boundary currents (Chapter 16).

The *vorticity equation*, derived from the curl of the momentum equation (McWilliams, 2006),

$$\frac{\partial \zeta}{\partial t} + \mathbf{u} \cdot \nabla \zeta = \frac{d\zeta}{dt} = -(f + \zeta)\nabla \cdot \mathbf{u}_H - \mathbf{u} \cdot \nabla f + \mathcal{V} \qquad (17.26)$$

relates the time rate of change of relative vorticity to the horizontal velocity divergence and spatial variations in f. The term $\mathcal{V}$ includes surface vorticity sources (e.g. wind stress curl) and frictional sinks. The vorticity equation contains the information about the departures from geostrophy for slowly evolving flow that are incorporated into the QG equations (Gill, 1982). QG models are sometimes formulated in terms of ζ and ψ; advection of ζ is computed from ψ using the Jacobian $\mathcal{J}[\psi, \zeta]$, after which a new streamfunction is generated by solving a boundary value problem from $\zeta = \nabla_h^2 \psi$.

17.2 Model grids and numerics

While there are numerous ocean GCMs in the literature, most derive in one way or another from a small set of widely used modeling frameworks or families. These include for example the Modular Ocean Model (MOM) from the NOAA Geophysical Fluid Dynamics Laboratory (GFDL), the Parallel Ocean Program (POP) from the Los Alamos National Laboratory, the Massachusetts Institute of Technology (MIT) ocean model, the Miami Isopycnic Model (MICOM) which is morphing into the Hybrid Coordinate Ocean Model (HYCOM), and the Regional Oceanic Model System (ROMS). Most of these codes are now community models, developed and tested by a large and often geographically dispersed user-group of scientists and engineers much in the same way as the open-source software movement. We illustrate modeling concepts by mentioning specific models as examples, but the reader should recognize that ocean GCMs are an ever-evolving set of tools. MOM1.0 and MOM2.0 may have the same name and come from the same developer group but have substantially different underlying code structure.

17.2.1 Vertical coordinates

The first step in building (or choosing) a 3D GCM is to decide on the vertical coordinate system, which alters in some cases the form of the equations to be solved. The three main types are the *z-coordinate* or *level model*, the *terrain-following model*, and the isopycnic or *isopycnal model* (Fig. 17.3). Each of the coordinate systems has its strengths and weaknesses and may be more appropriate for some applications than for others (Griffies *et al.*, 2000).

The z-coordinate system (z-models) is perhaps the simplest conceptually, and some of the earliest ocean GCMs were z-coordinate. Both MOM and POP are level models. As you would guess from the name, vertical grid surfaces z_k are at fixed depth levels and do not vary either horizontally or in time. The thickness of grid layers can vary with depth, and commonly more grid levels are packed into the upper ocean for higher resolution. In a depth–distance cross-section, the grid looks like stacks of rectangular blocks, and bottom topography has an unnatural, step-like character. Numerical discretization is straightforward and follows the approaches presented in earlier chapters for 1D and 2D advection–diffusion equations, and surface boundary layer parameterizations are relatively straightforward to implement.

The z-models, however, have problems accurately representing flow along bottom topography. This is especially troublesome for dense deep and intermediate water overflows (e.g. from Nordic Seas, the Antarctic shelf, or the Mediterranean) where there tends to be excessive artificial mixing with surrounding waters, resulting in model deep waters that are too light and too shallow in the water column. They also are not ideal for coastal regions where topography is a major factor governing currents. A variety of fixes have been created to get around topographic issues, ranging from partial and shaved bottom cells that modify the grid near topography to explicit overflow and bottom-boundary layer (BBL) parameterizations that circumvent the z-grid altogether.

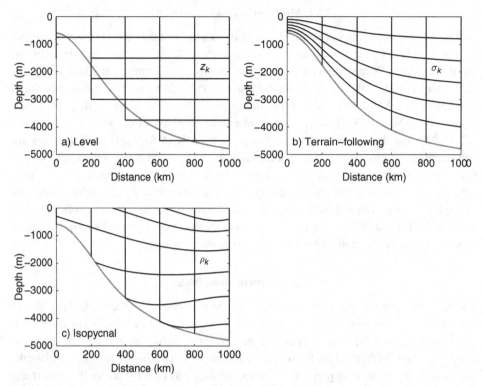

Figure 17.3 Schematic showing representative ocean GCM grids from three commonly used vertical coordinate systems: (a) level, (b) terrain-following and (c) isopycnal. Notice how the treatment of the model bathymetry differs in the three coordinate systems.

In the open ocean, z-models often do a poor job of representing flow and tracer properties in the stratified interior, which are oriented along isopycnal surfaces, not depth surfaces. In regions of sloping isopycnals near strong currents (e.g. Gulf Stream, Antarctic Circumpolar Current), horizontal diffusion (required for numerical stability) can induce artificial diabatic (i.e. cross-isopycnal) mixing, often termed the *Veronis effect* (Veronis, 1975; 1977). This arises because density horizons slope with respect to the grid cell layers, which are on a horizontal surface. The Veronis effect also induces artificial upwelling and downwelling circulation. Solutions to this problem are discussed below under sub-grid-scale parameterizations (Section 17.4).

As the name suggests, the depth of the vertical coordinate σ_k in terrain-following or σ-models, such as ROMS, varies horizontally (Fig. 17.3) and is a function of bottom depth H:

$$\sigma = \frac{z}{H} \tag{17.27}$$

Note that σ varies from 0 at the surface ($z = 0$) to 1 at the bottom ($z = H$). Because of the smooth representation of bottom topography, terrain-following models do a much better job for many coastal and regional applications, and σ-model grid cells have no side walls so we do not have to worry about side-wall friction issues. A drawback of σ-models is that horizontal pressure gradients are computationally more difficult to compute on constant σ surfaces, particularly in regions of steep topography (though some recent advances may resolve this issue). In fact all of the bookkeeping (conservation equations, flux divergences, etc.) becomes more complicated when you move away from z-coordinates, in part because the geometry (volumes and surface areas) of the cells varies.

In isopycnal models such as MICOM, the vertical coordinate system is discretized in terms of a fixed number of specified potential density layers ρ_k. Potential density referenced to the surface is used commonly for the thermocline, transitioning to deeper reference surfaces for intermediate and deep waters. The depth and thickness of ρ_k layers vary in space and also change with time as model integrations progress. The prognostic variables are potential temperature θ and layer thickness h (for any ρ_k layer S and θ covary and thus S can be diagnosed). Isopycnal coordinates are well suited for tracer transport in the stratified interior of the ocean, avoiding the pesky problems of artificial diapycnal diffusion mentioned above; for similar reasons ρ-models also do well at simulating deep-water overflows. A disadvantage is that ρ-layers naturally bunch up in the vertical in stratified regions and spread out in weakly stratified parts of the water column. Thus, isopycnal models have low effective resolution in the surface mixed layer and high-latitude deep convective mixing zones. Often a separate surface boundary layer parameterization (like the PWP model, Chapter 15) is incorporated, but there can be difficulties in computationally connecting the boundary layer with the ρ-layer stratified interior.

A recent trend is the development of hybrid models, such as HYCOM, that combine different vertical coordinate systems within a single model. The goals are to use the most appropriate coordinate system depending on regional topography and ocean physics and to transition smoothly between coordinate systems. So, for example, z-coordinates might be used for the surface mixed layer, ρ-coordinates in the open ocean interior, and σ-coordinates for continental margins and near the bottom.

17.2.2 Horizontal grid and resolution

The next step is to decide on the horizontal discrete grid. For 1D finite difference equations (Chapter 12), the grid is straightforward with tracer values assigned at the center of cells and velocities assigned at the interfaces between cells. More options with staggered grids arise as one moves to 2D and 3D. Variables of interest are u, v, w, θ, S, h, and horizontal streamfunction ψ. Arakawa (1966) established a widely used naming system for 2D structured grids (Fig. 17.4):

- *A-grid*: unstaggered grid with all variables evaluated at the same points;
- *B-grid*: θ, h, and ψ evaluated at grid-cell center and u and v at grid-cell corners;
- *C-grid*: θ and h at grid-cell center, u and v at mid-point of vertical and horizontal edges, respectively, and ψ at grid-cell corners;

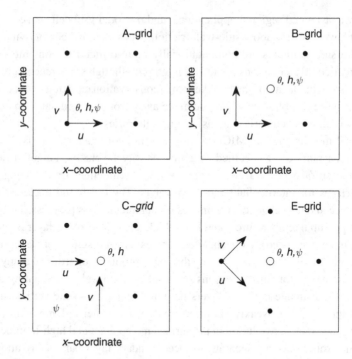

Figure 17.4 Schematic of structured 2D horizontal grids used in ocean models, marking the grid-cell locations where different variables are evaluated.

- *D-grid*: θ, h, and ψ at grid-cell corners and u and v at mid-point of vertical and horizontal edges, respectively (uncommon for ocean GCMs and not shown in Fig. 17.4);
- *E-grid*: a B-grid rotated by 45°.

Most ocean models utilize either a B-grid (mostly for coarse-resolution models) or a C-grid (mostly for high-resolution models), but there are examples of ocean models using each type of the Arakawa grids.

Velocities parallel to coastline and topographic side-boundaries in B-grid models are treated with no-slip boundary conditions, which makes it difficult to resolve flow through "narrow" straits (for a coarse global model even the Drake Passage can be effectively narrow). For example, a two-grid-point strait on a B-grid would have only a single non-zero, along-strait velocity at the center interface between the grid cells. The velocities applied to the tracers at the center of the cells to compute tracer transport would be an anemic half that size because of averaging with the zero no-slip values on either edge. Some numerical excavating and filling is often done in coarse models to smooth bottom topography, widen and deepen straits and even sink islands to better resolve the flow. Small-scale features, like the Straits of Gibraltar (14 km wide) separating the Mediterranean Sea from the North Atlantic, may be important oceanographically but difficult to capture in basin- to global-scale simulations with structured grids without excessive fudging of the topography. One approach might be to parameterize or specify the water exchange (i.e. artificially pushing water through a pipe between the Atlantic and the Mediterranean).

Horizontal grid resolution influences the character of ocean GCM simulations, sometimes quite dramatically. This is especially true in basin- to global-scale models where it is often too computationally expensive to fully resolve mesoscale eddies. *Eddy-resolving* models typically require horizontal resolution $\leq 0.1°$ latitude/longitude (roughly 10 km at mid-latitude) (Smith *et al.*, 2000). Weaker, eddy-like features appear at *eddy-permitting* resolutions (0.3–0.6°), and eddies are absent altogether from coarse, *non-eddy-resolving* simulations (1–3°). In coastal systems, eddy and filament scales can be even shorter, a few kilometers. You might consider simply always choosing a really fine-scale grid, but remember that computational costs increase quickly with improved resolution in a GCM. A factor of 2 reduction in horizontal grid size results in at least a factor of 8 increase in cost ($\times 2$ for the x and y directions and another $\times 2$ for the smaller time-step needed to maintain numerical stability; improvements in vertical resolution would only worsen the problem).

While you can't have really fine resolution everywhere, there are alternatives that concentrate resolution on locations of particular interest. These might include coastal and equatorial upwelling zones, river plumes, island wakes, and seamounts that exhibit small spatial scales and large physical, chemical, and biological gradients. Higher-resolution, regional sub-models can be nested within a larger spatial domain. *Nested models* come in two forms, one-way nesting where information passes in the form of lateral boundary conditions from the coarse-resolution model to the fine-resolution model, and two-way nesting where information also flows back from the finer-resolution model (a more challenging exercise). There can be multiple levels of nesting, and the fine-resolution domain can move with time to track features (e.g. a hurricane or a phytoplankton bloom). Another approach involves unstructured grids (e.g. finite elements) where horizontal resolution can be varied naturally over the domain.

One last grid issue has to do with the Arctic, which can be troublesome for global ocean simulations because in spherical coordinates the longitude grid lines converge at the pole. The small size of the grid cells leads to numerical instability (Section 12.4), and early models applied Fourier filtering in the Arctic to damp the resulting noise. A better solution is to use a distorted grid that displaces the polar singularity onto the land. The model dynamics and diagnostics with a generalized orthogonal curvilinear grid are of course somewhat more cumbersome, but standard software packages are available to address these issues.

17.2.3 Time-stepping and advection algorithms

The basic principles behind the numerics used in ocean GCMs will look familiar after reading the preceding chapters, but the details can often seem somewhat intimidating at first. Here we highlight some of the unique aspects, and the interested reader who wants a more in-depth view should see for example Haidvogel and Beckmann (1999), Griffies *et al.* (2000), and Griffies (2004). Be advised that computational methods (and even approaches to gridding) are evolving with time as more and more advanced methods from computational fluid dynamics and engineering (e.g. flow over an airplane wing) are being incorporated into ocean GCMs.

The horizontal velocity $\mathbf{u}_H$ in ocean GCMs is often decomposed into two components, the depth averaged flow or *barotropic velocity*, and the deviations from the depth average or *baroclinic velocity*. In most models, the faster evolving barotropic component and slower evolving baroclinic component are computed separately. Often the barotropic mode is computed from a 2D streamfunction ψ_{baro} of the depth averaged flow, which is a boundary value or elliptical problem usually requiring an iterative solution method. Because elliptical solvers need simultaneous information over the entire model grid, the barotropic velocity calculation can be a communication bottleneck for GCMs integrated on parallel computers. Alternatively, the barotropic mode can be solved for explicitly using short time steps. In many models, the horizontal velocity $\mathbf{u}_H$ is computed prognostically while the vertical velocity w is diagnosed using the continuity equation (17.7) by vertically integrating the divergence of the horizontal velocity terms starting from the bottom (where we must have $w = 0$).

Ocean GCMs use a variety of time-stepping schemes for the baroclinic velocity and tracer equations, with the *leap-frog* method being common in a number of models. The leap-frog method is a three time-level method where the rate of change of a tracer is computed in discretized form as:

$$C^{n+1} = C^{n-1} + 2\Delta t f(t^n, C^n) \tag{17.28}$$

where $f(t, C)$ includes all of the advective, diffusive and source/sink rate of change terms. The leap-frog method is more accurate than the simple Euler forward method but can introduce computational modes (i.e. unwanted numerical noise) because the odd time steps $(t^{n-1}, t^{n+1}, t^{n+3}, \ldots)$ are marched forward separately from the even time steps $(t^{n-2}, t^n, t^{n+2}, \ldots)$ (though the two timelines do communicate through the terms on the right-hand side). The computational noise is damped using either filtering or an occasional Euler forward step:

$$C^{n+1} = C^n + \Delta t f(t^n, C^n) \tag{17.29}$$

that blends information from the two timelines.

For many ocean modeling questions, we would like to know the steady-state solution in response to some specified surface forcing. Unfortunately the equilibrium timescale for the ocean thermocline is many decades to centuries and for deep waters several thousands of years. For many GCMs, it is intractable to integrate the model to equilibrium, particularly if we want to explore different values for physical parameterizations (e.g. mixing coefficients) or biogeochemical parameterizations (e.g. particle remineralization lengthscales). The maximum size of the time step Δt for the baroclinic momentum equation is constrained by stability limits, and some numerical acceleration techniques use longer time steps for the more slowly evolving temperature and salinity tracers to speed convergence. Other approaches go so far as using longer time steps for model deep waters compared with the thermocline. However, great care must be taken with any such numerical acceleration approaches, which can if you are not careful converge on the wrong solution. Advanced numerical approaches like *Newton–Krylov* methods potentially can reduce the time required to reach equilibrium by two orders of magnitude, and these methods are being applied experimentally to ocean biogeochemical models (e.g. Li and Primeau, 2008).

Modern ocean GCMs typically use some form of advanced tracer advection scheme in an attempt to limit the numerical errors associated with simple approaches such as first-order upwind methods. The somewhat inconsistent objectives are to limit artificial numerical diffusion and decrease the formation of spurious extrema (i.e. monotonicity) (of course, accurate and efficient algorithms are also desirable). See Rood (1987) for a good review of the basic concepts and approaches. Not too surprisingly given that there is no single perfect solution, the trade-offs have led to a plethora of advection algorithms. Common ones include third-order upwind, flux corrected, and semi-Lagrangian methods (Griffies *et al.*, 2000). While the details of these different schemes likely matter at some level, probably the best advice at this point is to be aware of the problems and use almost any method other than simply first-order upwinding (e.g. Hecht *et al.*, 1998).

17.3 Surface boundary conditions

You can build (or more likely download) a state-of-the-art ocean GCM, with all of the latest computational bells and whistles, but the quality of the solutions you'll generate also greatly depends upon your choice of surface forcing (Large *et al.*, 1997). Surface momentum fluxes establish the large-scale wind-driven circulation, control upwelling and Ekman pumping, and drive boundary layer turbulence and mixing. Buoyancy forcing governs water mass formations and upper ocean convections and restratification. The problem is similar to the boundary condition issues we faced for the 1D upper ocean model (Chapter 15), where we reviewed approaches for treating momentum, heat, and freshwater fluxes. The scope for a 3D GCM of course is considerably greater, requiring often a merging of ship-based, satellite, and atmospheric model data sets to get the required spatial and temporal coverage (Doney *et al.*, 2007). Here we focus on some unique issues for GCMs, including some of the added complications such as sea-ice and rivers.

Specifying the surface forcing for ocean GCMs is a non-trivial task. Direct measurements of turbulent heat and freshwater fluxes across the ocean–atmosphere interface are quite difficult, and the uncertainties in flux estimates were often as large as or larger than the actual fluxes for much of the ocean during the era of ocean GCM development (even today they remain substantial). Precipitation and even, to some degree, solar radiation fluxes also can be problematic particularly when they need to be specified for the whole globe. Also, even if the forcing were perfect, model errors could still cause the simulated ocean to drift away from reality. To avoid some of these issues, early ocean GCMs (and some idealized 3D models still) used fairly simple surface boundary conditions for heat and salt called *restoring boundary conditions*. Restoring terms tend to keep simulated surface properties, which we know well from observations, looking reasonable but as a result may hide blemishes in the model behavior in the simulated surface fluxes.

A restoring heat flux, for example, scales with the difference between the model θ_0 and observed sea surface temperatures θ_0^{obs}:

$$F_{\text{heat}} = \gamma(\theta_0^{\text{obs}} - \theta_0) \tag{17.30}$$

where γ is a proportionality coefficient with units of $\text{W m}^{-2}\,\text{K}^{-1}$. When simulated surface temperatures are too cold relative to data, a positive heat flux is generated acting to warm the model surface layer. Here we'll say that a positive heat flux transfers heat downward from the atmosphere to the ocean; there is no agreed sign convention for surface fluxes for ocean GCMs, and the problem of internal consistency gets even worse for coupled ocean–atmosphere models. The magnitude of γ is somewhat arbitrary but, unfortunately, can strongly influence models solutions.

Note an oddity of flux restoring that when the model temperature is skillful and matches the observations ($\theta_0^{\text{obs}} - \theta_0 = 0$) the surface heat flux also goes to zero, which doesn't make much sense for the large regions of the ocean that experience either substantial surface cooling (e.g. the Gulf Stream) or warming (e.g. the tropics). The next level of sophistication is:

$$F_{\text{heat}} = F_{\text{heat}}^{\text{obs}} + \gamma(\theta_0^{\text{obs}} - \theta_0) \tag{17.31}$$

where we add a specified heat flux term $F_{\text{heat}}^{\text{obs}}$ that could be from a ship-based flux climatology or an atmospheric reanalysis model. Then the simulation can support non-zero fluxes, and the restoring flux is (hopefully) a minor correction term to account for temperature errors arising from the combination of model dynamics and boundary forcing. Unfortunately the surface temperature errors are often large, especially on local scales near frontal regions, and these errors tend to grow because there are no feedback processes between the simulated SSTs and an externally specified heat flux.

An alternate approach involves using model SSTs, specified atmospheric conditions, and bulk turbulent flux formula. For sensible heat exchange (Large and Yeager, 2009):

$$F_{\text{sen}} = \rho c_p C_H(\theta_{\text{atm}} - \theta_0)\,|\mathbf{u}_{\text{atm}} - \mathbf{u}_0| \tag{17.32}$$

where ρ and c_p are the density and specific heat of seawater, C_H the sensible heat transfer coefficient, θ_{atm} the surface atmosphere temperature (typical $10\,\text{m}$), and the final term the magnitude of the difference between the atmosphere and surface ocean velocities. A similar equation could be written for latent heat flux:

$$F_{\text{lat}} = \rho L_v C_E[q_{\text{atm}} - q_{\text{sat}}(\theta_0)]\,|\mathbf{u}_{\text{atm}} - \mathbf{u}_0| \tag{17.33}$$

where L_v is the heat of vaporization, C_E the evaporative transfer coefficient, q_{atm} atmosphere specific humidity, and $q_{\text{sat}}(\theta_0)$ the saturated specific humidity above seawater with the model SST. The transfer coefficients depend on wind speed, waves, and atmospheric stability. The complete set of heat fluxes would also include terms for solar heating and net longwave flux.

Notice that the sensible and latent heat flux terms explicitly include model surface temperature, providing a damping feedback on model SSTs similar to a restoring term (Doney *et al.*, 1998). The effective restoring temperature, however, differs from the observed SST so that the model can match both heat fluxes and SSTs. While the bulk turbulent fluxes help limit model SST errors, they are not a panacea. In effect the atmosphere becomes an infinite reservoir of heat that can compensate for biases in the ocean model heat budget. This issue can be avoided by coupling together dynamic atmosphere and ocean models, but this raises a host of new problems when the fluxes being passed from the atmosphere model are different than what the ocean model wants and vice versa. As a general rule, individual component models that behave fine when integrated separately tend to misbehave when coupled together, at least at first until some tuning issues are resolved. Bulk turbulent fluxes can also be used to compute model evaporation rates, but freshwater flux schemes also require some form of explicit restoring on surface salinity because there are only indirect feedbacks of model salinity on computed evaporation rates. Uniform global forcing data sets are being constructed for ocean GCMs, for example through the international Coordinated Ocean-ice Reference Experiments (CORE) project (Large and Yeager, 2009). Considerable progress is also being made on developing coupled ocean–atmosphere models that generate internally consistent heat, freshwater, and momentum fluxes across the air–water interface (e.g. Doney *et al.*, 1998; Collins *et al.*, 2006).

The surface boundary conditions for net fresh water, the sum of precipitation and evaporation, also can be somewhat tricky because you are adding or removing mass (and volume) from the top model grid box. What then happens to the model sea level? Similar issues arise from net convergence or divergence from the wind-driven circulation. Actually there are two related concerns here, one dynamical and one associated with tracer balances. Many ocean GCMs assume a local constant volume for each model water column, which is enforced by a *rigid-lid approximation* where the barotropic flow is non-divergent and $w \equiv 0$ at the surface. Net mass divergence in flow that would drive surface displacements in the real ocean instead generates an artificial surface model pressure exerted on the rigid lid. The computed model surface pressure in turn contributes to the barotropic velocity calculation. The rigid-lid approximation is beneficial in that it filters out of the model solution some fast-evolving wave phenomena like surface gravity waves. But the approximation introduces its own problems and is also incompatible with models of, for example, tides or storm surge.

The next level of sophistication is a *free surface*, which prognostically tracks variations in surface elevation η:

$$\frac{\partial \eta}{\partial t} + \mathbf{u}_H \cdot \nabla \eta = w \qquad (17.34)$$

Free-surface models conserve ocean volume globally but not locally. There are issues when the size of the variations in η approaches the thickness of the surface layer. Finally, there are so-called *natural boundary conditions*, which incorporate surface elevation changes caused by the net freshwater flux across the ocean–atmosphere interface F_{fw} from precipitation minus evaporation:

$$\frac{\partial \eta}{\partial t} + \mathbf{u}_H \cdot \nabla \eta = w + F_{fw} \qquad (17.35)$$

Some authors argue that natural boundary conditions are more realistic and lead to substantially different circulation than free-surface models; however, natural boundary conditions have been explored in only a small number of experiments and are not yet implemented in most of the frequently used community ocean GCMs.

If in rigid-lid and free-surface models there is no actual freshwater transport between the ocean and atmosphere, how then do precipitation and evaporation influence model sea surface salinity? Instead a *virtual flux* of salt F_S is applied to the surface grid box, where F_S is related to the net freshwater flux F_{fw} and mean surface salinity S_0:

$$F_S = -\frac{S_0 F_{fw}}{1 - S_0} \qquad (17.36)$$

Salinity here is given in terms of the mass fraction of salt; that is, 35 PSU would be 0.035. Note the minus sign; a positive input of fresh water decreases surface salinity. For conservation purposes it is often best if S_0 is fixed in time (rather than varying in time and space). The same sort of issues apply to chemical tracers that have rather small vertical gradients, for example dissolved inorganic carbon (DIC) and alkalinity. The tendency term for the surface grid box for DIC then would include a surface flux term:

$$F_{DIC} = k\alpha_{CO2}(pCO^{atm} - pCO_2) - DIC_0 F_{fw} \qquad (17.37)$$

where the first term is the traditional air–sea gas exchange flux (Chapter 15) and the second term is the virtual DIC flux. For many model regions the virtual term is comparable to or larger than the air–sea gas exchange.

River inflow is another factor that needs to be considered for coastal and global GCMs. In many coastal regions, riverine freshwater inputs contribute to large spatial and temporal variations in near-shore salinity, and coastal currents are sometimes dominated by the resulting density-driven flow. For the largest rivers like the Amazon, Orinoco, Congo, Yangtze, and Ganges-Brahmaputra, the influence of the freshwater plume and low sea surface salinities can extend regionally. Global data sets of seasonal and interannual river discharge from the major gauged rivers (and even in aggregate the more numerous smaller rivers) are readily available for forcing ocean GCMs with freshwater inputs (Dai *et al.*, 2009). For coarse-resolution models that do not adequately resolve river mouths and coastal mixing process, the river freshwater input is often spread over a number of model grid cells rather than simply deposited as a point source. River nutrient models and databases are also available (Mayorga *et al.*, 2010), though often the reported nutrient levels reflect concentrations just upstream of the estuaries and near-shore coastal environmental, zones of intense biogeochemical transformations that can substantially modify net nutrient loading to the ocean (Fennel *et al.*, 2006). Also, there is the emerging realization of the potential importance of submarine groundwater discharge (SGD) to some tracers:

the actual net SGD water flux may be much smaller than the river flux, but because some chemical signatures are far more pronounced in groundwaters, this may be important for biogeochemical models of some substances.

Global ocean GCMs and many 3D regional ocean models extend into polar domains where sea-ice is a crucial factor in modulating air–sea momentum, heat, and freshwater fluxes (as well as, of course, trace gas exchange). Sea-ice also acts as an effective heat reservoir because of the substantial thermodynamic latent heat of formation for ice; on the converse side, heat transfer from the upper ocean to sea-ice can be a major factor governing ice distributions. Brine rejection during ice formation and meltwater inputs during ice melting strongly modulate polar surface salinities and ocean mixed layer dynamics. The bulk turbulent flux formula used for surface fluxes needs to be modified to account for the reduced fraction of open water, altered surface roughness, and transfer of heat directly through the sea-ice. Solar heating depends upon the thickness and albedo of the sea-ice, which can be strongly affected by snow cover and age (Uttal *et al.*, 2002). Often, specially constructed atmospheric forcing data sets are needed for polar regions because of inadequate sampling and data quality in the standard global forcing data sets for these environments. Also, the presence of sea-ice can profoundly affect air–sea gas exchange.

In some older global GCMs, surface forcing in ice-covered regions is approximated using the historical observed ice fraction, which can be derived for the last several decades from satellite data. Common but relatively crude simplifications include transferring the momentum from the wind stress directly into the ocean (basically ignoring the presence of the ice), assuming zero heat and water vapor fluxes in the ice-covered portion of a grid cell (diagnosed from the observed ice-fraction), and using the standard bulk formula for the open-water portion of the grid cell. Even then there can be inconsistencies between the atmospheric and sea-ice forcing and the ocean model, a good example being where the ocean model predicts cooling of the surface layer below the freezing point of water in nominally "ice-free" conditions. Of course a much better approach would be to incorporate prognostically the dynamics of sea-ice into the ocean GCM, and coupled ocean sea-ice models are becoming more common (Weatherly *et al.*, 1998; Maslowski and Lipscomb, 2003). Broadly speaking, sea-ice models consist of thermodynamic and dynamical components. The thermodynamic element computes the local formation and melting of sea-ice and the exchange of heat and mass from the ice to the ocean and atmosphere. The dynamical component computes ocean–atmosphere–air momentum transfers, the lateral redistribution of sea-ice due to wind stress and ocean currents, and the deformation of sea-ice due to lateral stress, which can lead to ice thinning and lead formation, or alternatively ice thickening and ridging (Briegleb *et al.*, 2004). Sea-ice models predict the spatial patterns of sea-ice fraction and thickness as a function of time, and some models track multiple sea-ice thickness classes within individual grid cells, which appears to improve some aspects of performance in terms of large-scale climate interactions. There are even efforts to incorporate the calving, transport, and melting of icebergs from glaciers and ice sheets.

17.4 Sub-grid-scale parameterizations

The size of a model's horizontal grid cells determines the scale of physical processes that can be resolved by the model (Chapter 9); the influence of any processes smaller than those scales will need to be treated through sub-grid-scale parameterizations. Ocean GCM simulations are particularly sensitive to the parameterizations of horizontal and vertical turbulent mixing (e.g. Danabasoglu *et al.*, 1994; Large *et al.*, 1997), which address the rectification of sub-grid-scale motion and transport back into the larger, resolved model fields. Previously we discussed how tracer or velocity fields can be decomposed into a resolved component (e.g. $\mathbf{u}$, C) and an unresolved (sub-grid-scale) turbulent anomaly (e.g. $\mathbf{u}'$, C') (Chapter 11) (here the overbar notation $\bar{\mathbf{u}}$ indicating the mean is dropped because it is already implicit in the resolved term, which typically reflects the mean over the grid cell). Turbulent anomalies are not random and influence the mean field through the covariances between turbulent anomalies, that is, Reynolds fluxes and stresses (e.g. $\overline{\mathbf{u}'\mathbf{u}'}$, $\overline{\mathbf{u}'C'}$). For a discrete model, additional terms equal to the divergence of Reynolds fluxes of the form $\nabla \cdot (\overline{\mathbf{u}'\mathbf{u}'})$ need to be added to the continuous forms of the governing equations written above for momentum and tracers (Section 17.1). Because $\mathbf{u}$ is a vector, there are multiple Reynolds stress components $(\overline{u'u'}, \overline{u'v'}, \overline{u'w'}, \ldots)$.

The problem of course is that almost by definition we don't have prognostic equations for the unresolved turbulent anomalies and Reynolds terms, and we need to parameterize the average effects of turbulent eddies in terms of the mean, resolved model fields. The simplest approach is through down-gradient eddy diffusion, resulting in what should by now be a familiar looking advection–diffusion form:

$$\frac{\partial C}{\partial t} + \mathbf{u} \cdot \nabla C - \nabla \cdot \kappa_H \nabla C - \frac{\partial}{\partial z}\kappa_V \frac{\partial}{\partial z}C = J_C \tag{17.38}$$

where κ_H and κ_V are the horizontal and vertical eddy tracer diffusivities, respectively. A similar equation can be constructed for momentum using eddy viscosities κ^M or in slightly more sophisticated form the 3D eddy viscosity tensor κ^M (Chapter 11). Many ocean GCMs use a variant of Equation (17.38), but, as we discuss below, other approaches are also in wide use.

17.4.1 Lateral mixing and mesoscale eddies

As mentioned in Chapter 11 and subsequent modeling chapters, turbulent mixing in the ocean occurs preferentially along isopycnal surfaces, and diapycnal diffusivities in the stratified ocean interior, $\mathcal{O}(10^{-5})\,\mathrm{m^2\,s^{-1}}$ are many orders of magnitude smaller than typical mesoscale isopycnal diffusivities, $\mathcal{O}(10^3)\,\mathrm{m^2\,s^{-1}}$. Since isopycnal surfaces are sloped, applying a simple horizontal diffusivity as in Equation (17.38) in a z-coordinate model can result in artificially enhanced diapycnal mixing. This mixing, which we mentioned above is termed the Veronis effect, is strongest in regions with steeply sloping isopycnals associated with western boundary currents and the Antarctic Circumpolar Current. Gent and McWilliams (1990) introduced a widely used parameterization to deal with horizontal mesoscale eddy mixing.

The Gent–McWilliams or *GM mixing scheme* has two parts, one that orients mixing along isopycnal surfaces and a second that addresses tracer transport caused by dynamical effects of mesoscale eddies. Ocean density fronts are unstable and tend to shed mesoscale eddies through a process called *baroclinic instability*. Eddy generation acts to flatten the isopycnals in the process converting potential energy to kinetic energy. Of course fronts don't disappear in the real ocean because there are other dynamics like wind- and buoyancy-forcing that act to maintain the fronts. But the resulting turbulence can greatly enhance lateral tracer transport. The integrated effects of mesoscale eddies can be mimicked by a down-gradient diffusion of isopycnal layer thickness, which can be cast in terms of an eddy-induced transport velocity $\mathbf{u}^*$.

In the original form of the GM scheme (Gent *et al.*, 1995), the rate of change for a generic tracer becomes:

$$\frac{\partial C}{\partial t} + (\mathbf{u} + \mathbf{u}^*) \cdot \nabla C - \nabla \cdot R(\kappa_I, C) - \frac{\partial}{\partial z} \kappa_V \frac{\partial}{\partial z} C = J_C \qquad (17.39)$$

The term $R(\kappa_I, C)$ is the diffusion oriented along the isopycnals (Redi, 1982), and κ_I is the isopycnal diffusivity. The horizontal and vertical eddy-induced transport velocities (sometimes referred to as the eddy-bolus velocities) are:

$$\mathbf{u}_H^* = -\frac{\partial}{\partial z} \kappa_B \frac{-\nabla \rho}{\partial \rho / \partial z} \qquad w^* = \nabla \cdot \kappa_B \frac{-\nabla \rho}{\partial \rho / \partial z} \qquad (17.40)$$

The term $-\nabla \rho / (\partial \rho / \partial z)$ is the projection of the isopycnal slope into the two horizontal directions. The bolus diffusivity κ_B is often set equal to κ_I. In many modern implementations, the eddy-induced velocity is rewritten in terms of a skew diffusion (Griffies, 1988). The GM scheme has many benefits for simulated ocean circulation patterns, water mass distributions, and transient tracer fields (e.g. Danabasoglu *et al.*, 1994; Danabasoglu and McWilliams, 1995), and GM is standard in most coarse-resolution z-coordinate models (and even in some eddy-permitting and eddy-resolving simulations).

The problem is somewhat more tricky for the equivalent horizontal viscosity term κ_H^M. Some form of viscosity or friction is needed to damp numerical noise and account for unresolved sub-grid-scale processes in the horizontal momentum equations (the turbulent cascade). Friction arises from the divergence of the Reynolds stresses including mean turbulent momentum fluxes (e.g. $\overline{u'u'}$) and shear stresses (e.g. $\overline{u'v'}$). In analogy with tracer eddy diffusivities, the simplest choice is *harmonic* or *Laplacian* friction:

$$\nabla \cdot (\overline{\mathbf{u}_H' \mathbf{u}_H'}) = -\kappa_{\text{Lap}}^M \nabla^2 \mathbf{u}_H \qquad (17.41)$$

where for clarity we assume a uniform horizontal viscosity κ_H^M. Because $\mathbf{u}_H$ is a vector, there are multiple Reynolds stress components (i.e. $\overline{u'u'}$, $\overline{u'v'}$, ...), and formally the viscosity term should be a tensor κ_H^M. Eddy-resolving simulations commonly use biharmonic friction:

$$\nabla \cdot (\overline{\mathbf{u}_H' \mathbf{u}_H'}) = \kappa_{\text{Bihar}}^M \nabla^4 \mathbf{u}_H \qquad (17.42)$$

which is more scale-selective and acts to damp variability strongly at the grid scale while not significantly affecting mesoscale eddies. Notice that the units for κ_{Lap}^M ($\text{m}^2\,\text{s}^{-1}$) differ from those for κ_{Bihar}^M ($\text{m}^4\,\text{s}^{-1}$).

There is no strong theoretical framework for assigning viscosity values. In principle, eddy viscosities should depend on the flow field and therefore should evolve in time and space. Often, however, models use specified viscosity values, either spatially uniform or smoothly varying with, for example, latitude. Some researchers have explored anisotropic viscosities (remember the tensor nature of κ_H^M) (e.g. Large et al., 2001). Smagorinsky (1963; 1993) proposed a commonly used adaptive scheme for horizontal viscosities:

$$\kappa_{\text{Smag}}^M = \gamma_{\text{Smag}} \Delta x \Delta y \sqrt{\left(\frac{\partial u}{\partial x} - \frac{\partial v}{\partial y}\right) + \left(\frac{\partial u}{\partial y} + \frac{\partial v}{\partial x}\right)} \qquad (17.43)$$

where γ_{Smag} is an adjustable parameter, $\Delta x \Delta y$ scales the viscosity to the size of the grid cell, and the term under the square root is the horizontal deformation rate. Without better guidance, one sensible approach then may be simply to set the viscosities as low as possible while still sufficient to minimize numerical instabilities.

There are several other types of sub-grid-scale parameterizations that could affect horizontal circulation and water mass properties in ocean GCMs. We won't go into details on these parameterizations, but you should be aware of their existence (see Haidvogel and Beckmann, 1999, and Griffies et al., 2000, for more in-depth discussion and references). Time-varying flow over bottom topography can induce time-mean barotropic velocities through eddy-topographic interactions (the *Neptune effect*). Model parameterizations for eddy-topographic interactions have been created based on statistical-mechanical and non-equilibrium thermodynamic arguments (Hollaway, 1992) but are not yet widely used in standard model configurations.

In the ocean, deep and intermediate waters are strongly influenced by the overflow of dense waters through sills and narrow straits, and from shelves down steep continental slopes. Important examples are Denmark Straits Overflow Water, Mediterranean Water, and Antarctic Bottom Water. Narrow, thin, rapidly flowing, dense overflows are either unresolved in large-scale models or rapidly diluted by excess mixing with ambient waters, resulting in too shallow and too light model outflows. The net effect of these bottom currents can be parameterized through *bottom-boundary layer* models (e.g. Beckmann and Doscher, 1997) and specialized *overflow models* (e.g. Price and Baringer, 1994). Bottom-boundary layer models are applied over the entire model domain of z-coordinate models and allow for direct downslope advection and lateral mixing between adjacent bottom cells even when the cells are at different depths. Overflow models are more similar to pipes embedded in the model at particularly critical locations. They transport dense water laterally and downslope external from the traditional model grid and are tuned typically to observations and process models for each specific overflow.

17.4.2 Vertical mixing and surface mixed layers

A second critical set of sub-grid-scale parameterizations involves the specification of vertical mixing. Vertical mixing, or more technically cross-isopycnal (diapycnal) mixing, governs so many aspects of an ocean GCM simulation, both physical and biogeochemical: sea surface temperature and air–sea fluxes; water mass properties (e.g. Large *et al.*, 1997; Doney *et al.*, 2004); the structure of the thermocline (e.g. Gnanadesikan, 1999); the strength of the thermohaline overturning circulation (e.g. Bryan, 1987; Jayne, 2009); nutrient supply and biological productivity (e.g. Najjar *et al.*, 2007); and ocean uptake and transport of chemical tracers (e.g. Dutay *et al.*, 2002; Matsumoto *et al.*, 2004). At the most basic level, we want any model parameterization to capture the anticipated high levels of diapycnal mixing in the surface mixed layer and low levels in the stratified interior. If we wanted to get fancy, we might even consider mixing in the bottom boundary and tidally driven mixing, both of which are important in coastal regions and perhaps also in the open ocean. We covered in some detail the basic issues behind diapycnal mixing through our exploration of 1D advection–diffusion models for the deep Pacific (Munk, 1966; Craig, 1969) (Chapter 13) and the Price–Weller–Pinkel (PWP) model (Price *et al.*, 1986) for the surface mixed layer (Chapter 15). Here we'll touch on some popular alternative parameterization approaches used in ocean GCMs (see also review by Large *et al.*, 1994).

The PWP model is a type of *bulk model* where a well-defined mixed layer is assumed, and the depth of the mixed layer is determined by a combination of wind and surface buoyancy forcing. The velocities and tracer properties are defined as uniform across the mixed layer, equivalent to infinitely fast mixing. The *Kraus–Turner model* (Kraus and Turner, 1967) is an important early example of a bulk model, and Kraus–Turner was used in many early ocean GCMs.

A second type of parameterization involves a family of *turbulence closure models*, developed from the pioneering work of Mellor and Yamada (1974; 1982), that predict turbulent flux over the entire water column, including the surface boundary layer. In contrast with bulk models, they do not define a mixed layer, a priori but well-mixed layers can appear near the surface because of large but finite mixing rates. The turbulence closure approach parameterizes turbulent fluxes in terms of the statistical moments of the three-dimensional eddy variability. So for example for the east–west velocity u, we have an equation that includes the spatial derivatives of unknown turbulent Reynolds stress terms:

$$\frac{\partial u}{\partial t} \cdots = -\left(\frac{\partial \overline{u'u'}}{\partial x} + \frac{\partial \overline{u'v'}}{\partial y} + \frac{\partial \overline{u'w'}}{\partial z} \right) \tag{17.44}$$

For clarity we have dropped the other terms. We could of course write a prognostic equation for the Reynolds stresses, for example:

$$\frac{\partial \overline{u'u'}}{\partial t} \cdots = -\left(\frac{\partial \overline{u'u'u'}}{\partial x} + \frac{\partial \overline{u'v'u'}}{\partial y} + \frac{\partial \overline{u'w'u'}}{\partial z} \right) \qquad (17.45)$$

Do you see the problem? Now we have new third-order moments the likes of $\overline{u'v'u'}$. In fact we could keep writing more and more equations, but there will always be new unknown moments one order higher than the one for which we are trying to solve (Stull, 1988).

Mellor and Yamada (1974; 1982) resolve the problem by prescribing higher-order moments in terms of known lower-order turbulent moments. That is, they specify a *closure* to the turbulence equations. In this case they keep the prognostic equations for key second-order moments while parameterizing any third-order and some second-order terms. It's a family of models because different choices can be made in terms of which second-order terms to keep versus parameterize. The most common closure approximations utilize either length-scale arguments or arguments based on the simulated turbulent kinetic energy TKE:

$$\text{TKE} = \frac{1}{2} \left(\overline{u'u'} + \overline{v'v'} + \overline{w'w'} \right) \qquad (17.46)$$

The model parameterizations track the vertical profile of TKE in response to surface winds and buoyancy forcing, internal wave-breaking, turbulent vertical transport, and dissipation. Mixing within turbulence closure models is always down-gradient and local, that is, between adjacent grid cells. Remote forcing may play a role but only through modifying the vertical distribution of TKE and other turbulent moments.

Another group of models, which are intermediate in complexity between bulk and turbulence closure models, utilize time-varying depth profiles of eddy diffusivity κ_V. Pacanowski and Philander (1981) present a good example of this model class where the eddy diffusivity and viscosity are a function of the gradient Richardson number. Another example is the *K-profile parameterization* or KPP model (Large *et al.*, 1994), which incorporates a non-local closure scheme in recognition that much of the mixing in the planetary boundary layer (PBL) is driven by the largest eddies that span the boundary layer. As described in its most basic form as in Doney *et al.* (1995), KPP computes a PBL depth h_{PBL} that depends on stratification, velocity shear, and surface buoyancy forcing, and h_{PBL} is computed from a bulk Richardson number. Turbulent fluxes in the boundary layer are formulated as a function of a depth-dependent diffusivity κ_V and a tracer specific, non-local or countergradient term γ_C:

$$\overline{w'C'} = -\kappa_V \left(\frac{\partial C}{\partial z} - \gamma_C \right) \qquad (17.47)$$

The shape of the PBL diffusivity profile κ_V is specified as a function of the distance z from the surface. The diffusivity is formulated to agree with the similarity theory of turbulence in the surface layer, and the magnitude of κ_V is determined by a turbulent velocity scale w_x and the depth of the boundary layer:

$$\kappa_V(z) = h_{\mathrm{PBL}} w_x \left(\frac{z}{h_{\mathrm{PBL}}} \left(1 - \frac{z}{h_{\mathrm{PBL}}} \right)^2 \right) \tag{17.48}$$

The value of $\kappa_V(z)$ reaches a maximum at a third of the PBL depth and tends toward zero at both the surface and h_{PBL}. The turbulent velocity scale w_x differs for momentum and scalars, and increases with surface wind stress and unstable surface buoyancy forcing (i.e. net cooling and evaporation). Below the PBL depth in the stratified interior KPP specifies the vertical diffusivity as a function of the gradient Richardson number, double-diffusion (e.g. salt-fingering), and a background internal-wave diffusivity.

The countergradient term γ_C scales with the tracer surface flux and arises only during unstable (convective) conditions. For strong convection, turbulence is asymmetric, dominated by narrow, rapidly sinking plumes in a background of more gradual upwelling. Under those spatially inhomogeneous conditions, the mean (resolved) vertical gradient $\partial C / \partial z$ is not reflective of the actual net turbulent fluxes, which can exist even when the mean field exhibits no vertical gradient.

Mixing schemes like KPP or turbulence closure models could be used as well to simulate shear driven mixing in bottom boundary layers. On continental shelves, the surface and bottom boundary layers often begin to merge, and tidal mixing is also essential often for capturing vertical exchange. Tides may even be a significant factor for mixing in deepwaters over rough topography. Simmons et al. (2004), for example, present an open-ocean parameterization of tidal mixing as applied to an ocean GCM.

17.5 Diagnostics and analyzing GCM output

A wide range of diagnostics are used to evaluate the skill and behavior of ocean GCMs, and here we present a few of the more common examples (e.g. Large et al., 1997; Doney et al., 2004, 2007). To be useful, a metric should be related to some fundamental aspect of ocean circulation (or biogeochemistry, biology, . . .). Preferably the metric should also be straightforward to measure in the real ocean, though, as we'll see, some common model diagnostics are not so easy to relate directly to observations (though they are still quite useful for intercomparing models). There are also test diagnostics for idealized situations that focus more on model numerics and the model's ability to capture basic geophysical fluid dynamical phenomena (Haidvogel and Beckmann, 1999). Given the complexity of an ocean GCM, we shouldn't expect an exact match of model output to data. A GCM is really a system of interconnected sub-models for dynamics, surface forcing, and various sub-grid-scale parameterizations. Even if each of the subcomponents has been tested individually, when you bring them together as a system the initial overall behavior can be disappointing, and additional tuning exercises are often needed for the integrated model. It is useful, therefore, to understand the difference between a relatively trivial model–data mismatch and a systematic bias that could reflect fundamental problems in surface forcing, parameterizations, or model structure.

Analyzing ocean GCMs starts with the same standard types of model–data products we've used before: model–data difference plots, mean bias, root mean square error, and spatial and temporal correlations. Other quite useful but less well-known model–data metrics include the reliability index RI and the modeling efficiency MEF (Stow *et al.*, 2009):

$$RI = \exp \sqrt{\frac{1}{N} \sum_i^N \left(\log \frac{C_i^{\text{obs}}}{C_i} \right)^2} \qquad MEF = \frac{\left(\sum_i^N (C_i^{\text{obs}} - \overline{C^{\text{obs}}})^2 - \sum_i^N (C_i^{\text{obs}} - C_i)^2 \right)}{\sum_i^N (C_i^{\text{obs}} - \overline{C^{\text{obs}}})^2}$$

(17.49)

As a logarithmically weighted mean of the ratios of observations to model values, one would expect an RI value close to 1 for good performance. The modeling efficiency measures how well a model predicts relative to the average of the observations $\overline{C^{\text{obs}}}$. A value near 1 indicates a close match between observations and model predictions. A value of 0 indicates that the model predicts individual observations no better than the average of the observations. Values less than 0 indicate that the observation average would be a better predictor than the model results.

Visualizing time-evolving, 3D model output is challenging with traditional 2D plotting methods (e.g. contour plots, property–property plots), and innovative approaches to solving this problem are discussed in Chapter 19. For visualization purposes, GCM properties are often contoured along different slices in space: depth surfaces, isopycnal surfaces, and depth–latitude and depth–longitude sections. Zonally averaged depth–latitude plots are often generated, either averaged globally or by basin. Temporally, model output is analyzed for annual and climatological means, seasonal cycles, and interannual variability. Taylor diagrams provide a compact method for comparing the correlations and RMS errors between model and observed fields as well as the amount of variance in the model versus the observations (Taylor, 2001). Empirical orthogonal functions are also used commonly to interpret spatial-temporal variability. Most model analysis concentrates on univariate metrics, in other words one model variable at a time; but there is also growing use of bivariate and multivariate metrics, which can be particularly useful when model and observed patterns are generally similar in shape but offset in space or time (Stow *et al.*, 2009).

17.5.1 Hydrography and air–sea fluxes

A good place to start one's analysis of ocean GCM output is with the distributions of temperature, salinity, and water masses (Large *et al.*, 1997; Doney *et al.*, 2004; 2007), which can be compared directly against the large historical hydrographic databases. Example databases include those from the National Ocean Data Center (NODC) (Levitus *et al.*,

1998), the World Ocean Circulation Experiment (WOCE), and the CLIVAR and Carbon Hydrographic Data Office (CCHDO). The WOCE program has also generated electronic atlases of ocean properties along sections, depth surfaces and isopycnal surfaces. Higher sampling density exists for many upper ocean physical data products from ship underway systems and expendable instruments like XBTs (expendable bathythermographs). Variations and trends in vertically integrated, upper ocean heat content on seasonal, interannual, and decadal timescales (Willis *et al.*, 2004) offer key measures of model performance and climate response (Gent *et al.*, 2006; Doney *et al.*, 2007). Extensive global sea surface temperature (SST) data sets and merged data products are available for the satellite era over the past several decades (Reynolds *et al.*, 2007). Seasonal variations of mixed layer depths modulate surface physical properties and the subduction of water masses into the ocean interior (as well as nutrient supply and phytoplankton growth rates; Doney *et al.*, 2009). Mixed layer depth estimates are available from observations (Boyer Montégut *et al.*, 2004) and operational ocean forecast models.

Air–sea heat and freshwater fluxes typically are less well known than sea-surface properties, but surface fluxes do provide another important metric for assessing model performance (Large and Yeager, 2009). Net surface heat and freshwater fluxes must be balanced by either net storage or advective divergence. Lateral heat and freshwater transports are common observational and model diagnostics (Wijffels *et al.*, 1992; Jayne and Marotzke, 2001). For example, the meridional heat transport $\mathcal{T}_{\text{heat}}(\phi)$ for an ocean basin as a function of latitude can be calculated by integrating the potential temperature times the north–south velocity from the bottom to the surface and from one edge of the basin to the other:

$$\mathcal{T}_{\text{heat}}(\phi) = \rho c_p \int_{z_{\text{bot}}}^{0} \int_{\lambda_{\text{west}}}^{\lambda_{\text{east}}} v\theta \, r_e \cos\phi \, d\lambda dz \qquad (17.50)$$

where geometric terms are included inside the integral to get the cross-section area correct. For a steady-state ocean, the meridional derivative of the transport $\partial \mathcal{T}_{\text{heat}}/\partial\phi$ equals the zonally integrated, net surface fluxes. Ocean heat and freshwater transports can be estimated from ocean hydrographic and velocity data (Section 18.3), air–sea flux products or in conjunction with atmospheric transport estimates. The atmospheric constraints arise, for example, because the total poleward ocean and atmosphere heat transport must balance the patterns of net planetary radiation flux to space.

Surface heat and freshwater fluxes also can be converted using Equations (17.13) and (17.36) into a surface buoyancy flux F_ρ (kg m^{-2} s^{-1}):

$$F_\rho = -\frac{\alpha F_{\text{heat}}}{c_p} - \frac{\rho_0 \beta S_0 F_{\text{fw}}}{1 - S_0} \qquad (17.51)$$

The surface buoyancy flux is the tendency for ocean–atmosphere exchange to modify upper ocean density. Model buoyancy fluxes can be interpreted in terms of spatial maps. Alternatively, buoyancy fluxes could be integrated over the surface area A_ρ for density classes $\Delta\rho$ to find the water mass transformation rates $\mathcal{F}(\rho)$ (in sverdrups or Sv, where $1\,\text{Sv} = 10^6\,\text{m}^3\,\text{s}^{-1}$):

$$\mathcal{F}(\rho) = \frac{1}{\Delta\rho} \sum A_\rho F_\rho \, \delta[(\rho + \Delta\rho) - \rho] \tag{17.52}$$

where the δ function is 1 for the interval ρ to $\rho + \Delta\rho$ and zero elsewhere (Speer and Tziperman, 1992; Doney *et al.*, 1998). Water mass transformation rates are the fluxes of water from one density class to another, and the net water mass formation rates are found from the difference $\mathcal{F}(\rho) - \mathcal{F}(\rho + \Delta\rho)$.

17.5.2 Ocean circulation diagnostics

A wide variety of observations and diagnostics are now used to constrain the velocity field and circulation of ocean GCMs. Velocities can be estimated indirectly from hydrographic data using the geostrophic relationship (Section 18.3). Direct velocities can be derived for the surface layer from drogued surface drifters (Ralph and Niiler, 1999; Fratantoni, 2001) and for the subsurface from Lagrangian floats (Davis, 1998; Lavender *et al.*, 2005). The global Argo profiling float network provides information on both subsurface velocities at the parking depth (typically 1000 db) and the heat content of the thermocline (and, with additional instruments, salinity and oxygen). With sufficient observations, eddy kinetic energy $(\overline{u'u'} + \overline{v'v'})/2$ can be estimated from drifter and float data. Current meter moorings are located in many key locations to constrain velocities and transport through straits and choke points and in the narrow, high-velocity regions of western boundary currents. Tropical current meter arrays (McPhaden *et al.*, 1998) are particularly useful because the geostrophic balance breaks down as f goes to zero at the equator.

A number of frequently used physical diagnostics reflect integrated properties of the simulated ocean circulation and physics. Sea surface height (SSH) anomalies arise from a variety of factors: wind-driven convergence or divergence, surface heat and freshwater fluxes, and anomalies in the subsurface ocean heat content and density field. SSH can be measured from satellites using altimetry, basically a fancy radar ranging technique for measuring the distance from the satellite to the ocean surface. Satellite SSH observations revolutionized physical oceanography by providing a global picture of eddy variability (Stammer, 1997) and ocean geostrophic circulation (Ducet *et al.*, 2000). The AVISO project offers readily accessible global and regional gridded SSH and geostrophic velocity products merging data from different satellite altimeters. Strictly speaking, altimeter SSH data provides information on the variations in ocean circulation, not on absolute values, because of issues with uncertainties in the Earth's geoid. Absolute velocities can be found by combining altimeters with bottom pressure observations, Lagrangian floats, and new satellite gravity measurements (Jayne, 2006; Willis and Fu, 2008).

The depth-averaged flow or barotropic circulation is a common diagnostic for ocean GCM simulations. Barotropic velocities can be cast in terms of a 2D streamfunction $\psi_{\text{barotropic}}$ in latitude and longitude and is typically presented in units of sverdrups. Plots

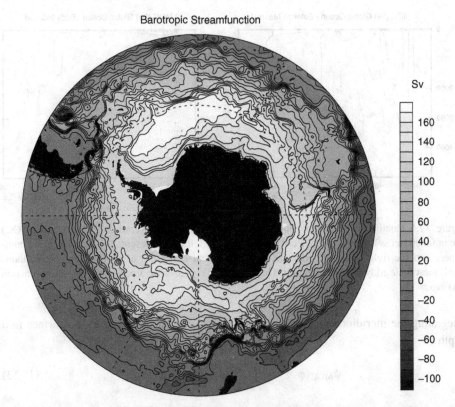

Barotropic Streamfunction

Figure 17.5 Contour plot of the depth-integrated, barotropic streamfunction $\psi_{\text{barotropic}}$ (in Sv) for the Southern Ocean from a global, eddy-permitting POP simulation (horizontal resolution 0.4°). Depth-integrated flow is oriented along lines of constant streamfunction and is proportional to the horizontal gradient in streamfunction (in Southern Hemisphere flow is to the left for positive gradient in $\psi_{\text{barotropic}}$). The dominant features are the fronts making up the eastward flowing Antarctic Circumpolar Current (ACC), which are often locked to topographic features.

of the time-mean barotropic streamfunction illustrate the main features and strength of the circulation field. In the Southern Ocean example shown in Fig. 17.5, the topographic steering of the Antarctic Circumpolar Current (ACC) is evident as are the regions of strongest flow marked by the bunching of the streamlines. Model barotropic transport also can be compared against the transport for a limited number of locations where we have solid observational estimates for the total flow, good examples being the Drake Passage and Florida Straits.

A second frequently used circulation metric, the meridional overturning circulation (MOC), also can be cast in terms of a 2D streamfunction, now in depth and latitude. The MOC streamfunction generally is presented either for an individual ocean basin or the globe. It can be calculated for a particular latitude and depth $\psi_{\text{MOC}}(\phi, z)$ by

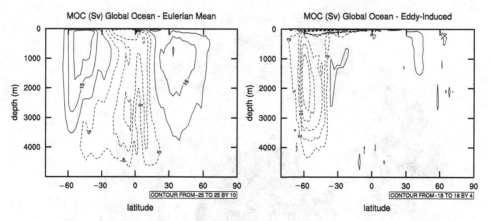

Figure 17.6 Contour plot of the zonally integrated meridional overturning circulation (MOC) streamfunction (Sv) from a global POP simulation. The left panel presents the Eulerian mean component and the right panel the eddy-induced component computed using the Gent–McWilliams (GM) mesoscale eddy parameterization. The circulation is clockwise around positive streamfunction maxima.

integrating the meridional velocities zonally across the basin and from the surface to a depth z:

$$\psi_{MOC}(\phi, z) = \int_z^0 \int_{\lambda_{west}}^{\lambda_{east}} v \, r_e \cos \phi \, d\lambda dz \qquad (17.53)$$

where the term $r_e \cos \phi$ is an areal geometric term. Figure 17.6 displays the global MOC stream function from an ocean GCM using the Gent–McWilliams mesoscale mixing parameterization, computed separately for the Eulerian mean velocity v and the eddy-induced velocity v^* (Equation (17.40)). The positive streamfunction values in the Eulerian mean in the Northern Hemisphere reflect the sinking and southward flow of North Atlantic Deep Water. Positive values in the Southern Hemisphere arise from the wind-driven upwelling in Southern Ocean near 60° S caused by the strong westerlies in the Southern Ocean and resulting northward ocean surface flow (remember the Coriolis force is to the left in the Southern Hemisphere). Substantial eddy-induced velocities appear only in the Southern Ocean associated with the steeply sloping isopycnals in the Antarctic Circumpolar Current and tend to partially cancel the Eulerian circulation.

Because sea surface height anomalies are now routinely monitored by satellite remote sensing, SSH anomalies are particularly useful for assessing model variability from eddies and fronts at the mesoscale to seasonal and interannual variations in ocean heat content. At the mesoscale, the dominant component of SSH anomalies is caused by variations in the depth of the main thermocline, where a positive SSH anomaly indicates a depression of the thermocline and vice versa. The warmer thermocline water is less dense, and it takes a thicker water column of lighter water (and thus elevated SSH) to create the same pressure at some depth below the thermocline (recall the hydrostatic relationship $p \approx -g \int \rho dz$,

Equation (17.21)). SSH anomalies are also linked to lateral pressure gradients in the upper ocean and, through geostrophy, to horizontal velocities (Equation (17.20)). Roughly speaking, upper ocean velocities follow SSH isolines, like wind direction tracking pressure contours on a weather map, where the flow direction is such that higher SSHs are to the right (in the Northern Hemisphere) and the velocities are largest for steep SSH gradients (Section 18.3.2).

The SSH η *steric* contributions (i.e. from temperature and salinity) can be computed by integrating fractional density anomalies from the base of the thermocline z_0 to the surface:

$$\eta' = \int_{z_0}^{0} (\alpha\theta' - \beta S')dz \qquad (17.54)$$

For a mean temperature appropriate to the tropical and mid-latitude upper thermocline ($\theta \sim 15\,°C$), a uniform $0.1\,°C$ anomaly over the upper 400 m leads to a ~ 1 cm positive SSH anomaly. Because α drops sharply with temperature, the steric SSH anomaly for the same temperature anomaly at polar latitude would be 3–4 times smaller. Salinity variations generally have a smaller effect on SSH variations except in polar latitudes.

Figure 17.7 contrasts the simulated SSH anomalies near the Gulf Stream from eddy-permitting and eddy-resolving cases using the same ocean GCM. In the simulations, the Gulf Stream is indicated by the sharp front in SSH separating warm subtropical waters to the south (high SSH) from cold slope waters to the north (low SSH), and Gulf Stream rings and mesoscale eddies appear as local maxima and minima. Not too surprisingly, the number and strength of eddies are weaker in the eddy-permitting case. More intriguingly, the location of the Gulf Stream shifts with horizontal resolution. In the eddy-permitting

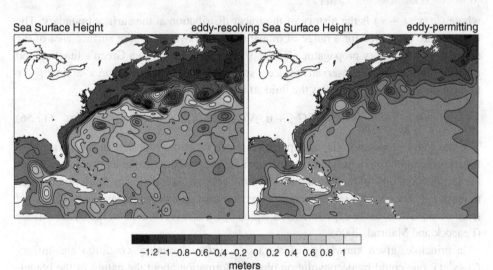

-1.2 -1 -0.8 -0.6 -0.4 -0.2 0 0.2 0.4 0.6 0.8 1

meters

Figure 17.7 Simulated sea surface height anomaly field for the western North Atlantic and Gulf Stream region from two POP simulations, (left) an eddy-resolving case (0.1° horizontal resolution) and (right) an eddy-permitting case (0.4° horizontal resolution).

case, the Gulf Stream leaves the coast far north of the observed separation point at Cape Hatteras, a common bias in coarse-resolution simulations. The take-home message is that just because eddy-like features arise in a simulation, this does not mean that these eddies are behaving correctly in a dynamical sense.

17.5.3 Transit time distributions

Another type of diagnostic, *transit time distribution* or TTD, bridges between physical circulation and chemical tracers. Based on model developments in atmospheric science, TTDs are built on the idea that a given fluid parcel may be made up of smaller parcels that traveled to the parcel's location from different sources and along different circulation pathways. Because tracer properties are often reset at the ocean surface by air–sea exchange and biology, we are especially interested in surface source regions and the time it takes to travel from the surface to the interior locations (i.e. transit times τ) (e.g. Hall and Haine, 2002). Any particular fluid parcel will contain a continuum of components characterized by a *transit time distribution*. In principle, one can attribute the distribution of some tracer property $C(\mathbf{x}, t)$ to surface boundary conditions, which are unique for each tracer, and physical redistribution processes, which are properties of the fluid flow and are the same for all passive tracers.

Assuming for the moment a *conservative* tracer (i.e. no *in situ* sinks or sources), this can be expressed in integral form as:

$$C(\mathbf{x}, t) = \int_{\text{area}} \int_{\tau} C_{\Omega}(\mathbf{x}', t - \tau) \, G_{\Omega}(\mathbf{x}, t | \mathbf{x}', t - \tau) \, d\tau \, d\mathbf{x}' \qquad (17.55)$$

where $C_{\Omega}(\mathbf{x}', t - \tau)$ is the history of the tracer distribution at the surface boundary. The integrals are computed for a range of transit times and over the entire surface area of the domain. The boundary propagator $G_{\Omega}(\mathbf{x}, t | \mathbf{x}', t - \tau)$ is related to a Green's function and describes how passive, conservative tracers spread from a surface location $\mathbf{x}'$ at time $t - \tau$ to location $\mathbf{x}$ at time t. It satisfies the fluid advection diffusion relation:

$$\frac{\partial G_{\Omega}}{\partial t} - \nabla \kappa \nabla G_{\Omega} + \mathbf{u} \cdot \nabla G_{\Omega} = \delta(\mathbf{x} - \mathbf{x}')\delta(t - t') \qquad (17.56)$$

The boundary propagator G_{Ω} is a pulse response function, reflected in the Dirac delta function terms on the right-hand side that are zero everywhere (and at all times) except for the location and time of the initial pulse. The integral of this function by definition is equal to unity $\int_{\text{area}} \delta(\mathbf{x} - \mathbf{x}')d\mathbf{x}' = 1$. Discrete forms of transit time distributions can be computed from ocean GCMs using experiments in which dye tracer is released on the ocean surface (Peacock and Maltrud, 2006).

In principle, given knowledge of a tracer's surface boundary condition and history $C_{\Omega}(\mathbf{x}', t)$, one could by deconvolution obtain information about the nature of the boundary propagator function G_{Ω} from subsurface tracer observations $C(\mathbf{x}, t)$. Such a process basically reveals "moments" of the transit time and/or pathways intrinsic in the circulation that are characteristic of the specific tracer used. Combining information from tracers with significantly differing boundary conditions may reveal additional moments. By way of example, observations of the evolving distribution of a linearly increasing (with time) transient tracer reveal a biased estimate of the mean transit time (sometimes referred to as Γ in the literature) between the sea surface and a location in the ocean interior. Combining such information with that from a different tracer (for example a pulsed tracer like bomb-tritium or radiocarbon) can be used to extract information regarding the second moment (the width) of the transit time distribution (called Δ). The former relates to the average advection, the latter relates to the mixing. See Waugh *et al.* (2003; 2004) and Jenkins (2004, and references therein) for a more in-depth discussion.

17.5.4 Tracer and biogeochemical diagnostics

The development of ocean biogeochemical models is synergistic with ocean field observations at many levels: experiments and process studies provide the conceptual framework for model formulation and parameter value estimation; process-studies, time-series and survey data provide essential constraints on model dynamics and for evaluating overall performance. Ocean biogeochemical modeling (Doney, 1999; Doney *et al.*, 2003; Sarmiento and Gruber, 2006) benefited greatly from the availability of field observations from the international Joint Global Ocean Flux Study (JGOFS) during the late 1980s and 1990s (Fasham *et al.*, 2001; Fasham, 2003). The Ocean Carbon-cycle Model Intercomparison Project (OCMIP), begun during the same era, constructed a standard suite of global ocean simulations for transient tracers, inorganic carbon system variables, and biogeochemical fields. Results from the program include systematic approaches for comparing model results against the field observations, especially the WOCE/JGOFS Global CO_2 Survey data and synthesis products from the GLobal Ocean Data Analysis Project (GLODAP) (Key *et al.*, 2004). The program identified observation-based tracer metrics (e.g. chlorofluorocarbons and radiocarbon) for evaluating and choosing among different model estimates of ocean uptake of anthropogenic CO_2 (Matsumoto *et al.*, 2004). Using relatively simple parameterizations of upper ocean biogeochemistry, OCMIP also established a baseline for assessing model skill against nutrient and oxygen distributions and export production estimates. Follow-on studies have utilized GLODAP data products, OCMIP model results, and new model variants to track horizontal transport and air–sea exchange of natural and anthropogenic CO_2 (Gruber *et al.*, 2009).

Parallel research on more biological and ecological modeling has been under way over the past two decades to improve model treatment of primary and new production, phytoplankton–zooplankton dynamics, export flux, and particle sinking and remineralization. The JGOFS process studies and time-series records have been essential in this regard, as has the advent of routine global satellite ocean color observations, beginning with Sea-viewing Wide Field-of-view Sensor (SeaWiFS) in 1997 (McClain *et al.*, 2004). The newest generation, 3D ocean biogeochemical models typically incorporate phytoplankton functional groups, multiple limiting nutrients, flexible elemental composition, and iron limitation (Moore *et al.*, 2004; Le Quéré *et al.*, 2005). Simulated plankton species are often aggregated into functional groups distinguished by size class, production of calcified or siliceous shells, and ability to carry out specific biogeochemical processes such as nitrogen fixation or dimethylsulfide production (Hood *et al.*, 2006). Doney *et al.* (2009) outline available evaluation data for marine ecosystem models that include: *in-situ* and satellite-derived surface ocean chlorophyll, primary productivity, phytoplankton growth rate and carbon biomass; large-scale climatologies of surface nutrients, underway pCO_2 data, and air–sea CO_2 and O_2 fluxes; and JGOFS time-series data. Additional biological evaluation data include data from repeat large-scale expeditions like the Atlantic Meridional Transect (AMT) (Aiken *et al.*, 2009) and new satellite data products from MODIS constraining specific plankton functional groups and plankton size structure. Other significant developments involve the application of high-resolution, mesoscale eddy simulations for regional coastal domains (Gruber *et al.*, 2006) and ocean basins (Oschlies *et al.*, 2000; McGillicuddy *et al.*, 2003).

For an individual investigator, the task of developing and evaluating these ever more complex physical, biogeochemical, and ecological simulations of the ocean is daunting. A number of international collaborations have sprung up to facilitate model–data evaluation and speed the model design cycle. This approach is encapsulated in projects on the physical side such as the CLIVAR Working Group for Ocean Model Development (WGOMD) and the Common Ocean-ice Reference Experiments (CORE); comparable efforts on the biological side include the Dynamic Green Ocean Project and the follow-on MARine Ecosystem Model Intercomparison Project (MAREMIP).

18

Inverse methods and assimilation techniques

> Reports of my assimilation have been greatly exaggerated.
> *Captain Jean-Luc Picard*

In Chapter 10 we introduced optimization techniques for finding unknown model parameters by minimizing a model–data cost function. Here we extend those concepts to generalized inverse and data assimilation methods. Inverse modeling actually covers a number of related numerical approaches and is applicable to a wide range of oceanographic problems, essentially any system where we want to interrogate data to help better constrain a model. Often we can measure the state of the ocean, say the temperature, density, inorganic carbon, or nutrient distributions, much better than we can determine the governing processes (in this case, circulation, mixing, air–sea fluxes, and biological uptake and release). Inverse techniques allow us to take advantage of the wealth of ocean data. With the rapid growth of satellite and ocean observing system data, inverse modeling will likely continue to grow in popularity.

The chapter starts with an introduction to the concepts behind linear inverse modeling (Section 18.1). Many ocean-related inverse problems involving tracer transport can be written as a set of linear equations, and we therefore focus in some detail on methods for solving under-determined linear systems (Section 18.2). To illustrate some of the basic ideas, we present an example case of computing the horizontal velocity field from geostrophic balance and tracer budgets using ocean hydrographic sections (Section 18.3). We follow with brief introductions to more advanced approaches such as variational data assimilation and Kalman filtering (Section 18.4).

18.1 Generalized inverse theory

Not all inverse problems are linear, but it is a useful place to start. In generalized linear inverse theory, one starts with the observations and, using a model as a framework, estimates model parameter values, forcing, and/or the initial conditions from which the system evolved. The terminology surrounding inverse methods can get pretty thick and thorny (and the notation is even worse), but the basic ideas are straightforward with a rich literature and

489

numerous ocean-related applications (e.g. Menke, 1984; Wunsch, 1996; Kasibhatla *et al.*, 2000; Bennett, 2002).

18.1.1 *Over- and under-determined systems*

Consider a set of linear model equations in matrix form:

$$\mathbf{Ap} = \mathbf{x} \tag{18.1}$$

where $\mathbf{x}$ ($N \times 1$) are the data, $\mathbf{p}$ ($M \times 1$) is the vector of model parameters we want to diagnose (generically called parameters but which could also include forcing and initial conditions), and $\mathbf{A}$ ($N \times M$) is the model design matrix or simply the "model". It has N rows, one model equation for each data point, and M columns, one for each parameter. The generalized solution to (18.1) is:

$$\hat{\mathbf{p}} = \mathbf{A}^{-g}\mathbf{x} \tag{18.2}$$

where $\hat{\mathbf{p}}$ is the vector of estimated parameter values and $\mathbf{A}^{-g}$ ($M \times N$) is the generalized inverse. We can also compute model estimates for each data point:

$$\hat{\mathbf{x}} = \mathbf{A}\hat{\mathbf{p}} \tag{18.3}$$

For a case where there are exactly as many equations as there are unknowns ($N = M$), we have what is referred to as an *even-determined* system. There is then a single, unique solution for the parameters that also exactly fits the data. An example would be fitting a straight line to two data points. Of course the matrix $\mathbf{A}$ cannot be rank deficient nor can it be poorly conditioned for this to work (Chapter 1). When $N > M$ there is "too much information" (i.e. more equations than unknowns), and we have an *over-determined* system of equations. In this case we can't uniquely fit all of the data, but we can determine the parameters that yield $\hat{\mathbf{x}}$ (estimates) closest to actual data ($\mathbf{x}$) in a least squares fashion, which was the topic from Chapter 3.

Unfortunately for most geophysical applications there are often more unknowns than equations ($N < M$), meaning there isn't enough information to solve uniquely for the model parameters. In fact there will be an entire family, an infinite number, of solutions (parameter values) that fit the data. Of course, we won't let that stop us. What we need to do is add some a priori information to constrain an "optimal" solution. Not unexpectedly, there is considerable "art" (some might say magic) to solving under-determined systems, and as we will see one of the preferred ways of finding a solution is our old friend singular value decomposition (SVD). Sometimes systems can be "mixed"; that is, some of the model parameters are over-determined by the data while others are left unconstrained. Most of the time when someone says "inverse approach" in a geophysical or geochemical framework, they are talking about solving under- or mixed-determined systems.

An illustration drawn from a problem in ocean science may be helpful here. Sound speed in seawater depends on temperature, and a method called ocean acoustic tomography was proposed by Walter Munk to monitor ocean warming due to climate change remotely

(Munk and Wunsch, 1979) (this was well before anyone envisioned the Argo profiling float array, when ships were the only feasible alternative; Munk, 2006). Consider the example where ocean sound travel time is measured across N paths, and we want to invert the data to find the sound speed in M water parcels. The N measurement paths could reflect both independent sound paths between different pairs of sound sources and receivers, and the multiple sound paths between a single pair that arise because of refraction and reflection off the surface and bottom. A set of simple schematics, adapted from Menke (1984), may help clarify these different cases (Fig. 18.1). Notice that even when $N \geq M$ the system may still be mixed- or under-determined. For example, we may have a case where several travel time observations were made for some parcels but none for others. Alternatively, we may only know the aggregate travel time data across multiple parcels, in which case we can't uniquely constrain sound speeds for separate parcels, only some combination. Of course, the situation will not always be as obvious or black and white for real world problems, as we discuss below.

18.1.2 Vector spaces, ranges, and the null spaces

Linear systems such as $\mathbf{Ap} = \mathbf{x}$ can be thought of as transforming or mapping vectors from parameter or model space $\mathcal{S}(\mathbf{p})$ to data space $\mathcal{S}(\mathbf{x})$. Conversely, the solution $\hat{\mathbf{p}} = \mathbf{A}^{-g}\mathbf{x}$

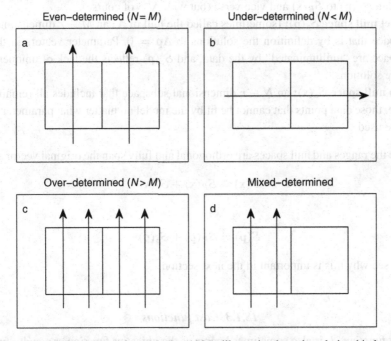

Figure 18.1 Schematics, adapted from Menke (1984), illustrating how the relationship between the number of observations N (arrows) and unknowns M (boxes) lead to (a) even-, (b) under-, (c) over-, and (d) mixed-determined systems.

maps vectors from the data space $S(\mathbf{x})$ to the model space $S(\mathbf{p})$. The Nth vector space is the collection of all vectors with N elements and is spanned by N linearly independent vectors. For example, the familiar three-dimensional space is a vector space spanned by the x, y, z coordinate axis (as well as any other independent vector triplets with three elements). Subsets of vectors within the vector space can make what are called subspaces, which are often of lower dimensionality. Again using the 3D example, subspaces could be planes or lines.

So, why do we care about vector subspaces (other than the joy of reading linear algebra books late at night)? It turns out that the linear transforms $\mathbf{A}$ and $\mathbf{A}^{-g}$ often do not completely span the model and data vector spaces for over- and under-determined systems. That is, no matter what set of model parameters you choose, you may never reach some data vectors and vice versa. The matrix pair $\mathbf{A}$ ($N \times M$) and $\mathbf{A}^{-g}$ ($M \times N$) has four relevant subspaces, the model and data *ranges* $S_R(\mathbf{p})$ and $S_R(\mathbf{x})$ and the model and data *null spaces* $S_o(\mathbf{p})$ and $S_o(\mathbf{x})$.

Data range $S_R(\mathbf{x})$: (sometimes called the range or column space of $\mathbf{A}$) an r dimensional subspace ($r \leq N$) that includes all $\mathbf{x}$ for which $\mathbf{A}\mathbf{p} = \mathbf{x}$ can be solved for any $\mathbf{p}$. $S_R(\mathbf{x})$ is the key to the existence of solutions.

Model range $S_R(\mathbf{p})$: an r dimensional subspace ($r \leq M$) that includes all $\mathbf{p}$ for which $\mathbf{A}^{-g}\mathbf{x} = \mathbf{p}$ can be solved for any $\mathbf{x}$. The linear model $\mathbf{A}$ transforms parameter vectors from $S_R(\mathbf{p})$ to $S_R(\mathbf{x})$ and vice versa (but with $\mathbf{A}^{-g}$ of course).

Model null space $S_o(\mathbf{p})$: (sometimes called the null of $\mathbf{A}$) an $M - r$ dimensional subspace that is by definition the solutions to $\mathbf{A}\mathbf{p} = 0$. Parameter vectors in the null space are "unilluminated" by the data, and $S_o(\mathbf{p})$ reflects the lack of uniqueness of the solutions.

Data null space $S_o(\mathbf{x})$: an $N - r$ dimensional subspace that includes all remaining $\mathbf{x}$, i.e. those data points that cannot be fit by the model no matter what parameter values are used.

Because the ranges and null spaces are orthogonal and fully span the original vector spaces:

$$S(\mathbf{x}) = S_R(\mathbf{x}) + S_o(\mathbf{x}) \tag{18.4}$$

and

$$S(\mathbf{p}) = S_R(\mathbf{p}) + S_o(\mathbf{p}) \tag{18.5}$$

We will see why this is important in the next section.

18.1.3 Cost functions

The *cost function*, J, is also sometimes known as the *objective function* or penalty function in inverse theory. The cost function is either minimized or maximized in inverse modeling, depending on the problem. For under-determined systems, the cost function provides a constraint for selecting a single parameter solution from the infinite number of possibilities that will fit the data. The cost function folds in all of the available information, beyond the model equations themselves, for solving the problem. This will certainly include the observations but may also involve a priori parameter estimates and other external constraints. For under-determined problems, this extra information beyond simply the data helps to close the problem. An example of an external constraint would be if you knew that some of the parameters must be positive definite, for example river inputs to the ocean, nutrient concentrations, or radioactive decay source. Others might involve linear combinations or non-local statements about the solution (e.g. mass conservation, geostrophy, chemical equilibrium).

The cost function for the linear system $\mathbf{Ap} = \mathbf{x}$ may take the general form:

$$ J = (\hat{\mathbf{x}} - \mathbf{x})' \, \omega_x^{-1} (\hat{\mathbf{x}} - \mathbf{x}) + \alpha(\hat{\mathbf{p}} - \mathbf{p}_\circ)' \, \omega_p^{-1} (\hat{\mathbf{p}} - \mathbf{p}_\circ) + \beta \sum_k^K c_k(\hat{\mathbf{p}}, \hat{\mathbf{x}}) \qquad (18.6) $$

The first term should be familiar from our earlier discussions of least squares and is simply the weighted model–data misfit, predicted minus observed. The weight matrix ω_x is $(N \times N)$ and is often (but not always; see Wunsch, 1996) a matrix of the data error covariance terms. The second term is the weighted length of the difference between the estimated parameters relative to a set of a priori values, $\mathbf{p}_\circ$, where ω_p is the parameter error covariance matrix. The a priori estimates or prior knowledge could be parameter estimates from previous experiments, models, or theory. Finally, the third term is the sum of external constraints c_k that can depend on both the data and parameters. The factors α and β are trade-off coefficients for adjusting the relative contributions of the different cost function terms. The cost function can take on other forms as well, of course. The power of the generalized inverse lies in the freedom of choice for the objective function.

Thinking about the meaning of Equation (18.6) in more detail, it should be worrisome, if not downright troubling, that we have included "part of the answer" into the cost function. Namely, we have added a-priori estimates or priors for the parameters $\mathbf{p}_\circ$ we are trying to estimate. The inverse and statistical communities tend to split into two distinct philosophical camps over this issue. Some argue that you should include all of the information at hand, which would of course incorporate parameter estimates derived from earlier experiments, the literature, conceptual models, etc. In fact some reject traditional null hypothesis testing altogether for a *Bayesian* approach. In Bayesian statistics, one starts off with a hypothesis with some (assumed) prior probability of the hypothesis being true. New observations are used to modify this probability, resulting in a new a-posteriori probability. Bayesians like to see many more shades of gray! Others are more skeptical, preferring to use very loose priors (or none at all).

Whether you stand on one side of this debate or the other, you should be cautious when building your cost function. It is quite easy to inadvertently overspecify the problem to the point that you simply get out the parameter set you put in. Conceptual model formulation and creating the cost function are where the science enters into inverse modeling, which can otherwise seem like (and often is) a giant mathematical exercise. No matter how complex and mathematically rigorous your computational machinery, the problem must be properly set up first.

18.2 Solving under-determined systems

18.2.1 Basic machinery

So how do we solve linear inverse problems? You might be tempted to think that the generalized inverse should be simply the matrix inverse, that is $\mathbf{A}^{-g} = \mathbf{A}^{-1}$. But recall that a true matrix inverse only exists for a square matrix, in this case the rare even-determined system ($N = M$). For the much more common purely under-determined system ($N < M$), the data do not uniquely constrain the parameters (though they might constrain combinations of parameters). There are thus an infinite number of solutions that satisfy the model equations and fit the data.

Analogous to the over-determined least-squares case (recall generalized least squares), we could choose a solution that minimizes the length of the model parameters:

$$J = \mathbf{p}'\mathbf{p} \tag{18.7}$$

subject to the constraint that the model solution exactly fits the data:

$$\mathbf{x} - \mathbf{A}\hat{\mathbf{p}} = 0 \tag{18.8}$$

This is called the *minimum length solution*. This is in essence what we accomplished using SVD in Section 1.4.2. Menke (1984) and Wunsch (1996) present the fully generalized solution minimizing the weighted (ω_p^{-1}), squared distance of the model parameters to some a priori estimate $\mathbf{p}_o$ as in Equation (18.6); the general equations can be incorporated by transforming the variables, but the simpler form will do for our purposes.

The conventional approach to solving (18.7) and (18.8) uses *Lagrange multipliers*, which are a handy way of adding implicit constraints. We minimize a new cost function J_μ:

$$J_\mu = \mathbf{p}'\mathbf{p} + 2\boldsymbol{\mu}'(\mathbf{x} - \mathbf{A}\mathbf{p}) \tag{18.9}$$

with respect to the parameters $\mathbf{p}$. At the solution when J_μ is at a minimum:

$$\frac{\partial J_\mu}{\partial \mathbf{p}} = 0 \tag{18.10}$$

Notice that we have added a second term to J_μ, a vector of length N equal to the prediction errors (which at the solution point are defined to be zero) times a vector of some "arbitrary" Lagrange multipliers $\boldsymbol{\mu}$. We then solve for $\mathbf{p}$ and $\boldsymbol{\mu}$ jointly. The parameter values $\mathbf{p}$ are not independent at the minimum of J_μ because of the constraint that the model exactly fit the data, but the sum of the two right-hand terms in Equation (18.9) is independent. We can then solve for $\partial J_\mu / \partial \mathbf{p} = 0$ separately for each individual parameter p_k just as we did in least squares. This is called a *strong constraint* because the model must be met exactly. The estimated parameters are then:

$$\hat{\mathbf{p}} = \mathbf{A}'[\mathbf{A}\mathbf{A}']^{-1}\mathbf{x} \tag{18.11}$$

Assuming that the inverse $[\mathbf{A}\mathbf{A}']^{-1}$ exists, we can solve for $\hat{\mathbf{p}}$. In this case $\mathbf{A}^{-g} = \mathbf{A}'[\mathbf{A}\mathbf{A}']^{-1}$.

The comparable set of matrix equations for the least squares solution of over-determined $(N > M)$ systems is found by minimizing a cost function equal to the prediction error:

$$J = (\mathbf{x} - \mathbf{A}\mathbf{p})'(\mathbf{x} - \mathbf{A}\mathbf{p}) \tag{18.12}$$

leading to:

$$\hat{\mathbf{p}} = [\mathbf{A}'\mathbf{A}]^{-1}\mathbf{A}'\mathbf{x} \tag{18.13}$$

For the mixed-determined case, a solution can found by minimizing a cost function with constraints on both data and parameters:

$$J = (\mathbf{x} - \mathbf{A}\mathbf{p})'(\mathbf{x} - \mathbf{A}\mathbf{p}) + \alpha \mathbf{p}'\mathbf{p} \tag{18.14}$$

leading to:

$$\hat{\mathbf{p}} = [\mathbf{A}'\mathbf{A} + \alpha \mathbf{I}_M]^{-1}\mathbf{A}'\mathbf{x} \tag{18.15}$$

where $\mathbf{I}_M$ is the $M \times M$ identity matrix. This solution is referred to as the damped or tapered least squares. The coefficient α balances how far the model parameters are allowed to drift from their priors against misfit the data points. Equation (18.15) is called a *weak constraint* because the model no longer exactly matches the data; but it is often the appropriate approach because the data have errors and the model may be imperfect.

18.2.2 *The SVD approach and the data and model resolution matrices*

The minimum length (18.11), least squares (18.13), and damped least squares (18.15) solutions can all be found using a single SVD equation (but of course!):

$$\hat{\mathbf{p}} = \mathbf{V}_r \mathbf{S}_r^{-1} \mathbf{U}_r' \mathbf{x} \tag{18.16}$$

Recall that in MATLAB the full sizes of the matrices returned by svd(A) are $\mathbf{U}$ ($N \times N$), $\mathbf{S}$ ($N \times M$), and $\mathbf{V}$ ($M \times M$). We will retain only the first r columns of the three matrices, denoted as $\mathbf{U}_r$, $\mathbf{V}_r$, and $\mathbf{S}_r$ because they are in the data and model ranges. Similar to standard SVD, if any of the eigenvalues are zero or very small (i.e. $\mathbf{A}$ is rank deficient as expected for a mixed-determined case) then r would be reduced further (Chapter 1). The discarded $\mathbf{U}$ columns are in the data null space $S_o(\mathbf{x})$ and the discarded $\mathbf{V}$ columns are in the model null space $S_o(\mathbf{p})$. For the pure under-determined case, $N = r < M$, $\mathbf{U}_r$ ($N \times N$),

$\mathbf{S}_r$ ($N \times N$), $\mathbf{V}_r$ ($M \times N$), and there will be $M - N$ model null eigenvectors and no data null eigenvectors.

The model matrix **A** can tell us a tremendous amount about the structure of the problem and the relationship of the model parameters and data. Useful quantities include the model resolution:

$$\mathbf{R}_p = \mathbf{A}^{-g}\mathbf{A} = \mathbf{V}_r\mathbf{V}'_r \tag{18.17}$$

and data resolution:

$$\mathbf{R}_x = \mathbf{A}\mathbf{A}^{-g} = \mathbf{U}_r\mathbf{U}'_r \tag{18.18}$$

The model resolution shows how the estimated model parameters are interrelated. If the parameters are independently resolved, the model resolution matrix will equal an identity matrix; off-diagonal elements reflect the fact that the data provide information only about combinations of parameters. Similarly, the data resolution matrix shows how well the data can be predicted by the model. The cartoons in Fig. 18.2 display two model/data resolution matrices. In the left panel, the properties are independently resolved while in the right panel they are much more interconnected. Another interesting matrix is the a-posteriori model parameter covariance matrix:

$$\hat{\boldsymbol{\omega}}_p = \sigma_x^2\mathbf{V}_r\mathbf{S}_r^{-2}\mathbf{V}'_r \tag{18.19}$$

where σ_x^2 is the observational measurement uncertainty. Note that we don't have to make any observations to calculate these results. Inverse modeling, therefore, is of particular use in designing experiments and observational networks.

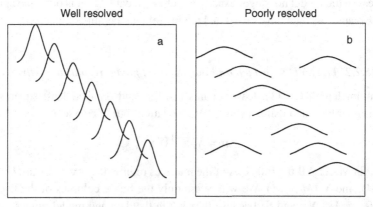

Figure 18.2 Schematics of well-resolved (a) and poorly resolved (b) model/data resolution matrices. The well-resolved case has large, non-zero values along the diagonal and negligibly small (or zero) values off the diagonal. In the poorly resolved case, the off diagonal elements are populated with substantial non-zero values, indicating strong covariation among multiple variables or data points in the inversion solutions.

18.2.3 An example calculation

A simple example may help clarify the situation. Consider an experiment like the sound speed experiment described in Section 18.1.1. In Fig. 18.3 we set up a mixed determined system with $M = 3$ unknowns ($\mathbf{p}$) and $N = 2$ measurements ($\mathbf{x}$). The corresponding model equations are:

$$p_1 = x_1 \tag{18.20}$$

and

$$0.5 p_2 + 0.5 p_3 = x_2 \tag{18.21}$$

or in matrix form:

$$A = \begin{bmatrix} 1.0 & 0 & 0 \\ 0 & 0.5 & 0.5 \end{bmatrix} \tag{18.22}$$

From Fig. 18.3 and the equations it is fairly obvious that p_1 is exactly determined by x_1 and that p_2 and p_3 are under-determined by x_2, but let's see what the SVD solution looks like. The singular value matrix:

$$S = \begin{bmatrix} 1.0 & 0 & 0 \\ 0 & 0.707 & 0 \end{bmatrix} \tag{18.23}$$

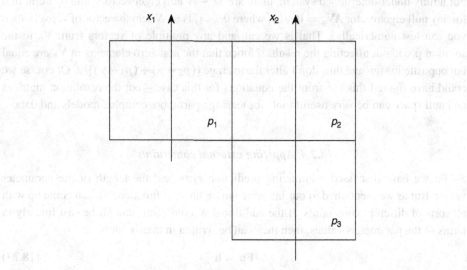

Figure 18.3 Schematic of a mixed-determined inverse problem. In this example, an ocean sound speed data set is collected along $N = 2$ paths but we are attempting to invert for $M = 3$ unknowns (in this case the temperatures within each box). The parameter p_1 is well determined by the measurement of x_1, but p_2 and p_3 are under-determined by the single measurement x_2; only the average of p_2 and p_3 can be determined.

has only two non-zero columns so the dimension r of the range of $\mathbf{A}$ is 2. Keeping only the first two columns of $\mathbf{U}$, $\mathbf{V}$, and $\mathbf{S}$, we have, as in Equation (18.16):

$$\mathbf{S}_r^{-1} = \begin{bmatrix} 1.0 & 0 \\ 0 & 1.414 \end{bmatrix} \quad \mathbf{U}_r = \begin{bmatrix} 1.0 & 0 \\ 0 & 1.0 \end{bmatrix} \quad \mathbf{V}_r = \begin{bmatrix} 1.0 & 0 \\ 0 & 0.707 \\ 0 & 0.707 \end{bmatrix} \quad (18.24)$$

The corresponding generalized inverse, model resolution, and data resolution matrices are:

$$\mathbf{A}^{-g} = \begin{bmatrix} 1.0 & 0 \\ 0 & 1.0 \\ 0 & 1.0 \end{bmatrix} \quad \mathbf{R}_p = \begin{bmatrix} 1.0 & 0 & 0 \\ 0 & 0.5 & 0.5 \\ 0 & 0.5 & 0.5 \end{bmatrix} \quad \mathbf{R}_x = \begin{bmatrix} 1.0 & 0 \\ 0 & 1.0 \end{bmatrix}$$

$$(18.25)$$

Without more information, the SVD solution suggests that our best "guess" is to assign the same value to the two under-determined parameters, p_2 and p_3, and the "data" information is spread equally over the two parameters in the model resolution matrix. This is a natural result of the SVD solution. We cannot recover the true values of p_2 and p_3 individually, and in fact there are an infinite number of possible solutions. This can be seen from the third column of $\mathbf{V}$, which is the eigenvector in the model null space:

$$\mathbf{V}_o = \begin{bmatrix} 0 \\ -0.707 \\ 0.707 \end{bmatrix} \quad (18.26)$$

For a fully under-determined system, there are $M - N$ null eigenvectors, and by definition for any null eigenvector $\mathbf{A}\mathbf{V}_o = \mathbf{0}_{(N\times1)}$ where $\mathbf{0}_{(N\times1)}$ is an N length vector of zeros (which you can test numerically). That is we can add any multiple of vectors from $\mathbf{V}_o$ to the solution $\hat{\mathbf{p}}$ without affecting the results. Notice that the non-zero elements in $\mathbf{V}_o$ are equal but opposite in sign and thus don't alter the average $[(p_2 + \gamma) + (p_3 - \gamma)]/2$. Of course we could have figured this out from the equations for this case – but the resolution matrices and null space can be very useful tools for teasing apart more complex models and data.

18.2.4 *Applying external constraints*

So far we have discussed minimizing prediction error and the length of the parameter vector. But as we mentioned in our introduction on the cost function, we can come up with all sorts of different constraints. If the additional K constraints can all be cast linearly in terms of the parameters values, then they can be written in matrix form as:

$$\mathbf{Fp} = \mathbf{h} \quad (18.27)$$

One possibility is that some subset of the parameters should sum to a particular value. This could arise, for example, if you wanted to constrain the total net northward ocean heat transport in a basin-scale ocean inversion based on atmospheric estimates, or if you wanted to constrain net community production integrated over the euphotic zone based on sediment trap data. Another example would be where you were fitting a straight line to data but trying to constrain the line to pass exactly through a particular point (z_0, h_0). This would look like $p_1 + p_2 z_0 = h_0$ or:

$$\mathbf{Fp} = [1 \quad z_0] \begin{bmatrix} p_1 \\ p_2 \end{bmatrix} = h_0 \tag{18.28}$$

where p_1 and p_2 are the unknown intercept and slope parameters.

There is a variety of ways to implement such constraints. They can be added to the original matrix:

$$\begin{bmatrix} \mathbf{A} \\ \mathbf{F} \end{bmatrix} \mathbf{p} = \begin{bmatrix} \mathbf{x} \\ \mathbf{h} \end{bmatrix} \quad \text{or} \quad \tilde{\mathbf{A}}\mathbf{p} = \tilde{\mathbf{x}} \tag{18.29}$$

We can then apply the SVD method to the combined matrix $\tilde{\mathbf{A}}$ where the weights on the constraints in ω_p^{-1} are set high to assure that the constraints will be closely met (so called weak constraint). Or alternatively we can use the Lagrange multiplier approach again to make sure the constraints are exactly met. After a bit of rearranging (see Menke, 1984), this leads to:

$$\begin{bmatrix} \mathbf{A'A} & \mathbf{F'} \\ \mathbf{F} & \mathbf{0}_K \end{bmatrix} \begin{bmatrix} \mathbf{p} \\ \mu \end{bmatrix} = \begin{bmatrix} \mathbf{A'x} \\ \mathbf{h} \end{bmatrix} \tag{18.30}$$

which again can be solved as $\tilde{\mathbf{A}}\tilde{\mathbf{p}} = \tilde{\mathbf{x}}$ using SVD.

In some situations, we might also want to specify inequality constraints, that is $\mathbf{Fp} > \mathbf{h}$. For an ocean inverse calculation, we might want to require a water mass to flow in one direction. Or we might want certain parameters to always be positive. This makes things a bit more complicated, but there are iterative methods that can do the job. We explored one such approach, the simplex method for solving linear programming problems, in Section 10.1.2.

As a final note, inverse problems can become computationally challenging as the size of the matrices grows. Sometimes the matrices we build have a lot of zeros in them. We call these *sparse matrices*. Fortunately, there are faster and more efficient algorithms available for manipulating and inverting many types of sparse matrices. The details of such algorithms are beyond the scope of this book, though you might start by examining some tools in MATLAB and other standard linear algebra packages.

18.3 Ocean hydrographic inversions

Unfortunately, the concepts behind generalized inverse modeling can be somewhat opaque when presented in generic form. To better illustrate inverse methods for under-determined systems, we discuss a specific example using ocean hydrographic sections, lateral density gradients, and geostrophic velocities. Ocean hydrographic sections have been collected for almost a century and continue to be a mainstay of oceanographic programs such as the

World Ocean Circulation Experiment (WOCE) and Climate Variability and Predictability
CLIVAR/CO_2 Repeat Hydrography Program.

18.3.1 Setting up the problem

Recall the mass balance equation for a tracer C in a moving ocean (Equation (17.9))
(neglecting turbulent mixing for the moment):

$$\frac{\partial C}{\partial t} + \mathbf{u} \cdot \nabla C = J_C \tag{18.31}$$

If we knew the ocean velocity field $\mathbf{u}$ we could estimate the often poorly known source/sink
terms J_C, at least for a steady-state ocean. But how do we find the large-scale $\mathbf{u}$ field?

There are not enough current meters to fully instrument the ocean, even for a single
basin, and ocean current meter data are often complicated by tides, mesoscale variability,
etc. A growing system of surface drifters and profiling Argo floats is now starting to provide
a more complete picture of ocean velocities, but only at two depths (the surface and 1000 m,
the parking depth of Argo floats). Sea surface height from satellite altimetry is invaluable
for understanding variability in ocean velocities, but it is more difficult to constrain the
mean velocity field. For many decades, the only alternative was to use the concept of
geostrophy, introduced in the last chapter, to solve for the horizontal velocity field $\mathbf{u}_H$
at right angles to the section from ocean density. The problem involves inverse modeling
because geostrophy provides information only on velocity shear (i.e. the vertical velocity
gradient), not absolute velocity. Thus there are unresolved reference velocities we need to
find by inversion using tracer budgets.

As illustrated in Fig. 18.4a, hydrographic sections consist of a series of stations where
the vertical profiles of temperature, salinity, and pressure are measured with a CTD (or
similar instrument). Density can be computed from the CTD data, and other tracers are
measured often either from continuous sensors on the CTD or from discrete water samples.
To avoid problems with mesoscale eddies, station spacing along the section (Fig. 18.4)
is typically ~50 km or better in the open ocean, with higher resolution near the coasts
and dynamical features such as the equator. The geostrophic velocity perpendicular to the
section can be computed in between station pairs from the lateral density gradient $\Delta\rho/\Delta x$
(note that for this section x refers to distance, not an element in the data vector as above).

18.3.2 Geostrophy and thermal wind

Geostrophy simply reflects the dominant balance of forces acting on ocean water parcels
on a rotating planet. Water parcels accelerate when the vector forces acting on the parcel
sum to a non-zero value. On small enough scales where rotation is not important, pressure
forces accelerate parcels creating a velocity parallel to the pressure gradient; water flows
directly from regions of high to low pressure (Fig. 18.5) (think about what happens if you
pile up water at one end of a bathtub and then release it suddenly). As scales increase

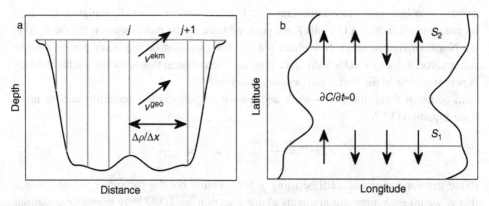

Figure 18.4 (a) Illustration of how hydrographic station data are used to compute the geostrophic velocities v^{geo} perpendicular to the section from the lateral density gradient $\Delta\rho/\Delta x$. Geostrophy constrains the vertical velocity shear, and a reference velocity v^{ref} is needed for each station pair (e.g. v_j^{ref}). The reference velocities can be estimated from ocean inverse calculations (b) using geostrophic velocities v^{geo}, Ekman velocities v^{ekm} and tracer budget balances (e.g. mass, nutrients, salt, heat) for ocean regions bounded by hydrographic sections (S_1 and S_2), assuming steady state.

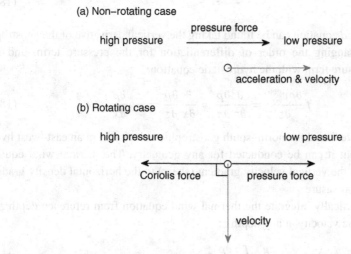

Figure 18.5 Schematic showing how pressure forces induce water parcel accelerations and velocities for non-rotating (a) and rotating (b) cases. Planetary rotation generates an apparent force, the Coriolis force, perpendicular to parcel motion and to the right in the Northern Hemisphere. Large-scale ocean circulation is approximately geostrophic where pressure and Coriolis forces balance and velocities are perpendicular to the horizontal pressure gradients.

and rotation becomes more important (i.e. Rossby number is small, Equation (17.22)), moving parcels induce a Coriolis force perpendicular to the parcel motion (to the right in the Northern Hemisphere). Now there are two forces acting on the water parcel, and the water parcel velocity rotates until there is no net vector force; this occurs when the velocity is perpendicular to the horizontal pressure gradient.

In equation form, the north–south and east–west geostrophic velocities can be taken from Equation (17.20):

$$\rho f v^{\text{geo}} = \frac{\partial p}{\partial x} \qquad \rho f u^{\text{geo}} = -\frac{\partial p}{\partial y} \tag{18.32}$$

(Note that for a moment we'll be using p for pressure not the parameter vector.) Notice that if we measure pressure gradients along a section we could only use these equations to calculate the component of the geostrophic velocity perpendicular to the section but not that along the section. It turns out that directly estimating pressure gradients along horizontal surfaces (or actually geopotential surfaces) from a ship is nearly impossible because we can only make depth measurements relative to the sea surface (there are new satellite approaches using altimetry and gravity measurements but this is a topic for a different day). We can, however, solve this problem by converting between pressure and density using the vertical hydrostatic balance (Equation (17.21)):

$$\frac{\partial p}{\partial z} = -g\rho \tag{18.33}$$

The *thermal wind* equation can be found taking the vertical derivative of the geostrophic relationship, rearranging the order of differentiation for the pressure term, and then replacing the pressure term with the hydrostatic equation:

$$f\frac{\partial \rho v^{\text{geo}}}{\partial z} = \frac{\partial}{\partial z}\frac{\partial p}{\partial x} = \frac{\partial}{\partial x}\frac{\partial p}{\partial z} = -g\frac{\partial \rho}{\partial x} \tag{18.34}$$

In this case we have found the north–south geostrophic velocity from an east–west hydrographic section, but it can be conducted for any geometry. The thermal wind equation allows us to relate the vertical velocity gradient (shear) to the horizontal density gradient, something we can measure.

We can then vertically integrate the thermal wind equation from reference depth z_0 to find the geostrophic velocity at any depth z:

$$v^{\text{geo}}(z) = -\frac{g}{\rho f}\int_{z_0}^{z}\frac{\partial \rho}{\partial x}dz + v^{\text{ref}} = v^{\text{tw}}(z) + v^{\text{ref}} \tag{18.35}$$

For ease of notation below, we define a thermal wind velocity v^{tw} as the geostrophic shear relative to the reference depth z_0. The unknown reference velocity v^{ref} arises as an integration constant for the vertical integral, and v^{ref} would change if we were to use a different z_0. A different reference velocity is needed for each station pair (i.e. $j \leftrightarrow j + 1$; $j + 1 \leftrightarrow j + 2$; etc.).

For the inversion we will also need a second velocity term, the Ekman velocity $\mathbf{u}^{\text{Ekm}}$. Surface wind stress drives an Ekman transport in the upper ocean. Because of the Earth's rotation, the depth-integrated Ekman transport is perpendicular to the surface wind direction, to the right in the Northern Hemisphere and to the left in the Southern Hemisphere:

$$F_u^{\text{Ekm}} = \frac{\tau_y}{f} \qquad F_v^{\text{Ekm}} = -\frac{\tau_x}{f} \qquad (18.36)$$

where τ_y and τ_x are the meridional and zonal components of the wind stress. The Ekman transport can be used directly in inverse calculations or an average Ekman velocity can be found by dividing through by an approximate Ekman depth z_{Ekm} and density.

18.3.3 Inverse box models

The next step is to build (and solve) an inverse box model in terms of the unknown v^{ref} values (Wunsch, 1996). This works best when you have two (or more) sections that bound ocean regions or boxes. The sections do not need to be strictly north–south or east–west, but typically we will want to use sections that extend across an entire basin (i.e. that nearly touch both coastlines) or that intersect other sections in the ocean interior (Fig. 18.4). The geostrophic velocities along two sections can be used to estimate the fluxes of mass (heat, salt, tracers, etc.) into and out of the ocean box confined by the sections. Assuming that the ocean is approximately in steady state (e.g. $\partial C/\partial t \approx 0$), then we can apply mass balance or mass conservation constraints to estimate the unknown v^{ref} values.

Let's begin with the mass balance for water and assuming one ocean box and two ocean sections. The sections are naturally discretized by the number of stations along the sections (say $M + 2$), resulting in M velocity profiles (one for each station pair along each section). CTD data are typically binned at 1 or 2 dbar intervals ($\sim$1 to 2 m), but for the box inversion we will divide the water column up into thicker vertical layers ($\mathcal{O}(10^2)$ m). Because there is usually more shear and steeper vertical tracer gradients in the upper ocean, we may want to use thinner layers in the thermocline and thicker layers in the deep. We can then sum the mass fluxes across the two sections by station pair j and layer k including both the geostrophic and Ekman contributions:

$$\sum_j \sum_k \left(v_{j,k}^{\text{tw}} + v_j^{\text{ref}} \right) \rho_{j,k} \, \Delta x_j \Delta z_k + \sum_j F_j^{\text{Ekm}} \rho_{j,0} \, \Delta x_j \approx 0 \qquad (18.37)$$

where Δx_j is the along-section distance for station pair j, and $\Delta x_j \Delta z_k$ is the cross-section area for station pair j and layer k. Depending on which section a station pair falls on, the geometry terms can have either a positive and negative sign to account for whether a positive velocity (mass flux) adds or subtracts material from the box.

Notice we've used the $\approx$ symbol rather than an equal sign; the mass balance is inexact because of errors in our measurements and in the model. We could equivalently write the equation as an equality by adding an error or noise term n. For example, geostrophy is an approximation and does not hold fully in strong western boundary currents. Aliasing of mass transport can arise from poorly resolved mesoscale eddies and narrow currents, a type of representation error. Also the ocean may not be in steady state, particularly for sections made over different time periods.

We can reorganize Equation (18.37) to group the unknowns on the left-hand side and the knowns on the right:

$$\sum_j \sum_k v_j^{\text{ref}} \rho_{j,k} \, \Delta x_j \Delta z_k + n^\rho = -\sum_j \sum_k v_{j,k}^{\text{tw}} \, \rho_{j,k} \, \Delta x_j \Delta z_k - \sum_j F_j^{\text{Ekm}} \rho_{j,0} \, \Delta x_j$$

$$(18.38)$$

which can be rewritten in a more familiar matrix form as:

$$\mathbf{A}_i \mathbf{v}^{\text{ref}} + n_i = x_i^{\text{obs}} \tag{18.39}$$

Here the "model" vector $\mathbf{A}_i$ $(1 \times M)$ consists of the measured densities $\rho_{j,k}$ and the specified station/layer areas $\Delta x_j \Delta z_k$, the parameter vector $\mathbf{v}^{\text{ref}}$ is our unknown reference velocities $(M \times 1)$, and the "data" value x_i^{obs} is a single scalar but one that depends on a whole lot of observations (all of the estimates of geostrophic velocity and Ekman mass transport). The addition of a formal noise term n_i does not alter the SVD solutions for $\mathbf{v}^{\text{ref}}$ (Equation (18.16)) while adding information on the model residual $\hat{\mathbf{n}} = \mathbf{x}^{\text{obs}} - \mathbf{A}\hat{\mathbf{v}}^{\text{ref}}$ (see Wunsch, 1996). We may also need to supply additional external information to x_i^{obs} to account for mass transport that is not captured by the hydrographic section. An example might be the mass transport across the Florida Straits, which can be estimated from an undersea cable between Florida and the Bahamas; this transport constraint would be added to open-ocean hydrographic section data across the subtropical North Atlantic. We now have a single constraint equation (18.38) in M unknowns (*sigh*, a truly under-determined system).

There are several ways to increase the number of constraints. We could add another hydrographic section, creating a second box and a second mass balance equation. More sections do add more information because now the choice of $\mathbf{v}^{\text{ref}}$ values is more limited by the need to balance what is leaving one box with what is entering another; but additional sections also add more unknown $\mathbf{v}^{\text{ref}}$. Keeping with two just sections and one box, we could develop additional mass balance constraints for individual density layers $(\Delta \rho_l)$, rather than just the whole water column. The argument is that away from the surface the transformation of water from one density class to another is driven only by weak diapycnal (cross-isopycnal) mixing (Chapter 11). Similarly, conservation equations can be written for quasi-conservative tracers by replacing the $\rho_{j,k}$ terms with tracer concentrations $C_{j,k}$ in Equation (18.38):

$$\sum_j \sum_k v_j^{\text{ref}} C_{j,k} \, \Delta x_j \Delta z_k + n^C = -\sum_j \sum_k v_{j,k}^{\text{tw}} \, C_{j,k} \, \Delta x_j \Delta z_k - \sum_j F_j^{\text{Ekm}} C_{j,0} \, \Delta x_j$$

$$(18.40)$$

What do we mean by a quasi-conservative tracer? Well, that depends and is part of the art (or science) of inverse modeling. In the past, researchers have used heat (away from the surface layer), salt, silicate, and combinations of nutrients and oxygen (e.g. the sum of $170PO_4 + O_2$ based on Redfield elemental ratios should compensate for organic matter respiration that releases phosphate while consuming dissolved oxygen). Ganachaud and Wunsch (2002; 2003) present a nice example of the box inverse approach based on hydrographic data from the World Ocean Circulation Experiment. The results of their analysis are global patterns of horizontal mass and tracer fluxes as well as the tracer divergences, which can tell us something about surface fluxes and biogeochemical transformations.

We could of course include non-conservative tracers and attempt also to solve for some of the unknown source/sink terms. Subsurface nutrient, inorganic carbon, alkalinity, and oxygen fields are all influenced by the decomposition and respiration of sinking particulate matter. For simplicity let us assume that the subsurface remineralization rate for tracer C scales with surface export flux E_0^C and a specified depth function $f(k)$:

$$J_k^C = E_0^C f(k) \qquad (18.41)$$

Then the conservation equation for C over an ocean box for a single layer k is:

$$\sum_j v_j^{\text{ref}} C_{j,k} \, \Delta x_j \Delta z_k + E_0^C f(k) \, V_k + n^C = - \sum_j v_{j,k}^{\text{tw}} C_{j,k} \, \Delta x_j \Delta z_k \qquad (18.42)$$

where V_k is the volume of layer k over the box. Combining multiple layers leads to the following linear inverse model:

$$\tilde{\mathbf{A}} \begin{bmatrix} \mathbf{v}^{\text{ref}} \\ E_0^C \end{bmatrix} + \mathbf{n} = \mathbf{x}^{\text{obs}} \qquad (18.43)$$

where we have now included E_0^C in the vector of unknowns to be solved for.

Not all ocean hydrographic stations fall along nice (relatively straight) sections. But it is not a large leap to see how the box inverse formulation can be transformed from sections to randomly distributed data using a grid of boxes and finite difference methods. Velocity estimates are needed for the horizontal faces of each box, and in a similar fashion geostrophic velocities (and unknown reference velocities) are constructed from the thermal wind relationship and lateral density gradients. For an example of the inner workings of such inverse model calculations see Schlitzer (2002) and other papers by the same author.

So far we have relied on geostrophy to compute the velocity field. But we can also utilize the circulation field from an ocean GCM and focus more on 3D ocean tracer transport. One approach involves adding unit concentrations or fluxes of a dye-like tracer to a few dozen different surface regions of an ocean GCM integration and then tracking the spread of the simulated tracers with time (or until steady state). A model matrix $\mathbf{A}$ can then be constructed relating regional surface conditions to subsurface tracer distributions as a function of time since the introduction of the tracers at the surface. Given an observed subsurface tracer field $\mathbf{x}^{\text{obs}}$, we can use linear inverse modeling $\mathbf{A}\mathbf{p}^{\text{flux}} = \mathbf{x}^{\text{obs}}$ to reconstruct surface fluxes $\mathbf{p}^{\text{flux}}$ and ocean tracer transport. This method has been successfully applied to steady-state tracers (e.g. oxygen, dissolved inorganic carbon, natural radiocarbon) and transient tracers (e.g. bomb-radiocarbon, tritium, anthropogenic carbon) (Mikaloff Fletcher *et al.*, 2006; 2007). The same basic approach, but perhaps with a somewhat higher level of mathematical sophistication, can be followed using transit time distributions (TTDs; Section 17.5.3) (Khatiwala *et al.*, 2009).

18.4 Data assimilation methods

A wide range of techniques fall under the rubric of *data assimilation*, including the linear inverse methods already discussed. They all address in one form or another the question of optimally combining models and data. What differentiates the approaches discussed below from the simple linear inverse systems discussed above is that the model equations can be significantly nonlinear and time-evolving. Data assimilation really comes into its own when the model becomes too complex and intricate to write down as a simple linear transformation. Two important application areas are (1) nonlinear parameter optimization problems where you would like to use your data to improve the model parameters for complicated models; and (2) time-evolving forecast problems, where current observations at time t_0 are incorporated to update the model state at t_0, which in turn is used as an initial condition for a model forecast at a future time t_n. The two problems are related because you can treat initial conditions as just other unknown model parameters.

18.4.1 Variational methods and parameter optimization

Variational methods minimize a cost function J by adjusting a set of control parameters $\mathbf{p}$ using information about the sensitivity of the cost function to parameter values $\partial J / \partial \mathbf{p}$. For nonlinear problems, variational approaches follow an iterative algorithm (Fig. 18.6). The first step of each iteration, labeled here as iteration j, involves either inputting an initial set of model parameters $\mathbf{p}_0$ or loading a parameter vector $\mathbf{p}_j$ updated from a previous iteration. The model "parameters" might be quite general, including initial conditions and boundary conditions as well as classical parameters such as mixing coefficients or biological growth rates. The forward model is then computed with these parameters leading to a new estimate of the model fields $\mathbf{x}_j$, where we somewhat expand the notation for $\mathbf{x}$ to include the time-evolving model variables that match the observations now denoted $\mathbf{x}^{\text{obs}}$.

A cost function is calculated comparing the model and observation fields $J(\mathbf{x}_j, \mathbf{x}^{\text{obs}})$. The next step involves finding the sensitivity of the cost function to the model parameters $\partial J / \partial \mathbf{p}$. The *adjoint method* is particularly efficient in this regard because it allows the model to be run "backwards" in time to search for the origins of model anomalies (but only if you have built an adjoint of your forward model, which we discuss below). The sensitivities are then used to estimate a new set of parameters $\mathbf{p}_{j+1}$, the methodological details of which vary from optimizer to optimizer (e.g. conjugate gradient, BFGS; Section 10.3). The cycle then begins again until at some point the model matches the data "closely enough"

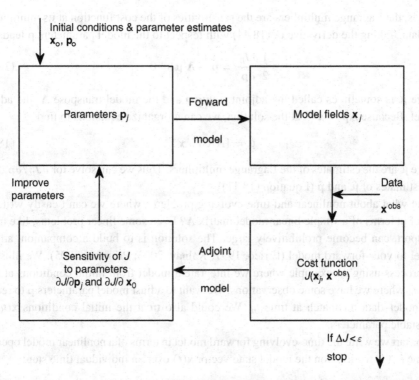

Figure 18.6 Schematic for a variational parameter optimization method using the model adjoint. An initial estimated set of parameters $\mathbf{p}_o$ and initial conditions $\mathbf{x}_0$ are used in the forward model to generate new model fields $\mathbf{x}_j$, which are then compared against the data $\mathbf{x}^{obs}$ with a cost function J. The adjoint model is used to compute the sensitivities of the cost function to parameters $\partial J/\partial \mathbf{p}_j$ (including initial conditions if desired, $\partial J/\partial \mathbf{x}_0$), which are then used in a optimization method to generate a new estimate of the parameters $\mathbf{p}_{j+1}$. The loop is repeated for the next iteration $j + 1$ until the improvement in the cost function ΔJ is sufficiently small.

$(\Delta J < \epsilon)$ and you stop (or you run out of computer time). Because many realistic models are nonlinear, the sensitivities depend on the model base state and thus change every cycle, which is why we can't just jump straight to the answer after the first pass.

Because the adjoint technique is used commonly in variational methods, we present the basics here starting with linear models (Wunsch, 1996). Recall from Section 18.2 that we introduced an additional term to the cost function equal to zero using Lagrange multipliers:

$$J_\mu = \mathbf{p}'\mathbf{p} + 2\mu'(\mathbf{x}^{obs} - \mathbf{A}\mathbf{p}) \tag{18.44}$$

A physical interpretation of the Lagrange multipliers is found by differentiating the cost function with respect to the data:

$$\frac{1}{2}\frac{\partial J_\mu}{\partial \mathbf{x}^{obs}} = \mu \tag{18.45}$$

That is, the Lagrange multipliers are the sensitivities of the cost function at its minimum to the data. Taking the derivative of (18.44) with respect to the model parameters $\mathbf{p}$ leads to:

$$\frac{1}{2}\frac{\partial J_\mu}{\partial \mathbf{p}} = \mathbf{p} - \mathbf{A}'\boldsymbol{\mu} = 0 \tag{18.46}$$

where $\boldsymbol{\mu}$ is sometimes called the adjoint solution and the model transpose $\mathbf{A}'$ the adjoint model. Because $\mathbf{Ap} = \mathbf{x}^{\mathrm{obs}}$ at the solution, we can reorganize equations to find:

$$\hat{\boldsymbol{\mu}} = [\mathbf{AA}']^{-1}\mathbf{x}^{\mathrm{obs}} \tag{18.47}$$

where $\hat{\boldsymbol{\mu}}$ are the estimates of the Lagrange multipliers. Thus we can solve for $\partial J_\mu/\partial \mathbf{p}$ using the estimates of $\hat{\boldsymbol{\mu}}$ and $\hat{\mathbf{p}}$ (Equation (18.11)).

But what about nonlinear and time-evolving problems where we can't easily write the model in terms of a simple linear model matrix $\mathbf{A}$? Even some linear problems, like tracer transport, can become prohibitively large. The solution is to build a companion, adjoint model to your forward model (Errico, 1997; Kalnay, 2003; Primeau, 2005). We illustrate the process using an example where we integrate a model from initial conditions at t_0 to time t_n where we have some observations. We want to adjust model parameters $\mathbf{p}$ to reduce the model–data mismatch at time t_n. We could also treat the initial conditions $\mathbf{x}(t_0)$ as adjustable parameters.

To start we write the time-evolving forward model in terms of a nonlinear model operator at time i $\mathcal{M}_i$ working on the model state vector $\mathbf{x}(t)$ over an individual time step:

$$\mathbf{x}(t_{i+1}) = \mathcal{M}_i[\mathbf{x}(t_i), \mathbf{p}] \tag{18.48}$$

where the index t_i stands for the time index, and $\mathbf{p}$ are the nonlinear model parameters. The model operator at time i $\mathcal{M}_i$, is time-dependent and changes with time as the simulation evolves. It also can be used to calculate the forward propagation of $\delta\mathbf{x}$, a perturbation to the state vector $\mathbf{x}$, as well as perturbations that arise because of incremental changes in model parameters $\delta\mathbf{p}$:

$$\delta\mathbf{x}(t_{i+1}) \approx \frac{\partial \mathcal{M}_i}{\partial \mathbf{x}}\delta\mathbf{x}(t_i) + \frac{\partial \mathcal{M}_i}{\partial \mathbf{p}}\delta\mathbf{p} \tag{18.49}$$

We define the tangent linear model as:

$$\mathbf{L}_x(t_i, t_{i+1}) = \frac{\partial \mathcal{M}_i}{\partial \mathbf{x}} \qquad \mathbf{L}_p(t_i, t_{i+1}) = \frac{\partial \mathcal{M}_i}{\partial \mathbf{p}} \tag{18.50}$$

$\mathbf{L} = [\mathbf{L}_x, \mathbf{L}_p]$ is basically a linearized version of $\mathcal{M}$ valid for small perturbations about the base state vector $\mathbf{x}(t)$, which evolves with time through a simulation.

To calculate the effects of a perturbation over multiple time-steps ($\delta\mathbf{x}(t_0) \rightarrow \delta\mathbf{x}(t_n)$), we need to use the product of the tangent linear models from the individual forward model time-steps:

$$\mathbf{L}(t_0, t_n) = \prod_{i=n-1}^{0} \mathbf{L}(t_i, t_{i+1}) = \mathbf{L}(t_{n-1}, t_n)\mathbf{L}(t_{n-2}, t_{n-1})\ldots\mathbf{L}(t_0, t_1) \tag{18.51}$$

Remember that the order of the matrix multiplication in Equation (18.51) matters. Because $\mathbf{L}$ is linear, we can compute the adjoint for an individual time step by simply taking the transpose of the linear tangent model, $\mathbf{L}'(t_{i+1}, t_i)$. For multiple time steps, the adjoint is:

$$\mathbf{L}'(t_n, t_0) = \prod_{i=0}^{n-1} \mathbf{L}'(t_{i+1}, t_i) = \mathbf{L}'(t_1, t_0)\mathbf{L}'(t_2, t_1) \ldots \mathbf{L}'(t_n, t_{n-1}) \qquad (18.52)$$

Notice that the time index for the adjoint moves backward in time.

Suppose we have a cost function J that depends on the model state at time t_n. We could conduct a systematic analysis by perturbing model parameters one by one to see how they affect the full model state at time t_n. For an individual perturbation:

$$\delta\mathbf{x}(t_n) = \mathbf{L}_p(t_0, t_n)\delta p_k \qquad (18.53)$$

where we have only changed the kth element in $\mathbf{p}$. This process is shown schematically in Fig. 18.7a for an example where a local perturbation to the model initial condition is transported and spread out over time by the model flow, leading to a large area of influence at t_n. The cost function sensitivity $\partial J / \partial p_k$ is:

$$\frac{\partial J}{\partial p_k} = \left(\frac{\partial J}{\partial \mathbf{x}(t_n)}\right)' \frac{\partial \mathbf{x}(t_n)}{\partial p_k} \qquad (18.54)$$

The partial derivative $\partial J / \partial \mathbf{x}$ can be computed analytically from our choice of cost function, and $\partial \mathbf{x} / \partial p_k$ can be estimated numerically from Equation (18.53). Equation (18.54) is quite generic and could be applied also to initial conditions and external forcing.

In a forward perturbation experiment, we get one piece of information on the cost function sensitivity, $\partial J / \partial p_k$, because J condenses all of the information on the numerous changes across the entire model state into a single number. We can build up a complete version of $\partial J / \partial \mathbf{p}$ through a series of perturbation experiments, but this can be rather laborious for a large number of parameters. This is where the model adjoint becomes handy.

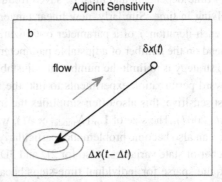

Figure 18.7 Schematics showing how altering a single model point by $\delta x(t)$ in a flow field influences other model points after a time interval of Δt downstream in a forward perturbation experiment (a) and upstream in an inverse adjoint sensitivity experiment (b).

Adjoint sensitivity experiments involve choosing a single metric (usually J) at time t_n and then integrating the adjoint model backwards in time to quantify the influence of earlier model states and model parameters on J. This is illustrated in Fig. 18.7b for the specific example of how the model–data misfit at a single location and time is affected by model conditions spread over a wider domain earlier in the integration. With the adjoint we can get the complete estimate of $\partial J/\partial \mathbf{p}$ in a single paired forward integration (to compute $\mathbf{L}$ and $\mathbf{L}'$) and backward adjoint integration.

To give a more concrete example, let's turn to an initial condition sensitivity experiment relevant to the next subsection on forecasting. We will also say that observations can be collected at multiple time points over the model integration, not just at the end at t_n. Therefore we define a more general cost function based on the model–data misfit at different times t_i:

$$J = \frac{1}{2} \sum_{i=0}^{n} \left[\left(\mathbf{x}(t_i) - \mathbf{x}^{\text{obs}}(t_i) \right)' \boldsymbol{\omega}_{\text{obs}}^{-1} \left(\mathbf{x}(t_i) - \mathbf{x}^{\text{obs}}(t_i) \right) \right] \qquad (18.55)$$

where $\mathbf{x}^{\text{obs}}(t_i)$ are the observations for time t_i and $\boldsymbol{\omega}_{\text{obs}}$ is the observational error covariance matrix. The cost function sensitivity is then:

$$\frac{\partial J}{\partial \mathbf{x}(t_0)} = \sum_{i=0}^{n} \left[\mathbf{L}'(t_i, t_0) \boldsymbol{\omega}_{\text{obs}}^{-1} \left(\mathbf{x}(t_i) - \mathbf{x}^{\text{obs}}(t_i) \right) \right] \qquad (18.56)$$

For each time point t_i with observations, we compute the model error $(\mathbf{x}(t_i) - \mathbf{x}_i^{\text{obs}})$ and then propagate that information backwards from time t_i to the initial time t_0 using the adjoint $\mathbf{L}'$.

Variational approaches often require substantial data storage capabilities to save the tangent linear and adjoint models as a function of time from the original forward model integration. Because the model operator $\mathcal{M}$ depends on model state, $\mathbf{L}$ and $\mathbf{L}'$ are typically time-dependent and must be saved frequently if the base model simulation is rapidly evolving in time. Similarly, new linear tangent and adjoint models need to be computed for each iteration j of a parameter optimization (Fig. 18.6). The matrices $\mathbf{L}$ and $\mathbf{L}'$ also depend on the number of adjustable parameters $\mathbf{p}$ (including adjustable initial conditions). One strategy is to limit the number of adjustable parameters by first testing the model with forward perturbation experiments to find the subset of parameters to which the model is most sensitive; this also often simplifies the interpretation of the optimization (Friedrichs *et al.*, 2007). The size of $\mathbf{L}_p$ ($N_{\text{model}} \times M$), where N_{model} is the number of state variables in $\mathbf{x}$, can also become problematic, particularly for a multi-dimensional model with a large number of state variables. Also for 2D and 3D tracer transport problems, the $\mathbf{L}$ matrix can be quite sparse for individual time-steps because the influence of a perturbation $\delta \mathbf{x}$ only spreads to adjacent grid points. For most medium to large applications, it is much easier to compute numerically only the required subset of $\mathbf{L}$ and $\mathbf{L}'$ on the fly in a forward (backward) time-evolving code rather than saving and manipulating large matrices to an external data file and solving the problem using matrix multiplication; more computations but less input/output and data storage.

For simple models, we can find the tangent linear and adjoint models analytically. Or more likely for modest sized problems we can code them by hand from the original nonlinear forward model using some simple rules dealing with linearization of the model equations and bookkeeping of information on the sensitivities to perturbations in model parameters and model state variables (Kalnay, 2003). There are now computer programs available that will "automatically" generate model code for the tangent linear and adjoint models, much like a compiler computes machine code. An example for FORTRAN codes is a web-based software system called the Tangent-linear and Adjoint Model Complier (TAMC) (Giering and Kaminski, 1998) and its successor the Transformation of Algorithms in FORTRAN (TAF). Adjoint compilers are not foolproof (though they are becoming much more sophisticated with time). They often work better for forward model codes structured in a particular fashion. For example, it is best to avoid boolean conditional statements (e.g. `if ... then`) because they result in binary forks in the code that are not easily differentiated (think of them as step functions). In many cases conditional statements can be replaced by smooth differentiable functions. Care also should be taken to limit the interdependence of variables between code loops (i.e. `i=1:n ... end`) and to avoid the redefinition of variables over the code. Finally, writing clean modular code is essential (at least if you want to be able to read the resulting adjoint code). As general advice, you should be familiar enough with the concepts behind linear tangents and adjoints and their construction to hand check (and numerically check) the output.

18.4.2 Analysis and forecast problems

Everyone is familiar with weather *forecasting* but perhaps not with the underlying numerical machinery – how do you optimally combine models and observations to better predict future behavior? The same mathematical tools are applied in *hindcast simulations* or *reanalysis*, retrospective studies that use historical observations to recreate past variations in oceanic and atmospheric circulation and properties. Ocean reanalysis provides a framework for interpreting past physical climate variability as well as variations and patterns in ocean biology and chemistry.

While the roots of data assimilation stem in large part from atmospheric numerical weather prediction (Kalnay, 2003), marine applications are now common, particularly for ocean circulation and physics. Ocean reanalysis efforts include the consortium Estimating the Circulation & Climate of the Ocean (ECCO) (Stammer *et al.*, 2004; Wunsch *et al.*, 2009) and the Simple Ocean Data Assimilation (SODA) analysis (Carton and Giese, 2008). In the first phase of ECCO, surface forcing was adjusted using variational methods so that the ocean model fields better match historical observations (e.g. temperature, salinity, sea surface height); the subsurface fields were not adjusted directly, and the solution to this *state estimation* problem is a dynamically consistent circulation and ocean hydrographic field over time. In contrast, SODA is a sequential scheme where observations are used

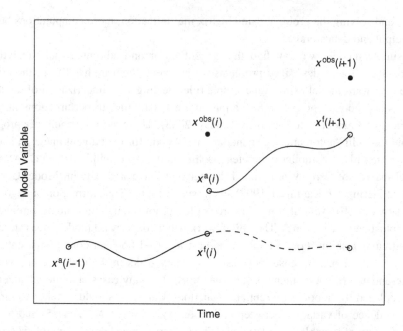

Figure 18.8 Schematic of how data assimilation is used to improve model forecasts. Beginning with initial conditions at time t_{i-1}, the model produces a forecast $x^f(i)$ (solid line) to time t_i. The model forecast and observations at time t_i are compared and then combined to produce an analysis $x^a(i)$, which is then used as the initial condition for the next forecast step to time t_{i+1}, leading to a better forecast than would have occurred following the original model trajectory forward in time to $i+1$ (dashed line).

directly to modify internal model fields over the historical time integration. Operational ocean forecast models include the UK Forecasting Ocean Assimilation Model (FOAM) (Storkey *et al.*, 2010), the French Mercator Ocean project, and the US Navy Coupled Ocean/Atmosphere Mesoscale Prediction System (COAMPS). Data assimilation is being applied increasingly to marine geochemical and biological problems as well (Gregg *et al.*, 2009).

Consider the schematic of a forecast/analysis cycle in Fig. 18.8. The cycle starts with an estimate of the model state at time t_{i-1}, referred to as the *analysis* $\mathbf{x}^a(t_{i-1})$. The *forecast step* then creates a model forecast for a future time t_i:

$$\mathbf{x}^f(t_i) = \mathcal{M}_{i-1}[\mathbf{x}^a(t_{i-1})] \qquad (18.57)$$

The *analysis step* merges the model forecast and new data from time t_i to produce a better estimate for the model state at time t_i:

$$\mathbf{x}^a(t_i) = \mathbf{x}^f(t_i) + \mathbf{W}_i \left(\mathbf{x}^{obs}(t_i) - \mathbf{x}^f(t_i) \right) \qquad (18.58)$$

where $\mathbf{W}_i$ are the *analysis weights* or gains that vary depending on the relative size of observational and model uncertainties. The cycle then repeats using the new analysis as the initial conditions for the next forecast step from $t_i \rightarrow t_{i+1}$. Because assimilation continually moves the model trajectory back toward the observations, the resulting predictions should be better than those from an unconstrained model integration. Of course the devil is in the details, and there are substantial differences among techniques in the treatment of model/data uncertainties and analysis weights.

The size of the adjustment in the analysis step (Equation (18.58)) depends on the *observational increment* or *innovation* $(\mathbf{x}^{obs} - \mathbf{x}^f)$ and analysis weights $\mathbf{W}_i$. Observations change the model trajectory most at locations and times with the largest model–data differences. Data often do not lie exactly on the model grid, and sometimes the available data variables differ from the state variables used in the model. For example, an atmosphere model predicts temperature and humidity but satellites measure radiances, or an ocean model tracks gridded temperature fields but acoustic arrays measure sound speed over long paths. Formally all data types can be assimilated by interpolating and converting the model variables to match the observations using the *observational operator* $\mathbf{H}$; the innovation would then be $(\mathbf{y}^{obs} - \mathbf{H}(\mathbf{x}^f))$, where we use a different notation $\mathbf{y}^{obs}$ to indicate that we are considering a broader suite of observations than simply the model state variables. For clarity below we keep with the simpler notation and neglect $\mathbf{H}$ (or mathematically set $\mathbf{H} = \mathbf{I}$), but if you need the complete equation forms see Kalnay (2003).

The analysis weights $\mathbf{W}_i$ can be computed in different ways. Some forecasting systems use purely statistical approaches like optimal interpolation, which we introduced in Section 7.3.1. Optimal interpolation attempts to minimize analysis errors at each grid point. The weights $\mathbf{W}^{OI}$ depend on prescribed observation and forecast error covariance matrices, ω_{obs} and ω_f, respectively:

$$\mathbf{W}^{OI} = \omega_f \left(\omega_{obs} + \omega_f \right)^{-1} \tag{18.59}$$

The observational error covariance matrix ω_{obs} is typically a diagonal matrix that is relatively straightforward if we understand the instrument systems; even here, though, there are issues about how well a measurement at a single location represents properties over larger scales like a model grid box (sometimes called the *representativeness error*). The forecast error covariances are more troublesome. We need to start somewhere (usually an educated guess at model skill), but over time better estimates of ω_f can be built by comparing model results against observations over many forecast cycles. The ability to routinely test forecast skill is a major difference between forecasting (e.g. weather) and future projections (e.g. climate change). The errors in the analysis:

$$\omega_a^{-1} = \omega_f^{-1} + \omega_{obs}^{-1} \tag{18.60}$$

are smaller than the errors in the forecast or observations alone.

Other forecast approaches are based on three-dimensional (x–y–z; *3D-Var*) and four-dimensional (x–y–z–t; *4D-Var*) variational methods. In 3D-Var, the analysis step is conducted by minimizing the following cost function:

$$J = \frac{1}{2} \left((\mathbf{x}^{\text{obs}} - \mathbf{x})' \, \boldsymbol{\omega}_{\text{obs}}^{-1} (\mathbf{x}^{\text{obs}} - \mathbf{x}) + (\mathbf{x} - \mathbf{x}^{\text{f}})' \, \boldsymbol{\omega}_{\text{f}}^{-1} (\mathbf{x} - \mathbf{x}^{\text{f}}) \right) \quad (18.61)$$

The two terms reflect the tension of adjusting the model toward the data but not too far away from the original forecast. Mathematically the effective analysis weights from 3D-Var are equivalent to those from optimal interpolation, but in practice 3D-Var involves an iterative procedure (Fig. 18.6). The 4D-Var method is a more sophisticated variant that takes into account observations from multiple time periods by minimizing a cost function such as:

$$J = \frac{1}{2} \left(\sum_{i=0}^{n} \left((\mathbf{x}_i^{\text{obs}} - \mathbf{x}_i)' \, \boldsymbol{\omega}_{\text{obs}}^{-1} (\mathbf{x}_i^{\text{obs}} - \mathbf{x}_i) \right) + (\mathbf{x}_0 - \mathbf{x}_0^{\text{f}})' \, \boldsymbol{\omega}_{\text{f}}^{-1} (\mathbf{x}_0 - \mathbf{x}_0^{\text{f}}) \right) \quad (18.62)$$

where i is the index for observational time intervals. In 4D-Var, model initial conditions are adjusted at time t_0 to create a dynamically consistent analysis at time t_n that is also consistent with the observations from $t_0 \rightarrow t_n$.

An alternative to variational methods is a family of techniques called *Kalman filtering* (Kalnay, 2003). The analysis/forecasting cycle is similar to optimal interpolation except now there are additional equations to track the time evolution of the forecast error covariance matrix $\boldsymbol{\omega}^{\text{f}}$ rather than assuming, a priori, a stationary structure. The forecast step for extended Kalman filtering is:

$$\mathbf{x}^{\text{f}}(t_i) = \mathcal{M}_{i-1}[\mathbf{x}^{\text{a}}(t_{i-1})] \quad (18.63)$$

$$\boldsymbol{\omega}^{\text{f}}(t_i) = \mathbf{L}(t_{i-1}, t_i) \boldsymbol{\omega}^{\text{a}}(t_{i-1}) \mathbf{L}'(t_i, t_{i-1}) + \mathbf{Q}(t_{i-1}) \quad (18.64)$$

where $\boldsymbol{\omega}^{\text{f}}$ depends on the analysis error $\boldsymbol{\omega}^{\text{a}}$ from the previous time step and a model noise matrix $\mathbf{Q}$ that needs to be specified. This followed by an analysis step:

$$\mathbf{x}^{\text{a}}(t_i) = \mathbf{x}^{\text{f}}(t_i) + \mathbf{K}_i \left(\mathbf{x}^{\text{obs}}(t_i) - \mathbf{x}^{\text{f}}(t_i) \right) \quad (18.65)$$

$$\boldsymbol{\omega}^{\text{a}}(t_i) = (\mathbf{I} - \mathbf{K}_i) \boldsymbol{\omega}^{\text{f}}(t_i) \quad (18.66)$$

The Kalman gain matrix $\mathbf{K}$:

$$\mathbf{K}_i = \boldsymbol{\omega}^{\text{f}}(t_i) \left(\boldsymbol{\omega}^{\text{obs}}(t_i) + \boldsymbol{\omega}^{\text{f}}(t_i) \right)^{-1} \quad (18.67)$$

is time-dependent and looks similar to $\mathbf{W}^{\text{OI}}$ from Equation (18.59).

Unfortunately the extended Kalman filter estimate for $\boldsymbol{\omega}^{\text{f}}$ (18.64) is quite expensive to compute for large models. The problem is the size of the $\mathbf{L}$ matrix, which scales with the number of state variables N_{model} in the model $\mathcal{M}$. In parameter optimization we could limit the number of adjustable parameters but now $\mathbf{L}$ is an unwieldy matrix roughly $N_{\text{model}} \times N_{\text{model}}$ in size, which makes Equation (18.64) approximately equivalent to integrating the model $\mathcal{O}(N_{\text{model}})$ times. For an ocean GCM this quickly becomes intractable. Several simplifications have been proposed. In *ensemble Kalman filtering*, a few dozen model forecasts with small offsets are integrated in parallel. The forecast error covariance is then estimated from the ensemble member spread:

$$\omega_\ell^{\mathrm{f}} \approx \frac{1}{\mathcal{L}-1} \sum_{\ell=1}^{\mathcal{L}} \left(\mathbf{x}_\ell^{\mathrm{f}} - \bar{\mathbf{x}}^{\mathrm{f}}\right) \left(\mathbf{x}_\ell^{\mathrm{f}} - \bar{\mathbf{x}}^{\mathrm{f}}\right)' \tag{18.68}$$

where ℓ is the ensemble index, $\mathcal{L}$ the number of ensemble members, and $\bar{\mathbf{x}}^{\mathrm{f}}$ the ensemble mean. While promising, this approach tends to underestimate the true forecast error covariance, and solving this is an active area of research.

19

Scientific visualization

It's not what you look at that matters, it's what you see.

Henry David Thoreau

19.1 Why scientific visualization?

Throughout this book we have used a number of MATLAB's graphical capabilities as tools to monitor the progression of our mathematical and numerical travails, to demonstrate some characteristic of our results, or to reveal underlying relationships in data. Our emphasis now will be on the basic process of *scientific visualization* and providing you with some advice on how to effectively use (and not abuse) the many tools available to you. You may think that scientific visualization is an easy and natural thing to do, especially given the relatively powerful and reasonably intuitive tools built into MATLAB and other "point and click" packages so readily available. However, in the many years that we have been attending conferences, reading journals, and perusing text-books, we have encountered some ghastly instances of computer graphics abuse (or more to the point, abuse of the poor viewer/reader). This is a shame, because invariably the presenter has worked hard, often under difficult circumstances, to acquire scientific data, execute a model, or discover an erstwhile hidden relationship . . . only to fail to communicate the final result effectively. After all, isn't *communication* the final end-product of all our scientific endeavors?

We could also regale you with the awe-inspiring size of today's huge data sets, but rest assured that tomorrow's will be even more impressive. While the advent of more data offers new opportunities, burrowing into such large data sets and coming out with useful, quantitative information can be a daunting challenge. With the continued growth in the size and complexity of numerical models, ever larger observational data sets, and the advent of computationally capable computers equipped with fast graphics engines, we have seen a proliferation of scientific visualization tools. These tools range from publicly available freeware packages (such as Generic Mapping Tools and Ocean Data View) through commonly used graphical programs (e.g. Excel, Origin), to mid-range commercial products (including MATLAB and Surfer), to even more sophisticated (and expensive) "high end" programs (like FlederMaus, AVS, and Iris Explorer). We will dwell mostly on using either MATLAB or when it's easier the public domain programs, although you may see fit to invest both financially and with time and effort in other tools. But we hope that in using them, you heed the advice we offer in this chapter.

We will give you some basic guidelines and tips to improve the way in which you communicate your science (both to yourself and to your colleagues), and to point out some basic pitfalls that you ought to avoid when working with scientific visualization. Please be aware that scientific visualization is by itself a burgeoning field of endeavor, and we shall only walk around its periphery while hoping to avoiding some of the bottomless chasms. But we hope at least to provide you with a minimum set of skills necessary to get your message across effectively. There are many texts that can serve as useful resources in this area. In particular, a series of books by Edward R. Tufte provide a compelling and entertaining guide to many aspects of scientific visualization. We've listed four of his books in our reference list, and you might be interested to know that he occasionally holds a workshop on the topic.

19.1.1 *Motivations for scientific visulization*

You use scientific visualization techniques in a variety circumstances, but ultimately for the same reason: communication. In the early stages of the endeavor, you will likely be communicating with yourself. That is, you may be monitoring model behavior or experiment progress, or after the fact exploring the nature of a data set. In the middle of a project, you may be sharing your ideas or discoveries with your collaborators, or convincing them (and yourself) of the robustness of the results, or perhaps building a case for changing experiment design. Near the end of your project you will likely be reaching a broader scientific audience, during a seminar, on a poster, or in a scientific journal article. Your graphics might be the key tool for convincing a reviewer of a proposal or a funding agency's program manager that your conclusions are sound, important, and worthy of further research. You may even be working with the popular press and communicating with the general public. Each of these activities places different demands on your graphical communication skills, but there are many common elements that deserve your attention and respect.

From the viewpoint of scientific visualization, we will make no distinction between observational data and output from numerical models (the latter sometimes called *computa*). We hope that over the course of this book, we have led you to the understanding that the two share similar underlying characteristics and limitations. Neither are truly absolute, but come with limitations such as statistical uncertainties or perhaps more fundamental ambiguities.

The principal objectives of scientific visualization may be thought of as:

- **Monitoring** the progression of some numerical computation (model), observation sequence, or experiment. Often the effective representation of the evolution of a system provides valuable insight into how it is (or is not) functioning. This provides an opportunity to correct problems interactively, or at least to understand and accommodate the limitations of the results.

- **Data exploration**, which may range from simply compacting a large data set into a more efficiently "browsed" format through examining data sets for patterns or relationships, often with preconceived ideas about what these relationships might be, but sometimes simply searching to see if there are any patterns in the data.

- **Feature extraction**, which in some respects characterizes the other end of the motivation spectrum, in that you know what you're looking for, and you want to show yourself and your collaborators this relationship in an informative and convincing manner.

- **Collaboration**, which, especially at a distance, can benefit from good tools for individuals to remotely and collaboratively interact with data. Well designed graphics can be a potent tool for efficiently communicating complex or volumous information.

- **Presentation graphics**, where you strive to illuminate the underlying functionality of the system, either for pedagogical reasons, to convince your audience of a basic "truth", or to bring them along some logical pathway toward some desired conclusion.

Many experts make a strong distinction between data exploration tools and presentation graphics. Their reasoning is that the latter is not true scientific visualization, but we would argue that there are more elements in common between data exploration and presentation than there are differences. The important point is that as a practicing scientist, you will be doing both. Effective use of presentation graphics in communicating the ultimate results of your work is as important as efficient utilization of scientific visualization tools in data exploration.

19.2 Data storage, manipulation, and access

Data management is a large and highly evolved field, and we won't delve into it here, except to say that large or complex data sets often require special strategies for efficient storage and effective access. For instance, the NCEP Reanalysis climatology (Kalnay *et al.*, 1996) used as boundary conditions for the one-dimensional upper ocean seasonal model in Chapter 15 was contained in netCDF files. A self-describing file structure is provided in netCDF files, which stands for NETwork Common Data Format. That is, it contains the built-in information necessary to identify the coordinates, dimensionality, organization, and scope of the data. In addition, the data are stored in as compact a form as possible while maintaining appropriate resolution. Prior to MATLAB release 2008b, there is no built-in support for reading netCDF files. If you have an older version, you can install a package authored by Chuck Denham, of the US Geological Survey, which we describe in Appendix A.

It's useful to be able to browse the contents of netCDF files. There is a MATLAB-based browser in the package mentioned above. In addition, a program named *Panoply*, available at `http://www.giss.nasa.gov/tools/panoply/`, is an intuitive and useful tool for browsing and visualizing geo-gridded and time-sliced netCDF files. As a Java-based program, it requires the Java run-time on your computer, which is also freely available on the web (e.g. at `http://www.java.com/en/download/manual.jsp`) if you don't already have it on your computer.

Metadata is an important concept to be aware of. In the case of model output, one needs to know certain information in addition to the model output. These include the date of the simulation, the model version, the choice of parameters, the boundary conditions, the resolution and nature of the algorithm, etc., as they have some impact on the nature and quality of the model results. With experimental data, details about the date and location of the sampling site, the sampling and analytical methodology, etc. are relevant: for example, what type of sampling device was used (Niskin bottle, *in situ* pump, or GoFLo bottle), were the samples filtered and/or preserved, was a photometric end-point used in a titration, what reagents were used? Which laboratory or investigator made the measurements? In addition, the analytical uncertainties may vary from measurement to measurement. The Biological and Chemical Oceanography Data Management Office (BCO-DMO at `http://www.bco-dmo.org/`) in the United States is an example of a metadata-cognisant data repository and management system that houses a number of data sets along with useful documentation and metadata online.

HDF, which is short for *Hierarchical Data Format*, a multi-object data format that contains both data and metadata, was created by NCSA (the National Center for Supercomputing Applications) for efficient storage and platform independent exchange of large data sets. In addition to a defined data format strategy, the HDF standard includes a collection of software utilities, a software library, and an application programming interface (API) specification. A complete description of the format and formal specifications is available from `http://www.hdfgroup.org`, and from `http://www.hdfgroup.org/products/hdf4_tools/` you can obtain useful utilities for file browsing, editing, and conversion. In addition, as of its most recent release, MATLAB has both high- and low-level access commands for HDF5 (the most recent version of HDF). There are also limited (low-level) capabilities in MATLAB for the earlier (HDF4) version.

In addition to these structured data files, many data collections such as WOCE/CLIVAR (World Ocean Circulation Experiment/Climate Variability) data sets at `http://whpo.ucsd.edu` also offer data formats arranged in "ASCII flat files" compatible with spreadsheets. They come with *data quality flags* that are simple integer indicators for data integrity. For example, a value of 2 indicates "no reported problems", while a value of 7 means the data is somehow questionable. Also in this instance, null (missing) data, either resulting when the property was not measured or because there was an analytical failure, are represented by a quality flag of 9, and an agreed on *null value* of -999 in the data column. Cognisant visualization programs, like *Ocean Data View v4.0* (see Section 19.5), can be used to effectively eliminate "flagged-as-bad" data points. It is important to use these flags to eliminate suspect data. We know of at least one instance where someone ignored them and eventually had to publish a correction to a *Nature* paper. How embarrassing!

More complex data sets may also come in the form of, or be suitable for storage as databases. Relational databases can be useful when there are numerous data types with varying degrees of resolution and inter-relationships. The BCO-DMO is built on a relational database foundation. For example, the entire CLIVAR/WOCE data set could be combined into a relational database that permits structured queries (using *SQL*, a *structured query language*) designed to select pertinent subsets of the data for subsequent visualization. The ability to navigate intelligently through large data sets is a form of *data mining*. Although data mining has traditionally been associated with business and commercial development (e.g. looking for market trends in business, purchasing patterns in supermarket sales), these tools can in principle be applied to scientific endeavors. Methodologies include classification, discriminant and cluster analysis, machine learning, and regression techniques (e.g. see Chapter 4). Publicly available programs such as *Weka*, *RapidMiner*, and *Scriptella ETL*, are written in *Java* and can be used on a variety of data files (including *XML*) and databases. XML is short for *Extensible Markup Language*, and is basically a language that resembles HTML (*Hypertext Markup Language*, the language used to create web pages), but is designed to wrap text-readable data files in a self-documenting framework.

19.2.1 A note on data gridding

Visualization, especially in multi-dimensional cases, requires data to be organized in a more-or-less regular grid. Fortunately, model simulations almost always produce output on a regular space-time grid, which greatly simplifies the visualization process. The same may be true for well-designed laboratory experiments. In real-world observations, however, data are usually distributed irregularly in space and/or time, and are often populated sparsely in one or more dimensions. It is tempting in visualization to interpolate sparse data onto very fine grids and to apply statistical smoothing to avoid discontinuities (jagged lines) and create visually appealing graphics objects, but danger lurks in apparently robust features that in fact arise from *spatial aliasing* (see Chapter 6). Be warned that although seductive imagery can be achieved this way, you must ask yourself whether you are communicating data or artwork.

Most visualization programs come with gridding tools, and MATLAB is no exception. In Chapter 7 we discussed the concept of gridding, mapping, and contouring of data, and made you aware of some of the challenges and subtleties associated with this activity. As a general rule, when gridding your data, choose a resolution that is appropriate to your data spacing and number. For example, if you have a section consisting of 30 stations with 36 depths, then a corresponding data grid might have around 1000 nodes. If it has an order of magnitude fewer or more, then you will be under- or over-gridding your data. With most "point-and-click" software packages, the built-in gridders tend to be robust, well-behaved algorithms that produce relatively smooth gridded products, but beware of the potential for intrinsic algorithmic effects. Spline routines, for example, can generate propagating anomalies (alternating minima and maxima) at the grid spacing. It is important to keep these issues in mind when you work with the resultant images because the manner in which you create gridded data can greatly influence the nature of the output. Given the sophistication of graphics technology, the images created can be sufficiently compelling that you may delude yourself that a gridding artifact is reality.

19.3 The perception of scientific data

Of the five senses, vision is undoubtedly the most information-intensive. The transformation of light reaching the eye into neuronal signals and subsequently to a mind-image of the scene being viewed is a complex process. It is not a simple "raster building" construction of an image, but rather a convoluted process of assembly of components, curves, and inter-relationships that builds an internal model of the scene being viewed. This is why "optical illusions" can be constructed that fool the eye. We mention this for two reasons. First, you don't want to create graphics that inadvertently fool the eye, i.e. that have unintended consequences. Second, you may wish to use the nature of visual perception to communicate information effectively, in the least ambiguous fashion. Not surprisingly, some treatments of scientific visualization start from, and concern themselves with, the physiology and psychology of human perception.

Although data sets are rarely restricted to one or two dimensions, the primary medium that we mostly deal with is two-dimensional, either the computer screen (we include the seminar room projection screen here) or the printed page. There are three-dimensional virtual reality immersive environments (holo-desks and CAVEs), but most of you won't have the opportunity to work with these extensively (at least not yet). Thus we must effectively use the two-dimensional medium to communicate multi-dimensional information. Strategies include 3D perspective plots, color-mapping, contouring, animations (i.e. using time as the third dimension), or a combination of two or more of these. Being able to move a 3D object on a computer screen provides important and intuitive visual cues as to the structure of the object.

The 3D perspective plots can involve simple line-drawing, or be further embellished with *scene rendering* methods, whereby object surfaces are created and manipulated in a virtual environment. These surfaces can be *texture mapped* (i.e. painted) with other variable information, including color and/or contours, and visual cues created with lighting models. The lighting models can be used to create shading that guide the eye to perceive relational structure associated with the object being viewed, and the object's reflectance (including the fraction of *specular reflection*) can be used to provide additional cues. Many of these features can be effectively manipulated in the more sophisticated software systems (e.g. AVS and IRIS Explorer) to provide remarkably vivid renditions of model output and multi-dimensional data sets. Effective rendering (displaying) of such objects is compute-intensive and requires higher-quality graphics engines (dedicated processors and memory). Clearly such tools are not only expensive, but also challenging to learn how to use. Moreover, it may be that unless the result is well designed and intuitive, the viewer will also need to be trained to fully understand the features being portrayed.

19.3.1 Color

Coloring your graphics can add another dimension to your presentation. You will have noticed that we have chosen (for reasons of economy) to restrict the graphics in this book to "black and white", or at least kept it to a gray scale. This is often the cheapest option in traditional (paper-based) publication; many journals charge extra for color reproduction. As we transition to web-based journal publications this no longer is a constraint. However, color, like any other tool, can be misleading if not used wisely. It is therefore important to understand it and how it is perceived.

Depending on the technology involved, there are three basic ways of referring to color. The one most closely associated with computer displays is the *RGB* or "red–green–blue" system. In this system colors are *additive*: that is, you construct colors by adding two or three of the end-member colors together. This works for displays where a triplet of small light emitters (LED, LCD, or phosphors) are sufficiently close together that the eye perceives them as a single light source, often referred to as a *pixel*. Any conceivable color is represented by a triplet of values corresponding to amount each of red, green, and blue respectively (hence the term RGB). In MATLAB the values range between 0 and 1, while in some systems they vary between 0 and 255 (corresponding to 8 bits full scale) for 24-bit colour systems. Using the MATLAB terminology, the color intensities are scaled between 0 and 1 such that [0,0,0] is black (no color at all), and [1,1,1] is white (the sum of all colors). You also have red = [1,0,0], blue = [0,1,0], and green = [0,0,1]. Other "terminal" binary mixtures include cyan =[1,1,0], magenta =[1,0,1], and yellow =[0,1,1]. You may notice that these color combinations are not what you're used to when mixing paints (more on that later with the CMYK system). Other colors can be arrived at by graduated mixtures of all three colors.

A useful prop for thinking about RGB colors is to use a *color cube*. We suggest you visit the GMT website at http://www.soest.hawaii.edu/gmt/gmt/examples/ex11/gmt_example_11.html (see also Section 19.5) and construct the RGB color cube. You will see the colors we just mentioned on the corners of the cube, and a fraction of the visible color spectrum associated with binary mixtures along the faces. Other colors, associated with ternary mixtures and not visible in the example, are embedded inside the cube. Bear in mind that color displays are not created equal. There are the old-fashioned CRT displays (rapidly going the way of the stegosaurus, we think), LCD, plasma, and more recently LED displays. Within each you'll find a range of quality, directly related to price, and performance. Even units from the same manufacturer with the same model number may differ significantly. Professional graphic artists often purchase expensive computer–display–printer–software systems that are calibrated (using Pantone calibration monitors) for optimal color fidelity. Even so, the entire gamut (color range) of visible colors cannot be accessed from RGB space. Some have to be left out.

A second system to specifying color is *HSV* or *hue-saturation-value*. Historically, this system evolved from the development of color television and the need for its broadcast signal to be "backward compatible" with the many black and white television sets still in use. In this three-coordinate system, you can also navigate color space, but in a different way. *Hue* describes the specific color (red, green, brown, orange), while *saturation* refers to its intensity (a low saturation tends toward pastel, a high saturation means more vivid), and *value* corresponds to brightness. Thus black and white televisions worked off the last variable alone and still presented a useful gray-scale picture. Typically one represents HSV color space in a conical or cylindrical map as an equivalent to the RGB color cube.

The *CMYK* color space (an abbreviation for *Cyan–Magenta–Yellow–blacK*) was developed for ink-based technology. As such, it is a *subtractive* color-mixing scheme that works very much like your paint mixing experience in kindergarten: cyan and yellow mix to produce green. In principle, black ink shouldn't be necessary as it can be produced with an appropriate mixture of all of the other colors, but having a dedicated black ink source saves a lot of ink in printing because many graphics rely extensively on black (including text). It also turns out that not all colors available in RGB space can be produced with CMYK combinations. Moreover, unless you have some very clever software and an expensive, calibrated computer display, what appears on the printed page depends on both the limitations of CMYK mapping and your printer technology. As a general rule, you'll find that printed colors tend to be more intense and darker, especially with blue hues, and that some experimentation will be necessary to achieve the desired effects.

Finally, we should mention that publishers may require that you submit your graphics, especially bitmapped files, with a specified minimum resolution and using a specific color model. Many high-end programs, like Adobe PhotoShop, allow you to change between color models. You will be warned as you do this that you may have colors in your image "outside of the target color gamut", which means that some colors will be changed. MATLAB also allows you to specify the color model when you use the print command.

We end this color discussion with a caveat; color is a useful tool, especially for codifying additional information in a two-dimensional plot. Like any tool, however, it can be misused, and you need to be aware not only of the physical limitations and characteristics of this tool, but also the perceptual and psychological aspects. People react to some colors instinctively, or the colors may "map" in some unexpected way. For example, red may represent passion or rage, light blue map to serenity, yellow to warmth, green to growth or fertility. Are these messages you want to transmit with your scientific data? We return to the subject of color in Section 19.6 to offer further advice.

19.3.2 Some specialty plots

Along the way, we've used a variety of different plotting strategies to look at our data and model results. For example, the power spectrum plots in Chapter 6 inform us about what frequency or wavelength band contributes the most to a noisy signal. We thought we'd mention a few other plots that have proved useful. There are obvious ones, like histogram and bar plots, but here are some that you may run across and that might bear looking into:

- **Cumulative probability plots** – We introduced you to the cumulative probability plot in Section 2.2.4 and just remind you that this is a useful tool for looking at population statistics and evaluating whether things are *normal*. We won't repeat the plot here, as you've seen it before.

- **Box and Whisker plots** – When you want to show something about the temporal or spatial evolution of the basic statistical distribution of a property, this is a useful non-parametric tool that goes beyond error bars. Rather than plotting the mean and standard error of each observation set, which presumes a normal distribution, with a box and whisker plot you plot five key values at each distribution: the lower extreme range, the lower quartile, the median, the upper quartile, and the upper extreme range. The extreme range is defined as 2.5 times the distance between the adjacent quartile and the median. The quartiles form the upper and lower ends of a box and the median is a horizontal line inside the box. The extreme ranges are drawn like error bars. Additionally, outliers are plotted as points (generally crosses). A potentially important aspect of this mode of presentation is that it is *non-parametric* in that it makes no assumptions about the underlying statistical distribution, while at the same time it reveals important information about its character. The viewer can immediately infer not only the basic trend of the median, but also the evolution of the spread in the data and variation in the *skew*, a measure of the distribution's asymmetry (see Section 2.2.2).

  An example box and whisker plot can be seen in Fig. 19.1, which shows the trend in the variation of 6-hourly NCEP wind speeds near Bermuda for monthly periods for 2005. While we could wax lyrical about the structure of the seasonal trends in the data, we'll just point out that to the trained eye there's a wealth of information in this kind of plot.

- **Waterfall plots** – When you want to characterize the temporal evolution of a one-dimensional profile (for example temperature), you can overlay the plots, but separate their values by some offset as a function of time. For example, in a two-dimensional case where you were documenting the onset of seasonal stratification, you could plot the first profile, then plot the next one offset by 0.1 degrees, and the third one by 0.2 degrees, etc. A slightly more "glamorous" approach is to plot the data in three dimensions as a mesh plot (see Fig. 19.2). In this instance, one uses the added spatial dimension to represent time, and obtains a ready visual impression of the temporal evolution of the profiles.

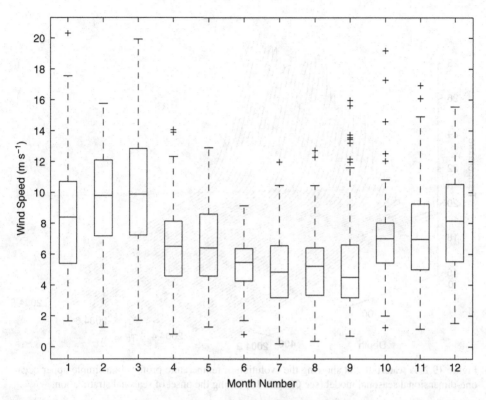

Figure 19.1 A box and whisker plot of monthly binned 6-hourly NCEP winds near Bermuda for 2005.

- **Hovmöller diagrams** – These are special versions of contour plots where data are contoured with one spatial dimension along the x-axis and time along the y-axis. This kind of plot is useful for showing time–space propagation of features. As an example, we show in Fig. 19.3 a Hovmöller plot of Topex Poseidon satellite radar altimetry data along $5°$ N in the Pacific (data obtained from NASA/JPL). Although the plot looks rather "busy", a clear trend virtually leaps out of the page at you. The contours consistently show an oblique orientation (sloping upward to the left in the diagram) that is indicative of a westward propagation of sea surface height anomaly. The slope is equivalent to about a 6-month basin-scale transit time, or a propagation velocity of about $1\,\mathrm{m\,s^{-1}}$. Moreover, if one makes a cross-sectional "cut" perpendicular to this trend, an approximately annual periodicity becomes evident associated with the troughs and crests in the sea surface height displacements. The full spatial data could be visualized as maps progressing in time through some kind of animation, but by reducing the dimensionality of the system the Hovmöller plot produces a clear quantitative view of underlying correlations.

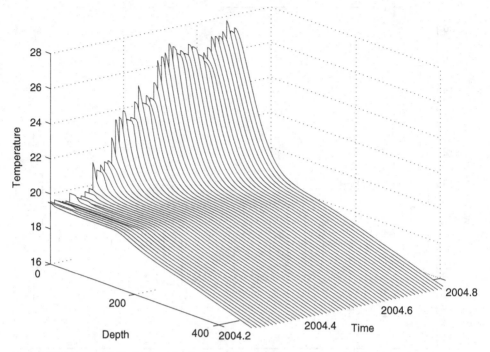

Figure 19.2 A waterfall plot showing the evolution of temperature profiles in a simple upper ocean one-dimensional seasonal model (see Chapter 15) during the onset of seasonal stratfication.

19.4 Using MATLAB to present scientific data

The early versions of MATLAB had rather primitive two-dimensional graphics capabilities, and the products were far from publication quality. This has changed. MATLAB now has many powerful data representation and visualization tools built into it, and for most of your needs, it constitutes an almost complete visualization environment. Most importantly it offers the advantage of being tightly coupled to your modeling and calculating tools. Throughout the course of this book, you have already gained experience in working with these tools – in fact most of the figures in this book were originally crafted using MATLAB.

We provide below a broadly categorized list of some of the more useful MATLAB tools, along with a brief descriptive sentence or two. You will already be familiar with a number of these. You can use the relatively extensive documentation associated with the doc command to gather more information as you need it.

19.4.1 Graphics control commands

There are some general graphics commands that manipulate graphics objects and displays. You can get a moderately complete listing with doc graphics. These include:

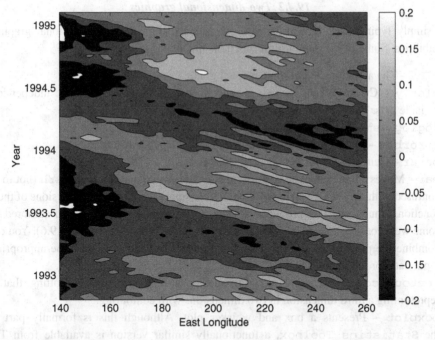

Figure 19.3 A Hovmöller plot of sea surface height anomaly from the Topex Poseidon satellite altimeter data set in the Pacific along 5° N between approximately 140° E and 100° W (260° E) over a three-year period in the mid 1990s. Contour interval is 0.10 m and negative anomalies are shown as dark (see grayscale bar on the right).

- figure – Create a separate graphic window, or if it already exists, display it and make it the focus of your script commands or keyboard.
- clf – Clear the current graphics figure.
- subplot – Create or make the focus a subplot within the current figure. This command can be used simply to place subplots in an array on the page, or can be used to specify the actual plot location and size.
- axis – Manipulate axis scaling and appearance.
- hold – Enables you to add subsequent graphics/plots to the current plot rather than clearing the plot each time.
- gca – Obtain the current graphics handle. This allows you to query (get) or change (set) specific characteristics of the graphics object.
- zoom – Allows you to zoom in on a plot.
- grid – Show a grid on the plot.
- title, xlabel, ylabel – Make a title or axis label.
- text – Place text on the plot at a specified location, or with a mouse-click (gtext).
- uicontrol – Create an interactive user interface control.
- showplottool – Brings up a useful "point and click" interface for visually modifying your graphs.

19.4.2 Two-dimensional graphics

By simply typing doc graph2d you can get a listing of two-dimensional graphics capabilities. Some of the more useful functions include:

- plot – The archetypical two-dimensional graph.
- plotyy – Creates a plot for two dependent variables against the same independent variable axis. The second variable axis is on the right.
- loglog, semilogx, semilogy – Produces log or semilog plots.
- errorbar – Makes a plot with error bars for the dependent variable.
- polar – Generates a plot in polar coordinates.
- bar – Makes a bar chart. You can also make a horizontal bar plot with barh (not to be confused with a dyslexic version of Planck's constant). There are "3D" versions of these functions, much like that produced by that popular spreadsheet program authored by some west coast company, but we advise against doing that (see Section 19.6). You can combine the preparation for a histograph with hist, which produces the appropriate input data for the bar plot.
- contour, contourf, clabel – Provides quasi-three-dimensional plotting, that is, representing a third dimension in two dimensions with contouring.
- boxplot – Presents a box and whisker plot. Although this is formally part of the Statistics Toolbox, a functionally similar version is available from The MathWorks website.

19.4.3 Three-dimensional graphics

MATLAB is relatively replete with three-dimensional graphics capabilities. Some of these are:

- plot3 – Plots lines and/or points in 3D space.
- contour3 – Produces a contoured volume in three dimensions.
- surf, surfc, mesh, meshc, meshz – Alternate representations of 3D data using colored surfaces or meshes.
- quiver – Allows you to plot a field of two-dimensional vectors by placing arrows at the grid points whose length and direction correspond to the vector value at that location.

There are a number of tools or controls specific to three-dimensional plots. You can use these to programmatically set up the presentation of your data, or perhaps to produce a crude animation (for example through progressive manipulation of camera angle) to highlight the structure of the data. These include:

- view – Sets the camera view angle and location.
- rotate3d – Enables you to manipulate the object interactively by rotation using your mouse.
- surfl, lighting – Modifies the lighting applied to the graphics object.

- diffuse, specular, material – Sets the nature of the object's surface reflectance. This can highlight (or hide) certain features of the object's shape.

19.4.4 Exporting graphics

Having labored extensively to produce your computational and graphical masterpiece, it would be nice to preserve it for posterity. You can do this by either saving the graphics to a file or printing it. Both involve the same basic issues. One of the key points to consider is whether the graphics will be saved in vector or bitmapped mode, and where possible we advise the former (see Section 19.6) as it preserves maximum resolution. In either case, the generic MATLAB instruction is print followed by a format specification (see the list below) and optionally a file name. If no file name is specified, the graphic is printed. The print or file format is prefaced with a "-d" such that the command print -djpeg MyGraph.jpg saves the graphic in jpeg (a compressed bitmap format) in the file MyGraph.jpg. File extensions are optional, but we advise that you follow convention such that they represent the content. Not doing so can lead to confusion (naming a bitmapped file as FileName.ps doesn't make it a postscript file, but your computer operating system will try to open the file up with a postscript interpreter and may become rather nasty with you). Here are some format options (we won't include the "-d"). The first four are vector graphics formats while the remainder are bitmapped.

- ps, ps2, psc, psc2 – Generates postscript output (a vector representation) in black and white or color ("c"). The "2" indicates "level 2" postscript, which is more capable and efficient. Early printers couldn't handle it.
- eps, eps2, epsc, epsc2 – Generates encapsulated postscript, which contains "wrapper" information that makes it suitable for inclusion in documents.
- pdf – Produces a postscript-like format (page description format) that is almost universally readable.
- ill – Generates vector graphics in a format readable by Adobe Illustrator.
- bmp – A kind of "raw" bitmapped format that is uncompressed, so preserves the resolution it was "painted at", but tends to produce large files.
- jpeg, tiff, png – These are compressed bitmapped formats. Depending on the nature of the image and the compression settings, some quality may be lost (e.g. lines may be blurred or sharp edges fuzzy), but files will be generally much smaller, which is useful for presentation on web pages.

There are numerous others, and we suggest you use doc print for more information.

19.5 Some non-MATLAB visualization tools

While MATLAB offers a relatively complete range of visualization tools that nicely integrate within the modeling environment and can produce publication-quality graphics, there are other tools that provide complementary capabilities. We list some that we have had good success with, but recognize that there are many, many others available. We also hasten to add that mentioning a commercially available product does not constitute an endorsement on our part.

- **Generic Mapping Tools**, also known affectionately as GMT, is a freeware package aimed at manipulation and presentation of spatially organized geophysical data. It was created by Paul Wessel and Walter Smith, and is available from `http://gmt.soest.hawaii.edu`. The package is somewhat large (especially if you select the high-resolution coastline options) but installation is relatively painless on most computer platforms. GMT is endowed with quite high-resolution coastline mapping, a wide selection of many mapping projections, extreme flexibility, and robust gridding capabilities. If this sounds too good to be true, there is a drawback. GMT is not a *WYSIWYG* (What You See is What You Get) "point and click" graphics program, but rather a collection of executable programs that you control via the computer operating system's command line interface. For example with a Windows PC, you would build a text file with the `.bat` extension, and run it from a Command Prompt Window. The programs are used to sequentially build color maps and gridded files, and to build a postscript output file (your graphics output). The behavior of these programs can be controlled with exquisite precision by numerous alphanumeric strings appended to command-line options, although it takes some experimentation and learning. Ample documentation and example scripts are provided with the package, and although a little daunting to learn, it is well worth the effort.
- **Ocean Data View** is a very capable, easily used, freely available visualization package created and maintained by Reiner Schlitzer. It interfaces nicely to a number of oceanographically important data sets (World Ocean Atlas, the WOCE hydrographic data set) and produces publication-quality profiles, sections and maps. Moreover, ODV contains a number of color maps that are commonly used in oceanography, thus producing graphics that are readily recognized by oceanographers. The program incorporates a number of features for data management and can import data in a number of formats. It is available for download at `odv.awi.de`. Unlike GMT, it has a "point and click" interface and allows interactive multi-variable analysis, with integrated tools for data gridding and interpolation, finding outliers, volumetric estimates, etc. It has built-in capabilities for mapping properties onto isosurfaces such as potential density anomaly horizons. It is a highly usable tool that can be invaluable for cruise preparation, on-board data analysis and publication.
- **FlederMaus, AVS, and IRIS Explorer** are examples of commercial software packages that allow the exploration and visualization of larger multi-dimensional data sets.

These programs are well suited to the analysis of model data arranged on some kind of regular grid. They have built in tools for *scene building*, whereby 3D *geometries* are constructed to represent aspects of the data. Once these geometries are assembled, they are *rendered* (converted to pixels on the computer screen) based on some presumed relationship between the geometric structure, one or more light sources, and the hypothetical *camera* with which you are viewing the scene. For example, you could represent a 3D scalar field (say of salinity in an ocean basin) by an *isosurface*, which is the 2D analog of a contour line. You can then texture map (paint) onto this isosurface a color-mapped representation of another 3D scalar field (say dissolved oxygen) along this surface. You can then represent a 3D vector field of water velocity with arrows emanating from the spatially distributed model grid points, with the arrow lengths proportional to velocity.

But given this "scene", how do you really comprehend its structure based on a 2D projection onto your computer screen? These visualization packages provide you with manipulation capabilities, allowing you to use your computer mouse to rotate the scene, zoom in (or out), or even "fly" over or through the scene. This imbues the visualized objects with full 3D characteristics and enables you to manipulate and examine the object as if it were in your hands. Moreover, most visualization programs offer the ability to control how you see the scene using lighting and reflectance models. Just as you might adjust the lighting in a photographic studio to bring out the best (or more robust) features in your subject, you can use lighting to control what you see and how you see it. Adjusting the nature of the surface reflectivity (e.g. specular versus diffuse reflection) tunes the observer's eye toward different characteristics of the shape of the surface. Another important visual cue is *perspective*, where the rendering software transforms the observed scene so that objects further away from the camera appear smaller.

Finally, one can construct *animations* from the recording of video sequences for demonstration purposes. Much as you gain insight into the scene's structure by manipulation, "automating" this manipulation allows you to reliably reproduce an optimal sequence of manipulations to efficiently make your point. These videos can then be displayed on the web, or perhaps as part of a scientific presentation or even a paper. One goes about doing this by using manual scene manipulation to arrive at a series of "snapshots" that serve as *keyframes* for the program. Each one represents a unique relationship between the scene and the *camera* (really your eye). Once you have finished collecting them, the program can use these keyframes to navigate smoothly from one keyframe to the next, adjusting the camera and scene orientation.

If used wisely, such tools are extremely effective in presenting compelling graphics. Bear in mind that not only do such software packages tend to be expensive, they usually demand high-end hardware, and take a lot of time to learn. However, they do have their place.

19.6 Advice on presentation graphics

There are some useful strategies that make for effective communication of your results. We have gleaned these from a combination of personal experience and reading the literature. There is some inevitable overlap between these points. Like any "rules", there may be times where you cannot avoid breaking some of them, but keeping this advice in mind may help you make better decisions about how to get your story accross.

- **Know your intended audience.** Always bear in mind the nature, motivation, and background of your audience. Is this a group of experts, collaborators, or readership of a specialty journal? If so, the use of jargon, abbreviations, and information content can be exploited to make the most efficient delivery of your message. For instance, a group of physical oceanographers may be comfortable with a color-keyed depth–latitude section of salinity with neutral density contours overlaid, along with hatched areas showing regions of low potential vorticity. However, if you are publishing in a general science journal or communicating to scientific non-experts, or even to the general public, you cannot make many assumptions about what the reader knows about the topic. Hence the graphics in that case will be less information-compact and more self-explanatory.

- **Communicate your message, not your "style".** Although we all want to produce attractive graphics, be aware that you don't want to distract the viewer from your message with pretty and unnecessary symbolism. We don't want to produce visual atrocities, and there is a certain elegance to simplicity. If the viewer walks away from your graph with only a memory of the pretty fish symbols you used to designate the daily primary production estimates at your time series site, and not the imprint of the lunar tidal cycle on it, you've failed.

- **Avoid strong, complex, or pictorial backgrounds.** Don't ever put an unnecessary background on a journal graphic, unless it's going into National Geographic. This rule applies more to PowerPoint slides and poster presentations than graphics, but it bears emphasis. Too often, presenters place colorful or complex pictures on the background of their graphics that make it difficult, distracting, and downright tiresome to read or assimilate the real information. We have witnessed virtually unreadable text or graphics because the presenter saw fit to put in a very colorful but irrelevant polarized light petrographic thin-section, or a pretty picture of penguins, as a background. The presenter never used or referred to the thin-section (or penguins) during the talk, but it was the only thing we remembered about the presentation (aside from the migraine it induced).

- **Maximize the data/ink ratio.** Tufte (2001) makes a compelling case for clarifying your message by reducing graphics clutter. You can do this by removing any marks, symbols or lines that are not directly related to conveying data. Figure 19.4 shows tidal displacement over a 1-month period at a location on the east coast of North America. The inclusion of hatch marks and strong grid lines in the upper plot was intended to provide the reader with good quantitative tools for evaluating the precise value of each of the data points, but in fact makes it difficult to assess the underlying patterns. Removal

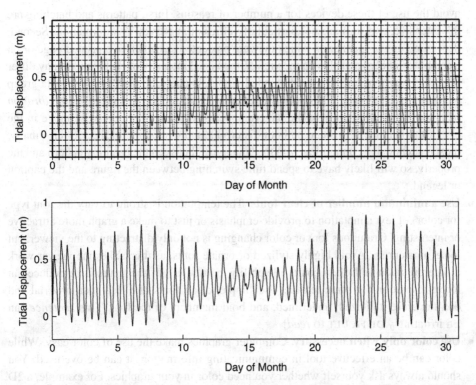

Figure 19.4 Two plots of tidal displacement at Woods Hole, MA for the month of May in 2005. Note the removal of excess ink (grid points) allows the reader to focus on the fundamental periodicity in the data.

of the hatches, lightening the axes, and slightly adjusting the ranges greatly clarifies the data patterns and allows additional useful annotation.

- **Don't use false perspective.** One of our personal gripes is the use of quasi-3D graphics when two dimensions are all that is needed. For example, the 3D bar charts used by certain popular spreadsheet programs ought to be avoided. Not only is it unnecessary, it is a waste of ink and distracts the viewer: the eye and brain must process and reject the extra lines and geometry. Sometimes, optical "inversions" can occur where the intended line-perspective appears reversed, leading to momentary confusion.

- **Avoid use of hatching or patterns.** Hatching or graphically embedded patterns have traditionally been used in geologic maps and figures. They have the advantage of conveying "type" information, and geologists have trained their eyes to quickly interpret conventional patterns (e.g. sandstone versus igneous outcrops). Such patterns are readily available in most graphics programs, but their use should be restricted to the representation of *nominal* data, that is, categorical data as opposed to *ordinal* and quantitative data (see Section 2.1.1). Outside of traditional geologic maps we think you should avoid the use of these devices for a number of reasons. First, patterns and hatching are ink-intensive, and thus are not particularly efficient at purveying information. Second, depending on the manner on which they are presented (whether on a printed page or on a computer screen) and also on the viewer, the visual intensity of each pattern may differ in such a way that undue and unintended emphasis could be placed on one data group or feature over another. Third, certain patterns can introduce so-called *Moiré vibration* (e.g. see Tufte, 2001), where the pattern appears to "dance" or otherwise move in the eye of the beholder. This can be both disturbing and distracting to the viewer. Finally, most readers will be unfamiliar with the "mapping" between the hatching pattern and the property, so will likely have to spend time switching between the figure and the caption or legend.

- **Use a minimum number of clear fonts.** The temptation is strong to vary the font type (or color) of text annotation to provide emphasis or just to make a graph more attractive or interesting. Gratuitous font or color changing is not only distracting to the viewer, but looks amateurish. Avoid highly stylized or ornate fonts, as this makes the viewer work at interpretation and often courts illegibility, particularly if the figure is reproduced at lower resolution for some reason. Use simple, clean fonts like Helvetica or Arial and don't overuse italicized, underlined, and bold members of the font. **This** *sentence* can be <u>irritatingly</u> DIFFICULT to *read*!

- **Use color only when necessary.** Computer graphics make the use of color easy. While color can be an effective tool in communicating information, it can be overused. You should always ask yourself whether you need color in your graphics. For example, a 2D line drawing or a simple contour map works much better in black and white, and may be easier to extract quantitative information from. Moreover, publishers often charge extra for color graphics.

- **If you use color, use an appropriate color map.** Bear in mind that different colors usually signify different things, and sometimes not what you intended (see Section 19.3.1). For example, the commonly used "rainbow" scale, which grades from blue/cyan through green, yellow, orange, and red/magenta has the unfortunate aspect that the colors in the middle of the spectrum (green–yellow–orange) tend to be the most vibrant and hence tend to be emphasized in the resultant image. Is this what you intended? The problem is compounded when you consider that a significant fraction of the readership may be color-blind. Light and Bartlein (2004) provide an (ahem!) enlightening comparison of how different color scales appear to such individuals. Their point is that the rainbow color scale performs far worse than simple two-hue diverging color scales. If multiple colors must be used, they recommend a modified spectral scheme. In fact, if a gray-scale or single hue (e.g. white to purplish-blue) color map will do, where the maximum corresponds to darkest (most intense hue) and minimum to lightest (white), use that.

For anomaly fields (i.e. containing positive and negative deviations from some mean state) we advise using a red–white–blue pattern (named "polar" in the GMT package, see Section 19.5), with blue associated with negative deviations and red with positive.

This color map is ideal for these circumstances because red and blue are more intuitively associated with positive and negative respectively, and the color saturation fades toward zero ink (white) for zero anomaly. GMT also offers the use of the *GEBCO* color map, which is particularly effective in presenting topographic maps. As an experiment, we suggest that you compare the effects of color map choice on an ocean basin relief map. Using the rainbow color map produces a rather surreal landscape image compared with GEBCO or grayscale.

- **Repetition is sometimes useful.** Although repetition is generally not encouraged in academia (have we mentioned this before?), it sometimes has its merits. An example where this may be warranted is the depiction of cyclical patterns, for example of seasonal warming in the subtropics. Figure 19.5 shows the annual temperature cycle in the upper ocean near Bermuda (upper panel). Repeating the data over a 2-year period (lower panel) gives a better view of the seasonal changes in temperature and mixed layer depth.

- **Use vector rather than bitmapped graphics files.** Be aware that the graphics file that you use to store your output, and that you send to the publisher or your colleagues, may be either bitmapped or vector graphics (or a combination of both). Most graphics programs, including MATLAB, let you choose output file format. While bitmapped graphics (really they're *images*) are fine for publishing on the web, vector graphics offer superior resolution in a generally smaller file. If you import a bitmapped file and then magnify

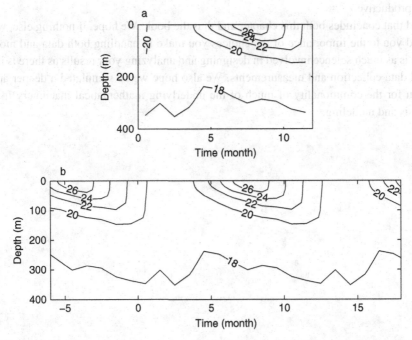

Figure 19.5 The annual temperature cycle at Bermuda shown (a) with no repeats and (b) with a half-year extension on both sides.

it, you eventually (depending on the resolution of the image) begin to see a "graininess" where once-smooth lines in, say, a letter become "steppy" or edges become fuzzy. This is not the case for vector graphics, where the detail and smoothness is preserved as resolution is increased. Examples of bitmapped files typically have `.gif`, `.tif`, `.jpg`, and `.png` extensions. You will typically view and edit these kinds of files with a program like *Adobe PhotoShop*. Vector graphic files include those with extension such as `.ps` (postscript), `.pdf` (page description format), and `.eps` (encapsulated postscript). Such files can be viewed and edited using a program like *Adobe Illustrator*. You should be warned that you may not be safe assuming files with these last three extensions are truly vector graphics, as these file formats can contain (i.e. act as a *wrapper* for) bitmapped graphics.

Although we've stressed scientific visualization as a form of communication, which is the final end-product of scientific research, we want you to view it as a necessary activity integrated within the practice of data analysis and modeling. Indeed, presenting and communicating your results (both to others and yourself) constitutes a link that closes the circle back to hypothesis construction, experiment design, data analysis, and numerical modeling and simulation. The very tools that you use for visualization – aside from the final rendering software – stem from the methodologies and techniques that are common to all of these areas. We hope that by familiarizing you with these methods, their strengths, limitations, and commonalities, your journey through this scientific process will be both easier and more productive.

And that concludes both this chapter and thus the book. We hope, if nothing else, we've altered you to the importance of the choices you make in handling both data and models. There is as much science involved in designing and analyzing your results as there is in the actual data collection and measurements. We also hope we've stimulated a deeper appreciation for the commonality of much of the underlying mathematical machinery in data analysis and modeling.

Appendix A
Hints and tricks

You know a conjurer gets no credit when once he has explained his trick; and if I show you too much of my method of working, you will come to the conclusion that I am a very ordinary individual after all.

Arthur Conan Doyle

A.1 Getting started with MATLAB

At the risk of being revealed as very ordinary indeed, we would like to show you some useful MATLAB and other numerical conjuring tricks, or at least give you some useful tips on how to get the most out of this powerful tool. We start by assuming that you have successfully installed MATLAB on your personal computer, or that you have found out how to access it on a shared system. The version you have access to might differ from the one we are currently using, but unless you are working with a particularly old version, small differences should not matter; the principles should remain the same.

Learning to use MATLAB is a lot like learning to use a bicycle. It seems very difficult at first, but after a while it comes so naturally that you wonder at those who cannot do it too. Like a bicycle, you learn to use it by *doing* it. There may be books on the theory of balancing on a bicycle or the gyroscopic physics of wheel motion, but all the reading in the world won't keep you upright without trying it and taking your lumps when you fall. Fortunately, the consequences of failure with MATLAB are far less painful, so get out there and ride!

When you start MATLAB you'll likely be presented with a compound window consisting of three sub-panes. The largest of the three is the *Command Window* where you enter your (ahem!) commands. Yes, Bertha, MATLAB is a "command-driven" program at its heart. This is primarily how you will be using it. Actually, "Command Window" is a bit of a misnomer, since it is really a "command and response window". That is, MATLAB presents you with results and information here as well. The other two, smaller panes are likely the *Workspace* (or *Current Directory*) and the *Command History* windows. More on these later.

Remember that MATLAB is a case-sensitive language. Variable and command names make a distinction between upper and lower case. So a variable named Alpha is not the same as a variable named `alpha` or `ALPHA`. This can lead to some confusion to those unfamiliar with this characteristic. Moreover, by convention all built-in MATLAB commands consist of all lower case letters. Bear in mind one historical (or is that hysterical?) quirk in MATLAB. When you see help documentation in the command window, the commands are fully capitalized for emphasis. Beware: the actual commands in usage should be in lower case.

　You enter a command in the command window by typing it and pressing the *Enter* key. If you're as bad a typist as we are, you may find it difficult to type more than two or three letters correctly without a mistake. Despair not, for if you have not pressed the enter key, you can use the arrow keys or the mouse to move the cursor back and forth in the line and make corrections. If you have pressed the *Enter* key and MATLAB has rudely responded with some sort of error message, you can recall the previous line with the *Up-Arrow* key, make your correction(s) and then press the *Enter* key again. As you execute your commands, whether correct or not, you will see them appear in the command history window in sequence. If you use the mouse and double click one of the entries in the command history window, it will be executed once again.

　The hard part about a command-driven program is that you have to know the commands to drive it. We introduce many of these commands as we go along in this textbook, but there are a few tools that you might find especially helpful. The ones that you should know about first are those that help you understand or look for other commands:

- The **help** command is the oldest and simplest-to-use help aid. If you simply enter `help` you'll get a long list of topics that you can get help for. The list is rather too long to fit in the window, so you'll have to use the scroll bar to move it up and down. If you prefer, you can turn on "more" by first entering *more on* (no, we're not calling you names, this is a hold-over from the UNIX days). When you enter `help` you will get just one "page" of information with `-more-` on the bottom. Pressing any key advances one line while pressing the space bar gives the next page. If you want stop at any time, press `Control-C` (hold the control key down while simultaneously pressing the `C` key). If you already know the command (aye, there's the rub!) but can't remember quite how to use it, you can ask for help on it. For example, if you want help with `rand` (a random number command) you would enter `help rand`. Most help output provides a description of what the command does and what arguments are required (and if some are optional), and provides examples. Often at the end of help text will be a short list of related commands. If they are blue and underlined, clicking on them gives you help for them. This is often a useful way to explore for new commands and functionality. Go ahead and discover!

- The **lookfor** command allows you to search for a term embedded in command documentation. For example, if you're looking for methods involving `decomposition` (breaking down a matrix into constituent parts), you would enter `lookfor decomposition`. Bear in mind that the search for this term may take a long time, but you will begin to see results appear in the command window. Since MATLAB may go on searching for some considerable time, you may want to terminate the proceedings with Control-C once you've got the answer you want. Once you have your prize, you can then use help to burrow deeper into the knowledge base. Be aware, also, that using too generic a term (for example lookfor matrix) may give you too much information.

- With the **doc** command MATLAB also has a more accessible and powerful tool. This provides you with a separate window containing a more intuitive interface that includes a structured tree (a visual table of contents) for navigating the documentation, a search capability, an index, and a list of demos (see next point). The largest part of the help window contains the help you requested. So for example you can use the table of contents to navigate to MATLAB, then down into some specific topic (say Cholesky Factorization) or Functions - By Category until you come to the command you are seeking. By clicking on this command, you raise the information in the help pane.

- By selecting the **demo** tab in the help window you will see a structured list of demos. The Getting Started with MATLAB group are a series of short video clips that you might find of value to get over the first hurdles. Subsequent items tend to be m-files (more on those in the next section) that guide you through some activity.

Finally, there are resources out there that we encourage you to explore. There are a number of MATLAB primers that are available (some for free, some at a price). The MathWorks website www.mathworks.com also contains a lot of useful information. Another is your colleagues: often you can learn how to get started with something by simply asking a more experienced practitioner (remember to return the favor when a less experienced newbie asks you something).

A.2 Good working practices

As we mentioned, MATLAB is a command-driven program. Sometimes you simply need to operate it that way to experiment your way through a problem; to see if something does or doesn't work. Other than in that kind of experimental mode, however, we tend to work systematically with m-files. These are text files that contain a sequence of MATLAB commands that it executes as a script. You typically build this file with the MATLAB editor, and you execute it by typing the name of the file in the command window. Note that these m-files by default end in ".m" (hence the name). To be executed, the file has to be in whatever directory MATLAB is working in, or somewhere in MATLAB's path.

MATLAB always has a current directory that is displayed near the top of the window. You can navigate to any directory on your computer, using the controls related to this display. One especially convenient feature is the drop-down list of previously occupied directories that you can use to get back to the last few things you were working on. We also advise that you set up a special, hierarchically organized directory structure in which you embed your individual projects. This makes it easier to find your work many weeks, months, or years later when you've forgotten what you've done and where you've put things.

Once you're situated in the correct directory, you can create or open an existing m-file with the command `edit foobar`, which opens or creates a file named `foobar.m` in the current directory. If you are among the emotionally insecure, you could equally well have commanded `edit foobar.m` but MATLAB knows that you're working with an m-file. This opens a separate window for the editor. If you have Luddite tendencies, you could use a simple text editor to create and work with your m-files, but the MATLAB editor offers a number of attractive features. These include *smart indenting*, some real-time syntax flagging (problems are underlined in squiggly red, just like a spell-checker in a word processor), signals for parenthesis and bracket imbalances, and color coding of keywords, comments, and strings. There are numerous other useful features for code debugging and testing that we encourage you to explore.

Bear in mind that m-files can call other m-files, either as *scripts* or as *functions*. The advantage of scripts is that they're simple; they just execute in sequence as if you typed them (flawlessly) into the command window. The disadvantage is that they must "know" what variables are being operated on. If you change the name of a variable elsewhere in your m-file script collage, it will break. Thus the scripts, although very useful, may not be portable to other problems. *Functions* are somewhat more complicated to construct; you must supply them with *input arguments* and have them return values. The key advantage, though, is that they are *encapsulated*, which makes them more robust and more useful in other situations. There are times and places for both types, and we will use them throughout the book.

So you'll find yourself with the main MATLAB window open on your screen, you'll be building and test-running an m-file in the editor, and you'll be checking out commands with the help window. Everyone has their own style, but you will likely use a "divide and conquer" strategy in your problem solving. Break your problem down into a sequence of steps (perhaps as few as two or three, or as many as a dozen or so) that you can construct, complete, and test individually. It's worthwhile convincing yourself that each individual component functions as intended before going on to the next step. Sometimes it is not possible to test each component completely, but the very exercise of segmenting your problem helps to better define your thought processes. In the old days, we used to construct diagrams called *flow charts* that schematized the logic flow from start to finish. This is not a bad habit to get into. Try sketching out on a piece of paper the way you would solve the problem. The exercise of representing branch (decision) points and procedural flows often clarifies how you would build your solution.

Unless you're terribly clever or the problem is trivially simple, you may not be able to do it "all in one sitting". Sometimes sleeping on a problem (putting this book under your pillow might help too) yields better insights. Working with m-files allows you to save your logic process for another day. As importantly (and we can't emphasize this enough) you should *document your work*. Put numerous comments in your m-files telling you (or the consumer of your work) how or why you've done this or that. Your reasoning may be blindingly obvious at the moment, but in a few weeks, you may struggle to get back to where you were mentally. Regular, informative comments are worth their weight in gold, but don't *over-comment* either, since it may be hard to find the important points in a morass of whimsical verbal wanderings. You need to strike a reasonable balance.

Sometimes the point at which you've arrived in your problem is so computationally time-consuming that you can't afford to run the calculation over and over again. Be aware that you can save your results (actually, saving your *workspace*) with the `save` command. This allows you to reconstitute the situation relatively quickly when you return. Bear in mind that in doing so, you must navigate back to the directory where you saved your workspace before you can recover it with the `load` command. Again, use `help` or `doc` to learn about these tools.

A.3 Doing it faster

Computers are getting ever faster and more powerful, doubling in capability every few years. Even so, what seems to happen is that the scale and scope of the numerical problems we tackle grow at least as fast as the speed and capacity of computers. All things being equal, though, we'd like to give a few suggestions on how to speed up your code. Most are simply "mechanical" in nature, but some may be more nuanced.

Preallocate your storage. Almost always, with a little effort and planning, you can figure out how big your storage vectors or matrices will be. It is possible to "grow" matrices by simply appending more and more data. For example, consider the timing that you obtain from the following code

```
tic                    % start the clock
  n=10000;             % start with a 10,000 element vector
  x=[];                % create an empty matrix to start
  for j=1:n            % step through the integers
      x=[x j^2];       % squaring each one and appending it to x
  end
toc                    % stop the clock and report elapsed time

clear all              % start again with a clean slate
tic                    % start the clock
  n=10000;             % start with a 10,000 element vector
  x=zeros(1,n);        % this time preallocate storage
  for j=1:length(x)    % step through the length of the vector
      x(j)= j^2;       % squaring each element
  end
toc                    % stop the clock and report elapsed time
```

Note that we've used the `tic` and `toc` statements to give us the elapsed time. This is a useful tool for timing operations. The speed improvement gained by preallocation is not much for short vectors (say a thousand elements long) because MATLAB and the operating system can be remarkably efficient for small blocks of data, but is worth it for large vectors or matrices. This is because every time you add elements to an existing matrix (or vector), MATLAB has to reallocate storage. It does this by asking the computer operating system to create a new storage space with the larger size, and then copying the old data (plus the new addition) into the new space. It turns out that the time required to do this goes roughly as the square of length of the vector. Thus it take approximately 100 times as long to accumulate a vector that is 10 times as big. By preallocating space, MATLAB (and your poor computer's operating system) has to do it only once. This results in huge time savings. Try the above as an m-file with various sized vectors (values of n) and you'll see what we mean.

Vectorizing your code always speeds things up. MATLAB is becoming ever cleverer about doing this for you "under the hood", but you can still confound it if you're not careful. That is, it is always a good thing to write your code in a way that allows vectorized computation. Vectorization often involves matrix-type (or array-type) operations, but can also include colon operations. Try the following code:

```
clear all              % start with a clean slate
tic                    % start the clock
  x=zeros(1,1000000);   % preallocate storage
  for j=1:length(x)    % step through the length of vector
     x(j)= j^2;        % square of integers
  end
toc                    % stop the clock and report elapsed time
clear all              % start all over again
tic                    % start the clock
  x=[1:1000000];        % this time assign as integer series
  x=x.^2;              % and square in vectorized fashion
toc                    % stop the clock and report elapsed time
```

You'll notice that we're doing 100 times as many computations in this example, and the time lapse increases approximately tenfold over the last example in the first instance. Note also that the vectorized code runs about 30 times faster than the non-vectorized code.

The bottom line is to avoid using loops unless you absolutely have to. In Chapters 12 and 16 we went as far as using off-set indices in our vectors and matrices to implement vectorized code. Although it looks obtuse, it really speeds up calculations. If you must use loops within loops, we recommend you embed the more "numerate" loop within the less numerate one. For example

```
for i=1:10
    for j=1:1000
     % do stuff
    end
end
```

works much more efficiently than

```
for j=1:1000
    for i=1:10
```

```
        % do stuff
    end
end
```

because in the second case you have to initialize and start a loop 1000 times rather than the first case where you do it only 10 times. Sometimes for specific reasons you have to logically do it the hard way, but if you can avoid it, do so.

A.4 Choose your algorithms wisely

Although the temptation is great, we urge you not to treat MATLAB functions as black boxes. When investigating what function call to use in a given project, you should always examine the documentation to find out what algorithm it uses. Often you can find out further information within the documentation system, or by consulting Press *et al. Numerical Recipes* (2007). Sometimes, a given function may employ different algorithms depending on the nature of the data you provide, how it's called, or how you set options. MATLAB functions tend to use the most efficient algorithms for the task at hand, but sometimes even this produces less than desirable results if you extend the tools into *terra incognita*. Being equipped with an understanding of the basic "dynamics" of an algorithm arms you against a lot of trouble.

Algorithms are often gauged by how they scale with the size of the problem. An algorithm that "scales as N" (where N is the number of data points) will take take 10 times longer for a data set that is 10 times larger. Not all algorithms scale as nicely as this. That is, an algorithm that works well on a small sample may not do nearly so well for a larger one. For example, say that you test your code by running it with a data set of 1000 points and it takes only 10 seconds. You may be shocked to find that running the same routine with a million points takes not 10 000 seconds (about 3 hours), but the better part of a year if it scales as N^2. This may be problematic if you need the calculation to complete a term project or present a paper at a meeting next month! A more elegant but more complicated algorithm that takes 30 seconds with the same smaller sample may take only a day if it scales as $N \log N$. As we discussed in Chapter 6, the discovery of the *fast Fourier transform*, which scales as $N \log N$, created a wholesale revolution in the technology of digital signal processing, and is used in everything from science, satellites, and cell-phones, to flat-screen televisions.

Paying attention to algorithm structure and efficiency, and in particular how it scales with sample size, is important. A detailed discussion of algorithms, except in the limited number of examples covered in the course of this book, is beyond our scope here. However, there are some useful questions that you might ask about the algorithms. For example, consider the following:

- How does the algorithm scale with N?
- Is the algorithm "smart"? For example, does it adjust step size dynamically?

- How are the data stored? In search algorithms, data structure (how data are stored) can significantly impact algorithm scaling and speed.
- Will the computing platform run into caching/paging issues? How big are memory and secondary caches?
- How accurately do you really need your answer? Relaxing convergence criteria can sometimes reap substantial efficiency gains, and even trigger the use of faster/coarser algorithms.

Algorithm design and selection is a venerable field of study, and there are many good books out there. A good, practical discussion is given by Skiena (2008), who offers an introduction to algorithms along with an entertaining and instructive collection of "war stories" that exemplify the value of choosing the right algorithm. The take-home lesson he gives is that even higher-powered supercomputers cannot compensate for sloppy thinking and poor design.

A.5 Automating tasks

Sometimes you need to do something repetitive, for example performing a data extraction from a large number of monthly data files, or running a numerical model over a range of parameter values. You will see examples of this throughout this book. You can readily do this by embedding your tasks within one or more `for` loops, and using the `sprintf` command to construct file names in some logical manner. For example, consider the following:

```
for i=1:100           % do this 100 times
    % do some numerical calculation using i as a key parameter
    fname=['TheDataFile' sprintf('%.0f',i) '.dat'];
    fp=fopen(fname,'w');
    % write some data to this file using the file pointer "fp"
    fclose(fp);           % close the file
end
```

where we have concatenated a string in the vector `fname` that encodes the value of `i`. For example, when $i = 11$ we would have `fname = TheDataFile11.dat`. We encourage you to look at how `sprintf` is used to create strings.

The same can be done for loading or saving native MATLAB mat-files. You might plan on doing this because storing the data in native, binary format is more efficient, both in terms of storage and speed. In order to use the `save` and `load` commands, you have to use the rather cumbersome `eval` command. What this basically involves is that you have to encode the entire command as a string (as if you were typing it into the command window) and then execute it using `eval`. Here's an example:

```
for i=1:100          % do this 100 times
    % do some numerical calculation using i as a key parameter
    myCommand =['save TheDataFile' sprintf('%.0f',i) '.mat'];
```

```
      eval(myCommand)
end
```

which looks kind of strange (well, it did to us), but is potentially powerful. This doesn't have to be restricted to saving and loading files; you can imagine all sorts of things you could do with this, can't you?

You can do the same for reading files, assuming that you know in advance that the files exist. Which brings us to a small dilemma; what if we know there are some files in a directory that we want to read, but they may not be all there? For example, maybe a ship went out to collect a time series station once a month, but because of bad weather, an oceanographer's strike, or a malfunctioning CTD, the data were not taken. You want a program to read all the files in a specific directory. Here's how you do it using the `dir` command :

```
cd 'C:\Data\TimeSeries'   % change to the data directory of interest
theFiles=dir('*.dat');    % get a list of the "dat" files
for i=1:length(theFiles)  % do this for each of the files
    fp=fopen(theFiles(i).name,'r');    % open the file for reading
    % read in your data from the file
    fclose(fp);
end
```

Note that the `dir` command returns a *structure array* that contains information about the contents of the specified directory.[1] You access the relevant parts of the data structure with the `.` operator. For example, `theFiles(i).name` is a string containing the file name and `theFiles(i).date` contains its date. Try `doc dir` to learn more about this handy command. The `*` is a wild-card operator that allows us to select all file names that match the pattern. Notice also that if there are other files that don't end in `.dat` or sub-directories in the directory of interest, they will be ignored. Also, if there are no `.dat` files in the directory, the length of `theFiles` will be zero, so the code above will run without error (that is, the `for` loop won't execute at all).

A.6 Graphical tricks

As we've mentioned in Chapter 19, MATLAB does a fine job of producing quick and serviceable graphics, and can with some cajoling produce publication-quality plots. To improve on the "default" behavior, we need to use *handle graphics*. Admittedly, handle graphics can be confusing to newcomers to MATLAB, but they are in fact quite simple. Most graphics functions in MATLAB can be used "straight", but often quietly return one or more handles to the graphical output they produce. A handle is actually just an internal number that MATLAB uses to associate actions with a graphics object. For example, the `plot` command may be used with the `set` command operated this way:

---

[1] In truth, the variable `theFiles` is just a vector of pointers to data structures. To learn about structure arrays, type `doc struct` in the MATLAB command window.

```
h=plot(x,y,'.');
set(h,'MarkerSize',20,'MarkerEdgeColor','k','MarkerFaceColor','g');
```

produces moderately large, filled circles with a green interior and a black edge. (We think that's kind of stylish.) MATLAB actually allows you to do this same thing in shorthand, with the following structure:

```
plot(x,y,'.','MarkerSize',20,'MarkerEdgeColor','k','MarkerFaceColor','g');
```

which means that you don't have to deal with those messy handles. But there are still reasons that you might continue to use them.

If you were to look at the value of h in this instance, it would be some number like 159.0022 that means nothing to you and something to MATLAB. But how do you know what options are available? You can (gasp!) use the documentation (doc plot), or you can simply do the following:

```
h=plot(x,y,'.');
set(h)
```

which gives you an exhaustive list of your options.

Supposing you had a more complex plot, for example with:

```
h=plot(x,y,'.',w,z);
```

Now if you looked at h again, you would see that it would contain two numbers (it would be a vector of length 2), with each number corresponding to a separate graphical element. The first would represent the symbols (plotted for y as a function of x) and the second would point to the line segments associated with z plotted as a function of w. You could then alter each one separately and according to its own properties:

```
h=plot(x,y,'.',w,z);
set(h(1),'MarkerSize',20,'MarkerEdgeColor','k','MarkerFaceColor','g');
set(h(2),'LineWidth',4,'Color','r');
```

There is also a pre-defined handle that you're already familiar with: gca. This points to the axes. Again, typing set(gca) will generate a huge list of possible options, but you've already seen some of these:

```
plot(x,y,'.',w,z);
set(gca,'LineWidth',2,'FontSize',16,'FontWeight','bold','YDir','Reverse');
```

contains some examples. You can change tick sizes, tick labels, etc., as well.

You can also change tick labels "manually". For example, if plotting a seasonal cycle, you might do this:

```
plot(t,y);
mNames =['Jan'; 'Feb'; 'Mar'; 'Apr'; 'May'; 'Jun'; ...
            'Jul'; 'Aug'; 'Sep'; 'Oct'; 'Nov'; 'Dec'];
set(gca,'XTickLabels',mNames);
```

If an option requires some string-type option (e.g. FontWeight), you can get a list of those options by typing set(gca,'FontWeight'). Bear in mind, this doesn't work for all properties.

Further, you might be interested in using the *Plot Edit Mode* on your figure. Having used your MATLAB m-file to generate a plot, you can use the figure's toolbar to enter into Plot Edit Mode and interactively edit your plot, adding annotations and changing a number of aspects. This is often a useful approach if you're doing something for the first (and perhaps only) time. You do have the ability to undo and redo things in this mode, but long, complex operations can be tedious or difficult to replicate, so often it is better to code it up with handle graphics.

Finally, we mention one thing that MATLAB doesn't always do very well or gracefully, namely *contour labeling*. You can choose what contour levels will be contoured, but where the labels are placed is another matter. The default choice is certainly serviceable, but often not very pretty. When you think of it, it's a really tough problem to predict where a label should be placed to make the graph clear and useful. Here, you might try:

```
[C,h]=contour(X,Y,Z,cLev);
hc=clabel(C,h,'manual');
set(hc,'fontsize',16);
```

which contours a three-dimensional data set (**X,Y**, and **Z**) according to some pre-defined contour levels contained in the vector cLev. This returns a contour matrix **C** and a handle vector pointing to the graphics objects. You then stipulate a manual contour label mode. What happens is that window focus is set to the plot, and you use the cursor (cross-hairs) to select which contour is labelled where by "point and click". When finished, you press the *Enter* key. The clabel command also returns a handle that can be used to change aspects of the labels, for example their font size.

A.7 Plotting oceanographic sections

Sometimes it is useful in oceanography to plot property values in *sections*, that is, with horizontal distance along the x-axis, and depth (positive downward) along the y-axis. This allows one to interpret property distributions in terms of water masses and their circulation. Programs such as GMT (Generic Mapping Tools) and ODV (Ocean Data View, see Chapter 19) are well suited to doing such tasks, but sometimes it's convenient to do the same within MATLAB. Further, if the depth of the section plot is sufficient, it is often useful to represent the ocean bottom as a "blacked" (or "grayed") out area so that contours extrapolated below the seafloor are not interpreted as real features. We will show you by example how you might accomplish this.

We begin by obtaining a data set from the *CLIVAR and Carbon Hydrographic Data Office* (CCHDO) website.[2] In this instance we obtain bottle data from the CLIVAR

---

[2] http://whpo.ucsd.edu/

Repeat Hydrography section A20, occupied along 52° W in the North Atlantic during 2003. The file downloaded is a text file with the name a20_2003a_hy1.csv that contains station data as well as hydrographic properties. We import it into the spreadsheet program Excel to identify the columns of interest, and export it as a spreadsheet a20_2003a_hy1.csu for future report generation. The file has a "header" documenting the cruise information and data history, but in its "body" contains the measurement data organized with one column per measurement type and with one row per water sample. Station-related data, such as location and water depth, are repeated on each line as appropriate.

You will find the file is quite wide, having 98 columns. As we are going to plot the silicate distribution as function of latitude and pressure, the columns of interest include latitude (column 10), CTD pressure (column 13), and silicate (column 23). Additional columns include water depth (column 12) and the silicate data quality flag (column 24). We begin by reading the data file into MATLAB with:

```
d=xlsread('a20_2003a_hy1.csu');      % read the data in
```

This will take some time on your computer, because MATLAB has to translate this from the comma separated value text format. But if you look carefully, you'll notice that the data array is not 98 columns wide, but rather 96! Examining the file again, you'll see that the first two columns contain alphanumeric (mixed letters and numbers) characters, and since MATLAB doesn't know how to convert them to numbers, it ignores those columns. This, of course, means that we need to offset our column choices by 2. Hence to slim down our choice of data, we extract from the data array two separate arrays, one containing bottom depths and the other containing the data of interest:

```
b=d(:,[8 10]);          % latitude, water depth
d=d(:,[8 11 21 22]);    % lat, pres, sil, flag
```

From this reduced data set, we now extract an array containing the bottom depths:

```
b=unique(b,'rows');               % latitude, water depth
```

which produces a much-shortened list of station locations and water depths arranged in ascending latitude order. Now to make this useful for masking out the ocean bottom, we need to close this "curve" to make it a polygon that encompasses the bottom part of the chart. If we want to plot the section from 6.5 to 43.5° N and from the surface to 6000 meters depth, we do the following:

```
x=[6.5:43.5];           % the latitude range we're plotting
y=[0:50:6000]';         % the depth range we're plotting
ulc=[min(x) min(y)]';   % upper left corner of plot
llc=[min(x) max(y)]';   % lower left corner of plot
```

```
lrc=[max(x) max(y)]';        % lower right corner of plot
urc=[max(x) min(y)]';        % upper right corner
b=[llc ulc b' urc lrc]';     % append corners to polygon
```

Note that the *y*-vector must be columnar for gridding purposes later, and we've made the plot vertices little column vectors to append to b in the correct orientation to draw the bottom mask in a clockwise fashion.

Now we must recognize that sometimes the silicate data are suspect or missing. If missing, the data value defaults to a "null value" of −999, but if it is suspect, we need to rely on the quality flag. WOCE protocols dictate that a flag value of 2 indicates a good quality measurement, and otherwise it is missing or suspect. Thus we filter our data using the following:

```
d=d(d(:,4)==2,:);        % accept only good quality data
```

which shortens our data set a little more. Finally, we are in a position to grid and present our data with the following:

```
[X,Y,Z]=griddata(d(:,1),d(:,2),d(:,3),x,y);  % grid the data
colormap('gray')                   % select a tasteful colormap
contourf(X,Y,Z)                    % contour plot
```

Now this doesn't help us very much because it is upside down and has no topographic masking. So we add:

```
set(gca,'ydir','reverse','fontsize',14);   % orient pressure pos-downward
xlabel('Latitude'); ylabel('Pressure')     % label axes
hold on                                     % hold plot so that we can...
fill(b(:,1),b(:,2),0.75*ones(1,3));         % mask out the ocean bottom
hold off                                    % and cancel the hold
```

This produces a serviceable section plot.

A.8 Reading and writing data

Getting data from other sources into MATLAB can be challenging. If the data constitute a single matrix and the program can export data in an organized fashion into plain-text format, then it is easy to import into MATLAB using the load command. However, for larger data sets this isn't particularly efficient, either because the file size balloons up beyond belief, or it takes a horrendous amount of time to read (and translate) it into MATLAB. Moreover, it may be rather tedious to "translate" a directory full of, let's say, 2027 files into separate text files before starting.

Another concern you may have is whether something might be lost in translation. That is, the conversion to plain text may round or truncate data in a way that destroys potentially valuable information. For example, your favorite spreadsheet program may export your temperature and salinity data with a default of two decimal places when you really need three (or perhaps four) to represent abyssal ocean data properly. Thus you need to pay attention to this detail.

Excel spreadsheet files can be read into MATLAB in native format with the `xlsread` command, but don't expect to break any world speed records; the translation process is slow! You can also use `xlswrite` (guess what that does?). Check them out with `doc`. We will only discuss reading spreadsheets here and leave the writing part for your further investigation.

You can select specific worksheets within a file as well. Recall, for example, we did this in the previous section:

```
d=xlsread('a20_2003a_hy1.xls','Hydro');        % read the data in
```

where we have read in the worksheet named `Hydro` from the spreadsheet file. Note that the arguments are entered as strings, and that you get an error if that worksheet doesn't exist in the file, or the file doesn't exist or has an incorrect format. If you want interactive control, using `d = xlsread('theFile.xls',-1);` will raise Excel in a separate window allowing you to interactively select the worksheet and even a subset of the data to input.

Also, be aware that MATLAB will by default ignore any "outer" rows and/or columns that contain only alphanumeric data. Thus if your worksheet contains 21 rows and 11 columns, but the top two rows are labels and the last column contains comments, then the resultant matrix will be 19 rows and 10 columns. Any non-numeric data contained within the matrix will be represented by `NaN`. You can also read in the non-numeric data into a text array with:

```
[d,txt]=xlsread('theFile.xls','theSheet');        % read the data in
```

As we mentioned in Chapters 15 and 19, a lot of model and climatological data are stored in a self-describing format called *netCDF*. This format is very efficient but is complicated to translate without the proper tools. Fortunately the most recent versions of MATLAB (from release 2008b and later) have this capability built in, but if you have an earlier version, please read on.

To read netCDF files into older versions of MATLAB you need to install a special interface called MEXNC.[3] This is a one-time task, but it is rather complicated. It involves downloading a special file and some m-files to make the system work. We suggest you visit `http://mexcdf.sourceforge.net` and follow the links to download the appropriate tools.

Accessing a netCDF file requires you know what is inside it. There are a number of "netCDF browser programs" out there, most free. We use *Panoply* or Intel's *Array Viewer*. There are even MATLAB programs. Regardless, once you know what's inside, you can extract a portion of a data matrix with (for example):

---

[3] This was previously called MEXCDF and was pioneered by Chuck Denham.

```
nc = netcdf(fname,'nowrite');              % obtain pointer to file
myData = nc{'uflx',1}(:,30,[121:188]);     % subset uflx data set
myMatrix=reshape(myData,length(myData),68);% convert to 2D matrix
result = close(nc);                        % close the file
```

where we've had to reshape the multi-dimensional matrix extracted to suit our more modest needs. Note that we had to know that the variable `uflux` existed in the file, and what the data elements or ranges corresponded to in terms of the other variables (in this case: time, latitude, and longitude).

References

Aiken, J., Y. Pradhan, R. Barlow *et al.*, 2009, Phytoplankton pigments and functional types in the Atlantic Ocean: A decadal assessment, 1995–2005, *Deep-Sea Res. II*, **56**(15), 899–917.

Alvarez, A., V. Bertram, and L. Gualdesi, 2009, Hull hydrodynamic optimization of autonomous underwater vehicles operating at snorkeling depth, *Ocean Eng.*, **36**(1), 105–112.

Anderson, L. A. and J. L. Sarmiento, 1994, Redfield ratios of remineralization determined by nutrient data-analysis, *Glob. Biogeochem. Cycles*, **8**(1), 65–80.

Arakawa, A., 1966, Computational design for long-term numerical integrations of the equations of fluid motion: Two-dimensional incompressible flow, Part 1, *J. Comput. Phys.*, **1**, 119–143.

Armi, L. and D. Haidvogel, 1982, Effects of variable and anisotropic diffusivities in a steady state diffusion model, *J. Phys. Oceanogr.* **12**, 785–794.

Athias, V., P. Mazzega, and C. Jeandel, 2000, Selecting a global optimization method to estimate the oceanic particle cycling rate constants, *J. Mar. Res.*, **58**(5), 675–707.

Beckmann, A. and R. Doscher, 1997. A method for improved representation of dense water spreading over topography in geopotential-coordinate models, *J. Phys. Oceanogr.*, **27**, 581–591.

Bennett, A. F., 2002, *Inverse Modeling of the Ocean and Atmosphere*, New York: Cambridge University Press.

Berner, R. A., 1980, *Early Diagenesis*, Princeton, NJ: Princeton University Press.

Bevington, P. R. and D. K. Robinson, 2003, *Data Reduction and Error Analysis for the Physical Sciences*, 3rd edn, New York, NY: McGraw-Hill Inc.

Bishop, C. M., 1995, *Neural Networks for Pattern Recognition*, Oxford, UK: Oxford University Press.

Boas, M. L., 2005, *Mathematical Methods in the Physical Sciences*, 3rd edn, New York, NY: John Wiley & Sons.

Boyer Montégut, C., G. Madec, A. S. Fischer, A. Lazar, and D. Iudicone, 2004, Mixed layer depth over the global ocean: an examination of profile data and a profile-based climatology, *J. Geophys. Res.*, **109**, C12003.

Briegleb, B. P., C. M. Bitz, E. C. Hunke *et al.*, 2004, Scientific description of the sea ice component in the Community Climate System Model, Version Three, Tech. Rep. NCAR-TN-463STR, Boulder, CO: National Center for Atmospheric Research.

Bretherton, F. P., R. E. Davis, and C. B. Fandry, 1976, A technique for objective analysis and design of oceanographic experiments applied to MODE-73, *Deep-Sea Res.*, **23**, 559–582.

Bretherton, C. S., Smith, C., and Wallace, J. M., 1992. An intercomparison of methods for finding coupled patterns in climate data. *J. Clim.*, **5**, 541–560.

Broecker, W.S. and T.-H. Peng, 1982, *Tracers in the Sea*. Palisades, NY: ELDIGIO Press.

Bryan, K., 1969, A numerical method for the study of the circulation of the world ocean, *J. Comput. Phys.*, **4**, 347–376.

Bryan, F., 1987, Parameter sensitivity of primitive equation ocean general circulation models, *J. Phys. Oceanogr.*, **17**, 970–985.

Canellas, B., S. Balle, J. Tintore, and A. Orfila, 2010, Wave height prediction in the Western Mediterranean using genetic algorithms, *Ocean Eng.*, **37**(8–9), 742–748.

Carton, J. A. and B. S. Giese, 2008, A reanalysis of ocean climate using Simple Ocean Data Assimilation (SODA), *Mon. Wea. Rev.*, **136**, 2999–3017.

Case, T. J., 1999, *An Illustrated Guide to Theoretical Ecology*, Oxford, UK: Oxford University Press.

Chatfield, C., 2004, *The Analysis of Time Series: An Introduction*, 6th edn, Boca Raton, FL: Chapman & Hall/CRC Press Texts in Statistical Science.

Chilés, J.-P. and P. Delfiner, 1999, *Geostatistics, Modeling Spatial Uncertainty*, New York: John Wiley & Sons.

Collins, W. D., M. Blackmon, C. M. Bitz *et al.*, 2006, The Community Climate System Model: CCSM3, *J. Clim.*, **19**, 2122–2143.

Cox, M. D., 1984, A primitive equation, 3-dimensional model of the ocean. GFDL Ocean Group Technical Report No. 1. Available from Geophysical Fluid Dynamics Laboratory, Princeton, New Jersey, 08542.

Craig, H., 1969, Abyssal carbon and radiocarbon in the Pacific, *J. Geophys. Res.*, **74**, 5491–5506.

Cressie, N. A. C., 1993, *Statistics for Spatial Data*, New York: Wiley-Interscience.

Cressie, N. A. C. and D. M. Hawkins, 1980, Robust estimation of the variogram, I, *J. Int. Assoc. Math. Geol.*, **12**, 115–125.

Dai, A., T. T. Qian, K. E. Trenberth, and J. D. Milliman, 2009, Changes in continental freshwater discharge from 1948 to 2004, *J. Clim.*, **22**, 2773–2792.

Danabasoglu, G., J. C. McWilliams, and P. R. Gent, 2004, The role of mesoscale tracer transports in the global ocean circulation, *Science*, **264**, 1123–1126.

Davis, J. C., 2002, *Statistics and Data Analysis in Geology*, 3rd edn, New York: John Wiley and Sons.

Davis, R. E., 1998, Preliminary results from directly measuring middepth circulation in the tropical and South Pacific, *J. Geophys. Res. Oceans*, **103**, 24619–24639.

Davis, T. A. and K. Sigmon, 2004, *MATLAB Primer*, 7th edn, Boca Raton, FL: Chapman and Hall/CRC.

deBoor, C., 1978, *A Practical Guide to Splines*, New York: Springer-Verlag.

Derbyshire, J., 2003, *Prime Obsession, Bernhard Riemann and the Greatest Unsolved Problem in Mathematics*, New York: Plume Book, member of Penguin Group (USA).

Deuser, W. G., 1986, Seasonal and interannual variations in deep-water particle fluxes in the Sargasso Sea and their relation to surface hydrography, *Deep-Sea Res.*, **33** (2), 225–246.

Doney, S. C., 1996. A synoptic atmospheric surface forcing data set and physical upper ocean model for the U.S. JGOFS Bermuda Atlantic Time-series Study Site. *J. Geophys. Res.*, **101**(C10), 25615–25634.

Doney, S. C., 1999, Major challenges confronting marine biogeochemical modeling, *Glob. Biogeochem. Cycles*, **13**, 705–714.

Doney, S. C., R. G. Najjar, and S. Stewart, 1995, Photochemistry, mixing, and diurnal cycles in the upper ocean, *J. Mar. Res.*, **53**, 341–369.

Doney, S. C., W. G. Large, and F. O. Bryan, 1998, Surface ocean fluxes and water-mass transformation rates in the coupled NCAR Climate System Model, *J. Clim.*, **11**, 1420–1441.

Doney, S. C., K. Lindsay, and J. K. Moore, 2003, Global ocean carbon cycle modeling in *Ocean Biogeochemistry*, ed. M. Fasham, Springer, 217–238.

Doney, S. C., K. Lindsay, K. Caldeira *et al.*, 2004, Evaluating global ocean carbon models: the importance of realistic physics, *Glob. Biogeochem. Cycles*, **18**, GB3017, doi:10.1029/2003GB002150.

Doney, S. C., S. Yeager, G. Danabasoglu, W. G. Large, and J. C. McWilliams, 2007, Mechanisms governing interannual variability of upper ocean temperature in a global hindcast simulation, *J. Phys. Oceanogr.*, **37**, 1918–1938.

Doney, S. C., I. Lima, J. K. Moore *et al.*, 2009, Skill metrics for confronting global upper ocean ecosystem-biogeochemistry models against field and remote sensing data, *J. Mar. Sys.*, **76**, 95–112, doi:10.1016/j.jmarsys.2008.05.015.

Ducet, N., P. Y. Le Traon, and G. Reverdin, 2000, Global high-resolution mapping of ocean circulation from TOPEX/Poseidon and ERS-1 and-2, *J. Geophys. Res. Oceans*, **105**, 19477–19498.

Dutay, J.-C., J. L. Bullister, S. C. Doney *et al.*, 2002, Evaluation of ocean model ventilation with CFC-11: comparison of 13 global ocean models, *Ocean Modelling*, **4**, 89–120.

Eckart, C. and G. Young, 1936, The approximation of one matrix by another of lower rank, *Psychometrika*, **1**, 3, 211–218.

Emerson, S. R. and J. I. Hedges, 2008, *Chemical Oceanography and the Marine Carbon Cycle*. Cambridge, UK: Cambridge University Press.

Emery, W. J. and R. E. Thompson, 1998, *Data Analysis Methods in Physical Oceanography*, New York: Pergamon Press, Elsevier Science Inc.

Errico, R. M., 1997, What is an adjoint model? *Bull. Am. Met. Soc.*, **78**, 2577–2591.

Fallat, M. R., S. E. Dosso, and P. L. Nielsen, 2004, An investigation of algorithm-induced variability in geoacoustic inversion, *IEEE J. Ocean. Eng.*, **29**(1), 78–87.

Fasham, M. (ed.), 2003, *Ocean Biogeochemistry*, New York: Springer.

Fasham, M. J. R., B. M. Balino, M. C. Bowles *et al.*, 2001, A new vision of ocean biogeochemistry after a decade of the Joint Global Ocean Flux Study (JGOFS), *AMBIO*, Spec. Iss. **10**, 4–31.

Fennel, W. and T. Neumann, 2004, *Introduction to the Modelling of Marine Ecosystems*, Elsevier Oceanography Series, 72, San Diego, CA: Elsevier, Inc.

Fennel, K., J. Wilkin, J. Levin *et al.*, 2006, Nitrogen cycling in the Middle Atlantic Bight: Results from a three-dimensional model and implications for the North Atlantic nitrogen budget, *Glob. Biogeochem. Cycles*, **20**, GB3007.

Fratantoni, D. M., 2001, North Atlantic surface circulation during the 1990's observed with satellite-tracked drifters, *J. Geophys. Res. Oceans*, **106**, 22067–22093.

Friedrichs, M. A. M., J. A. Dusenberry, L. A. Anderson *et al.*, 2007, Assessment of skill and portability in regional marine biogeochemical models: the role of multiple planktonic groups, *J. Geophys. Res. Oceans*, **112**, C08001, doi:10.1029/2006 JC003852.

Ganachaud, A. and C. Wunsch, 2002, Oceanic nutrient and oxygen transports and bounds on export production during the World Ocean Circulation Experiment, *Global Biogeochem. Cycles*, **16**(4), 1057, doi:10.1029/2000GB001333.

Ganachaud, A. and C. Wunsch, 2003, Large-scale ocean heat and freshwater transport during the World Ocean Circulation Experiment, *J. Clim.*, **16**, 696–705.

Garcia, H. E., R. A. Locarnini, T. P. Boyer, and J. I. Antonov, 2006, *NOAA Atlas NESDIS 64, World Ocean Atlas 2005*, Volume 4, *Nutrients (Phosphate, Nitrate, Silicate)*, ed. S. Levitus, Washington, DC: US Government Printing Office.

Gargett, A. E. 1984, Vertical eddy diffusivity in the ocean interior, *J. Mar. Res.*, **42**, 359–393.

Garrett, C., 1983. On the initial streakiness of a dispersing tracer in two- and three-dimensional turbulence, *Dyn. Atmos. Oceans*, **7**, 265–277.

Garrett, C., 1989, A mixing length interpretation of fluctuations in passive scalar concentration in homogeneous turbulence, *J. Geophys. Res.*, **94**, 9710–9712.

Garrett, C., 2006, Turbulent dispersion in the ocean, *Prog. Oceanogr.*, **70**, 113–125.

Gent, P. R. and J. C. McWilliams, 1990: Isopycnal mixing in ocean circulation models, *J. Phys. Oceanogr.*, **20**, 150–155.

Gent, P. R., J. Willebrand, T. McDougall, and J. C. McWilliams, 1995, Parameterizing eddy-induced tracer transports in ocean circulation models, *J. Phys. Oceanogr.*, **25**, 463–474.

Gent, P. R., F. O. Bryan, G. Danabasoglu *et al.*, 2006, Ocean chlorofluorocarbon and heat uptake during the twentieth century in the CCSM3, *J. Clim.*, **19**, 2366–2381.

Giering, R. and T. Kaminski, 1998, Recipes for adjoint code construction, *ACM Trans. Math. Software*, **24**(4), 437–474.

Gill, A. E., 1982, *Atmosphere–Ocean Dynamics*, New York: Academic Press.

Gill, P. E., W. Murray, and M. H. Wright, 1981, *Practical Optimization*, New York, NY: Academic Press, Inc.

Glover, D. M., S. C. Doney, A. J. Mariano, R. H. Evans, and S. J. McCue, 2002, Mesoscale variability in time-series data: Satellite-based estimates for the U.S. JGOFS Bermuda Atlantic Time-Series Study (BATS) site, *J. Geophys. Res. Oceans*, **107**(C8), 3092, doi:10.1029/2000JC000589.

Gnanadesikan, A., 1999, A simple predictive model for the structure of the oceanic pycnocline, *Science*, **283**, 2077–2079.

Gordon, A. D., 1999, *Classification*, 2nd edn. Boca Raton, FL: Chapman and Hall.

Gradshteyn, I. S. and I. M. Ryzhik, 2000, *Table of Integrals, Series, and Products*, 6th edn, San Diego, CA: Academic Press.

Gregg, 1987, Diapycnal mixing in the thermocline: A review, *J. Geophys. Res.*, **92**, 5249–5286.

Gregg, W. W., M. A. M. Friedrichs, A. R. Robinson *et al.*, 2009, Skill assessment in ocean biological data assimilation, *J. Mar. Syst.*, **76**, 16–33.

Griffies, S., 2004, *Fundamentals of Ocean Climate Models*, Princeton NJ: Princeton University Press.

Griffies, S. M., 1998, The Gent–McWilliams skew-flux, *J. Phys. Oceanogr.*, **28**, 831–841.

Griffies, S. M., C. Böning, F. O. Bryan *et al.*, 2000, Developments in ocean climate modelling, *Ocean Modelling*, **2**, 123–192.

Gruber, N., H. Frenzel, S. C. Doney *et al.*, 2006: Eddy-resolving simulation of plankton ecosystem dynamics in the California Current System, *Deep-Sea Res. I*, **53**, 1483–1516.

Gruber, N., M. Gloor, S. E. Mikaloff Fletcher *et al.*, 2009, Oceanic sources, sinks, and transport of atmospheric CO_2, *Glob. Biogeochem. Cycles*, **23**, GB1005, doi:10.1029/2008GB003349.

Haidvogel, D. B. and A. Beckmann, 1999, *Numerical Ocean Circulation Modeling*, London, UK: Imperial College Press.

Hall, T. M. and T. W. N. Haine, 2002, On ocean transport diagnostics: the idealized age tracer and the age spectrum, *J. Phys. Oceanogr.*, **32**(6), 1987–1991.

Hasselmann, K., 1976, Stochastic climate models, Part I. Theory, *Tellus*, **28**(6), 473–485.

Hastings, A., 1997, *Population Biology, Concepts and Models*, New York, NY: Springer-Verlag.

Hecht, M. W., W. R. Holland, and P. J. Rasch, 1995, Upwind-weighted advection schemes for ocean tracer transport: an evaluation in a passive tracer context, *J. Geophys. Res.*, **100** (C10), 20763–20778.

Hecht, M. W., F. O. Bryan, and W. R. Holland, 1998, A consideration of tracer advection schemes in a primitive equation ocean model, *J. Geophys. Res. Oceans*, **103**(C2), 3301–3321.

Hollander, M. and D. A. Wolfe, 1999, *Nonparametric Statistical Methods*, 2nd edn, New York: John Wiley and Sons.

Holloway, G., 1992, Representing topographic stress for large-scale ocean models, *J. Phys. Oceanogr.*, **22**, 1033–1046.

Hood, R. R., E. A. Laws, R. A. Armstrong *et al.*, 2006, Pelagic functional group modeling: progress, challenges and prospects, *Deep-Sea Res. II*, **53**(5–7), 459–512.

Huse, G. and O. Fiksen, 2010, Modelling encounter rates and distribution of mobile predators and prey, *Progr. Oceanogr.*, **84**(1– 2), Spec. Iss. SI, 93–104.

Jayne, S. R., 2006, Circulation of the North Atlantic Ocean from altimetry and the Gravity Recovery and Climate Experiment geoid, *J. Geophys. Res.*, **111**, C03005, doi:10.1029/2005JC003128.

Jayne, S. R., 2009, The impact of abyssal mixing parameterizations in an ocean general circulation model, *J. Phys. Oceanogr.*, **39**, 1756–1775.

Jayne, S. R. and J. Marotzke, 2001, The dynamics of ocean heat transport variability, *Rev. Geophys.*, **39**, 385–411.

Jenkins, W. J., 1980, Tritium and ^{3}He in the Sargasso Sea, *J. Mar. Res.* **38**, 533–569.

Jenkins, W. J., 1998, Studying thermocline ventilation and circulation using tritium and ^{3}He. *J. Geophys. Res.*, **103**(C8): 15817–15831.

Jenkins, W. J., 2004, Tracers of ocean mixing. In *Treatise on Geochemistry*, Volume 6, *The Oceans and Marine Geochemistry*, ed. H. D. Holland and K. K. Turekian. Amsterdam: Elsevier, pp. 223–246.

Jenkins, W. J. and J. C. Goldman, 1985, Seasonal oxygen cycling and primary production in the Sargasso Sea. *J. Mar. Res.*, **43**, 465–491.

Jenkins, W. J. and D. W. R. Wallace, 1992, Tracer based inferences of new primary production in the sea. In *Primary Production and Biogeochemical Cycles in the Sea*, ed. P. G. Falkowski and A. D. Woodhead. New York: Plenum, pp. 299–316.

Jenkins, G. M. and D. G. Watts, 1968, *Spectral Analysis and its Applications*, Oakland, CA: Holden-Day.

Jerlov, N. G., 1968, *Optical Oceanography*, Amsterdam, Elsevier Publishing Co.

Jumars, P. A., J. H. Trowbridge, E. Boss, and L. Karp-Boss, 2009, Turbulence–plankton interactions: a new cartoon. *Mar. Ecol.* **30**, 133–150.

Kalnay, E. Kanamitsu, M. and Kistler, R., 1996, The NCEP/NCAR 40-year reanalysis project. *Bull. Am. Met. Soc.*, **77**, p437–471.

Kalnay, E., 2003, *Atmospheric Modeling, Data Assimilation, and Predictability*, New York: Cambridge University Press.

Karstensen, J. and Tomczak, M., 1998, Age determination of mixed water masses using CFC and oxygen data, *J. Geophys. Res.* **103**, 18599–18609.

Kasibhatla, P., Heimann, M. and Rayner, P., eds., 2000, *Inverse Methods in Global Biogeochemical Cycles*, Washington, DC: American Geophysical Union.

Keeling, R. F., 1993, On the role of large bubbles in air-sea gas exchange and supersaturation in the ocean. *J. Mar. Res.*, **51**, 237–271.

Kent, J. T., G. S. Watson, and T. C. Onstott, 1990, Fitting straight lines and planes with an application to radiometric dating, *Earth Planet. Sci. Lett.*, **97**, 1–17.

Key, R. M., A. Kozyr, C. L. Sabine *et al.*, 2004, A global ocean carbon climatology: results from GLODAP, *Glob. Biogeochem. Cycles*, **18**, GB4031.

Khatiwala, S., F. Primeau, and T. Hall, 2009, Reconstruction of the history of anthropogenic CO_2 concentrations in the ocean, *Nature*, **462**, 346–349.

Kraus, E. B. and J. S. Turner, 1967, A one-dimensional model of the seasonal thermocline II. The general theory and its consequences, *Tellus*, **19**, 98–106.

Large, W. G. and S. G. Yeager, 2009, The global climatology of an interannually varying air–sea flux data set, *Clim. Dynam.*, **33**, 341–364.

Large, W. G., J. C. McWilliams, and S. C. Doney, 1994, Oceanic vertical mixing: A review and a model with a nonlocal boundary layer parameterization, *Rev. Geophys.*, **32**, 363–403.

Large, W. G., G. Danabasoglu, S. C. Doney, and J. C. McWilliams, 1997, Sensitivity to surface forcing and boundary layer mixing in a global ocean model: annual-mean climatology, *J. Phys. Oceanogr.*, **27**, 2418–2447.

Large, W. G., G. Danabasoglu, J. C. McWilliams, P. R. Gent, and F. O. Bryan, 2001, Equatorial circulation of a global ocean climate model with anisotropic horizontal viscosity, *J. Phys. Oceanogr.*, **31**, 518–536.

Lavender, K. L., W. B. Owens, and R. E. Davis, 2005, The mid-depth circulation of the subpolar North Atlantic Ocean as measured by subsurface floats, *Deep-Sea Res. I*, **52**, 767–785.

Ledwell, J. R., A. J. Watson, and C. S. Law, 1998, Mixing of a tracer in the pycnocline, *J. Geophys. Res.*, **103**, 21499–21529.

Le Quéré, C., S. P. Harrison, I. C. Prentice *et al.*, 2005, Ecosystem dynamics based on plankton functional types for global ocean biogeochemistry models, *Glob. Change Biol.*, **11**, 2016–2040.

Levitus, S., T. P. Boyer, M. E. Conkright *et al.*, 1998, *NOAA Atlas NESDIS* 18, *World Ocean Database 1998*, Volume 1, *Introduction*, Washington, DC: US Government Printing Office.

Levitus, S., J. Antonov, and T. Boyer, 2005, Warming of the world ocean, 1955–2003, *Geophys. Res. Lett.*, **32**, L02604, doi:10.1029/2004GL021592.

Levy, M., P. Klein, and A.-M. Treguier, 2001, Impact of sub-mesoscale physics on production and subduction of phytoplankton in an oligotrophic regime. *J. Mar. Res.* **59**, 535–565.

Li, X. W. and F. W. Primeau, 2008, A fast Newton–Krylov solver for seasonally varying global ocean biogeochemistry models, *Ocean Modelling*, **23**, 13–20.

Light, A. and P. J. Bartlein, 2004, The end of the rainbow? Color schemes for improved data graphics, *EOS Trans. AGU*, **40**, 385, 391.

Lomb, N. R., 1976, Least-squares frequency analysis of unequally spaced data, *Astrophys. Space Sci.*, **39**, 447–462.

Lotka, A. J., 1920, Undamped oscillations derived from the law of mass action. *J. Am. Chem. Soc.*, **42**, 1595–1599.

Lotka, A. J. 1925, *Elements of Physical Biology*, Baltimore, MD: Williams and Wilkins Co.

MacMahon, D., 2006, Half-life evaluations for ^{3}H, ^{90}Sr, and ^{90}Y. *Appl. Rad. Isotopes*, **64**, 1417–1419.

Mallat, S., 1999, *A Wavelet Tour of Signal Processing*, San Diego, CA: Academic Press.

Marcotte, D., 1991, Cokriging with MATLAB, *Comp. Geosci.*, **17**(9), 1265–1280.

Marcotte, D., 1996, Fast variogram computation with FFT, *Comp. Geosci.*, **22**(10), 1175–1186.

Martens, C. and Berner, R., 1977, Interstitial water chemistry of anoxic Long Island Sound sediments. 1. Dissolved gases, *Limnol. Oceanogr.*, **22**(1), 10–25.

Maslowski, W. and W. H. Lipscomb, 2003, High resolution simulations of Arctic sea ice, 1979–1993, *Polar Res.*, **22**, 67–74.

Matear, R. J., 1995, Parameter optimization and analysis of ecosystem models using simulated annealing — a case-study at station-P, *J. Mar. Res.*, **53**(4), 571–607.

MathWorks, The, 2009, *MATLAB 7 Function Reference*, Natick, MA: The Mathworks, Inc.

MathWorks, The, 2010, *Optimization Toolbox 5, Users Guide*, Natick, MA: The Mathworks, Inc.

Matsumoto, K., J. L. Sarmiento, R. M. Key 2004, Evaluation of ocean carbon cycle models with data-based metrics, *Geophys. Res. Lett.*, **31**, L07303, doi:10.1029/2003GL018970.

Mayorga, E., S. P. Seitzinger, J. A. Harrison *et al.*, 2010, Global Nutrient Export from WaterSheds 2 (NEWS 2): Model development and implementation, *Environ. Modelling Software*, **25**, 837–853.

McClain, C. R., G. C. Feldman, and S. B. Hooker, 2004, An overview of the SeaWiFS project and strategies for producing a climate research quality global ocean bio-optical time series, *Deep Sea Res. II*, **51**, 5–42.

McGillicuddy, D. J., Jr., L. A. Anderson, S. C. Doney, and M. E. Maltrud, 2003, Eddy-driven sources and sinks of nutrients in the upper ocean: Results from a 0.1 resolution model of the North Atlantic, *Glob. Biogeochem. Cycles*, **17**(2), 1035, doi:10.1029/2002GB001987.

McPhaden, M. J., A. J. Busalacchi, R. Cheney *et al.*, 1998, The tropical ocean global atmosphere observing system: A decade of progress, *J. Geophys. Res. Oceans*, **103**, 14169–14240.

McWilliams, J. C., 2006, *Fundamentals of Geophysical Fluid Dynamics*, Cambridge, UK: Cambridge University Press.

Mellor, G. L. and T. Yamada, 1974, A hierarchy of turbulence closure models for planetary boundary layers, *J. Atmos. Sci.*, **31**, 1791–1806.

Mellor, G. L. and T. Yamada, 1982, Development of a turbulence closure model for geophysical fluid problems, *Rev. Geophys. Space Phys.*, **20**, 851–875.

Menke, W., 1984, *Geophysical Data Analysis: Discrete Inverse Theory*, New York: Academic Press, Inc.

Metropolis, N., A. Rosenbluth, M. Rosenbluth, A. Teller, and E. Teller, 1953, Equations of state calculations by fast computing machines, *J. Chem. Phys.*, **21**, 1087–1092.

Mikaloff Fletcher, S. E., N. Gruber, A. R. Jacobson *et al.*, 2006, Inverse estimates of anthropogenic CO_2 uptake, transport, and storage by the ocean, *Glob. Biogeochem. Cycles*, **20**, GB2002.

Mikaloff Fletcher, S. E., N. Gruber, A. R. Jacobson *et al.*, 2007: Inverse estimates of the oceanic sources and sinks of natural CO_2 and their implied oceanic transport, *Glob. Biogeochem. Cycles*, **21**, GB1010.

Monahan, E. C. and T. Torgersen, 1991. Enhancement of air–sea gas exchange by oceanic whitecapping. In *Air–Water Mass Transfer, Second International Symposium on Gas Transfer at Water Surfaces*, New York: ASCE, pp. 608–617.

Moore, J. K., S. C. Doney, and K. Lindsay, 2004, Upper ocean ecosystem dynamics and iron cycling in a global 3-D model, *Glob. Biogeochem. Cycles*, **18**, GB4028, 10.1029/2004GB002220.

Munk, W., 1966, Abyssal recipes, *Deep-Sea Res.* **13**, 707–730.

Munk, M., 2006, Ocean acoustic tomography from a stormy start to an uncertain future In *Physical Oceanography Developments Since 1950*, ed. M. Jochum and R. Murtugudde. New York, NY: Springer, pp. 119–138.

Munk, W. and C. Wunsch, 1979, Ocean acoustic tomography: a scheme for large scale monitoring, *Deep-Sea Res. A*, **26**, 123–161.

Murray, J. D., 1993, *Mathematical Biology*, 2nd Corrected edn Berlin: Springer-Verlag.

Nabney, I. T., 2002, *NETLAB: Algorithms for Pattern Recognition*, Advances in Pattern Recognition, London, UK: Springer.

Najjar, R. G., X. Jin, F. Louanchi *et al.*, 2007, Impact of circulation on export production, dissolved organic matter and dissolved oxygen in the ocean: Results from Phase II of the Ocean Carbon-cycle Model Intercomparison Project (OCMIP-2), *Glob. Biogeochem. Cycles*, **21**, GB3007, doi:10.1029/2006GB002857.

Nelder, J. A. and R. Mead, 1965, A simplex method for function minimization, *Comput. J.*, **7**, 308–313.

Nightingale, P. D., P. S. Liss, and P. Schlosser, 2000, Measurements of air–sea gas transfer during an open ocean algal bloom, *Geophys. Res. Lett.*, **27**(14), 2117–2120.

Nocedal, J. and S. J. Wright, 1999, *Numerical Optimization*, Springer Series in Operations Research, New York, NY: Springer-Verlag.

Okubo, A., 1971, Oceanic diffusion diagrams, *Deep-Sea Res.*, **18**, 789–802.

Oschlies, A., W. Koeve and V. Garcon, 2000, An eddy-permitting coupled physical–biological model of the North Atlantic 2. Ecosystem dynamics and comparison with satellite and JGOFS local studies data, *Glob. Biogeochem. Cycles*, **14**, 499–523.

Pacanowski, R. C. and S. G. H. Philander, 1981, Parameterization of vertical mixing in numerical models of tropical oceans, *J. Phys. Oceanogr.*, **11**, 1443–1451.

Paulson, C. A. and J. J. Simpson, 1977, Irradiance measurements in the upper ocean. *J. Phys. Oceanogr.*, **7**, 952–956.

Pavia, E. G. and J. J. O'Brien, 1986, Weibull statistics of wind speed over the ocean, *J. Clim. Appl. Oceanogr.* **25**, 324–332.

Peacock, S. and M. E. Maltrud, 2006, Transit-time distributions in a global ocean model, *J. Phys. Oceanogr.*, **36**, 474–495.

Pedlosky, J. C., 1992, *Geophysical Fluid Dynamics*, 2nd edn, Berlin: Springer-Verlag.

Preisendorfer, R. W., 1988, *Principal Component Analyses in Meteorology and Oceanography*, Amsterdam: Elsevier.

Press, W. H., S. A. Teukolsky, W. T. Vetterling, and B. P. Flannery, 2007, *Numerical Recipes*, 3rd edn, New York: Cambridge University Press.

Price, J. F., 2003, Dimensional analysis of models and data set, *Am. J. Phys.* **71**, 437–447.

Price, J. F. and M. Baringer, 1994, Outflows and deep water production by marginal seas, *Prog. Oceanogr.*, **33**, 161–200.

Price, J. F, R. A. Weller and R. Pinkel, 1986, Diurnal cycling: observations and models of the upper ocean response to diurnal heating, cooling and wind mixing, *J. Geophys. Res.*, **91**, 8411–8427.

Priestley, M. B., 1981, *Spectral Analysis and Time Series*, New York: Academic Press.

Primeau, F., 2005, Characterizing transport between the surface mixed layer and the ocean interior with a forward and adjoint global ocean transport model, *J. Phys. Oceanogr.*, **35**(4), 545–564.

Qiu, B. and R. X. Huang, 1995, Ventilation of the North Atlantic and North Pacific: subduction versus obduction, *J. Phys. Oceanogr.*, **25**, 2374–2390.

Ralph, E. A. and P. P. Niiler, 1999, Wind-driven currents in the tropical Pacific, *J. Phys. Oceanogr.*, **29**, 2121–2129.

Record, N. R., A. J. Pershing, J. A. Runge *et al.*, 2010, Improving ecological forecasts of copepod community dynamics using genetic algorithms, *J. Mar. Syst.*, **82**(3), 96–110.

Redi, M. H., 1982, Oceanic isopycnal mixing by coordinate rotation, *J. Phys. Oceanogr.*, **12**, 1154–1158.

Reed, B. C., 1992, Linear least-squares fits with errors in both coordinates. II: Comments on parameter variances, *Am. J. Phys.*, **60**, 59–62.

Reynolds, R. W., T. M. Smith, C. Liu *et al.*, 2007, Daily high-resolution-blended analyses for sea surface temperature, *J. Clim.*, **20**, 5473–5496.

Richman, M. B., 1985, Rotation of principal components. *J. Climatol.*, **6**, 293–335.

Roache, P. J., 1972, On artificial viscosity, *J. Comput. Phys.*, **10** (2), 169–184.

Roache, P. J., 1998, *Fundamentals of Computational Fluid Dynamics*, Albuquerque, NM: Hermosa Publishers.

Rood, R. B., 1987, Numerical algorithms and their role in atmospheric transport and chemistry problems, *Rev. Geophys.*, **25** (1) 71–100.

Roughgarden, J., 1998, *Primer of Ecological Theory*, Upper Saddle River, NJ: Prentice Hall.

Ruth, C., Well, R. and Roether, W., 2000. Primordial ^{3}He in South Atlantic deep waters from souces on the Mid-Atlantic Ridge. *Deep-Sea Res. I*, **47**, 1059–1075.

Sarmiento, J. L. and N. Gruber, 2006, *Ocean Biogeochemical Dynamics*, Princeton, NJ: Princeton University Press.

Sarmiento, J. L., H. W. Feely, W. S. Moore, A. E. Bainbridge, and W. S. Broecker, 1976, The relationship between vertical eddy diffusion and buoyancy gradient in the deep sea, *Earth Planet. Sci. Lett.*, **32**, 357–370.

Schey, H. M., 2005, *Div, Grad, Curl, and All That: An Informal Text on Vector Calculus*, 4th edn, New York: W. W. Norton.

Schlitzer, R., 2002, Carbon export fluxes in the Southern Ocean: Results from inverse modeling and comparison with satellite based estimates, *Deep-Sea Res. II*, **49**, 1623–1644.

Schöter, J. and C. Wunsch, 1986, Solution of nonlinear finite difference ocean models by optimization methods with sensitivity and observational strategy analysis, *J. Phys. Oceanogr.*, **16**(11), 1855–1874.

Shampine, L. F. and M. W. Reichelt, 1997, The MATLAB ODE Suite, *SIAM J. Sci. Comput.*, **18**, 1–22.

Shulenberger, E. and J. L. Reid 1981, *Deep-Sea Res. A*, **28**, 901–920.

Simmons, H. L., S. R. Jayne, L. C. St Laurent, and A. J. Weaver, 2004, Tidally driven mixing in a numerical model of the ocean general circulation, *Ocean Modelling*, **6**, 245–263.

Skiena, S. S., 2008, *The Algorithm Design Manual*, 2nd edn, London: Springer-Verlag.

Smagorinsky, J., 1963, General circulation experiments with the primitive equations: I. The basic experiment, *Mon. Wea. Rev.*, **91**, 99–164.

Smagorinsky, J., 1993, Some historical remarks on the use of nonlinear viscosities. In *Large Eddy Simulation of Complex Engineering and Geophysical Flows*, ed. B. Galperin and S. A. Orszag. Cambridge: Cambridge University Press.

Smith, R. D., M. E. Maltrud, F. O. Bryan, and M. W. Hecht, 2000, Numerical simulation of the North Atlantic Ocean at 1/10°, *J. Phys. Oceanogr.*, **30**, 1532–1561.

Smith, W. H. F. and P. Wessel, 1990: Gridding with continuous curvature splines, *Geophysics*, **55**, 293–305.

Spall, J. C., 2003, *Introduction to Stochastic Search and Optimization*, Wiley Inter-Science Series in Discrete Mathematics and Optimization, Hoboken, NJ: Wiley.

Speer, K. and E. Tziperman, 1992, Rates of water mass formation in the North Atlantic Ocean, *J. Phys. Oceanogr.*, **22**, 93–104.

Stammer, D., 1997, Global characteristics of ocean variability estimated from regional TOPEX/POSEIDON altimeter measurements, *J. Phys. Oceanogr.*, **27**, 1743–1769.

Stammer, D., K. Ueyoshi, A. Kohl *et al.*, 2004, Estimating air-sea fluxes of heat, freshwater, and momentum through global ocean data assimilation, *J. Geophys. Res. Oceans*, **109**, C05023.

Stanley, R. H. R., 2007, A determination of air–sea gas exchange processes and upper ocean biological production from five noble gases and tritiugenic helium-3. Unpublished Ph.D. thesis, MIT–WHOI Joint Program in Chemical Oceanography, Woods Hole, MA.

Stanley, R. H. R., W. J. Jenkins, and S. Doney, 2006, Quantifying seasonal air-sea gas exchange processes using a noble gas time-series: a design experiment, *J. Mar. Res.*, **64**, 267–295.

Stanley, R. H. R., W. J. Jenkins, D. E. Lott III, and S. C. Doney, 2009, Noble gas constraints on air–sea gas exchange and bubble fluxes, *J. Geophys. Res. Ocean*, **114**, C11020, doi:10.1029/2009JC005396.

Stoer, J. and R. Bulirsch, 2002, *Introduction to Numerical Analysis*, 3rd edn, New York: Springer.

Stokes, G. G., 1922, On the effect of the internal friction of fluids on the motion of pendulums. In *Mathematical and Physical Papers*, reprinted from the *Transactions of the Cambridge Philosophical Society*, 1850, Vol. IX. p. [8], Cambridge University Press, pp. 1–141.

Stommel, H., 1948, The westward intensification of wind-driven ocean currents. *EOS Trans. AGU*, **29**, 202–206.

Storkey, D., E. W. Blockley, R. Furner *et al.*, 2010, Forecasting the ocean state using NEMO: The new FOAM system, *J. Operational Oceanogr.*, **3**, 3–15.

Stow, C. A., J. Jolliff, D. J. McGillicuddy Jr. *et al.*, 2009, Skill assessment for coupled biological/physical models of marine systems, *J. Mar. Syst.*, **76**, 4–15, doi:10.1016/j.jmarsys.2008.03.011.

Strang, G., 1986, *Introduction to Applied Mathematics*, Wellesley, MA: Wellesley-Cambridge Press.

Strang, G., 2005, *Linear Algebra and its Applications*, 4th edn, Pacific Grove, CA: Brooks Cole.

Strang, G. and T. Nguyen, 1997, *Wavelets and Filter Banks*, Wellesley, MA: Wellesley-Cambridge Press.

Stull, R. B., 1988, *An Introduction to Boundary Layer Meteorology*, Boston, MA: Kluwer: Academic.

Stumm, W. and J. J. Morgan, 1996, *Aquatic Chemistry*, 3rd edn, New York: Wiley-Interscience.

Sundermeyer, M. A. and J. F. Price, 1998, Lateral mixing in the North Atlantic Tracer Release Experiment: observations and numerical simulations of Lagrangian particles and a passive tracer, *J. Geophys. Res.*, **103**, 21481–21497.

Taylor, B., 1715, *Methodus incrementorum directa et inversa (Direct and Indirect Methods of Incrementation)*, reprinted by Gale ECCO (2010), (Latin Edition).

Taylor, K. E., 2001, Summarizing multiple aspects of model performance in a single diagram, *J. Geophys. Res.*, **106**, 7183–7192.

Tennekes, H. and J. L. Lumley, 1972, *A First Course in Turbulence*. Cambridge, MA: MIT Press.

Thomas, G. B. Jr., M. D. Weir, and J. R. Hass, 2009, *Thomas' Calculus*, 12th edn, Reading, MA: Addison-Wesley.

Thurstone, L. L., 1935, *The Vectors of the Mind*, Chicago, IL: University of Chicago Press.

Tomczak, M. and D. G. B. Large, 1989, Optimum multiparameter analysis of mixing in the thermocline of the Eastern Indian Ocean. *J. Geophys. Res.*, **94**, 16141–16149.

Torrence, C. and G. P. Compo, 1998, A practical guide to wavelet analysis, *Bull. Am. Met. Soc.*, **79**(1), 61–78.

Tufte, E. R., 1990, *Envisioning Information*, Cheshire, CT: Graphics Press.

Tufte, E. R., 1997, *Visual Explanations*, Cheshire, CT: Graphics Press.

Tufte, E. R., 2001, *The Visual Display of Quantitative Information*, 2nd edn, Cheshire, CT: Graphics Press.

Tufte, E. R., 2006, *Beautiful Evidence*, Cheshire CT: Graphics Press.

Uttal, T., J. A. Curry, M. G. McPhee *et al.*, 2002, Surface heat budget of the Arctic Ocean, *Bull. Am. Met. Soc.*, **83**, 255–275.

van der Pol, B. and J. van der Mark, 1927, Frequency demultiplication, *Nature*, **120**, 363–364.

Venkataraman, P., 2002, *Applied Optimization with MATLAB Programming*, New York: Wiley Interscience.

Veronis, G., 1975, The role of models in tracer studies. In *Numerical Models of Ocean Circulation*, Washington, DC: National Academy of Sciences, pp. 133–146.

Veronis, G., 1977, Use of tracers in circulation studies. In *The Sea*, Volume 6, *Marine Modeling*, ed. E. D. Goldberg, I. N. McCave, J. J. O'Brien, and J. H. Steele. New York: Wiley-Interscience, pp. 169–188.

Vinogradov, S. V., R. M. Ponte, P. Heimbach, and C. Wunsch, 2008, The mean seasonal cycle in sea level estimated from a data-constrained general circulation model, *J. Geophys. Res.*, **113**, C03032, doi:10.1029/2007JC004496.

Volterra, V., 1926, Variazionie fluttuazioni del numero d'individui in specie animali con-viventi. *Mem. Acad. Lincei.*, **2**, 31–113. (Variations and fluctuations of a number of individuals in animal species living together. Translation in: R.N. Chapman, 1931, *Animal Ecology*, New York: McGraw Hill, pp. 409–448.)

Wanninkhof, R., 1992, Relationship between wind speed and gas exchange over the ocean. *J. Geophys. Res.*, **97**, 7373–7382.

Wanninkhof, R. and S. C. Doney, 2005, Cruise Report CLIVAR A16S 2005, http://www.aoml.noaa.gov/ocd/gcc/a16s/

Wanninkhof, R. and W. R. McGillis, 1999, A cubic relationship between air–sea CO_2 exchange and wind speed, *Geophys. Res. Lett.*, **26**(13), 1889–1892.

Ward, B. A., M. A. M. Friedrichs, T. R. Anderson and A. Oschlies, 2010, Parameter optimi-sation techniques and the problem of underdetermination in marine biogeochemical models, *J. Mar. Syst.*, **81**(1–2), 34–43.

Waugh, D. W., T. M. Hall, and T. W. N. Haine, 2003, Relationships among tracer ages, *J. Geophys. Res.*, **108**(C5), 3138, doi:10.1029/2002JC001325.

Waugh, D. W., T. W. N. Haine, and T. M. Hall, 2004, Transport times and anthropogenic carbon in the subpolar North Atlantic Ocean, *Deep-Sea Res. I*, **51**, 1475–1491.

Weatherly, J. W., B. P. Briegleb, W. G. Large, and J. A. Maslanik, 1998, Sea ice and polar climate in the NCAR CSM, *J. Clim*, **11**, 1472–1486.

Welty, J. R., C. E. Wicks, R. E. Wilson, and G. L. Rorrer, 2007, *Fundamentals of Momentum, Heat and Mass Transfer*, 5th edn, Hoboken, NJ: Wiley.

Wijffels, S. E., R. W. Schmitt, H. L. Bryden, and A. Stigebrandt, 1992, Transport of freshwater by the oceans, *J. Phys. Oceanogr.*, **22**, 155–162.

Willis, J. K. and L. L. Fu, 2008, Combining altimeter and subsurface float data to estimate the time-averaged circulation in the upper ocean, *J. Geophys. Res. Oceans*, **113**, C12017.

Willis, J. K., D. Roemmich, and B. Cornuelle, 2004, Interannual variability in upper ocean heat content, temperature, and thermosteric expansion on global scales, *J. Geophys. Res. Oceans*, **109**, C12036.

Wunsch, C., 1996, *The Ocean Circulation Inverse Problem*, New York: Cambridge University Press.

Wunsch, C., P. Heimbach, R. Ponte, and I. Fukumori, 2009, The global general circulation of the ocean estimated by the ECCO-consortium, *Oceanography*, **22**, 88–103.

York, D., 1966, Least-squares fitting of straight lines, *Canad. J. Phys.*, **44**, 1079–1086.

York, D., 1969, Least squares fitting of a straight line with correlated errors, *Earth Planet. Sci. Lett.*, **5**, 320–324.

York, D., N. M. Evensen, M. L. Martinez, and J. D. B. Delgado, 2004, Unified equations for the slope, intercept, and standard errors of the best straight line, *Am. J. Phys.*, **72**, 367–375.

Young, W. R., P. B. Rhines, and C. J. R. Garrett, 1982, Shear-flow dispersion, internal waves and horizontal mixing in the ocean, *J. Phys. Oceanogr.*, **12**, 515–527.

Zwillinger, D., 2002, *CRC Standard Mathematical Tables and Formulae*, 31st edn, Boca Raton, FL: Chapman & Hall/CRC.

Index

Printed in the United States
By Bookmasters